MICROBES
AND SOCIETY

The Jones & Bartlett Topics in Biology Series

We are pleased to offer a series of full length textbooks designed specifically for your special topics courses in biology. Our goal is to supply comprehensive texts that will introduce non-science majors to the wonders of biology. With coverage of topics in the news, emerging diseases, and important advances in biotechnology, students will enjoy learning and relating science to current events.

AIDS: The Biological Basis, Fourth Edition
Benjamin S. Weeks, PhD, Adelphi University, I. Edward Alcamo, PhD, formerly of State University of New York at Farmingdale

Cancer: The Biological Basis
Benjamin S. Weeks, PhD, Adelphi University
Coming Soon!

Human Embryonic Stem Cells, Second Edition
Ann Kiessling, PhD, Harvard Medical School, Boston, Massachusetts, Scott C. Anderson

The Microbial Challenge: Human-Microbe Interactions
Robert Krasner, PhD, Providence College

Microbes and Society, Second Edition
Benjamin S. Weeks, PhD, Adelphi University, I. Edward Alcamo, PhD, formerly of State University of New York at Farmingdale

Related Titles in Microbiology

20th Century Microbe Hunters
Robert Krasner, PhD, Providence College

Alcamo's Fundamentals of Microbiology, Eighth Edition
Jeffrey C. Pommerville, PhD, Glendale Community College

Alcamo's Laboratory Fundamentals of Microbiology, Eighth Edition
Jeffrey C. Pommerville, PhD, Glendale Community College

Principles of Modern Microbiology
Mark Wheelis, PhD, University of California, Davis

Guide to Infectious Diseases by Body System
Jeffrey C. Pommerville, PhD, Glendale Community College

Understanding Viruses
Teri Shors, PhD, University of Wisconsin—Oshkosh

MICROBES AND SOCIETY

SECOND EDITION

BENJAMIN S. WEEKS, PhD

*Department of Biology and Environmental Sciences Program,
Adelphi University*

I. EDWARD ALCAMO, PhD

JONES AND BARTLETT PUBLISHERS

Sudbury, Massachusetts

BOSTON   TORONTO   LONDON   SINGAPORE

WORLD HEADQUARTERS
Jones and Bartlett Publishers
40 Tall Pine Drive
Sudbury, MA 01776
978-443-5000
info@jbpub.com
www.jbpub.com

Jones and Bartlett Publishers Canada
6339 Ormindale Way
Mississauga, ON L5V 1J2
CANADA

Jones and Bartlett Publishers
International
Barb House, Barb Mews
London W6 7PA
UK

Jones and Bartlett's books and products are available through most bookstores and online booksellers. To contact Jones and Bartlett Publishers directly, call 1-800-832-0034, fax 978-443-8000, or visit our website www.jbpub.com.

Substantial discounts on bulk quantities of Jones and Bartlett's publications are available to corporations, professional associations, and other qualified organizations. For details and specific discount information, contact the special sales department at Jones and Bartlett via the above contact information or send an email to specialsales@jbpub.com.

PRODUCTION CREDITS
Chief Executive Officer: Clayton Jones
Chief Operating Officer: Don W. Jones, Jr.
President, Higher Education Professional Publishing: Robert W. Holland, Jr.
V.P., Design and Production: Anne Spencer
V.P., Sales and Marketing: William Kane
V.P., Manufacturing and Inventory Control: Therese Connell
Executive Editor, Science: Cathleen Sether
Acquisitions Editor, Science: Shoshanna Grossman
Managing Editor, Science: Dean W. DeChambeau
Associate Editor, Science: Molly Steinbach
Editorial Assistant, Science: Briana Gardell
Senior Production Editor: Louis C. Bruno, Jr.
Production Editor: Dan Stone
Marketing Manager: Andrea DeFronzo
Text and Cover Design: Anne Spencer
Illustrations: Elizabeth Morales
Photo Research Manager and Photographer: Kimberly L. Potvin
Associate Photo Researcher and Photographer: Christine McKeen
Composition: Shepherd, Inc.
Printing and Binding: Courier Kendallville
Cover Printing: Courier Kendallville
Cover Photo: (Cheese) © Robert Harding Picture Library Ltd/Alamy Images
(Insert) © Phototake Inc./Alamy Images

Library of Congress Cataloging-in-Publication Data
Weeks, Benjamin S.
  Microbes and society / Benjamin S. Weeks. — 2nd ed.
    p. cm.
  Includes bibliographical references and index.
  ISBN-13: 978-0-7637-4649-0
  ISBN-10: 0-7637-4649-5
  1. Microbiology. I. Title.
  QR41.2.A434 2008
  616.9'041—dc22
  6048   2007023500

Printed in the United States of America
12 11 10 09 08   10 9 8 7 6 5 4 3 2

About the Cover
French cheese makers wrap Roquefort cheese in the natural caves of Roquefort-sur-Soulzona. The cheese is made from the milk of a single breed of sheep, the Lacaune, and seeded with *Penicillium roqueforti* spores. The mold, which grows naturally in the caves, gives the cheese its bluish color and distinctive aroma. Inset: A scanning electron micrograph of *P. roqueforti* hyphae with conidiophores (magnified 205x).

Dedication

I would like to dedicate this book to its first author, Professor I. Edward Alcamo, whom I never met in person, but met as a student of his writing and now through the honor of continuing his work. Here's to you, Professor Alcamo.

I would also like to dedicate this book to my wife, Melissa, who has provided me with undying love and support, and my children, Samuel, Hayden, Jessica and David, who have provided me with endless joy and pride.

Brief Contents

Contents

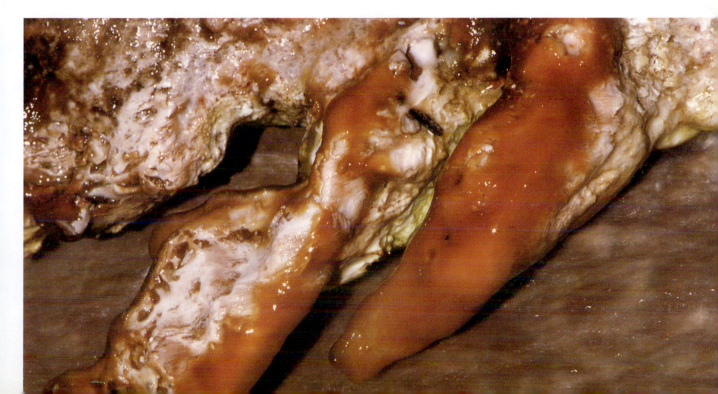

Preface

Rarely does a day go by when a story about microbes does not appear in a newspaper, magazine, or TV news show. What we see on these shows can influence our view of microbes, and each of us may have a different vision. Perhaps we see them as fearsome "germs" lurking about waiting to do us harm. Or perhaps we have heard that they are the biological factories in which the products of biotechnology are made. We may know them as producers of dairy products such as yogurt and sour cream or of fine wines and cheeses. Perhaps our vision is clouded by thoughts of microbes as agents of bioterrorism. Our vision of microbes likely comes from limited knowledge of them. If so, then we are about to broaden our spectrum because this book will take us on an extended tour of the microbial world. We'll see the remarkable variety among microbes; we'll experience the myriad ways they help maintain life on Earth; and we'll explore the marvelous jobs they perform for us and the not-so-marvelous ways they cause illness. In the end we may even discover that our new view of microbes is quite unlike the limited view we had before.

Audience

Microbes and Society, Second Edition is geared to the informed citizen of the 21st century. It discusses such topics as the place of microbes in ecology and the environment, the uses of microbes in biotechnology, the role of microbes in food production, and the numerous other ways that microbes contribute to the quality of our lives. The book also explores bioterrorism, examines the problem of antibiotic resistance, and surveys several microbial diseases of history and contemporary times. Students will find that understanding microbes will help them do well in such fields as business, sociology, food science, pharmaceutical and health sciences, economics, and agriculture. The book assumes little or no science background, and it should accommodate one-quarter or one-semester courses.

Objectives

The 21st century is destined to be the Century of Biology. In future decades we can anticipate new products of genetic engineering, new ways of preserving and protecting our environment, new methods in agriculture, and new technologies not yet even in the idea stage. And microbes are at the center of all of these. They are the hammers and nails of genetic engineering, the worker bees for purifying polluted water, the sources of imaginative insecticides and pesticides, and the jumping off points for futuristic technologies. Knowing the microbes is essential to knowing the future. And knowing the microbes is the first major objective of this book.

What of today? Consider these thoughts: Rarely does a day go by when we do not enjoy a "microbial food;" each time we put out the garbage, we assume that microbes will break it down; whenever we take a breath, we inhale the oxygen that microbes have put into the atmosphere; and each time we cover a sneeze, we try to stop a microbial disease from spreading. All of which brings us to this book's second major objective: understanding the places that microbes occupy in our day-to-day existence.

And what would the present and future be without the past? So the third major objective is showing how microbes have had a significant impact on history. We shall study, for example, how microbes changed the course of Western civilization, how microbes helped Alexander the Great conquer Asia, how microbes influenced the ways cultures arose, and how microbes made much of the current work in genetic engineering possible. Few groups of organisms have such a rich and powerful place in history.

I hope you will enjoy your education in microbiology and come to understand the influence of microbes on our society today, in the past, and in the future.

What's New?

The second edition of *Microbes and Society* contains updates, additions, reorganization, and new features. For example, this second edition includes information on recent *E. Coli* outbreaks, the emergence of antibiotic resistant bacteria, new updated epidemiology on rates of disease, and a furtherance on the uses of bacteria in biotechnology. Additionally, this second edition contains a new Chapter 3 on the "Molecules of the Cell: The Building Blocks of Life." In keeping with the spirit of this book, this material is presented on the introductory level and presented with clarity and simplicity. By providing this new chapter, both the student and professor can explore the remaining chapters of the book with a deeper scientific foundation. The second edition has also been slightly reorganized so that the chapters flow between related topics more directly. A glossary and pronunciation guide have also been added to assist the reader. In all, the second edition is an up-to-date, in-depth exploration of the many facets of microbiology and how microorganisms touch our lives, in ways both good and bad, on a daily basis. This book maintains a scientific foundation and basis while also bringing to life the impact of microbes on society.

Organization

The second edition of *Microbes and Society* contains two parts. In Part I, we introduce the microbial world over the span of nine chapters. Some of the chapters explore the bacteria, viruses, fungi, protozoa, and other microbes; and other chapters describe how these microbes grow and reproduce, the unique genetic patterns they display, and the methods used to control them.

Part II moves to the practical applications of microbiology. We visit a restaurant for a microbial meal, we wander through a research facility and see microbes at work, we stop at various locations in the environment and observe microbes acting on our behalf, and we examine their places in disease. The bottom line is that microbes are relevant.

Must the chapters be studied in sequence? Absolutely not. We understand that time constraints often prevent courses from using the entire book, so we invite instructors and students to pick those topics that fit best. To encourage flexibility, each chapter has been written independently of the others, and each section in a chapter stands alone. Instructors may, therefore, design their own approach to microbiology according to their students' needs.

Special Features

Approaching a course in microbiology can be an anxious experience. There are new insights to learn, new concepts to master, and an entirely new vocabulary to memorize. To smooth over the bumps, we have incorporated several features that should help increase the comfort level of *Microbes and Society*.

Each chapter begins with a section titled "Looking Ahead" to let students know what they should take away from the chapter. The reading then opens with an engaging story to set a tone for the pages that follow. Key terms in the chapter have been boldfaced to draw the attention of readers, and pronunciations of difficult terms are presented in the margins. Boxes in each chapter ("A Closer Look") encourage a moment of relief from the rigors of study and present an historical insight, an interesting aside, or a current research direction. The figures are presented in full color, and special attention has been given to setting them close to their text reference. The chapter concludes with a list of the key terms for review and a set of thought questions that provide challenging opportunities to apply what has been learned.

Students may note that all chapters are about the same length. This was done purposefully because we wanted to provide a symmetrical framework in which students can learn. Each chapter has several sections and numerous smaller subsections to accommodate limited study times. Even the paragraphs are about the same size (there should be a rhythm in reading). The bottom line is that we're hoping to provide a thorough and balanced presentation of microbiology within an enjoyable context.

Ancillaries to Accompany *Microbes And Society, Second Edition*

To assist you in teaching this course and supplying your students with the best in teaching aids, Jones and Bartlett Publishers has prepared a complete ancillary package available to all adopters of *Microbes and Society, Second Edition*. Additional information and review copies of any of the following items are available through your Jones and Bartlett Sales Representative.

For the Instructor

Instructor's ToolKit CD-ROM

Compatible with Windows and Macintosh platforms, the Instructor's ToolKit CD-ROM provides adopters with the following traditional ancillaries.

The **Test Bank**, prepared by the author, is available as text files. The test bank contains over 1,200 questions. An additional set of test bank files is formatted for your own online courses using WebCT and Blackboard.

The **PowerPoint® Lecture Outline Slides** presentation package provides lecture notes, graphs, and images for each chapter of *Microbes and Society, Second Edition*. Instructors with the Microsoft PowerPoint software can customize the outlines, art, and order of presentation. The PowerPoint files have also been prepared in HTML format for use in online course management systems.

The **PowerPoint Image Bank** provides the illustrations, photographs, and tables (to which Jones and Bartlett Publishers holds the copyright or has permission to reprint digitally) inserted into PowerPoint slides. With the Microsoft PowerPoint program, you can quickly and easily copy individual images into your existing lecture slides. If you do not own a copy of Microsoft PowerPoint or a compatible software program, a Microsoft PowerPoint Viewer is included on the CD-ROM.

The answers to the "Questions to Consider" are also on the CD-ROM.

For the Student

We have developed a website (http://microbiology.jbpub.com/book/microbes) exclusively for *Microbes and Society, Second Edition*. The site contains a free student study guide that includes chapter outlines, flash cards, and quizzes. The site also provides research and reference links, and links to microbiology in the news.

Laboratory Fundamentals of Microbiology, Eighth Edition, is a series of over 30 multi-part laboratory exercises providing basic training in the handling of microorganisms and reinforcing ideas and concepts described in the textbook.

Encounters with Microbiology brings together "Vital Signs" articles from *Discover Magazine* in which health professionals use their knowledge of microbiology in their medical cases.

Guide to Infectious Diseases by Body System, by Jeffrey C. Pommerville, Glendale Community College, is an excellent tool for learning about microbial diseases. Each of the fifteen body system units presents a brief introduction to the anatomical system and the bacterial, viral, fungal, or parasitic organism capable of infecting the system.

20th Century Microbe Hunters, by Robert Krasner, Providence College, offers a dramatic portrayal of the achievements and lives of microbiologists such as Charles J. Nicolle (typhus epidemic), Barry Marshall and J. Robin Warren (*Helicobacter pylori*), Luc Montagnier and Robert Gallo (HIV), and Donald R. Hopkins (Guinea worm).

How Pathogenic Viruses Work, by Lauren Sompayrac, is a concise summary of the basics of virology written in an understandable and entertaining manner. The book is composed of nine lectures covering the essential elements of virus-host interactions with descriptive graphics, helpful mnemonic tactics for retaining the concepts, and brief lecture reviews. This is an ideal text for medical, science, and nursing students who want a review, or simple explanation, of virology.

Acknowledgments

Professor Alcamo, along with those at Jones and Bartlett Publishers, envisioned the need for a book on the topic of microbiology for the non-science major. I am honored and grateful to have been given the opportunity by Cathleen Sether to continue Ed Alcamo's work by updating and extending the material for the second edition of this book. I want to thank the rest of my "family" at Jones and Bartlett Publishers, Daniel Stone, Shoshanna Grossman, Molly Steinbach, Dean DeChambeau and Lou Bruno for their encouragement and support through the process of bringing this second edition together. I would also like to thank Donald G. Lindmark of Cleveland State University and Linda Bruslind of Oregon State University who took valuable time to offer their suggestions for improving this edition.

A Note From the Author

I completed my doctoral training in 1988, investigating mechanisms through which environmental pollutants can cause disease in humans. My real interest in microbiology did not take shape until 1997 with my arrival at Adelphi University. With training in virology and immunology, I was asked to lead the microbiology course for our nursing majors, a course that focuses on the infections and diseases that these "nasty" bacteria cause in humans. One of my favorite laboratory exercises for the nursing microbiology course is to have the students wash their laboratory bench top with detergents and disinfectants and then sample the cleaned bench top immediately, and over several hours, for the presence of bacteria. Invariably more than half of the class finds bacteria still present even at time zero with increases in cultivable bacteria over the following hours. The students are also amazed, as am I despite repeated experience, that at least one student each semester will culture a bacteria from the environment that is resistant to all antibiotics tested against it. My interest in bacterial samples from the environment quickly turned to the soil. My students and I were imagining all the antibiotic resistant bacteria in the ground that school children must be playing and rolling around in. However, this line of investigation actually led me to the "other side" of microbiology and back home to my roots as an environmental scientist. Bacteria are not just infectious agents that cause disease; the bacteria in the soil, as decomposers, play a vital role in the cycle of life. I went digging for germs, but I found a treasure! One to be cared for.

While my own research and recent reports raised concerns that pesticides could cause neurodegenerative disease in people, my concerns were now also with the effects of insecticides and their mixtures on the bacteria that make the soil their home. Disruption of soil micro-organisms with pesticides and other pollutants could have a devastating effect on soil health and crop production as well as disturb a fundamental balance in the ecosystem. Indeed, my laboratory is demonstrating that soil bacteria are susceptible to environmental pollutants such as pesticides and while bacteria can cause disease, poisoning bacteria in the environment may not only be the equivalent of poisoning ourselves, but is also ecologically and environmentally unsound.

In addition to having a valuable position in our ecosystem, bacteria are also valuable in the production of foods and beverages and are also vital in the production of vaccines and other medicines, including antibiotics. Further, bacteria have proved to be an essential tool for molecular biologists who are engaged in projects that range from gene therapy to genetically modified produce. Because of the real value of bacteria to the quality of human life, as well as their ability to cause disease, I saw the need for a course in microbiology for the non-science major. Microbiology embodies the beautiful and ugly, the simple and complex, and the big and the small of life and in this regard is a fascinating, useful and approachable topic for non-science majors to learn as well as being the same for scientists and professors as a tool to teach non-science majors.

In the development of my microbiology course for non-science majors, "Microbiology: The Biological Basis," I searched for an appropriate book and found Professor Ed Alcamo's, *Microbes and Society*. I was delighted to find that Professor Alcamo had prepared a book that covered the good and bad of microbiology with a solid infusion of science. I found *Microbes and Society* to be in most ways an ideal companion for my course. While I never met Professor Alcamo, I came to know his genuine interest in educating people about AIDS and HIV through his book *AIDS: The Biological Basis,* of which I have been honored to become a part. It is therefore a great honor to have the opportunity to continue to develop *Microbes and Society*, with a second edition built upon the foundation of Professor Alcamo's work.

Microbes and Society

I

The Microbial World

In areas of Great Britain, milkmen still visit homes regularly and deliver bottles of fresh, pasteurized milk. Unfortunately, magpies, crows, and other birds arrive soon thereafter and use their strong beaks to peck through the foil caps covering the bottles. They then help themselves to the milk. In doing so, they often transmit species of bacteria that cause intestinal illness. Unsuspecting families soon suffer the indignity and discomfort of diarrhea, cramps, and nausea. One British scientist has whimsically suggested: "Battered Brits better beware bacteria-bearing birds."

But we should not castigate all the microbes for the ills that a few species cause. In that same milkbox, for example, are a variety of dairy products that owe their existence to microbes. For instance, both the sour cream and buttermilk have been produced by harmless species of bacteria. The yogurt would remain condensed milk without the help of two types of bacteria. And those blue streaks in the Roquefort cheese are harmless fungi that have turned milk curds into a protein-rich and perfectly safe cheese we can sprinkle on a salad.

Indeed, the microbes are a fascinating group of living things. Unappreciated by the vast majority of people, microbes bring about changes that are both mind-boggling and awe-inspiring. For example, microbes are responsible for putting most of the oxygen we breathe into the atmosphere; they trap the nitrogen that all living things need to make proteins; they are waste decomposers *par excellence*; and they are used in industrial plants to make such diverse things as perfumes, growth supplements, and chocolate-covered cherries. Unfortunately, they are also responsible for the profound changes that occur during the course of infectious disease.

Microbes live in a world that we can scarcely comprehend. Many species survive quite well in oxygen-free environments; others live 6 miles below the ocean surface along boiling water vents; and still others thrive in acidic conditions that would dissolve a handkerchief. Moreover, some microbes have the ability to pick up genes from their environment and incorporate those genes into their hereditary material, and some can exchange genes with other microbes. Both of these processes are completely unknown in other living things. They point up some of the extraordinary processes taking place in microbes, processes that set the microbes apart from other living things and make knowledge of microbiology almost indispensable to all segments of society.

And so in Part I of *Microbes and Society,* we get to know the microbes. In the first six chapters, we explore how their world was discovered (Chapter 1) and how they fit into the scheme of living things (Chapter 2). In Chapters 3 and 4, we review the basic structure of microbial cells and highlight the roles of microbes in the discoveries surrounding DNA. Next, we look at the remarkable diversity they exhibit (Chapters 5 to 8). Here we get a flavor of the microbial world as we explore the unique characteristics of different groups of microbes and the influence they exert on society. Then, in the last three chapters of Part I, we see how microbes grow (Chapter 9), we survey their extraordinary genetics (Chapter 10), and we study some methods by which they can be controlled (Chapter 11). The pages of Part I contain some eye-opening concepts that will tune you into a microbial world unknown to most of your colleagues. And once we have an understanding of microbes and *their* world, we can better appreciate how they influence *our* world. But that's a story for Part II.

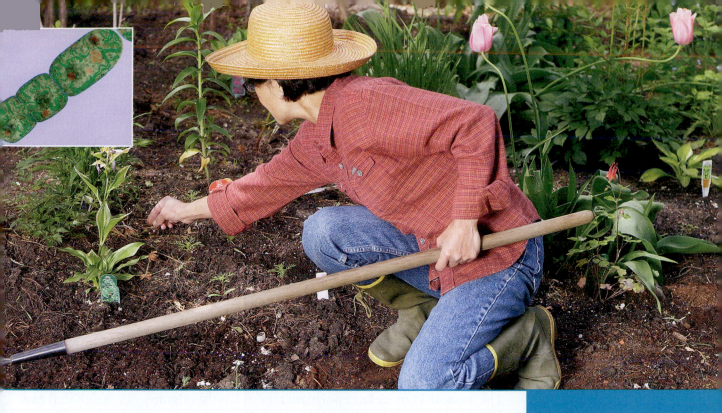

The Microbial World: Surprising and Stunning

1

Looking Ahead

We share our world with thousands of species of plants and animals that we can see and thousands of species of microbes that we cannot see. This microbial world is both surprising and stunning—surprising because it contains such a wealth of different forms of life, and stunning because we scarcely understand how those life forms affect our own world. This chapter begins our trek into the microbial world with some insights into an invisible realm.

On completing this chapter, you should be able to . . .

- appreciate how the microbial world affects our world in such areas as agriculture, industry, research, and the environment.
- provide numerous examples of how microbes contribute substantially to the quality of our lives.
- discuss the origins of microbiology and how the origins are related to the studies on spontaneous generation.
- describe the contributions of Louis Pasteur and Robert Koch to the development of the science of microbiology.

Perhaps you've heard the saying "there are more stars in the heavens than all the grains of sand on Earth." The same may be said of microbes and grains of soil, and without microbes life would be scarcely possible. The microbe in the inset is the cyanobacterium *Nostoc,* which can take nitrogen from the air and transfer it to the soil where it is essential for plant growth.

- provide thumbnail sketches of bacteria, viruses, protozoa, and other members of the microbial world.
- understand the diversity of microbes and explain why an appreciation of microbiology encourages an appreciation of life.

Each year, a group of pilgrims gather in the English countryside outside the village of Eyam and pay homage to the townsfolk who three and a half centuries before gave their lives so that others might live. The pilgrims bow their heads and remember what happened that fateful year.

Bubonic plague had erupted in Eyam during the spring of 1666, and before long, many had fled to the countryside. But most of the townsfolk realized that by doing so they would probably spread the plague to nearby communities. They were confronted not only by the terrible disease but also by the moral dilemma they faced. Then the village rector made a passionate plea that the remainder of the townsfolk stay, and after much soul-searching, they reluctantly decided to do so and take their chances. They marked off the village limits with a circle of stones, and the neighboring villagers brought food and other supplies to the self-quarantined group. In the end, 259 of Eyam's 350 people died of the plague.

The memorial service has a poignant moment as the pilgrims join their hands and somberly recite a poem whose roots trace to that period:

> *Ring-a-ring of rosies*
> *A pocketful of posies*
> *Achoo! Achoo!*
> *We all fall down.*

There is no laughter in the group; indeed, some are moved to tears. The ring of rosies refers to the rose-shaped splotches that surrounded the neck and shoulders of plague victims. Posies were flowers that people tucked into their pockets hoping to ward off the evil spirits. "Achoo" refers to the fits of sneezing that accompany plague; and the last line, the saddest of all, refers to the death that claimed so many plague victims.

Disease has always left people thunderstruck with terror. Before the late 1800s, however, that fear was compounded by ignorance because no one was really sure what causes disease, much less how to deal with it. As we shall see later in this chapter, the answers would not start coming until the late 1800s, when Louis Pasteur and Robert Koch solidified the link between infectious disease and microscopic forms of life. These life forms include bacteria, viruses, fungi, protozoa, and other groups of organisms invisible to the unaided eye. Many scientists refer to them as "microorganisms," but we shall use the equally acceptable and more simplified term microbes. This term was coined in 1879 by the French scientist Charles E. Sedillot. It implies any living thing that must be magnified to be seen.

To be sure, most of us are inclined to think of disease when we think of microbes. And this connection is probably justified because microbes cause much misery and pain. But microbes are also responsible for much that adds quality to our lives. For instance, microbes break down the remains of everything that dies and recycle the essential elements so that vital nutrients can be regenerated. Moreover, the very air we breathe is a product of microbial chemistry because certain microbes perform photosynthesis (the process by which light energy is converted to chemical energy) in the vast expanses of the forests and oceans, where they generate the oxygen that sustains us. Closer to home, many species of microbes live in our mouth, skin, intestines, respiratory tract, and other body systems, where they prevent the multiplication of organisms that might cause disease. Microbes are also responsible for the final forms of many foods we eat, including fermented dairy products such as yogurt, buttermilk, and sour cream. We shall embellish on these and other ways in which microbes influence humans and society as we begin our journey through the microbial world.

Why Microbes Matter

If you took a pinch of rich soil and placed it in the palm of your hand, you would come face-to-face with an estimated billion microbes—bacteria, viruses, fungi, protozoa, and numerous other microscopic forms of life, shown in FIGURE 1.1 . To appreciate the extraordinarily small size of the microbes, consider this: If you were to count the microbes in a pinch of soil at a rate of one per second without stopping, it would take over 33 *years* to complete your counting.

But we are not here to count microbes. Rather, we are here to appreciate them, and so we should ask: What are the microbes doing in the soil, and why are they there? To be sure, some are disease-causing (pathogenic) microbes, probably in transit from one living thing to another. Others, however, are probably quite benign. By their sheer numbers, they control the pathogens in the soil and maintain nature's balance in the environment. Many species capture energy from the sun and store that energy in the form of sugar molecules; to do this, they use the chemistry of photosynthesis, the same chemistry that adds

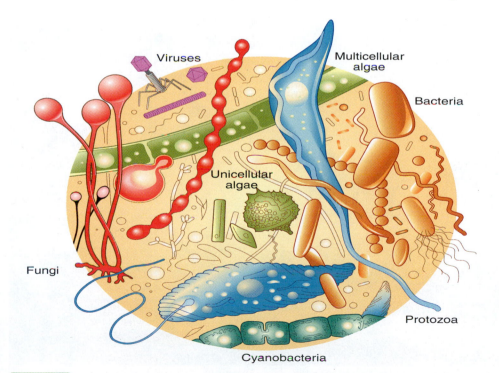

FIGURE 1.1 **A Microbial Menagerie.** A menagerie of microbes as conceived by an artist. Most of these organisms and many others are present in a pinch of rich soil. All are discussed in later chapters of this book.

oxygen to our atmosphere. Other species are decomposers, the great recyclers of nature that return carbon, nitrogen, sulfur, and other elements to the atmosphere so that they can be used to renew the chemical molecules essential to life's continuance.

And these are but a few of the ways that microbes benefit society. Some other examples are presented in the following paragraphs.

Down on the Farm

A cow rests peacefully under a tree on a sunlit day in an open meadow and chews its cud. The scene is calm and reassuring and makes us feel that all is well in the world (FIGURE 1.2). We might even be inclined to think of Beethoven's beautiful Sixth Symphony—*The Pastorale*—with its endless visions of the charm and grace of the countryside.

But in the gut of the cow, microbes are busily at work. They have mixed with grasses in the first stomach of the cow (the rumen) and have begun the conversion of carbohydrate to protein, a conversion that will turn parts of the plants the cow eats into its muscle and dairy products. Soon the cow will regurgitate the material from the rumen, chew it a bit to mix in some saliva, then reswallow the mixture for more fermentation. At the end of the line, we can expect hamburgers and dairy products, as we explore further in Chapter 15.

In other areas of agriculture, humans use microbes as insecticides and pesticides. One bacterium called *Bacillus thuringiensis* is sprayed onto plants, where its poisonous toxins kill the caterpillar forms of many agricultural pests. Biotechnologists have even extracted from these bacteria the genes that encode their toxins and biochemically in-

■ *B. thuringiensis*
B. thur-in-jē-en'sis

FIGURE 1.2 **Microbial Factories.** These cows lazily enjoying the afternoon sun are in fact microbial factories. Within the cow's first stomach (the rumen), various species of microbes are breaking down grasses and plants and assisting the conversion of carbohydrates to proteins of the cow's muscle and dairy products. The cow chews the regurgitated mass, the cud, to help the microbes make the essential conversions.

serted those genes into plant cells. Remarkably, the plants then produce the insecticidal toxins as they grow in the field. Most of the soybean plants currently in American fields contain these bacterial genes. And this astounding accomplishment is only the beginning of what promises to be an agricultural revolution in the decades ahead.

In the Industrial Plant

On an industrial scale, microbes are cultivated in huge batches in building-sized tanks, where they produce chemicals such as lactic, acetic, and other organic acids, as well as the amino acids used to construct proteins. The citric acid in a bottle of soda, for example, is produced by fungi. And bacterial proteins produced in industrial quantities are used in laundry detergents, as well as in adhesives, and in baking, as we explore in Chapter 14. In our foods, microbial proteins are used to tenderize meat, to clarify fruit juices, and to soften the centers of chocolate-covered cherries. Working like miniature chemical factories, the microbes churn out industrial quantities of vitamins (especially B vitamins) that eventually find their way into the tablets we take to promote good health. Even some perfumes contain products of microbial origin—musk oil, for instance—is the product of the chemistry in a fungus.

Although most natural food flavorings are produced from plants by traditional processes, new industrial methods have made it possible to use microbes to convert relatively cheap starting materials into higher-value flavor and aroma additives. These additives are then used in foods and beverages as well as in cosmetics and other consumer items. Examples are the peach and coconut flavoring agents belonging to a chemical group called lactones. Although lactones can be generated from the fats in sweet potatoes, these fats are expensive to modify and occur only in low concentrations in the potatoes. To circumvent these problems, microbiologists use fungi to convert less expensive and more available fats into compounds that other microbes can easily transform to lactones. Transitions like these were studied a century ago by one of the first industrial microbiologists, whose life is explored in A Closer Look on page 10.

At the Hub of Biotechnology

Perhaps the most sophisticated contemporary use of microbes is in biotechnology. In this industry, various species of bacteria and yeasts are used as living factories to produce such products as insulin for diabetics, blood-clotting Factor VIII for hemophiliacs, and human growth hormone for undersized individuals. Microbes, especially viruses, are also used to ferry genes into tissue cells during gene therapy, as we note later in this chapter. Other viruses have been genetically altered to increase disease resistance in plants and to kill insect pests in the environment, as Chapter 16 explains. And some yeasts have been modified to produce viral proteins that can be used in vaccines, such as the one for hepatitis B.

There is even a bacterial cement that researchers have used to fix cracks in concrete blocks. Biochemists begin with species of bacteria that digest a nitrogen-containing compound called urea (the major waste product of urine). They mix the bacteria with sand, calcium chloride, and urea. When the bacteria break down the urea, they produce carbon dioxide and ammonia, which reacts with water to form ammonium hydroxide. The latter reacts with the carbon dioxide and the calcium chloride to produce calcium carbonate, which crystallizes to form limestone and fill the cracks in the concrete. One observer has whimsically suggested that science has finally found a "concrete use for bacteria."

A CLOSER LOOK
Of Microbes and Cherry Trees

During the 1880s, while medical microbiologists were searching out the causes of infectious disease, a Japanese biochemist was busy acquiring enormous experience and wealth as one of the first industrial microbiologists. His name was Jokichi Takamini.

In the 1870s, Takamini left his native Japan to study engineering in the West. By the 1890s, he had married an American and was an expert on the enzymes of a common mold named *Aspergillus* (a-spèr-jil'lus). In one of his first successes, Takamini isolated and purified an *Aspergillus* enzyme, and he used it to help make whiskey at his fermentation plant in Peoria, Illinois. Local brewers turned against his innovative methods, however, so he patented the enzyme and licensed it to a pharmaceutical company. The company mixed it with peppermint and sold it as a digestive aid. Marketed as "takadiastase," the enzyme became the Alka-Seltzer of the 1890s.

Takamini next moved to New York, where he and an assistant set out to isolate the active principle (the hormone) in adrenal glands. For months, they separated, precipitated, dissolved, purified, and repurified in their hunt for the elusive substance. Then one night, his assistant, too tired to wash the glassware, left it by the sink and went home. The next day, a glass containing extracts from adrenal gland tissue was lined with crystals of epinephrine (adrenaline). The isolation was a success. Takamini's patent rights to epinephrine yielded wealth beyond his imagination.

In later years, Takamini became a patron of the arts and a philanthropist. On August 30, 1909, the city of Tokyo announced a gift to the people of the United States: The city would be honored to adorn the Tidal Basin in Washington, D.C. with dozens of Japanese cherry trees. In the decades that followed, the cherry trees grew to become a splendid attraction in the nation's capital and an annual harbinger of spring. What few people know is that the gift was funded anonymously by Jokichi Takamini.

With Most Meals

Rarely does a meal go by when we do not rely on microbes for something on our plate or in our glass (as **FIGURE 1.3** illustrates). Sausages, for example, owe their tangy taste to microbial activity. The food industry also depends on microbes to produce sauerkraut, pickles, and vinegar. Many dairy products and cheeses result from microbial action on milk. Virtually all types of bread depend on microbes for their taste and spongy textures. The wine and beer at the hub of the fermentation industries could not be produced without microbes. Even the coffee we drink depends in part on microbes for its flavor and aroma. Chapter 12 explores these and other fascinating uses for microbes in food production.

On the negative side, foods are among the principal vehicles for transporting pathogenic microbes among individuals. Chicken, eggs, and other poultry products, for example, are notorious for their populations of *Salmonella*, which can infect individuals who fail to cook these foods properly. In some cases, individuals unknowingly contaminate uncooked foods. For example, they use a knife to cut up raw chicken and then, without washing it, use the same knife to cut up salad fixings. Scientists are working to resolve the *Salmonella* problem, as A Closer Look on page 11 points out. In the meantime, public health officials advise thorough cooking of poultry products. And when preparing eggs, the operative expression from health agencies is "Scramble or gamble."

■ Salmonella
säl-mōn-el'lä

FIGURE 1.3 **An Array of Foods and Beverages Produced by Microbes.** Society is dependent on microbes for many of its foods. Fermented foods and dairy products are examples.

A CLOSER LOOK

"Sorry, No Vacancy"

In the old days, chicks hatched from their eggs, scrambled to their feet, and nestled under their mother's wing until it was safe to come out. By staying close to their mother hen, the chicks received protection, caring . . . and bacteria. Bacteria? Yes, bacteria—hundreds of strains of harmless organisms that entered the chicks' guts and prevented *Salmonella* species from causing infection. With all the harmless enterococci, lactobacilli, and other strains of bacteria lurking about, there simply was no room for *Salmonella*. The chicks remained healthy.

But mother hen is gone. The high-tech chicken farms of today use machines to remove the eggs from the hens as they are produced. The chicks hatch and develop without ever seeing what made them. To be sure, that is sad, but microbiologically speaking, the sadness is compounded by the absence of the harmless bacteria from the chick's gut. This absence encourages the chicks to acquire and develop *Salmonella* infection. And when they become fully grown chickens, they pass the *Salmonella* on to humans, where it causes severe intestinal disease.

As of 1998, there was an answer. That year, the Food and Drug Administration approved a spray called Preempt that showers chicks with a mix of 29 species of harmless bacteria. The chicks pick the bacteria off their feathers and swallow them. As the harmless bacteria set up housekeeping in a chick's gut, they compete with and exclude the dangerous ones. (Scientists call the process "competitive exclusion.") Numerous research experiments show that the harmless bacteria crowd out the *Salmonella* species and significantly reduce or completely eliminate them. Once again, science has replaced mother hen.

In the Environment

Although many people think of microbes as water pollutants, the more modern view is that microbes can be effectively used to alleviate pollution. This immensely appealing way of putting microbes to work for society is called bioremediation. When an oil spill occurs, for instance, technologists add nitrogen, phosphorus, and other mineral nutrients to the water and encourage microbes to grow and gorge themselves on the petroleum (as Chapter 16 details). Oil in areas treated this way clears away faster than that in areas not treated. Bioremediation is also used to eliminate other environmental pollutants, such as the waste products of explosives, as well as cleaning agents and radioactive compounds.

Additionally, microbes remain the prime factor in sewage treatment. In some municipalities, waste is piped into oxidation lagoons, where microbes digest the organic matter and completely convert the complex compounds into simple ones that can be recycled. In larger cities, waste treatment plants rely on microbial chemistry to handle the massive amounts of sewage and garbage generated daily. Various stages of treatment address various types of waste, and soon the ugly profusion of grime, gunk, and grot is converted to chemical fertilizers that can be used as soil enhancers for growing crops.

In the Pharmaceutical Lab

Most microbiologists are dreamers, and those who work in pharmaceutical laboratories are among the more imaginative ones. These individuals continue the search for new medicines to combat the emerging diseases of our era. For example, when a certain author was growing up in the Bronx, doctors had not yet heard of Lyme disease, Legionnaire's disease, or AIDS. In today's world, each of these diseases is a serious health problem, and pharmaceutical scientists are hard at work searching for effective treatments and vaccines. They try modifications of existing drugs while exploring new approaches for innovative therapies, as we study in Chapter 15.

And the approaches they use often boggle the imagination. For example, it is now possible to extract the genes from the virus that causes genital herpes and insert those genes into cells isolated from the human immune system. Using the new genes in conjunction with their own genes, the cells then synthesize antibodies, a series of highly specific proteins that bind to and destroy the herpes virus in the body. But there is more: Biotechnologists can also obtain from the cells the complete set of antibody genes and transfer them to soybean plants. Sown in the fields, the soybean plants produce antiherpes antibodies. (Scientists whimsically refer to these antibodies as "plantibodies.") Indeed, it may some day be possible to fight genital herpes merely by eating a soybean product.

New diagnostic tests are also in the future. Pharmaceutical scientists (FIGURE 1.4) are now producing gene probes, collections of small fragments of deoxyribonucleic acid (DNA), the material of which genes are made. Gene probes seek out and unite with the complementary DNA of a microbe much like a left hand meets its complementary right hand when one claps. Once the match is made, a biochemical signal is sent, and the technologist knows that a specific microbe is present. This remarkable technology even makes it possible to diagnose a disease without cultivating the responsible pathogen. And the probes can also be used to find the DNA fingerprint of a microbial contaminant (rather than the microbe itself) in an environmental sample.

FIGURE 1.4 **The Research Lab.** Scientists continue to use microbes as centerpieces of their research to develop imaginative and novel pharmaceutical products.

The explosive developments in pharmaceutical microbiology are paralleled by equally explosive research in other arenas. For example, scientists are exploring the hydrothermal vents at the bottoms of the oceans and identifying strange and unusual microbes; they are looking for evidence of life on Mars by determining whether microbes could survive conditions on that planet; they are discovering new clues to evolution by studying the DNA of microbial forms that have been in existence for eons; and they are developing new concepts of the richness of life on Earth by focusing on microbial diversity. To be sure, it is a wonderful time to be studying microbiology.

However, we must pause before we get too deeply into the study of microbes, for, as one philosopher has said, to understand where you are going, you must know where you have been. And so before we launch into our study of microbiology, we shall survey its origins; for our roots give us the strength to grow upward and outward.

The Roots of Microbiology

The microbial world was virtually unknown until the mid-1600s, when an English scientist named Robert Hooke became fascinated with a newly developed instrument, the microscope, and wrote about his observations. Although Hooke is best remembered for his descriptions of minute compartments in slices of cork (which he named "cells"), he also reported the microscopic details of threadlike fungi, referring to them as "elongated stalks." Hooke's curiosity led him to study the fungus that infects rose plants, and his illustrations of its threads were among the first descriptions of a microbe. But Hooke could scarcely imagine what others would find in this invisible world.

Anton van Leeuwenhoek

The individual best remembered for bringing microbes to the world's attention is a Dutch merchant named Anton van Leeuwenhoek, pictured in FIGURE 1.5 . In the 1670s, van Leeuwenhoek developed the skill of grinding lenses for the purpose of magnifying and inspecting cloth. Soon he was using the lenses to satisfy his own curiosity. Squinting through a lens, van Leeuwenhoek studied the eye of an insect, the scales of a frog's skin, and the intricate details of muscle cells. In 1673, while peering into a drop of pond water, he came upon microscopic forms of life, darting back and forth and rolling and tumbling in this microcosm of life. He dubbed the microbes "animalcules" (he assumed they were tiny animals). They at first delighted him, then amazed him with their variety, and finally perplexed him as he pondered their meaning. Today, we recognize that the microbes he saw were protozoa. To the best of our knowledge, few had seen them before.

Van Leeuwenhoek excitedly communicated his findings to a group of English scientists called the Royal Society. The members of the society encouraged van Leeuwenhoek to continue his work, and over the next 40 years, he wrote a long series of letters describing the new microscopic forms he observed. His letter of September 17, 1683 is particularly noteworthy because it apparently contains the first known descriptions of bacteria. In other letters, he described sperm cells and drew representations of microscopic yeast cells he gathered from the bottom of a beer vat. Van Leeuwenhoek became one of the most famous individuals of his time, and he hosted royalty and heads of state, who came to peer into his lenses. When he died in 1723, he was 90 years old, in itself an achievement for those times.

■ Leeuwenhoek
  LAY-wen-hoke

(a)

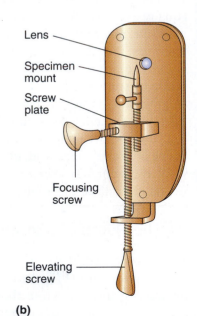

(b)

FIGURE 1.5 **Anton van Leeuwenhoek.** (a) Van Leeuwenhoek at work in his study. Using a primitive microscope, van Leeuwenhoek was able to achieve magnifications of over 200 times and describe various biological specimens, including numerous types of microbes. (b) Details of van Leeuwenhoek's lens system. The object is placed on the point of the specimen mount. The mount is adjusted by turning the focusing screw and the elevating screw. Light is reflected from the specimen through the lens, thereby magnifying the specimen.

Louis Pasteur

After the passing of van Leeuwenhoek, interest in microbes gradually waned for at least two reasons: Scientists lacked the technology for learning about microbes; and most people believed that microbes were mere curiosities with little or no effect on society. In the 1850s, however, both of these concepts changed with the work of Louis Pasteur and his contemporaries. Pasteur, the renowned French scientist, called attention to the microbes as possible agents of infectious disease. When he was proven correct, the technology for studying microbes rapidly developed.

Pasteur, shown in FIGURE 1.6 , believed that the discoveries of science should have practical applications, and in 1857, he seized the opportunity to try and unravel the mystery of why French wines were turning sour. The prevailing theory held that fermentation results from the purely chemical breakdown of grape juice to alcohol. No living thing seemed to be involved. But Pasteur's microscope consistently revealed that wine contained large numbers of tiny microbes known as yeast cells. Moreover, he noticed that sour wines contained populations of the barely visible bacteria described by van Leeuwenhoek. In a classic series of experiments, Pasteur boiled several flasks of grape juice and removed all traces of yeast cells from the flasks; he then set the juice aside to see if it would ferment. Nothing happened. Next, he carefully added pure yeast cells back into the flasks, and soon the fermentation was proceeding normally. Moreover, he found that if he used heat to remove all bacteria from the grape juice, the wine would not turn sour; it would not "get sick."

Pasteur's work shook the scientific community because it showed that microscopic yeast cells and bacteria are tiny, living factories where important chemical changes take place; indeed, scientists began to wonder if bacteria could also make people sick. In 1857, Pasteur published a short paper on the bacterial souring of milk, and he implied that microbes might be related to human illness. In so doing, he set down the foundation for

(a)   (b)

FIGURE 1.6 **Louis Pasteur.** Two portraits of Louis Pasteur, one of the founders of the science of microbiology. (**a**) As a young man while studying chemistry at the École Normale Supèrieure. (**b**) As a scientist working in a laboratory in his home. In his right hand, Pasteur holds a flask of broth previously exposed to microbes in the air and now cloudy with growth. In his left hand, he is holding a swan-neck flask, in which the broth is clear because airborne microscopic organisms could not enter through the neck of the flask.

the germ theory of disease, a fundamental tenet holding that microbes play significant roles in the development of infectious disease. Pasteur also recommended using heat to control bacterial contamination. Acceptance of his technique, known as pasteurization, eventually ended the sour wine problem and made Pasteur famous.

Pasteur's interest in microbes grew as he learned more about them. He found bacteria in soil, water, and air, and, importantly, in the blood of disease victims. He reasoned that microbes might enter the blood from the environment; if so, perhaps they could be killed in the environment and their spread halted. However, many other scientists had a different opinion: They believed that bacteria arose spontaneously in a diseased patient; therefore, disease was inevitable and had nothing to do with bacteria or any microbes. This belief was called spontaneous generation. Pasteur had to discredit spontaneous generation to salvage his germ theory of disease.

In an elegant series of experiments, Pasteur prepared nutrient-rich broth in a series of swan-neck flasks (so named because their S-shaped necks resembled those of swans). Pasteur boiled the broth in the flasks, thereby destroying all microbes; then he left the flasks open to the air. However, the S-shaped neck trapped any microbes in the air and prevented their entry into the flasks. Thus, when the flasks were set aside to incubate in a warm environment, no microbes appeared in the broth (advocates of spontaneous generation assumed that microbes would arise, since air and the so-called life force were being allowed in). Moreover, when the neck was cut off a flask and microbes from the air dropped into the broth, the broth soon became cloudy with microbial growth. It was clear that the microbes were in the air and were not spontaneously arising from nutrients in the broth. Pasteur's classic experiment is illustrated in **FIGURE 1.7**.

Pasteur's work brought to an end a long and tenacious debate on spontaneous generation begun two centuries earlier. Moreover, his work showed that microbes could cause disease. But Pasteur was stymied by his inability to obtain a pure culture by cultivating one type of bacterium apart from other types. In an effort to help French industry once again, he turned his attention to pébrine, a disease of silkworms. In 1865, he identified a protozoan infesting the silkworms and, by separating healthy silkworms from diseased ones and their food, managed to quell the spread of the disease. Still the definitive proof eluded him.

Robert Koch

Although Pasteur failed to relate a specific organism to a specific human disease, his work stimulated others to investigate the association of microbes with disease. Among them was Robert Koch, a country doctor from East Prussia (now part of Germany). Koch's primary interest was anthrax, a deadly blood disease in cattle and sheep (now known as a weapon of bioterrorists).

Koch was determined to learn all he could about anthrax. In 1875, in a makeshift laboratory in his home, he injected mice with the blood of sheep that had died from anthrax. He then performed meticulous autopsies and noted that the same symptoms appeared regularly in the mice. Next, he magnified a blood specimen under his microscope and observed rod-shaped bacteria, all apparently of the same type. Determined to cultivate the bacteria, he visited the local butcher and obtained the eye of a cow, from which he obtained the clear fluid in the inner chamber. Then, while peering intently through his microscope, he successfully separated a few anthrax bacteria from the sheep blood and added them to the clear fluid. Next, he watched for hours as the bacteria elongated, multiplied, formed tangled threads, and reverted to highly

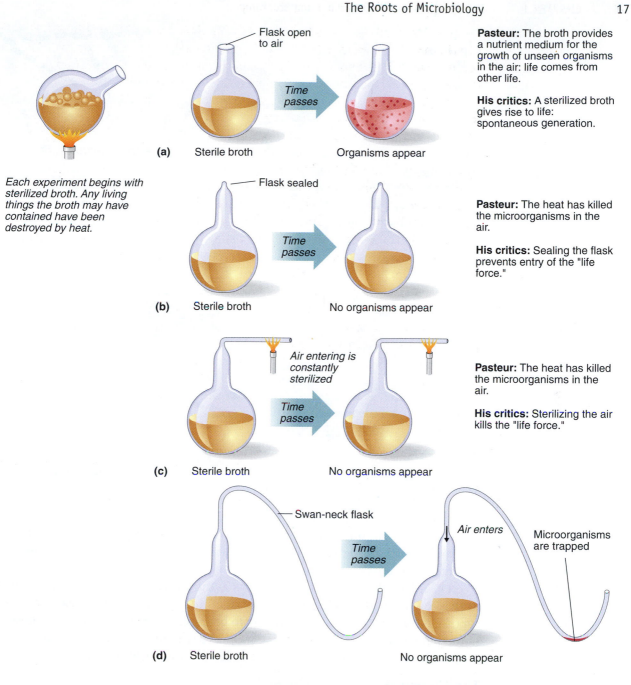

Each experiment begins with sterilized broth. Any living things the broth may have contained have been destroyed by heat.

(a) Sterile broth — Flask open to air — *Time passes* — Organisms appear

Pasteur: The broth provides a nutrient medium for the growth of unseen organisms in the air: life comes from other life.

His critics: A sterilized broth gives rise to life: spontaneous generation.

(b) Sterile broth — Flask sealed — *Time passes* — No organisms appear

Pasteur: The heat has killed the microorganisms in the air.

His critics: Sealing the flask prevents entry of the "life force."

(c) Sterile broth — Air entering is constantly sterilized — *Time passes* — No organisms appear

Pasteur: The heat has killed the microorganisms in the air.

His critics: Sterilizing the air kills the "life force."

(d) Sterile broth — Swan-neck flask — *Time passes* — Air enters — Microorganisms are trapped — No organisms appear

Pasteur: No living thing will appear in the flask because microorganisms will not be able to reach the broth.

His critics: If the "life force" has free access to the flask, life will appear, given enough time.

Some days later the flask is still free of any living thing. Pasteur has disproved the doctrine of spontaneous generation.

FIGURE 1.7 **Pasteur and the Spontaneous Generation Controversy.** Louis Pasteur engaged in a series of experiments to show that spontaneous generation is not a valid doctrine. Pasteur's experiments and the objections of his critics are displayed. (**a**) When a flask of sterilized broth is left open to the air, organisms appear. (**b**) When a flask of sterilized broth is boiled and sealed, no living things appear. (**c**) When air entering a flask of sterilized broth is heated with a flame, no living things appear. (**d**) Broth sterilized in a swan-neck flask is left open to the air. The curvature of the neck traps dust particles and microbes, preventing them from reaching the broth.

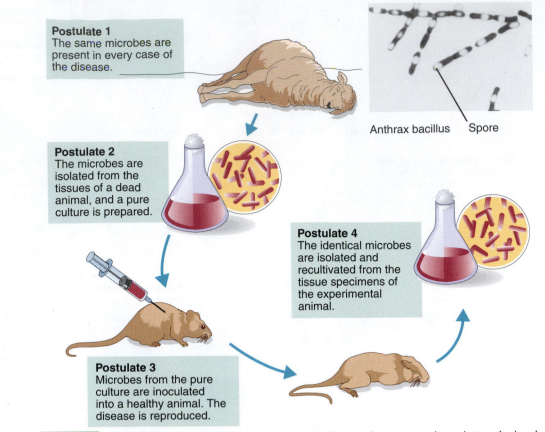

Postulate 1
The same microbes are present in every case of the disease.

Anthrax bacillus   Spore

Postulate 2
The microbes are isolated from the tissues of a dead animal, and a pure culture is prepared.

Postulate 4
The identical microbes are isolated and recultivated from the tissue specimens of the experimental animal.

Postulate 3
Microbes from the pure culture are inoculated into a healthy animal. The disease is reproduced.

FIGURE 1.8 **A Demonstration of Koch's Postulates.** Koch's postulates are used to relate a single microbe to a single disease. The photo shows the rods of the anthrax bacillus as Koch observed them. Many rods are swollen with spores.

resistant bodies called *spores*. At this point, Koch took some spores on a sliver of wood and injected them into healthy mice. Several hours later, the symptoms of anthrax appeared. Koch excitedly autopsied the mice and found their blood swarming with anthrax bacteria. The cycle was complete.

A year later, Koch presented his work at the University of Breslau. The scientists were astonished. Here was the verification of the germ theory of disease that had escaped Pasteur. Koch's procedures came to be known as *Koch's postulates*. These techniques, illustrated in **FIGURE 1.8**, were adopted as a guide for relating specific microbes to specific diseases. They are still used today for this purpose.

The Golden Age of Microbiology

The science of microbiology blossomed during a period of about 60 years, now referred to as the Golden Age of Microbiology. This period began in 1857 with Pasteur's proposal of the germ theory of disease, and it continued into the twentieth century, until the advent of World War I. During these years, numerous branches of microbiology were established, and the foundations were laid for the maturing process that has led to the modern science.

During the Golden Age, a lively competition sprang up between France and Germany to see which country would lead the newly emerging field of microbiology. For example, Koch presented his work in 1876, and within 2 years, Pasteur reported that some

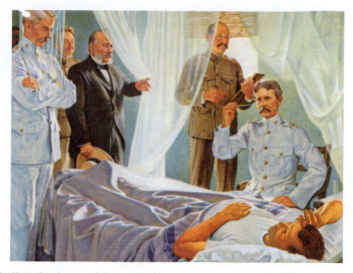

FIGURE 1.9 **Finding the Cause of Yellow Fever.** A painting by Robert Thom depicting members of the 1900 yellow fever commission in Cuba at the bedside of Private John Kissinger. Kissinger allowed himself to be bitten by mosquitoes after they had bitten yellow fever patients. When asked why he and others were participating in the hazardous experiments, Kissinger replied: "We volunteer solely for the cause of humanity and in the interest of science." Left to right: Major W. C. Gorgas, Aristides Agaramonte, Carlos J. Finlay, James Carroll, and Walter Reed.

types of bacteria were temperature-sensitive: Chickens did not acquire anthrax at their normal body temperature of 42° C but did so when the animals were cooled down to 37° C. Then, in 1880, Pasteur found that he could enfeeble the bacteria associated with chicken cholera and inject them into healthy chickens, where they protected the animals against a lethal dose of the bacteria. This principle is the basis for today's vaccines.

Pasteur's experiments encouraged France to take the microbiological lead. However, Koch soon isolated the microbe that causes tuberculosis, and his coworkers were the first to cultivate the bacteria that cause typhoid fever and diphtheria. Before long, news came from France that scientists in Pasteur's lab had linked diphtheria to a chemical poison, a toxin, which is produced by the bacterium that the Germans had discovered. In later years, Koch's coworker Emil von Behring successfully treated diphtheria in patients by injecting antitoxin, a preparation of blood proteins obtained from animals immunized against diphtheria. For his work, von Behring was awarded the first Nobel Prize in Physiology or Medicine. A Closer Look on page 20 describes the history of this prestigious award.

By the turn of the century, microbiology had become international in scope, moving far beyond France and Germany: Ronald Ross, an English physician working in the Far East, proved that mosquitoes transmit the microbes that cause malaria; another Englishman, David Bruce, showed that tsetse flies transmit sleeping sickness and opened the African continent to colonization; and the Japanese investigator Masaki Ogata reported that rat fleas transmit bubonic plague, thereby solving the centuries-old mystery of how plague spread. American microbiologists were represented by Howard Taylor Ricketts, who located the agent of Rocky Mountain spotted fever in the human bloodstream and demonstrated its transmission via ticks; and by Walter Reed, who led a contingent to Cuba to pinpoint mosquitoes as the insects that transmit yellow fever. His discovery led to the mosquito eradication programs that made possible the building of the Panama Canal. Reed and his group are portrayed in **FIGURE 1.9** .

A CLOSER LOOK
The Nobel Prize

The Nobel Prizes are among the world's most venerated awards. They were first conceived as a gesture of peace by a man whose discovery had unintentionally added to the destructive forces of warfare.

Alfred Bernhard Nobel was the third son of a Swedish munitions expert. As a young engineer, he developed an interest in nitroglycerine, an oily substance 25 times more explosive than gunpowder. In 1863, Nobel obtained a patent for a detonator of mercury fulminate, and within 4 years, he used it to ignite solid nitroglycerine mixed with a type of sandy clay. The mixture was called dynamite, from the Greek *dynamis* meaning "power."

Dynamite had a clear advantage over other explosives because it could be transported easily and handled with less fear. It became an overnight success and was adapted to applications in mining, tunnel construction, and bridge and road building. Before long, dynamite was being used in armaments on the battlefield.

Nobel soon amassed a fortune through the control of several European companies that produced dynamite. However, toward the end of his life, he became a pacifist and began speaking out against the use of his explosive in warfare. In 26 lines of his handwritten will, Nobel directed that the bulk of his estate should be used to award prizes that would promote peace, friendship, and service to humanity.

After his death in 1896, the governments of Sweden and Norway established Nobel Prizes in five categories: chemistry, physics, physiology or medicine, literature, and peace. A sixth category, economics, was added in 1969. Every year, Nobel laureates assemble in Oslo or Stockholm on December 10, the anniversary of Nobel's death. Each laureate receives a medallion, a scroll, and all or part of a cash award currently valued at about $1 million per category.

The first Nobel Prize winners were announced in 1901. Among the recipients were Wilhelm K. Roentgen, the discoverer of X rays; Jean Henri Durant, the founder of the Red Cross; and Emil von Behring, the developer of the diphtheria antitoxin.

Amid the burgeoning interest in microbes, other scientists devoted research to their environmental importance. The Russian scientist Sergius Winogradsky, for example, discovered that certain bacteria use carbon dioxide to synthesize sugars, much as plants do in the process of photosynthesis. And Martinus Beijerinck, a Dutch investigator, isolated bacteria that trap nitrogen in the soil and make it available to plants for use in constructing amino acids and proteins. It was clear that interest in microbes was reaching far beyond their importance in medicine.

Into the Twentieth Century

With the advent of World War I, scientists turned to research on blood products and vaccines to treat and prevent war-related infections. In the postwar years, scientists researched the key roles that microbes play in industrial processes. For example, food microbiologists identified the processes by which microbes manufacture numerous dairy products.

Work with viruses expanded greatly during the 1930s and 1940s with the invention of the electron microscope. An ordinary light microscope permits magnifications of 1,000 times, while the electron microscope permits magnifications of 100,000 times. The mysterious viruses could now be visualized for the first time, and an intense period of research in virology ensued (Chapter 5). Also about that time, antibiotics were discovered, and physicians were presented with previously unimagined therapies for established cases of disease.

In the period after World War II, large sums of money became available for biochemical research, and scientists worked out the key processes by which microbes synthesize proteins. These processes are centered in the hereditary units called genes, and soon the discipline of microbial genetics was in full flower. Elegant applications of this research were realized in the 1970s, with the advent of genetic engineering and biotechnology. Scientists learned to isolate genes, manipulate and splice them, and control the biosynthetic processes that genes oversee in microbial cells.

With the seminal discoveries in gene research, the Age of Molecular Genetics dawned, and today we are experiencing the fruits of that age. We live in a world where bacteria are engineered to destroy pollutants, where yeasts are reconfigured to produce human proteins, and where viruses carry therapeutic genes into patients with inherited diseases. Nor does the harvest end here. Scientists continue to dream of genetically modified plants that are rich enough in protein to resolve the world's shortage of this nutrient; they have given us new tests that dramatically reduce the time to diagnose disease; and they are just beginning to ponder the implications of knowing the intricate details of the human genome (Chapter 4). Truly, they have changed the face of science.

And the microbe stands at the center of all these discoveries and advances. Microbiology is a very hot topic in today's world, and you are fortunate to be studying this discipline of biology. Microbiologists range from the research scientist to the food technologist to the environmental scientist; they include the individuals who study industrial microbiology, agriculture, and medicine. Microbiology penetrates virtually every aspect of our existence, and as we shall see time and time again, microbes are essential elements for improving the quality of our society and our lives.

The Microbial World

The microbial world includes a diverse variety of organisms that inhabit most environments on Earth. This world is the subject matter of the discipline of microbiology and this book. The microbes are extremely diverse, as the following paragraphs will illustrate.

Bacteria

Among the best known of all microbes are the bacteria. Not only are the bacteria important in research and clinical medicine, but they are also vital to many beneficial activities that take place in the environment. They recycle the elements in dead and decaying matter, purify many of our waterways, and produce numerous foods we eat. A collection of bacteria magnified thousands of times with an electron microscope is shown in FIGURE 1.10 .

Bacteria have existed on Earth for approximately 3.5 billion years. Over this extraordinarily long period of time, they have evolved to occupy every conceivable niche on Earth. For instance, various species live in the outer reaches of the atmosphere, at the bottoms of the oceans, in the frigid valleys of the Antarctic continent, and in the scalding hot deserts of Africa. No other organisms of any kind (including humans) have adapted so thoroughly to Earth's varied conditions.

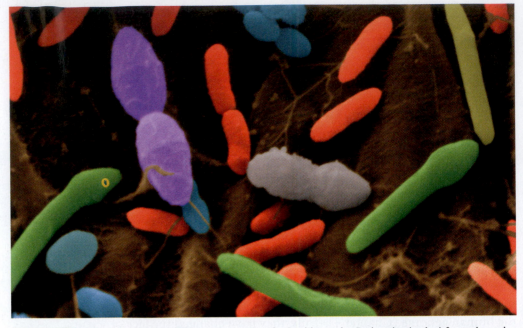

FIGURE 1.10 **Mixed Bacteria.** A false color image of mixed bacteria. Rod and spherical forms in various sizes are visible in this view. The bacteria shown here have been color enhanced to help distinguish them. These bacteria are not this color normally.

As we discuss in Chapter 5, most bacteria fit into three broad groups according to their shapes. There are rod-shaped bacteria, spherical bacteria, and spiral bacteria. Many bacterial species are heterotrophic—that is, they acquire their food from organic matter. Some species are autotrophic—they can synthesize their own food from basic elements. Most species of bacteria utilize oxygen in their chemistry, but the bacteria existing in oxygen-poor environments are also important. In tightly compacted landfills, for instance, such bacteria decay the garbage and recycle the elements for reuse by other organisms.

As we mentioned earlier, many species of bacteria contribute to society in the food industry, in the industrial plant, on the farm, and in the genetic engineering lab. In addition, bacteria make their presence felt in the environment. Certain species, for example, live on the roots of pod-bearing plants called legumes. Here, they extract nitrogen from the air and make it available to the plants. The plants then use the nitrogen to construct the proteins they need, which are also needed by animals, including humans.

Unfortunately, many species of bacteria are involved in disease. Using their considerable powers of reproduction and their ability to overcome body defenses, bacteria can infect vital tissues or organ systems and bring on illness and death. Moreover, certain bacterial species produce toxins that interfere with physiological processes in the body. For example, the toxin produced by the botulism microbe interrupts the passage of impulses between nerve and muscle cells. Inhibition of muscle contraction follows, and in many situations, paralysis and death ensue. We study bacterial disease in Chapter 18.

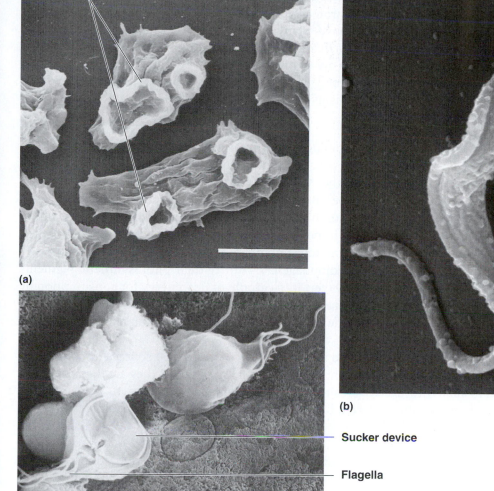

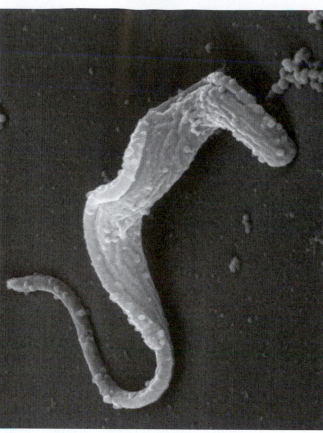

(a)

Adhesive disk

(b)

Sucker device

Flagella

(c)

 **Three Species of Protozoa.** Electron micrographs of three species of protozoa illustrating differences in shape and size. (**a**) The amoeba *Naegleria fowleri,* a cause of meningitis in humans. Note the irregular shape of this organism. (**b**) A protozoan called a trypanosome. The elongated shape and hairlike appendage (flagellum) are apparent. This organism causes African sleeping sickness. (**c**) Another flagellated protozoan, *Giardia lamblia* (×8062). The organism has a flat shape with multiple flagella extending toward the rear. The protozoan on the left shows the sucker device on its lower surface for holding fast to tissue. *Giardia* causes diarrhea in humans.

Protozoa

Protozoa are a diverse group of microbes that differ substantially from the bacteria and viruses in their structural composition. For example, protozoal cells have nuclei, while bacterial cells do not. In addition, protozoal cells have in their cytoplasm a series of ultramicroscopic cellular bodies. Partly for these reasons, protozoa are classified separately from the bacteria and viruses. The diversity of protozoa is illustrated in FIGURE 1.11 .

Protozoa have been used for decades as research tools. For example, the flagellar motion exhibited by these microbes is quite similar to that of human sperm cells, and the information gleaned from protozoal studies can be used to help scientists better understand how sperm cells move. Many species of protozoa live freely in the environment and break down the remains of plants and animals and recycle their components. And certain protozoa are photosynthetic—they trap the sun's energy and transform it into the chemical energy of carbohydrates (sugars). These protozoa serve as food sources for other organisms in Earth's waters.

Algae

The term algae is not a formal biological term; rather, it implies a large group of photosynthetic organisms, some of which are microscopic and relatively simple. The simple algae are plantlike in that they contain photosynthetic pigment molecules, but they are considered microbes because they are single-celled (unicellular) organisms.

Several freshwater species of unicellular algae exist, but most species are found in marine environments. Two important algal groups, the diatoms and dinoflagellates, inhabit the oceans in astronomical numbers, where they form the bases of many food chains. The diatoms and dinoflagellates utilize their photosynthetic pigments to trap sunlight's energy and convert it to carbohydrates' energy for consumption by other organisms. Scientists estimate that the microscopic algae of the oceans trap more energy, use more carbon dioxide, and produce more oxygen than all the land plants and other algae combined.

Simple algae have also been investigated as possible food sources for expanding world populations. The algae are cultivated in massive tanks; they are then dried and added to such things as ice cream and yogurt to enhance the nutrient content. Algae used as food are often referred to as *single cell protein* (or SCP). Chapter 7 explores these microbes in depth.

Algae do not infect humans; however, algae can be the cause of serious poisoning. Shellfish and fish bioaccumulate agal products that can be toxic to humans. When people eat the contaminated seafood, the algal toxins can cause serious neurological dysfunction. Freezing and cooking will not neutralize these toxins, so contaminated food must be avoided. In addition to harming humans, algal blooms can cause massive fish kills (see page 157).

Fungi

Fungi are among the major decomposers of organic matter on Earth. They are distinguished from other microbes by their physical structure and by the way they obtain nutrients: They secrete enzymes into the environment and break down the nearby organic matter. Then, they absorb the molecular particles of matter through their cell membranes. Other microbes take up small molecules directly from the environment.

Although their cells are microscopic, fungi are considered by many biologists to be multicellular organisms. This is because the body of a fungus consists primarily of cells joined in long, tangled filaments, as pictured under the electron microscope in FIGURE 1.12 . These filaments, called hyphae, form networks that are visible to the unaided eye; the network is called a mycelium (the term "mold" is commonly used for the mycelium of a fungus). Other species of fungi consist of single cells lacking a

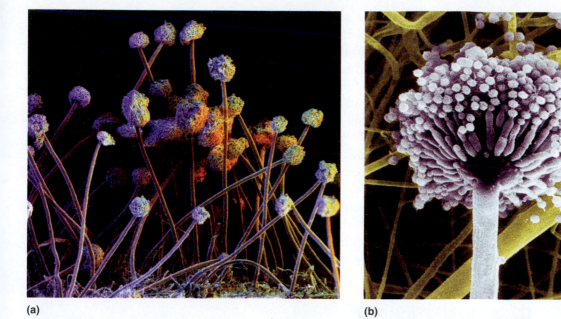

(a)   (b)

FIGURE 1.12 **The Fungus *Aspergillus niger*.** (**a**) A scanning electron micrograph showing the moldlike phase of *Aspergillus niger*. Many stalklike conidiophores are present within the mycelium. Conidiophores contain conidia, the unprotected spores of the fungus. This photograph demonstrates the three-dimensional image possible with the scanning electron microscope. (**b**) A close-up view of one conidiophore with a mass of conidia.

filamentous arrangement. These single-celled fungi, known as *yeasts*, are central to the bread and spirits industries, as we describe in Chapters 8 and 12.

Instances abound where fungi benefit society. For example, one mold called *Aspergillus* is used to produce the alcoholic rice beverage called sake, and another mold named *Claviceps* is cultivated to produce drugs that help relieve migraine headaches. Unfortunately, many species of fungi are agricultural pests, causing grave economic damage to farmers. Indeed, a funguslike microbe called a water mold caused the great Irish potato famine, a disaster that changed the face of two continents. Chapter 8 chronicles that story.

■ *Aspergillus*
a-spěr-jil'lus

■ *Claviceps*
kla'vi-seps

Fungi also can cause serious disease in humans, other animals, and plants. In humans and animals, fungi can infect the skin, mucous membranes, urinary tract, bone, liver, spleen, etc., and can obviously be the cause of serious disease. Fungi can infect plants, including crops. Because the same group of fungi can infect both humans and plants, crop infection with fungi is a potential source of infection for people.

Viruses

Although most people are familiar with the word "virus," few people are aware of the characteristics of a virus. Viruses consist of a fragment of nucleic acid encased in a coating of protein. In some cases, a membranous envelope encloses the protein coat. Viruses do not grow; they produce no waste products; they display none of the chemical reactions we associate with living things; and they are unable to reproduce independently of a host cell.

But viruses do reproduce actively within their host cells, and in doing so, they use the chemical machinery of the cells for their own purposes. Some minutes or hours later, hundreds of new viruses exit from each cell, often leaving in their wake disintegrated host cells. As this wave of cell destruction spreads, the tissue suffers damage and the symptoms of disease ensue. We define and discuss the characteristics of viruses in more depth in Chapter 6.

Despite their role in causing infectious disease (e.g., hepatitis, influenza, AIDS), viruses perform some valuable services to medical science. For example, the virus that normally causes common colds has been genetically altered to make it less infectious, then programmed to carry the genes that relieve the symptoms of cystic fibrosis. Patients with cystic fibrosis have damaged genes that encourage mucus buildup in their airways and, consequently, difficulty in breathing. The genetically altered cold virus ferries correct copies of the genes into the respiratory tract cells, and the proteins encoded by the genes relieve the mucus accumulation. Soon, normal breathing is restored. We shall see many more instances of how microbes benefit us as we move through the chapters ahead.

A FINAL THOUGHT

It should be clear from this chapter that microbes, despite their involvement in disease, contribute substantially to the quality of life in our society. Rather than gush with enthusiasm on the positive roles they play, I prefer to paraphrase several concepts of applied microbiology set down by the late industrial microbiologist David Perlman of the University of Wisconsin. In a 1980 publication, Perlman wrote:

1. The microbe is always right, your friend, and a sensitive partner;
2. There are no stupid microbes;
3. Microbes can and will do anything;
4. Microbes are smarter, wiser, and more energetic than chemists, engineers, and others; and
5. If you take care of your microbial friends, they will take care of your future.

QUESTIONS TO CONSIDER

1. In her biography of Louis Pasteur, Patrice Debré describes Pasteur's 1857 paper on lactose fermentation of milk as "the birth certificate of microbiology." Why do you suppose she thinks so highly of the paper? Further, she writes that as a result of the "Pasteurian revolution," medicine could no longer do without science, and hospitals could no longer be mere hospices. What does she have in mind?
2. "Microbes? All they do is make you sick!" From your introduction to microbes in this chapter, how might you counter this argument?
3. Every now and then in science, a seminal experiment sets off a barrage of studies that lead to the discovery of an important principle. Which experiment do you believe was the spark that ultimately led to our understanding of the germ theory of disease?
4. If you were to ask someone to describe a microbe, he or she might think of a dot under a microscope. However, the microbial world is quite varied, and each of its members is unique. Although your experience in microbiology is somewhat limited at this

juncture, the information in this chapter should give you some insight into the micro-bial world. How, then, would you now describe a microbe?

5. This chapter noted two reasons why interest in microbes ebbed after the death of van Leeuwenhoek. Can you think of any other reasons?

6. The poet John Donne once wrote: "No man is an island, entire of itself." This maxim applies not only to humans, but to all living things in the natural world. What are some roles microbes play in the interrelationships among living things?

7. Our world is somewhat "germ-phobic." The media cover new outbreaks of disease, we eagerly await new antibacterial medicines, and we hear of new ways to "fight germs." But suppose there were no microbes to contend with. What do you suppose life would be like?

KEY TERMS

Informative facts are necessary for the expression of every concept, and the information for a concept is founded in a set of key terms. The following terms form the basis for the concepts of this chapter. On completing the chapter, you should be able to explain and/or define each one:

algae
anthrax
antibodies
Anton van Leeuwenhoek
bacteria
bioremediation
biotechnology
fermentation
fungi

gene probes
germ theory of disease
Golden Age of Microbiology
Louis Pasteur
microbe
protozoa
Robert Koch
viruses

http://microbiology.jbpub.com/book/microbes

The site features **eLearning,** an online review area that provides quizzes and other tools to help you study for your class. You can also follow useful links for in-depth informa-tion, read more stories of microbiology, or just find out the latest microbiology news.

2

Microbes in Perspective: Of Collectors and Classifiers

Life in the microbial world comes in many forms and variations. Microbiologists identify the species they encounter by using a familiar classification system based on cell structure and properties. Here, a lab technician has identified the fuzz in the Petri dish as *Fusarium moniliforme,* a fungus shown in the false-color scanning electron micrograph.

■ Looking Ahead

Despite their incredibly small size, microbes occupy extensive and well-established places in the world of living things. Furthermore, their names, chemical makeups, and other characteristics conform to the principles that apply to all life forms, as we shall note in these pages.

On completing this chapter, you should be able to . . .

- recognize the work of four scientists who contributed to the development of classification methods for microbes and other living things.

- note how microbes were first introduced to the biological classification scheme and define the places they occupy in modern systems of classification.

- distinguish the characteristics that separate prokaryotes from eukaryotes and indicate where different types of microbes fit into these two categories.

- understand how microbes are assigned their binomial names and how the names should be expressed in writing.

- recognize the units of measurement used for microbes and specify how various microbes relate to one another in size.

- describe some principles of microscopy and name several types of microscopes used to view microbes.

Carolus Linnaeus was in trouble. The ship from Africa was pulling into port, and soon the Swedish botanist would be confronted with new plants, new animals, and new problems. "What shall we call this one?" he would be asked. "Or this one?" "Or that one?" This nomenclature thing had gotten thoroughly out of hand. To be sure, the biological world needed some order, but who had appointed him king of nomenclature? It was exasperating, to say the least!

The year was 1740, and Carl von Linne, known as Carolus Linnaeus to his scientific friends, was in the midst of naming all the known plants and animals on Earth. The museums were full of specimens to be named, and people were bringing new specimens to him almost daily (he soon learned to dread the sight of arriving ships).

But Linnaeus was performing a valuable service. Before he popularized his nomenclature scheme in 1735, confusion was rampant—scientists could not agree on what to call many organisms, and various individuals coined different names to both designate and describe an organism. Linnaeus' great simplifying answer was to use the classification scheme he devised and select out of that scheme the name of the category (the genus) to which an organism belonged. Then he added a descriptive adjective, and the combination became the organism's binomial (double) name.

Now, in 1740, even as the ship was tying up at port, Linnaeus was working feverishly lest other scientists give duplicate names to different organisms. In a monumental task of linguistic invention, Linnaeus (**FIGURE 2.1**) would spend the next few years ransacking his Latin for enough terms to make up thousands of binomial names. Some names he took from an organism's manner of growth, others from the discoverer, others from classical heroes, and still others from vernacular names. Any parent who has had to name a child can only begin to imagine what he was up against.

In a 1753 book on plants, Linnaeus supplied binomial names for over 5900 different plants known at that time. In his tenth edition of *Systema Naturae* (1759), he

(a)   (b)   (c)   (d)

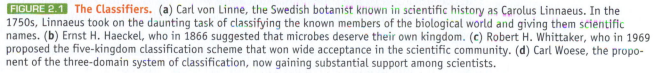

FIGURE 2.1 **The Classifiers.** (**a**) Carl von Linne, the Swedish botanist known in scientific history as Carolus Linnaeus. In the 1750s, Linnaeus took on the daunting task of classifying the known members of the biological world and giving them scientific names. (**b**) Ernst H. Haeckel, who in 1866 suggested that microbes deserve their own kingdom. (**c**) Robert H. Whittaker, who in 1969 proposed the five-kingdom classification scheme that won wide acceptance in the scientific community. (**d**) Carl Woese, the proponent of the three-domain system of classification, now gaining substantial support among scientists.

extended the scheme to animals. During the following decades, his binomial names were adopted by European scientists and were being introduced around the world. And just in time, because explorers kept returning to Europe from distant corners of the globe with newly discovered animals and plants (such as penguins, tobacco plants, potatoes, manatees, and kangaroos). In the end, Linnaeus even added to this frenzy: He inspired an unprecedented worldwide program of specimen hunting, and he sent his students around the globe in search of new life forms.

Carolus Linnaeus is just one of the collectors and classifiers that we shall encounter in this chapter as we fit microbes into the scheme of living things and see how they are related to members of the visible world. Numerous scientists have contributed to the contemporary view of where microbes fit in, and they have considerably modified our perspective of the tiniest forms of life. Knowledge of microbes has come a long way since the time of Linnaeus, who revealed his relative ignorance about them by grouping them apart from other living things under the heading Vermes (as in vermin) in a category he called Chaos (as in confusion). Linnaeus could hardly anticipate what the future held.

Microbes in the Biosphere

Microbes (with the notable exception of viruses) have a set of characteristics that are common to all members of the world of living things, the so-called biosphere. Among these characteristics are the ingestion and assimilation of nutrients for growth, the excretion of waste matter, and the abilities to reproduce independently, to adapt to environmental changes, and to react to stimuli.

And, like other living things, microbes are placed into groups according to certain established rules of taxonomy, the science of classification. Taxonomic rules bring order to the microbial world and place microbes into categories in which they can be studied as they relate to other microscopic and macroscopic members of the biosphere. (Indeed, biologists generally tend to be compulsive classifiers, placing all the species of the world into neat pigeonholes from which they can be retrieved at will, and microbiologists are no different from their colleagues.)

Classification

Before the invention of the microscope, biologists had little difficulty classifying living things. It was obvious that all living things were either animals or plants, and as early as Aristotle's time (fourth century B.C.), scientists recognized two major kingdoms of living things: the Animalia and the Plantae. It was easy to distinguish the two groups because animals moved about, while plants were rigid and immobile.

After van Leeuwenhoek's reports of microbes, it became clear that these organisms should be incorporated into the biological kingdoms. But scientists were unsure exactly where the microbes belonged. As we have noted, Carolus Linnaeus, the most eminent biologist of his day, was among the uncertain ones.

The classification of microbes remained somewhat chaotic until 1866, when the German naturalist Ernst H. Haeckel disturbed the tidiness of the plant and animal kingdoms by proposing a new system that separated the microbes. Haeckel was upset with the practice of classifying mushrooms with plants because mushrooms are fungi and lack the green pigment chlorophyll used in photosynthesis. Moreover, by his time, sci-

entists had found a host of "in-between" microbes, including many novel types of protozoa, slime molds, single-celled algae, and bacteria. Haeckel therefore coined the term
protist for a microbe, and he placed all fungi and other protists in a new kingdom known
as Protista (the Greek word *protistos* means "primitive" or "first"). Protista soon came
to include fungi, protozoa, algae, bacteria, and virtually all organisms that share plant
and animal characteristics but are not plants or animals (Chapter 7). The fungi would
later be moved to their own kingdom (Chapter 8).

Haeckel's suggestion met with mixed success. Some biologists continued to believe
that microbes were either plants or animals, and they continued to classify protozoa as
animals, and bacteria, algae, and fungi as plants. Indeed, through the mid-1900s, if you
wished to study bacteria you often had to register in a university's department of botany.

Prokaryotes and Eukaryotes

In the 1940s, the electron microscope became available, and scientists realized that
many species of microbes possess characteristics considerably unlike those of plants
and animals. For example, bacterial cells have no nucleus, the cellular body where
the hereditary information is stored; by contrast, plant and animal cells have nuclei.
In addition, developments in biochemistry indicated that some chemical constituents
of microbes were different from those of plants and animals. Researchers learned, for
instance, that the cell walls of fungi contain a complex sugar called chitin, while plant
cell walls lack this chemical substance.

In their continuing effort to give the natural world some semblance of order, biologists decided in the 1940s that all living things should be classified as either prokaryotes or eukaryotes, as described below. This decision was based on observations made
with the high-magnification electron microscopes as well as on studies in biochemistry and genetics. The division of all living things into prokaryotes and eukaryotes is
of great significance because the terms are used consistently in all sciences, especially
in microbiology. In this book, we shall examine such prokaryotes as primitive bacteria (the archaea), as well as traditional bacteria and their subgroups, the chlamydiae,
rickettsiae, and mycoplasmas; and we shall discuss such eukaryotes as protozoa, fungi,
and simple algae. FIGURE 2.2 compares the cells of prokaryotes and eukaryotes.

The principal difference between prokaryotes and eukaryotes is the presence of
a nucleus in a eukaryotic cell and its absence in a prokaryotic cell. In addition, eukaryotes have in their cytoplasm a series of complex membrane-enclosed cellular
bodies called organelles. Organelles include mitochondria, Golgi bodies, chloroplasts,
lysosomes, and the endoplasmic reticulum: Mitochondria are organelles where the
energy from sugars is released for cell use; chloroplasts contain the chlorophyll pigments used in photosynthesis; the Golgi bodies are sites of protein processing; and
the endoplasmic reticulum is a set of membranes in the cytoplasm. Prokaryotic cells
have no organelles, but both prokaryotes and eukaryotes have ribosomes, the ultramicroscopic bodies in the cytoplasm where proteins are constructed. Ribosomes contain the organic substance ribonucleic acid (RNA) together with some proteins.
Prokaryotic ribosomes are smaller than eukaryotic ribosomes.

Other differences exist between prokaryotes and eukaryotes. The cells of prokaryotes generally have their DNA organized in a single chromosome in a closed loop,
while the cells of eukaryotes have their DNA in multiple chromosomes in threadlike
strands. In eukaryotes, cell reproduction occurs by the complex process of mitosis

■ chlamydiae
kla-mi′dē-ä

■ rickettsiae
ri-ket′-sē-ä

■ mycoplasmas
mī-kō-plaz′mä

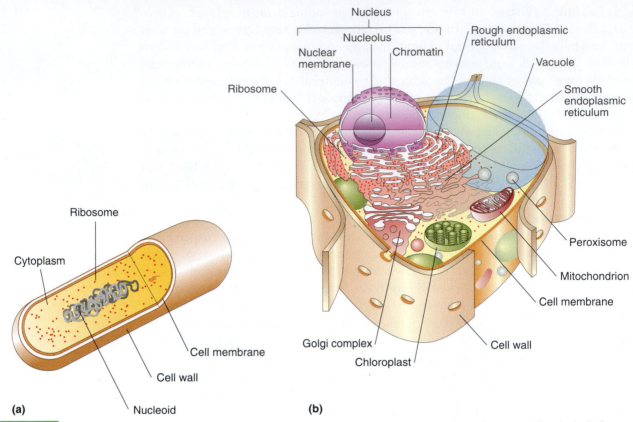

FIGURE 2.2 **Prokaryotic and Eukaryotic Cells.** (**a**) A stylized bacterial cell as an example of a prokaryotic cell. Relatively few structures are seen. (**b**) A plant cell as a typical eukaryotic cell. Note the variety and relatively large size of the cellular features, many of which are discussed in the text. The presence of a nucleus is the primary feature distinguishing the eukaryotic cell from the prokaryotic cell.

(in this process, the chromosomes duplicate and align themselves in the center of the cell before moving to opposite poles; then, the cell divides into two daughter cells). In prokaryotic cells, reproduction occurs by the more simple process of chromosome duplication followed by cell fission, as **TABLE 2.1** indicates. In addition to the fungi, protozoa, and algae, eukaryotes include the plants and animals, including humans. (The author has a friend who describes himself as a "card-carrying eukaryote.")

The Five-Kingdoms

As the twentieth century progressed, advances in cell biology and interest in evolutionary biology led scientists to question the three-kingdom classification scheme. Then, in 1969, Robert H. Whittaker of Cornell University proposed a system that has gained wide acceptance in the scientific community. Further expanded in succeeding years by Lynn Margulis of the University of Massachusetts, the system recognizes five kingdoms of living things: Monera, Protista, Fungi, Animalia, and Plantae.

In the five-kingdom system, bacteria are classified in the kingdom Monera (also called Procaryotae by bacteriologists). Bacteria differ significantly from members of the other four kingdoms in the cellular details described earlier. They are believed to be the oldest forms of life on Earth. All the Monera are prokaryotes.

TABLE 2.1   A Comparison of Prokaryotes and Eukaryotes

Characteristic	Prokaryotes	Eukaryotes
Nucleus	Absent	Present with nuclear membrane
Organelles	Absent	Present in a variety of forms
DNA structure	Single closed loop	Multiple chromosomes in nucleus
	Almost naked strand with very little protein	Structural protein (histone) associated with DNA
Chlorophyll	When present, dissolved in cytoplasmic membranes	When present, dissolved in chloroplast membranes
Ribosomes	Smaller than eukaryotic ribosomes	Larger than prokaryotic ribosomes
	Free in cytoplasm	Free or bound to membranes
Cell walls	Generally present	Present in some types, absent in others
	Complex chemical composition	Complex chemical composition
Flagella	Rotating movement	Whipping movement
Cilia	Absent	In some cells
Reproduction	Usually by fission	By mitosis
	Sexual reproduction unusual	Sexual reproduction usual
Examples	Bacteria, rickettsiae, chlamydiae, cyanobacteria	Fungi, protozoa, plants, animals, humans (all other organisms)

The second kingdom is Protista (Haeckel's term). It includes eukaryotes such as protozoa, slime molds, and single-celled algae. Members of this kingdom generally have flagella at some time in their lives, and even though some larger forms consist of colonies of cells, the tissue level of organization is lacking. Many members of the kingdom Protista are considered "taxonomic misfits" because they do not appear to belong to other kingdoms. Protists share certain characteristics with plants and animals, and some species appear to be ancestors of those more complex life forms. Indeed, the protists are thought to have emerged from the Monera over many eons of time, and they are believed to be the basic evolutionary stock for the next three kingdoms.

In Whittaker's system, shown in **FIGURE 2.3**, the kingdom **Fungi** includes fungi and yeasts, both of which are nonpigmented eukaryotic organisms having cell walls with unique chemical constituents. Although the fungi usually occur as long, branching chains (filaments) of cells, the cytoplasm mingles among adjacent cells, and the true multicellularity characteristic of plants and animals does not exist (although some biologists debate this point). In their feeding patterns, fungi absorb dissolved nutrients from their surrounding environment. Moreover, some fungi live together with plants, and one fungus apparently produces the same valuable chemicals as the tree it inhabits, as A Closer Look on page 35 relates.

The final two kingdoms, Plantae and Animalia, are the traditional plants and animals. Plants use the chemical process of photosynthesis to synthesize their energy-rich

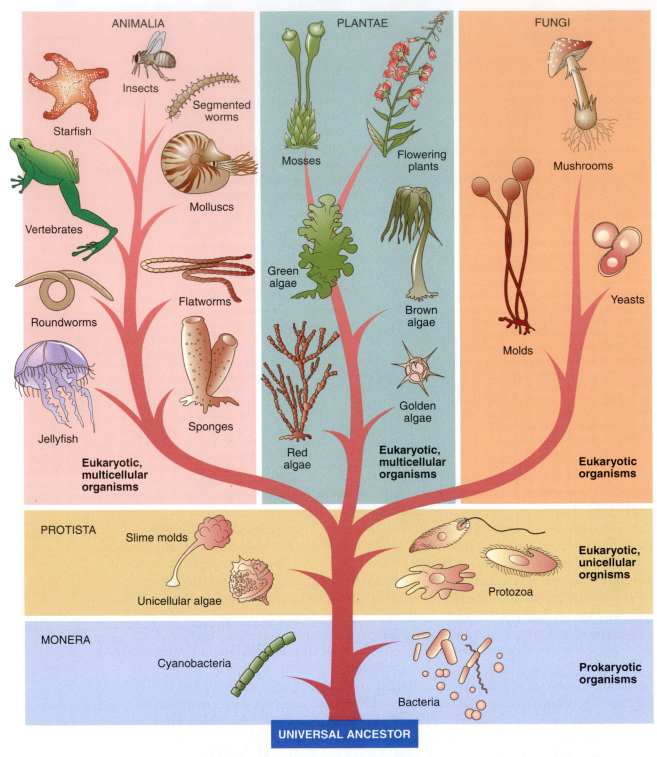

ANIMALIA

Insects

Segmented worms

Starfish

Molluscs

Vertebrates

Flatworms

Roundworms

Jellyfish

Sponges

Eukaryotic, multicellular organisms

PLANTAE

Mosses

Flowering plants

Green algae

Brown algae

Red algae

Golden algae

Eukaryotic, multicellular organisms

FUNGI

Mushrooms

Yeasts

Molds

Eukaryotic organisms

PROTISTA

Slime molds

Unicellular algae

Protozoa

Eukaryotic, unicellular orgnisms

MONERA

Cyanobacteria

Bacteria

Prokaryotic organisms

UNIVERSAL ANCESTOR

FIGURE 2.3 **The Five-Kingdom System of Classification.** Devised by Robert H. Whittaker, this system implies an evolutionary lineage, beginning with the Monera and extending to the Protista. Certain of the Protista are believed to be ancestors of the Plantae, Fungi, and Animalia. Divergence at each level is based on the mode of nutrition: photosynthesis, absorption, or ingestion. Unicellular or multicellular organization is also a key feature in the system, although the fungi are not truly multicellular.

A CLOSER LOOK
Taxol, Tumors, and Twists

In 1989, researchers at Johns Hopkins University discovered that taxol, a chemical derived from yew trees, could greatly reduce the size of tumors in women suffering from ovarian cancer. Two years later, in January 1991, the Food and Drug Administration approved taxol for treating ovarian cancer, while noting that the drug might be useful for breast, head, and neck tumors. Unfortunately, the exhilaration that accompanied approval of the new treatment was counterbalanced by the cost of the drug (about $1000 per treatment cycle) and the fear that the yew tree might be over-farmed to provide bark for the drug.

Then, in 1993, a new twist was added to the taxol story. In April, Montana researchers discovered growing within the bark of the yew a fungus that produces taxol on its own. Plant pathologist Gary Strobel and chemist Andrea Stierle, both from Montana State University, led the research effort. Under Strobel's intuitive direction, Stierle searched the Montana woods for local yews *(Taxus pacifica)* that would yield taxol. After she found one such yew, they went a step further and isolated a fungus from within the folds of the yew's bark. The fungus continued to produce taxol even after removal from its host plant. They named the fungus *Taxomyces andreanae* (Andrea's taxus-fungus).

Although *T. andreanae's* yield of taxol is low, the potential for increasing the yield is great. For example, huge fermentation tanks can be used to produce enormous amounts of the fungus and much larger amounts of the drug. Moreover, genetic engineering techniques can be used to pinpoint and clone the taxol-producing genes, then transfer them to high-yield vector organisms such as bacteria. Apparently the drug companies believe that these and other approaches can work. Months before their scientific paper appeared in print, the Montana researchers had secured a patent on the fungal production of taxol and were being courted by numerous drug companies. The fungus's future appears bright.

Taxomyces and andreanae. (**a**) The fungus *Taxomyces andreanae* showing its hyphal strands and fruiting bodies with spores (×2300). (**b**) Strobel and Stierle with the yews from which the fungus was obtained.

foods, while animals ingest their foods, then use digestive enzymes to break complex food particles into absorbable fragments.

You will note that Whittaker's five kingdoms do not include the viruses. This omission is intentional because the viruses are not considered cellular entities. Indeed, to many biologists, the viruses are not living things because they display no growth patterns and do not metabolize nutrients or reproduce independently. Thus, they are not placed in a kingdom of living things.

The Three Domains

The view that five kingdoms represent the natural lines of division among living things has been further modified by the development of the three domains (also called super-kingdoms). First developed in the 1980s by Carl Woese and his coworkers at the University of Illinois, the three-domain system is based on contemporary techniques in molecular biology and biochemistry. It also encompasses new knowledge about a group of bacteria called archaebacteria (also known as archaea in the plural form and archaeon in the singular). Archaebacteria (archaea) are prokaryotic forms with unique biochemical properties, as we note below (and explore in Chapter 3).

In Woese's three-domain system, illustrated in **FIGURE 2.4**, the first domain includes the archaea and is appropriately termed Archaea; the second encompasses all the remaining traditional bacteria (or "eubacteria") and is logically called Eubacteria; and the third domain includes the remaining four kingdoms of Whittaker and is called Eukarya (as in eukaryotes). Archaea and Eubacteria differ substantially in at least three ways: the form of ribonucleic acid (RNA) present in their ribosomes, the composition of their cell walls, and their sensitivity to certain antibiotics.

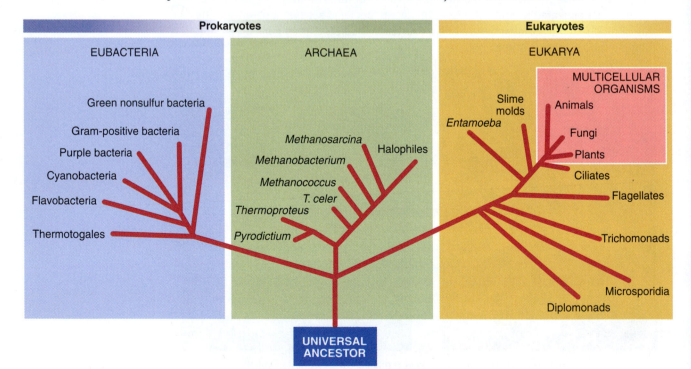

FIGURE 2.4   The Three-Domain System. Fundamental differences in genetic endowments are the basis for the three domains encompassing all organisms on Earth. The relationships are determined from the sequences of nitrogenous bases in ribosomal RNA. The line length between any two groups is proportional to their genetic differences.

Scientists were initially reluctant to accept the three-domain system of classification, and many deemed it a threat to the tenet that all living things are either prokaryotes or eukaryotes. But then, in 1996, molecular biologist Craig Venter and his coworkers deciphered the composition of the DNA of the archaeon *Methanococcus jannaschii*. The results demonstrated that almost two-thirds of the genes of *Methanococcus* are different from those of a common eubacterium. The scientists also found that certain proteins involved in DNA replication and RNA synthesis in archaea have no counterpart in the eubacterium. With persuasive evidence like this, the three-domain system has won new converts.

■ Methanococcus
jannaschii
meth-a-nō-kok′kus jan-
nä′shē-ē

Naming and Measuring Microbes

The system of names used for biological specimens is called nomenclature. It is closely allied to the system of classification because the scientific names are obtained from the classification scheme. This concept was first used by Carolus Linnaeus, the biologist who applied Latin names to the plants and animals in the mid-1700s, as described earlier in this chapter.

Linnaeus' classification scheme is still in use today. In the scheme, the fundamental rank is the species (pl., species). A species is a group of closely related organisms. In microbiology, two organisms are considered to be the same species if their hereditary material (their DNA) is 70 percent identical or if experienced researchers in the field proclaim the members of a group to be a species by mutual consent. By comparison, in the world of plants and animals, a species is a group of organisms that interbreed with one another and produce new members, which are similar to the parents and are fertile.

In the system of classification, a group of species are gathered together to form a genus (pl., genera). For example, a housecat, a lion, and a tiger belong to separate species, but they are classified together in the genus *Felis* because they have similar characteristics. Furthermore, a group of genera comprise a family. For instance, lions (genus *Felis*) and panthers (genus *Pantheria*) are categorized in the family Felidae. Families are organized into orders, and orders are brought together in a class. Various classes comprise a phylum for animals and animal-like microbes, or they make up a division for plants and plantlike microbes (e.g., fungi and single-celled algae). The phyla or divisions are grouped together in a kingdom, such as the animal kingdom, and then, as we have discussed, a domain.

Binomial and Common Names

In the system of nomenclature, a microbe (or any other organism) has a binomial name consisting of the name of the genus to which the organism belongs and a modifying adjective. The modifying adjective is called the species modifier or the specific epithet. For instance, a bacterium normally found in the human colon is called *Escherichia coli*. The first part of the binomial name, *Escherichia*, is the name of the genus to which the organism belongs. It is derived from the name of the scientist, Theodor Escherich, who first reported the microbe's existence in 1888. The second part of the binomial, *coli*, is an adjective derived from the word "colon," where the bacterium was first found by Escherich. (The same rules apply to humans, who are technically known as *Homo sapiens*.) TABLE 2.2 lists some other microbial names and their origins.

The correct way to express the binomial for a microbe (or for any organism) is to capitalize the first letter of the genus name and to write the remainder of the genus name

■ Escherichia coli
esh-ėr-ē′kē-ä kō′lī
(or kō′lē)

TABLE 2.2	How the Microbes Acquire Their Names
Name of Organism	**Meaning of Name**
Streptococcus lactis	A chain of spheres; produces lactic acid in milk
Escherichia coli	Escherich's bacillus; found in the colon
Treponema pallidum	A turning organism (spiral); pale
Entamoeba histolytica	An intestinal amoeba; digests (lyses) tissue
Salmonella typhi	Salmon's bacillus; causes typhoid fever
Salmonella typhimurium	Salmon's bacillus; causes typhoid fever in rodents
Staphylococcus aureus	A grapelike cluster of spheres; forms a golden pigment
Haemophilus ducreyi	A blood-loving organism; named after Ducrey
Neisseria gonorrhoeae	Neisser's bacterium; causes gonorrhea
Trichomonas vaginalis	A flagellated cell; causes a vaginal infection
Sarcina lutea	A bundle of bacteria; produces a yellow pigment
Saccharomyces cerevisiae	A sugar fungus; ferments beer

and the species modifier in lowercase letters. The binomial should always be italicized (if this is not possible, it should be underlined). After the full name has been introduced in a piece of writing, the name can be abbreviated by writing the first letter of the genus name and the full species modifier. Thus, *Escherichia coli* can be abbreviated as *E. coli*.

It is important to remember that the species is a concept, while the binomial name is merely a title given to the species. In this sense, a bacteriologist conceptualizes *Escherichia coli* as a bacterium having certain characteristics. The correct name for the species is the binomial *Escherichia coli*. It is incorrect to say that *Escherichia* is the genus and *coli* is the species.

In addition to the binomial name, many microbes have a common name that is frequently used in conversation. For instance, the bacterium that causes bacterial pneumonia, *Streptococcus pneumoniae,* is commonly referred to as the pneumococcus; and the bacterium that causes bacterial meningitis, *Neisseria meningitidis,* is often called the meningococcus. In addition, a microbial species may have various strains or subspecies whose identifiers are added to the binomial. An example is *Escherichia coli* O157:H7. During the 1990s, this strain of *E. coli* emerged as the cause of numerous outbreaks of bloody diarrhea.

Special problems exist in classifying microbes and defining a microbial species. Because microbes do not interbreed like plants and animals, it is not possible to define a microbial species based on reproductive patterns. Nor are there many traces of microbes in the fossil record because their cells break down quickly after death and the contents dissolve. Therefore, systems of microbial classification tend to be artificial, and the simplest and most direct way of defining a microbial species is based on similarities in the DNA. Other criteria used are similarities in structure, biochemistry, physiology, and immunology.

Microbial Measurements

One of the defining features of microbes is their extremely small size. Microbes are not measured in inches, millimeters, or other well-known units, but in much smaller and less familiar units. The unit most often used for microbial measurements is the

■ Streptococcus
pneumoniae
strep-tō-kok'kus
nü-mŏ'nē-ī

■ Neisseria meningitidis
nī-se'rē-ä me-nin
ji'ti-dis

■ Staphylococcus aureus
staf-i-lō-kok'kus ô-rē-us

micrometer (sometimes referred to as the micron). A micrometer is a millionth of a meter. The abbreviation for a micrometer is expressed by using the Greek letter mu (written as μ) together with the letter m. Thus, the length of a typical bacterium might be expressed as 5 micrometers, or 5 μm.

To conceptualize the extraordinarily small size of a micrometer, consider that comparing a micrometer to an inch is like comparing a fly to New York's Empire State Building. Put another way, 1 million micrometers is equivalent to a meter (approximately a yard). A bacterium such as *Staphylococcus aureus* (the common "staph"), has a diameter of roughly 1 micrometer. Therefore, 1 million staphylococci lying side-by-side would occupy the space taken up by a meter. If you were to attempt to count the staphylococci in a meter's space (at a rate of one bacterium per second), you would be counting for a considerable period of time—about 11 days!

The smallest bacteria measure about 0.15 μm in diameter. These bacteria are the mycoplasmas. Slightly larger than the mycoplasmas are the chlamydiae, with a diameter of about 0.25 μm. Then come the rickettsiae, about 0.45 μm in diameter, as **FIGURE 2.5** illustrates. These three groups of microbes are so small that lab workers cannot normally see them with a light microscope. Among the bacteria that can be seen

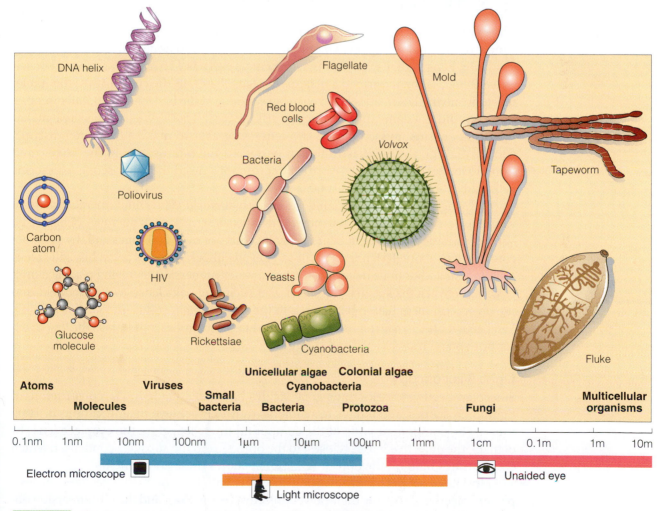

FIGURE 2.5 Size Comparisons Among Various Living Things and Molecules.

with a light microscope, the smallest measure about 0.5 µm in diameter (or width if they are rods), and the sizes range upward from there, with the largest having widths of up to 10 µm. The length of these bacteria can be as much as 25 µm (although some immense bacteria have been reported in the recent literature, as we note in Chapter 5).

Most other species of microbes are also measured in micrometers. For example, yeast cells are approximately 5 µm in diameter, and the cells of molds may be 25 µm long or longer, with varying widths. Protozoa may be as large as 100 µm (0.1 mm), or about the size of the period at the end of this sentence. Simple algae are also relatively large microbes; many exceed 100 µm in size.

At the opposite end of the microbial scale are the viruses and the subcellular organelles found in many microbes. These objects are measured in nanometers. A nanometer (nm) is equivalent to one-billionth of a meter, or one-thousandth of a micrometer. Among the smallest viruses are the polioviruses, about 25 nm in diameter. Among the larger viruses are the poxviruses, which are about 250 nm in diameter. Mitochondria, chloroplasts, and other organelles are also measured in nanometers.

Microscopy: Seeing the Unseen

The existence of microbes was largely a matter of speculation until the 1600s, for the simple reason that seeing them required a microscope and none of decent quality were available. Lens systems were known to exist as early as 1267, when the eminent scientist Roger Bacon described a magnifying lens. Later, in the early 1300s, the Italian inventor Salvino D'Armati is believed to have produced the first spectacles. Microscopes, however, did not come into existence until the early 1600s, when a spectacle maker named Zacharius Janssen placed two lenses together to make a crude microscope. Galileo Galilei, the great astronomer, perfected the microscope in the 1620s, and Robert Hooke, the imaginative British microscopist, used the microscope to describe cells and other objects in the 1660s.

The stage was set for revealing the microbial world. The Dutch merchant Anton van Leeuwenhoek was the first to provide detailed descriptions of microbes (Chapter 1). Van Leeuwenhoek's lenses obtained magnifications of over 200 times and, by some accounts, up to about 400 times. Part of van Leeuwenhoek's success lay in his skill at lens grinding. In addition, he devised a system of water immersion involving placing a drop of water between the lens and the object. Van Leeuwenhoek's lenses were handheld, and the objects on them had to be exposed to the sun to obtain sufficient light for viewing. Van Leeuwenhoek may not have been the first to see microbes, but he was the first to describe them with such detail and clarity.

Light Microscopy

Since van Leeuwenhoek's time, great strides have been made in the construction of light microscopes, and today's instruments routinely achieve magnifications of 1000 times. The component parts of the compound microscope are the eyepiece lens (or ocular), the objective lenses (closest to the object), and the substage condenser, which concentrates light on the object. FIGURE 2.6a shows these components.

Most microscopes have a revolving nosepiece with three or more objectives: the low-power objective (10×), the high-power objective (40 or 45×), and the oil-immersion ob-

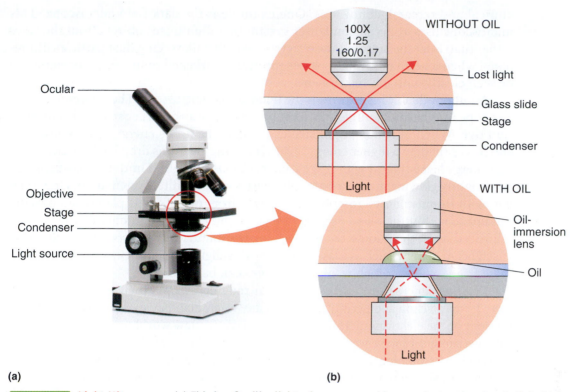

FIGURE 2.6 **Light Microscopy.** (**a**) This is a familiar light microscope used in many instructional and clinical laboratories. Note the important features of the microscope that contribute to the visualization of the object. (**b**) Aspects of oil-immersion microscopy. When light rays enter the air they bend (solid arrows) and miss the objective lens. However, they remain on a straight line (dashed arrows) in the oil. This pathway leads them directly into the lens. The clarity of the object increases with increased light.

jective (97 or 100×). The objective lens magnifies the object and creates an image in the tube of the microscope. The eyepiece then uses this image as an object, remagnifies it, and forms a final image in space below the stage of the microscope. This is the image that is seen by the observer. The total magnification achieved by the instrument is thus the magnification of the objective multiplied by the magnification of the eyepiece. For example, the low-power objective (10×) used with a 10× eyepiece lens yields a total magnification of 100×.

Because the oil-immersion objective must be placed extremely close to the microscope slide, it is usually very difficult to obtain sufficient light for viewing. Also, as light rays pass through the top surface of the microscope slide, they bend and miss the exceptionally small opening of the oil-immersion objective, as **FIGURE 2.6b** shows. The amount of light can be increased considerably by placing a drop of oil in the gap between the oil-immersion objective and the slide. The oil, known as immersion oil, has the same refractive index (or light-bending ability) as glass, and light coming out of the glass slide does not bend away as it does in air. Rather, the light continues on a straight path into the objective, and the object is illuminated sufficiently to be seen clearly.

Other Types of Microscopy

While the light microscope is considered the standard tool of microbiology, scientists use other types of microscopes for viewing particular types of microbes under clinical and research conditions. One example is the dark-field microscope. This microscope uses a special condenser system that illuminates objects from the sides rather than from the bottom. The effect is somewhat like seeing dust particles illuminated when a beam of sunlight passes through a darkened room. Spiral bacteria can be seen clearly with this instrument.

Another valuable instrument is the fluorescence microscope. The fluorescence microscope uses beams of ultraviolet light to illuminate an object coated with a fluorescent dye. When ultraviolet light strikes the dye, the light's energy excites electrons and forces them to a higher energy level. Then, the electrons drop back to their original energy levels, emitting the excess energy as visible light. The dyed object appears as a brightly glowing image, whose color varies with the filter system used. One problem with fluorescent microscopy is that light from out-of-focus planes can obscure the image within the plane of focus. Most recently, confocal microscopy has vastly improved upon the fluorescent microscope by illuminating the specimen with a focused beam of light, such as a laser. By doing this, light from unfocused planes is eliminated and extremely high resolution images can be obtained.

The light microscope and other specialized microscopes increase the lower limits of human vision and permit us to see many microbial species. However, the development of the electron microscope opened a whole new world to scientists because this instrument represented a quantum leap beyond the capabilities of the light microscope. Microbiologists could now see the viruses, an entire group of microbes that had previously been invisible, and they could visualize the finer structures of other microbes, such as the ultramicroscopic flagella of bacteria.

The first electron microscope can be traced to Ernst Ruska, a German scientist working in the 1930s. Ruska designed the microscope based on his earlier observation that a beam of electrons passing through a vacuum can be bent by a magnetic field. He set up an evacuated tube and used a 60,000-volt pulse of electricity to activate an electron source. Then, using magnets, he focused the beam on a sample of microbes. Acting similar to a beam of light, the electrons bounced off the microbes and created an image on a monitor. A photograph of this preliminary image created a final image at a magnification of many thousands of times.

Two types of electron microscope are in widespread use today: the transmission electron microscope (TEM) and the scanning electron microscope (SEM). The TEM takes pictures of microbes sliced into pieces, while the SEM permits us to see the surfaces of microbes. In technical terms, the TEM uses electrons that deflect off the object and form an image, while the SEM knocks loose showers of electrons from the surface of the object. The final magnification possible with the TEM is approximately 20,000,000 times, while the SEM yields images that are magnified about 100,000 times. **FIGURE 2.7** shows photographs of the bacterium *Pseudomonas* taken with both types of electron microscope.

■ Pseudomonas
sū-dō-mō′näs

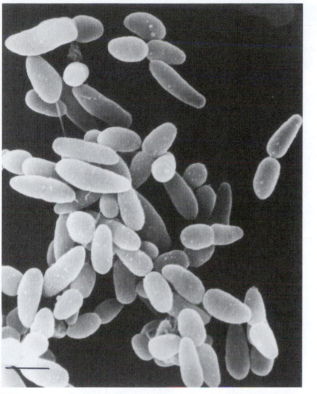

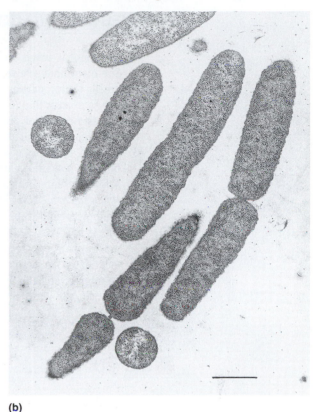

(a)   (b)

FIGURE 2.7 **Scanning and Transmission Electron Microscopy Compared.** The bacterium *Pseudomonas aeruginosa* as seen with two types of electron microscopy. (**a**) A view of whole cells seen with a scanning electron microscope. (Bar = 1.0 μm.) (**b**) A view of sectioned cells seen with a transmission electron microscope. (Bar = 0.5 μm.) The difference in perspective with the two microscopes is clear.

◼ A FINAL THOUGHT

The diversity of life forms has astounded scientists for centuries, and the challenge of characterizing and categorizing organism (taxonomy) has been the focus of many careers. The challenge of naming and categorizing this vast diversity is nowhere more complicated than it is with the microbial world. When it comes to microscopic life forms, which are often unicellular, the scientist is dependant on various forms of microscopy and means of detecting cellular behavior. Recent advances in microscopy and cell theory have led to the development of a domain system to describe the various forms of microscopic life and have challenged our thinking about the basic nature of life. The history of taxonomy, from Carolus Linnaeus to Charles Darwin to Carl Woese, is a magnificent study of life forms, and we encourage you in this Chapter to gain an appreciation of the diversity of microorganisms, how they can be viewed, and the fundamental differences in cell architecture between the domains.

▮ QUESTIONS TO CONSIDER

1. A student is asked on an examination to write a description of the protozoa. She blanks out. However, she remembers that protozoa are eukaryotes, and she recalls the properties of eukaryotes. What information about eukaryotes can she use to answer the question?

2. A local newspaper once contained an article about "the famous bacterial ecoli." How many things can you find wrong in this phrase?

3. In 1987, in a respected scientific journal, an author wrote, "Linnaeus gave each life form two Latin names, the first denoting its genus and the second its species." A few lines later, the author wrote, "Man was given his own genus and species *Homo sapiens*." What is conceptually and technically wrong with both statements?

4. Biologists tend to be collectors and compulsive classifiers. Why do you think this is so? Also, which classifier mentioned in this chapter do you think had the most impact on the science of his day?

5. Microbes have been described as the most chemically diverse, the most adaptable, and the most ubiquitous organisms on Earth. From this chapter, what can you add to this list of "mosts"?

▮ KEY TERMS

Informative facts are necessary for the expression of every concept, and the information for a concept is founded in a set of key terms. The following terms form the basis for the concepts of this chapter. On completing the chapter, you should be able to explain and/or define each one:

Archaea
Carl Woese
Carolus Linnaeus
dark-field microscope
domain
Ernst Haeckel
Ernst Ruska
Eubacteria
eukaryote
fluorescence microscope
genus
immersion oil

micrometer
Monera
nanometer
nomenclature
objective lens
prokaryotes
Protista
Robert Whittaker
scanning electron microscope (SEM)
species
taxonomy
transmission electron microscope (TEM)

▮ http://microbiology.jbpub.com/book/microbes

The site features **eLearning,** an online review area that provides quizzes and other tools to help you study for your class. You can also follow useful links for in-depth information, read more stories of microbiology or just find out the latest microbiology news.

Molecules of the Cell: The Building Blocks of Life

3

The molecules of life have their origin in a primordial soup in which organic molecules began to form. Later, these molecules developed the ability to replicate and formed cells. The major molecules that serve as the building blocks of the cells are shown in the inset. These are nucleic acids, proteins, lipids and carbohydrates.

▮ Looking Ahead

Now that we have discussed the basic structure of microbial cells, it is important to understand the molecules that compose cellular structures and the roles of these molecules in the cell function. One of the most important features of the cell is that it is composed mostly of water and that the molecules of life function in an aqueous environment.

On completing this chapter, you should be able to . . .

- understand and appreciate the role of water in the formation and function of the molecules of the cell.

- understand the role of polymerization and dehydration synthesis in the formation of the major macromolecules of the cell.

- know what the names are of the major macromolecules of the cell are and understand their composition and function.

- recognize that polysaccharides are not only an important source of sugar, but also a significant component of the cell wall.

- appreciate the role of membrane phospholipids in accompanying an aqueous cytoplasm and environment.

- recognize that proteins carry out most of the functions of the cells by serving as enzymes, molecular motors and signals.
- understand that DNA is a large molecule that contains information in the form of polymerized nucleotides.

The basic units of matter are known as atoms. Atoms have three basic components: negatively charged electrons, positively charged protons, and uncharged neutrons. Protons and neutrons possess most of the mass of an atom and associate closely to form a nucleus while the electrons orbit the nucleus. There are well over 100 different types of atoms that differ on the basis of the number of protons and neutrons in the nucleus and the number of electrons that orbit the nucleus. In a very real way, the atoms of this world are like a box of Legos® insofar as atoms of different sizes and shapes can snap together to form bonds, provided that they "fit." When this happens, the electrons of the atoms share the orbit around both nuclei and a molecule is formed. So, molecules are simply two or more atoms "snapped" together by their electrons.

Water is an example of a molecule that is made up of three atoms—two hydrogen (H) and one oxygen (O)—and is written as H_2O. Water is a very important molecule of life. Not only is water the most abundant molecule in the cell, but it is believed that the first cell evolved in the ocean approximately 3.5 billion years ago. Two other molecules available around at that time are thought to be methane (CH_4, one carbon atom and four atoms of hydrogen) and ammonia (NH_3, three hydrogen atoms bound to an atom of nitrogen, N). At the University of Chicago in 1953, Stanley L. Miller and Harold C. Urey conducted an experiment with these molecules (H_2O, CH_4, and NH_3), using sparks to mimic lightening and an abundance of water to mimic the ocean. At the end of one week of continuous operation, Miller and Urey observed the formation of amino acids, which are essential molecules for the cell. Although these experiments suggest that the molecules of life could form in the ocean, it is still difficult to imagine how the first cells overcame the diluting effect of the water by bringing all of the required molecules close together in space and capturing them within a membrane. After all, what is a cell? A cell can be thought of as a narrow confining room such as in a prison cell, or a small enclosed cavity or space, such as a compartment.

The capture of organic molecules, the molecules of life, into a concentrated area, within a membrane bound compartment, permitted the chemical reactions of life to take place at a reasonable rate, something that would not have happened with the molecules floating freely. There is no life without membranes, for it is the membrane that defines the cell and keeps the contents of the cell concentrated together, preventing them from spilling and protecting them from the diluting effect of water. Of course, when the membrane of the cell formed, water was also captured within the cell, much like an ocean within an ocean, except that the ocean inside the cell (the "cytoplasm") is more highly concentrated with organic molecules. This means that the cell membrane must accommodate a watery interior and a watery exterior.

Again, water is the most abundant molecule of the cell. Approximately 70% of the mass of a cell is water, but it is a small molecule. Unlike water, molecules of the cell that mediate the reactions of life are significantly larger than three atoms. Indeed, the molecules of life are hundreds to billions of atoms in size. For this reason, they are considered the four major macromolecules of the cell: carbohydrates, lipids, proteins, and nucleic acids. Atoms that comprise each of these four macromolecules are essentially the same: hydrogen (H), carbon (C), oxygen (O), and nitrogen (N). Carbohydrates and lipids contain only C–H–O (the dashes between the letter represent the "bonding" of the atoms) while proteins and nucleic acids also contain nitrogen, C–H–O–N. Lipids and nucleic acids may also contain small amounts of phosphorous (P) and proteins may include a small percentage of sulfur (S). Of course, as we shall see, for each different macromolecule, these atoms are "snapped" together in different arrangements. Because these four major macromolecules account for the all of the cell structures and for all cell functions—including metabolism, cell movement, and cell growth—it is essential to explore the nature of these molecules, their function, and formation.

The Molecules of Microbes

In Chapter 2, we looked at microbes in the context of other organisms by showing where they fit into biological classification systems, by illustrating that they are named in the same manner as other organisms, and by studying their size relative to that of other living things. We continue that theme in this chapter by discussing the chemical molecules that make up the microbes.

All living organism must control a flow of energy. Energy is released when bonds between atoms are broken, so when microbes eat, bonds in the nutrient molecules are broken and energy is released. Microbes capture this energy and then use it to make bonds, the bond between atoms that are organized to that organism's self-configuration. Thus, microbes eat food, break the bonds in the food molecules, and capture the energy to synthesize self-molecules. A true transformation! The name given to this all-encompassing process of chemical activity is "metabolism." In this chapter, we will discuss the synthesis of self-macromolecules, the carbohydrates, lipids, proteins, and nucleic acids. The chemical breakdown of food that provides the energy for this synthesis will be discussed in detail in Chapter 9.

Carbohydrates

Carbohydrates are organic compounds containing carbon, hydrogen, and oxygen, generally in a ratio of one atom of carbon to two atoms of hydrogen to one atom of oxygen. Thus, the basic formula unit for a carbohydrate is CH_2O.

Polysaccharides

Carbohydrates vary from extremely large molecules to relatively small ones. The large molecules are chains composed of hundreds and thousands of subunits consisting of the small molecules. Such a large molecule is called a "polymer." The smallest carbohydrates are the monosaccharides ("single sugars") and disaccharides ("double sugars"), both names derived from the Latin word saccharon, for "sugar." Complex carbohydrates are called polysaccharides ("many sugars"). Carbohydrates generally serve as energy sources in cells and are found in several microbial structures, such as the cell walls of all bacteria and the capsules surrounding the cells of many bacterial species.

A monosaccharide may contain three to seven carbon atoms. Among the most significant monosaccharides are the five-carbon sugars (pentoses) called ribose and deoxyribose and the six-carbon sugars (hexoses) known as glucose, fructose, and galactose

A CLOSER LOOK

"Not Without My Beano®!"

Some people would not dare sit down to a meal of corned beef and cabbage without a knife, fork, soda bread—and, of course, their Beano®. Nor would they have Brussels sprouts with their steak or broccoli with their fried chicken unless they were sure their Beano was nearby. Eating pasta y fagiola ("pasta fahzoole") without Beano? No thanks!

To the scientist, there is really no mystery: Beano is the trade name for an enzyme preparation from the mold *Aspergillus* (a-spér-jil'lus). The enzyme breaks down galactose, a monosaccharide in beans, cabbage, broccoli, Brussels sprouts, and other "strong tasting" vegetables and high-fiber foods. Normally, galactose is broken down by the body's natural enzyme (alpha-galactosidase). But in the absence of the enzyme, bacteria in the large intestine digest the galactose. Unfortunately, they do so at a heavy price: gas (flatulence), bloating, embarrassment—and an unwillingness to go back for seconds.

Enter Beano. All it takes is a couple of tablets or drops of liquid with the first bites of food (it tastes somewhat like soy sauce). Then the fungal enzyme takes over and breaks down the galactose, leaving none for the bacteria. And leaving a happy memory of the meal (or so says the manufacturer).

(A Closer Look above). These hexoses have the same numbers of carbon, hydrogen, and oxygen atoms, and they all have the same chemical formula: $C_6H_{12}O_6$. However, their atoms are arranged differently. Molecules such as these are called isomers of one another. Glucose serves as the basic energy supply for most living creatures. (We discuss more of its chemistry, including its production in photosynthesis, in Chapter 7.) Fructose and galactose are components of disaccharides as we discuss next.

Dehydration Synthesis

Monosaccharides act as the fundamental building blocks for larger carbohydrate molecules and as sources of energy for cellular processes. These monosaccharides can be polymerized in a chemical reaction that involves the removal of H_2O from the sugars. When water is lost during the synthesis of a molecule, it is known as dehydration synthesis (FIGURE 3.1a). Microbes often break down a polysaccharide to obtain the constituent monosaccharides by adding water back to the polysaccharide. The breakdown of a molecule by adding water is known as "hydrolysis." Both dehydration synthesis and hydrolysis of polysaccharides involve the action of cellular enzymes. Enzymes are protein molecules that digest or rearrange the components of organic substances while themselves remaining unchanged.

Combinations of two monosaccharide molecules are called disaccharides. Among the commonly encountered disaccharides is maltose, also known as malt sugar. This disaccharide is found in cereal grains such as barley. It is fermented by yeast cells in the absence of oxygen to produce the alcohol in beer (microbial fermentations are explored in Chapters 7 and 10). Another well-known disaccharide is lactose, the principal carbohydrate in milk. This carbohydrate is a combination of a glucose molecule and a galactose molecule. It is chemically changed to lactic acid by certain species of bacteria. The acid causes the milk to become sour, yielding yogurt, buttermilk, and sour cream (depending on the starting material). A third disaccharide, sucrose, is a combination of a glucose molecule and a fructose molecule. It is commonly known as table sugar.

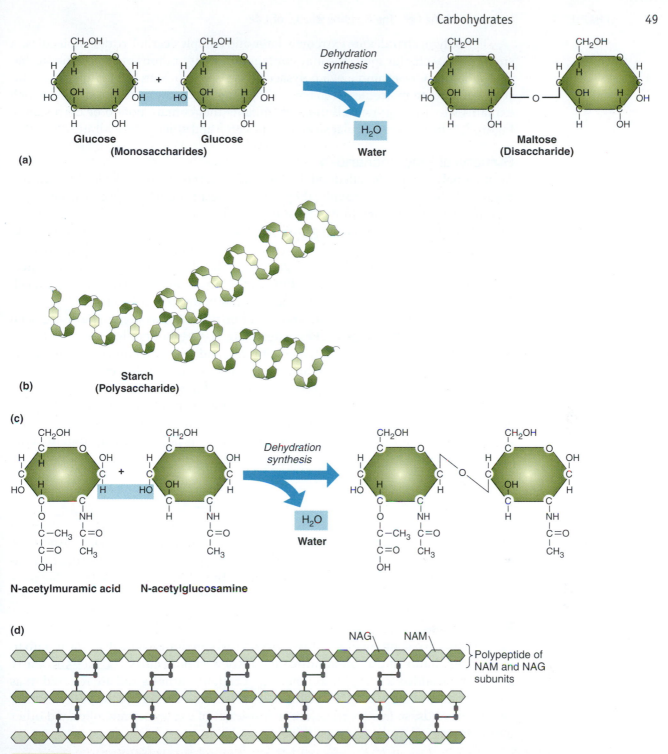

FIGURE 3.1 **Monosaccharides, Disaccharides, and a Polysaccharide.** (**a**) Glucose is a monosaccharide that can polymerize with more glucose in a dehydration synthesis (where water is removed from the sugars) to form maltose, a disaccharide. (**b**) The further polymerization of glucose lease to the formation of a polysaccharide. Note that each little hexagon is glucose. Glucose-based polysaccharides can form branches with one another to form starch, a storage form of glucose. (**c**) N-acetyl muramic acid (NAM) and N-acetyl glucosamine (NAG) are modified sugars that can polymerize to form peptidoglycan. (**d**) Peptidoglycans can be held together by side chains to form the cell wall of some types of bacteria.

The polysaccharides are extremely large and complex carbohydrate molecules. A single polysaccharide molecule may contain hundreds or thousands of monosaccharide subunits. One example of a polysaccharide is starch. Starch is composed exclusively of glucose molecules. It is favored by fungi in their growth processes, which explains why fungi grow well in environments that contain potato or corn starch. Figure 3.1 shows the molecular structures of starch and other carbohydrate molecules.

Structural Polysaccharides

Another polysaccharide called cellulose is a major part of plant cell walls. Because humans lack the necessary enzyme, they cannot digest cellulose to its constituent glucose molecules (and therefore excrete the cellulose as roughage). However, certain species of bacteria and protozoa synthesize the enzyme and can perform this breakdown. One species of protozoa lives in the intestines of termites and allows them to digest plant cell walls and obtain the glucose molecules for use as an energy source. Fungal cell walls are composed of cellulose, too. In addition to cellulose, fungal cell walls contain chitin, a polysaccharide also found in insect exoskeleton.

Moreover, the cell wall of bacteria contains polymers of sugars that have been modified to contain nitrogen. These sugars are N-acetylmuramic acid (NAM) and N-acetylglucosamine (NAG); polymers of NAM and NAG are known as peptidoglycan (Figure 3.1). Peptidoglycan is an important component of bacterial cell walls. In some cases, bacteria have multiple layers of peptidoglycan chains that serve as the cell wall and (in these cases) the adjacent peptidogylcans are stabilized and held together by peptide side chains (polymers of amino acids; Figure 3.1). As we will see in Chapter 5, Gram's stain can be used to determine the peptidoglycan content of the bacterial cell wall.

Lipids

Like carbohydrates, lipids contain only carbon, hydrogen, and oxygen, but there are usually more hydrogen atoms in a lipid than in a carbohydrate. Based on their chemical composition, lipids may be subdivided into three different groups: fats, phospholipids, and steroids.

Triglycerides

Fats contain two components, a three-carbon carbohydrate called glycerol and up to three long chains of carbon atoms called fatty acid molecules (FIGURE 3.2). Fatty acids can be added to the glycerol in a dehydration synthesis and are referred to as fatty acid tails (Figure 3.2). The fatty acid molecules contain much energy in their chemical bonds, so fats are therefore rich sources of energy for microbes and other living things.

Phospholipids

Phospholipids are phosphorus-containing lipids that possess a phosphate group (PO_4) in the place of a fatty acid (FIGURE 3.3). The phosphate group has a negative charge, while the fatty acid tails are not charged. Because this charge is found on one side of the molecule, it is called "polar." The negative charge on the phosphate

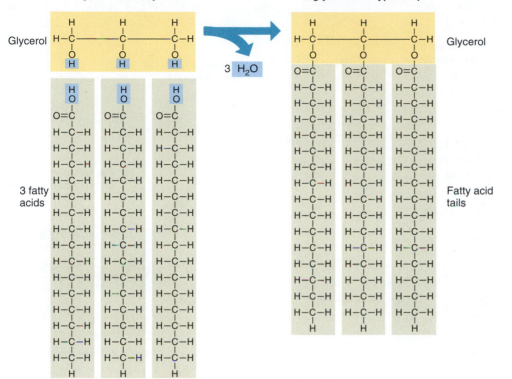

Glycerol + 3 fatty acids

Triglyceride: a type of lipid

FIGURE 3.2 **Glycerol and Fatty Acids Combine to Form Triglyceride, Fat.** Glycerol is a three-carbon molecule with each carbon possessing a hydroxyl (OH) group. Fatty acids are long carbon chains with a carboxyl group (COOH) on the terminal carbon. Three fatty acids can combine at each of the OH groups of the glycerol in a dehydration reaction in which the fatty acids bind to the glycerol to form fat, a type of lipid.

allows this polar portion of the molecule to associate with water, so it is known as a hydrophilic (water loving) polar head group (Figure 3.3). Conversely, the fatty acid tails of the phospholipid contain no charges and are not able to accommodate the charges on the water molecule, so the fatty acid tails are hydrophobic (water fearing). This gives the phospholipid the property of being amphipathic, which means that one portion is hydrophobic and another portion is hydrophilic. The amphipathic property of the phospholipid is the basis on which the cell was able to form a membrane in an aqueous environment. By forming a double-layer (or bilayer), phospholipids are able to accommodate a watery exterior with polar head groups toward the cell environment, and accommodate the watery interior of the cell with polar head groups toward the cytoplasm (Figure 3.3). Such a configuration also allows the hydrophobic fatty acid tails to associate with one another and not be exposed to water (Figure 3.3).

Steroids are lipids, but have a limited role in the metabolism of microbes. Notably, they are present in the cell walls of fungi (where they are the focus of certain antifungal agents). In addition, a fungus called *Rhizopus stolonifer* has industrial significance because it is used in the production of human steroids such as cortisone (Chapter 13).

■ Rhizopus stolonifer
rī′zo-pus stō-lon-i-fer

(a) Molecular model of a phospholipid

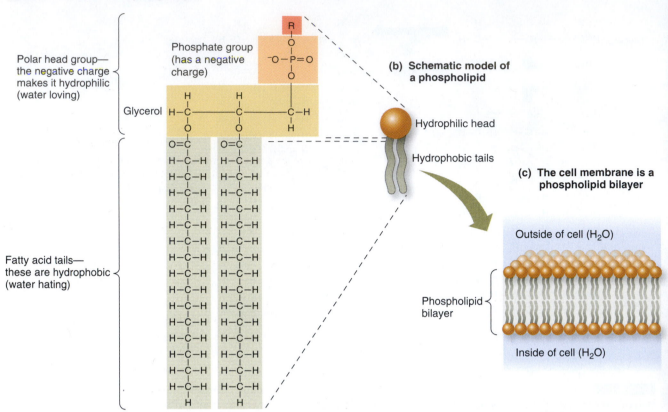

FIGURE 3.3 **Phospholipid, the Lipid of Cell Membranes.** (**a**) Phospholipids are composed of glycerol and fatty acid tails and a charged phosphate group. The charge makes this region of the protein polar and hydrophilic. (**b**) A schematic drawing of a phospholipid with the glycerol and polar head group shown as a circle with the fatty acid tail drawn as line. (**c**) The membrane of the cell is a phospholipid bilayer. This allows the hydrophilic polar head groups to associate with the watery exterior and interior of the cell.

Proteins

Proteins are important components of microbial structures, as are enzymes. Enzymes catalyze the chemical reactions of metabolism and spur the chemical processes that release energy for cell use. In bacteria, proteins are used to construct structures such as whiplike flagella, hairlike pili, and the cell membrane. Viruses are composed of nucleic acids enclosed in coats of protein, and much of the cytoplasm's substance in protozoa, fungi, and algae is protein.

Amino Acids

All proteins are polymers of subunits called amino acids, whose fundamental structure is shown in **FIGURE 3.4**. This structure always includes at least one amino group (NH_2) and one organic acid or carboxyl group (COOH). The major elements in amino acids are carbon, hydrogen, oxygen, and nitrogen; in some cases, sulfur is also present. Each amino acid has a central carbon between the amino and carboxyl group. Amino acids vary from one another on the basis of which atoms are attached to this central carbon. These atoms are known as the "R-group" and can be as simple as a hydrogen in the case of glycine, or involve up to 21 additional in the case of tryptophan, or result in charged

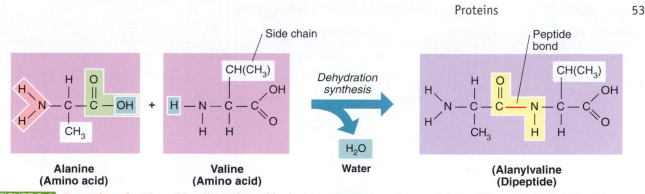

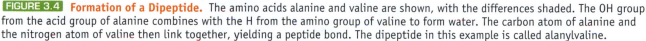

Alanine   Valine   Water   (Alanylvaline
(Amino acid)   (Amino acid)   (Dipeptide)

FIGURE 3.4 **Formation of a Dipeptide.** The amino acids alanine and valine are shown, with the differences shaded. The OH group from the acid group of alanine combines with the H from the amino group of valine to form water. The carbon atom of alanine and the nitrogen atom of valine then link together, yielding a peptide bond. The dipeptide in this example is called alanylvaline.

R-groups as seen with lysine and glutamic acid (**FIGURE 3.5**). Up to 20 different R-groups are on amino acids, which mean that there are 20 different amino acids available for use in the synthesis of proteins. The number of amino acids in a single protein may vary from a very few (in which case, the small protein is called a peptide) to thousands.

   To form a protein from amino acids, a reaction called a dehydration synthesis takes place. In this reaction, the components of water (–H and –OH) are removed from adjacent amino acid molecules, and a bond forms between the organic acid group of one amino acid and the amino group of the second amino acid. This bond linking two amino acids is referred to as a peptide bond. By forming successive peptide bonds, more and more amino acids can be added to the growing chain. Thus, an extraordinary variety of proteins can be formed from the 20 available amino acids. The genetic information for placing the correct amino acids in the correct sequence in the protein is provided by the DNA of the microbe, as we discuss at length in Chapter 12. Indeed, how the genetic code is controlled to produce proteins in microbes is a key underpinning of the field of biotechnology.

Protein Structure

Proteins carry out most of the activities of the cell, and therefore different proteins take on a wide range of structures to accommodate their range of functions. Protein structure can be defined on four levels (**FIGURE 3.6**). The first or primary structure refers to the specific sequence of amino acids in the protein polymer. This sequence is unique to each protein. The R-groups in the amino acids incorporated in the primary sequence are able to interact with one another. For example, sometimes hydrogen, when bound to an oxygen or nitrogen, can have a slight positive charge; if this occurs on an R-group (Figure 3.5), an attraction, known as a "hydrogen bond," can form between that R-group and a negative charge on the R-group of another amino acid. Other R-group associations are also possible. These R-group interactions depend on and vary with the primary amino acid sequences and in some cases cause the protein to coil (α-helix; Figure 3.6). Hydrogen bonded R-groups, if located further away on the primary sequence, also can lead to the formation of a β-pleated sheet. The α-helix and β-pleated sheet are secondary structures that arise from the primary sequence (Figure 3.6). The third, or tertiary structure of proteins results from combinations of two or more secondary structures to give an overall shape to a protein (Figure 3.6). When combinations of α-helices and β-pleated sheets

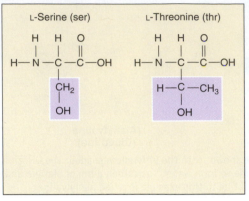

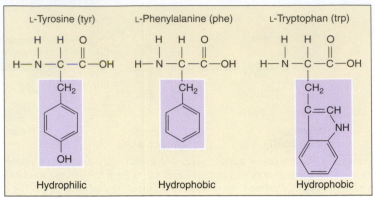

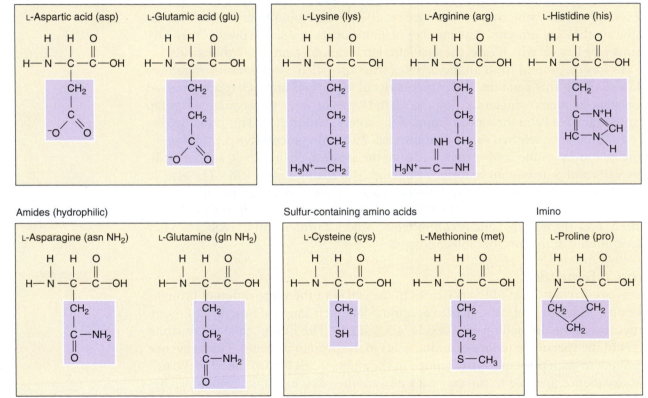

FIGURE 3.5 **Structure of the Amino Acids.** All amino acids have the same basic structure, but each vary by the R-group, which is a set of variable atoms attached to the central carbon between the amino (NH_2) and carboxyl (COOH) group of the amino acid. Shaded in this figure, we can see that the R-groups are sometimes charged or have hydrophobic and hydrophilic properties. Interactions between these shaded R-groups give rise to secondary protein structures.

combine, the protein can have an overall globular structure such as that seen with hemoglobin or immunoglobulin. The tertiary structure of a protein also can form a long fibrous structure such as that seen in collages. Globular and fibrous structures are examples of tertiary protein structure. Lastly, not all proteins can carry out their functions alone. Many proteins must work together in an associated group; these proteins are known as "polypeptides." For example, a protein that is composed of

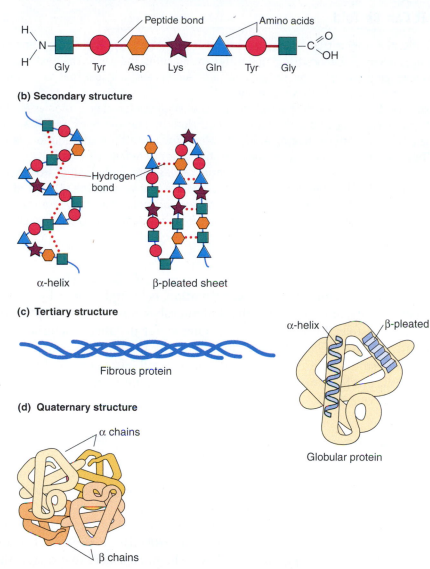

(a) Primary structure: polypeptide chain

Peptide bond · Amino acids

H — N — Gly — Tyr — Asp — Lys — Gln — Tyr — Gly — C

(b) Secondary structure

Hydrogen bond

α-helix   β-pleated sheet

(c) Tertiary structure

Fibrous protein

α-helix   β-pleated

Globular protein

(d) Quaternary structure

α chains

β chains

FIGURE 3.6 **Primary, Secondary, Tertiary, and Quaternary Protein Structures** (**a**) The primary protein sequence refers to the sequence of amino acids. (**b**) Interactions between R-groups on the amino acids can cause changes in shape of the amino acid polymer and are the secondary protein structures. (**c**) Combinations of secondary structures give the protein its overall shape, which is known as a tertiary structure. (**d**) Some proteins require more than one polypeptide to function, and the polypeptide configuration of these proteins is known as the quaternary structure.

multiple polypeptide subunits is hemoglobin. Hemoglobin actually has four polypeptide chains, two pair of the same polypeptide. Therefore, hemoglobin has a quaternary structure we call a "tetramer," with two α-chain subunits and two β-chain subunits (Figure 3.6).

It is important to note that the hydrogen bond and other association between R-groups are relatively weak. These weak chemical forces that hold a protein in its secondary and tertiary structures are easily disrupted with the addition of heat or chemicals that will cause the protein to unravel, thereby reverting to its primary

A CLOSER LOOK
Now It Can Be Told

Apparently, the story is true, at least part of it: There was a Little Miss Muffet; and at one point she was probably sitting on a tuffet eating curds and whey. The next part may have taken place in the poet's imagination: "Along came a spider that sat down beside her, and frightened Miss Muffet away."

In the late 1500s, there lived a well known English physician named Thomas Muffet. Muffet had several avocations, among them nutrition. In a famous book on this subject, he advised that fresh cheese, still in its whey, was a healthful addition to one's diet. Thus, we know that curds and whey (fresh cheese) were familiar to him.

Muffet had a daughter named Patience. Patience was apparently fond of sitting on a stool-like tuffet heeding her father's nutritional advice. But what of the spider and the rhyme? It seems that Thomas Muffet had two other avocations: the study of arthropods (with a specialty in spiders) and poetry. This is where imagination may have taken over from fact.

structure. This process, referred to as denaturation, is brought about by heat, antiseptics, disinfectants, or other agents that kill microbes. Sometimes, denaturation is commercially valuable: To produce cottage cheese, for instance, bacteria are incubated in milk, where the bacterial enzymes denature the milk proteins and curdle the milk. The result is milk curds (cottage cheese) and leftover milk fluid (whey). These two milk products are the main subjects of a famous tale, as revealed in A Closer Look on page 56.

■ Nucleic Acids

Nucleic acids are the fourth major group of organic compounds found in microbes. Like proteins, nucleic acids are generally composed of hundreds of subunits; these subunits are not amino acids, however, but are nucleotides.

The two important nucleic acids in microbes and all other living things are deoxyribonucleic acid (DNA) and ribonucleic acid (RNA). DNA is the nucleic acid of which chromosomes are composed, and RNA is the nucleic acid that carries the genetic message from the DNA to the site of protein synthesis.

Each nucleotide subunit of a nucleic acid is composed of three major elements: a sugar molecule, a phosphate group (PO_4), and a nitrogen-containing molecule with basic properties; the latter molecule is known as a nitrogenous base, or simply, a base. The sugar in DNA is the five-carbon carbohydrate deoxyribose, while in RNA it is the five-carbon carbohydrate ribose. Deoxyribose and ribose molecules are similar except that the former contains one fewer oxygen atom (hence, deoxyribose).

The phosphate group found in nucleic acids is derived from phosphoric acid (H_3PO_4). This group binds the sugars to one another in both RNA and DNA. The chain of alternating sugar and phosphate subunits forms the stem of the nucleic acid molecule, or the "backbone," as it is known colloquially.

In both DNA and RNA molecules, the bases are attached to the sugar molecules and extend out from the backbone somewhat like the teeth on a comb. The four bases in DNA are adenine, cytosine, guanine, and thymine; in RNA, they are adenine, guanine, cytosine, and uracil. (Note that DNA has thymine but no uracil, and RNA has uracil but no thymine.) Adenine and guanine are double-ring molecules called "purines," while cytosine, thymine, and uracil are "pyrimidines."

To visualize a DNA molecule, picture a ladder. In the molecule, two sugar-phosphate backbones make up the sides of the ladder, and the rungs (the steps) of the ladder are composed of the bases. On one side of each rung is a purine molecule, and on the other side is a pyrimidine molecule. That is, the purine and pyrimidine molecules project from the ladder's sides toward the middle. An adenine molecule lies opposite a thymine molecule (and vice versa), and a guanine molecule lies opposite a cytosine molecule (and vice versa). The ladder is twisted to form a double helix, the so-called spiral staircase shown in **FIGURE 3.7** . Microbes and all other living things have their DNA in this form. The DNA functions as the hereditary material and provides the genetic codes for the proper placement of amino acids in proteins. How this is accomplished is discussed in Chapter 12.

Like proteins, nucleic acids cannot be denatured without injuring the microbe or killing it. For example, ultraviolet radiation damages DNA by binding together adjacent thymine or cytosine units; thus, it can be used to lower the microbial population on an environmental surface. Chemicals such as formaldehyde alter the nucleic acids of viruses, and they are used to produce viral vaccines. Moreover, certain antibiotics interfere with nucleic acid activity, thereby killing bacteria. We shall encounter many other instances where tampering with nucleic acids or with the other key organic substances of a microbe leads to its destruction. Indeed, we can gain a fuller appreciation of a microbe's activity when we understand its chemistry.

■ A FINAL THOUGHT

We could probably talk about microbes without talking about their chemistry, but it would be like trying to describe a Big Mac® without knowing what a hamburger is. I realize that to some people, the word "chemistry" is equivalent to "root canal," but I also know that chemical compounds are the nuts and bolts of all living things.

In later chapters, we shall discuss milk products containing "carbohydrates," membranes composed of "lipids," antibodies consisting of "proteins," genes composed of "nucleic acids," and a host of other concepts that include a smattering of chemistry. To understand how yeasts cause bread to rise, we must understand the chemistry of the process; it takes a background in chemistry to appreciate nitrogen recycling; and to explain genetic engineering to our friends, we must know a bit of the chemistry behind the process. Viral replication is centered in chemistry; the production of yogurt is a chemical process; and the process of disinfection is based in chemistry.

I recommend that you give the chemistry section of this chapter a careful reading. In succeeding chapters, I shall try to make your investment of time worthwhile.

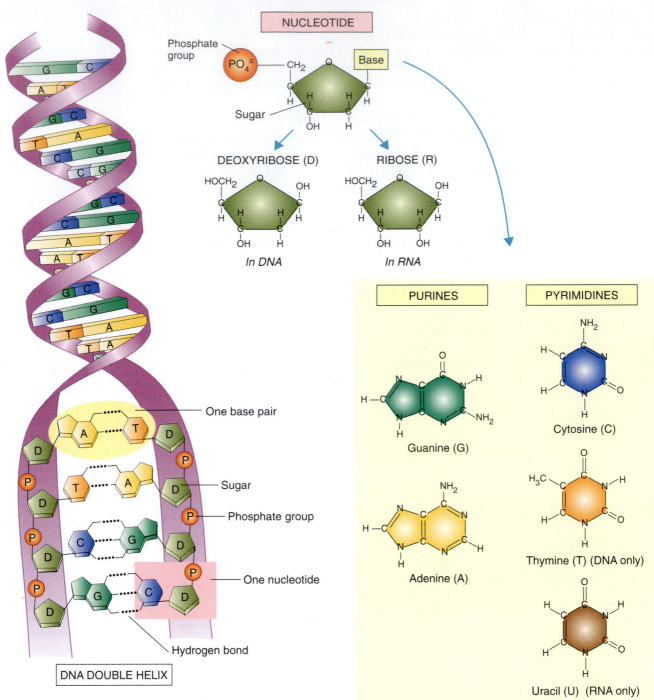

NUCLEOTIDE

Phosphate group

PO₄ — CH₂

Base

Sugar

DEOXYRIBOSE (D)
In DNA

RIBOSE (R)
In RNA

PURINES

Guanine (G)

Adenine (A)

PYRIMIDINES

Cytosine (C)

Thymine (T) (DNA only)

Uracil (U) (RNA only)

One base pair

Sugar

Phosphate group

One nucleotide

Hydrogen bond

DNA DOUBLE HELIX

FIGURE 3.7 **Nucleic Acid Components.** The pentose sugars in nucleotides are ribose and deoxyribose. These two small molecules are identical except for the lack of an oxygen atom at one carbon atom in deoxyribose. The nitrogenous bases include adenine and guanine, which are large purine molecules, and thymine, cytosine, and uracil, which are smaller pyrimidine molecules. Note the similarities in the structures of these bases and the differences in their inside groups.

QUESTIONS TO CONSIDER

1. Polysaccharides are important sources of sugar for energy; however, polysaccharides are also important structural components of the cell. Give an example of at least two structural polysaccharides. Which sugars are polymerized to form these polysaccharides?

2. The cell membrane is made of phospholipids. What unique properties of phospholipids and what unique properties of our world make this molecule uniquely adept at forming cell membranes?

3. Suppose you had the option of destroying one type of organic compound in a bacterium as a way of eliminating the microbe. Which type of compound would you choose and why?

4. If proteins are all polymers of amino acids, then how can different proteins have different shapes and take on different functions?

5. Oxygen comprises about 65% of the weight of a living organism. This means that a 120-pound person contains 78 pounds of oxygen. How can this be?

6. Which do you think formed first? Nucleic acids or proteins?

KEY TERMS

Informative facts are necessary for the expression of every concept, and the information for a concept is founded in a set of key terms. The following terms form the basis for the concepts of this chapter. On completing the chapter, you should be able to explain and/or define each one:

amino acid
carbohydrate
cellulose
dehydration synthesis
denaturation
deoxyribonucleic acid (DNA)
enzymes
fat
fatty acid
glycerol
hydrophilic
hydrophobic
isomer

lipid
nucleic acid
peptide
peptide bond
peptidoglycan
phospholipid
polysaccharide
primary protein structure
protein
ribonucleic acid (RNA)
secondary protein structure
tertiary protein structure

http://microbiology.jbpub.com/book/microbes

The site features eLearning, an online review area that provides quizzes and other tools to help you study for your class. You can also follow useful links for in-depth information, read more stories of microbiology, or just find out the latest microbiology news.

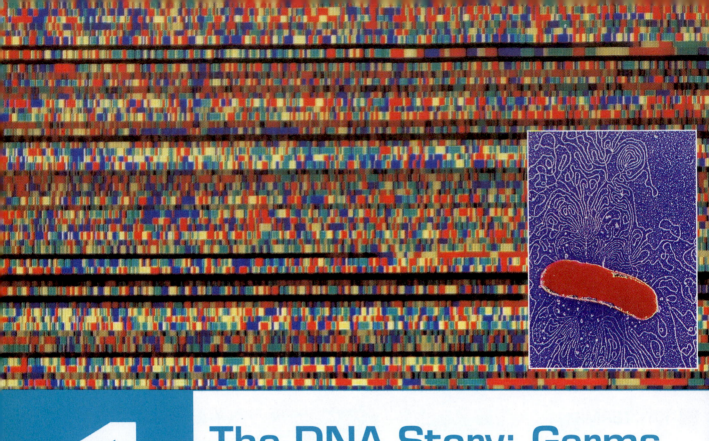

4

The DNA Story: Germs, Genes, and Genomics

E.coli, shown in the smaller, colored micrograph, doesn't get credit very often for its role in our current understanding of human DNA, but the early work to understand microbial genetics used E. coli. When the first attempts to create recombinant DNA were made, E.coli was again the organism used. Ultimately, the discoveries made in those experiments helped open the way to the astounding feat of mapping the entire human genome. The larger picture shows a sequence of base pairs forming part of the human genetic code.

Looking Ahead

Microbes have made vital contributions to our understanding of living things by serving as the nuts and bolts of biochemical research. Nowhere is this more evident than in the awe-inspiring advances in genetic engineering and biotechnology. This chapter explores the background of these modern technologies.

On completing this chapter, you should be able to . . .

- trace the landmark research of Mendel, Morgan, and other scientists of the past 150 years, which laid the foundation for recognizing DNA as the hereditary substance in microbes and other living things.

- summarize the thought processes of Watson, Crick, and their contemporaries as they resolved the puzzle of DNA's structure.

- explain how the sequence of bases in the DNA molecule serves as a genetic code and how this code is transcribed to a messenger RNA molecule and then translated to a sequence of amino acids in a protein molecule.

- describe the biochemistry involved in the regulation of gene activity using the operon model.

- discuss the methods used to identify the sequence of bases in the genomes of living things.

- explore the applications resulting from the sequencing of the genomes of microbes and humans.
- follow developments in biotechnology as they are reported in the scientific literature and the press.

In past centuries, it was customary to explain inheritance by saying, "It's all in the blood." People believed that children received blood from their parents and that something in the blood was responsible for the blending of parental characteristics. Such expressions as "blood relations" and "blood lines" reflect this belief.

But in the late 1800s and the early 1900s, the blood basis of heredity was challenged and then discarded. In its place, scientists focused on the role of nucleic acids organized as functioning units called genes. As the new science of genetics evolved, researchers guessed that the nucleic acids of genes control heredity by specifying which proteins are produced in cells, since proteins appeared to be the stuff of cellular enzymes and structural elements. But even this view of heredity seemed questionable because there was so little nucleic acid in a cell.

Nevertheless, scientists pushed on, and in the second half of the 1900s, their perseverance was rewarded. As we shall see in this chapter, one kind of nucleic acid called deoxyribonucleic acid (DNA) turned out to be the molecule of heredity. What biologists learned about DNA tested their wildest imaginations, for DNA turned out to be the jumpoff point for one of the central principles of biological science: how heredity and protein synthesis work. This principle ranks with the germ theory of disease, the cell theory, and the theory of evolution. How DNA controls heredity and protein synthesis helps explain how cells are built, how microbes move, how enzymes differ, how we come to look like our parents, how evolution occurs, and so much more.

And at the center of the spate of discoveries about DNA were the microbes. Microbes were the first living things for which the new theories were postulated, and they provided the testing grounds for verifying the newly discovered processes. Furthermore, microbes were the sources of the chemical nuts and bolts that made it possible to replicate the natural processes in test tubes. As models for other forms of life, microbes served as the sources for cascades of discoveries about the secrets of heredity.

But it did not end there. As the years passed, scientists learned how to control the processes of heredity and protein synthesis to benefit our day-to-day lives. At that time, controlling a natural process was unheard of in science—scientists were supposed to observe nature at work and ask probing questions to further human understanding of natural phenomena. Scientists simply did not alter organisms for society's benefit. And yet, that is exactly what scientists learned to do, for understanding heredity in terms of DNA and protein synthesis opened the door to genetic engineering and biotechnology. The scientific process soon became a practical technology. And what a technology it was: In this and the next chapter, we shall see how manipulating DNA has yielded products and processes well beyond the hopes of scientists. Success has led to success, and microbiologists have been in the vanguard of this progress.

To be sure, we are getting ahead of our story. To understand the great revolution taking place in gene technology, we must understand its roots, and to do that, we shall discuss the thought processes leading to a DNA-centered concept of heredity. Then we shall explore how the genetic message in DNA yields the biochemical message in protein and see how scientists have used their understanding of DNA to develop the new discipline of genomics. It is a thrilling story and one that is current as today's headlines. Hop aboard for the ride.

The Roots of DNA Research

Today's students discuss DNA as if scientists always knew about it. Over the years, the term DNA has become part of everyday speech (there is even a perfume named DNA), and DNA is as familiar as "Big Mac®." Of course, it was not always that way. Scientists spent decades figuring out the significance of DNA and the processes it controls. In this section, we shall examine some of their work.

Mendel and Morgan

Scientists were disputing the blood theory of inheritance even before they realized that semen contains no blood. But if blood was not the substance of heredity, then what was? This question would be answered by numerous experiments, beginning with those performed by Gregor Mendel, an Austrian monk (FIGURE 4.1). In the 1860s, Mendel worked with pea plants and laid the groundwork for the intensive study of genetics that would later blossom into the science of biotechnology.

Mendel studied seven characteristics in his plants during an extensive series of breeding experiments. His experiments indicated that somewhere in their reproductive cells plants have "factors" that control their heredity. The factors occur in pairs, he concluded, and they express themselves as traits in the plants and their offspring. Mendel surmised that the factors of a pair are transmitted to offspring independently of one another and that certain factors dominate over others. From his work, he developed a theory of inheritance completely at odds with the blood theory. Mendel's

(a)

(b)

FIGURE 4.1   **At the Roots of DNA and Genetics.** (**a**) Gregor Mendel, the Austrian monk who established the principles of genetics through meticulous experiments with pea plants. (**b**) Thomas Hunt Morgan, whose exhaustive experiments related an inherited characteristic (eye color) to a fruit fly chromosome and called attention to this cellular component as a carrier of inherited traits.

work provided the first inkling that inherited factors (now known as genes) are discrete units transmitted in reproductive cells.

Unfortunately, scientists of the time paid little attention to Mendel's work or its meaning. However, forty years later, in 1900, three European botanists working independently of each other reported experimental results that verified Mendel's work. Within a few short years, scientists tentatively connected Mendel's factors to cell parts called chromosomes. Chromosomes (literally, "colored bodies") are threadlike strands of nucleic acid floating freely in the cytoplasm of prokaryotic cells and contained within the nucleus of eukaryotic cells. However, scientists were uncertain whether chromosomes contained the units of heredity until Thomas Hunt Morgan performed landmark experiments with fruit flies in 1910. Morgan determined that a single chromosome determines sex as well as eye pigmentation in fruit flies; in doing so he provided statistical evidence relating eye color to chromosomes, placing the chromosomal theory of inheritance on a firmer footing.

At Morgan's time, geneticists believed that only a part of a chromosome specifies a trait because there simply were not enough chromosomes in cells to account for all an individual's traits. Scientists suggested that each chromosome might be divisible into smaller hereditary units, like beads on a string, and before long the concept of the gene emerged. The term "gene" was suggested by a Scandinavian researcher, Willard Johannsen. He used the word to mean *origin*, because "gene" is derived from the Greek *gena* meaning "birth" or "descent." It stood to reason that if genes are the hereditary units on a chromosome, then a first step in learning the chemical composition of the genes would be learning the chemical composition of the chromosome; and this is precisely what researchers next set out to do.

Focus on DNA

During the first quarter of the 1900s, the attention of scientists was drawn to the major material in the cell nucleus: deoxyribonucleic acid (DNA). DNA had been first reported in 1869 by the Swiss researcher Johann Fredrich Meischer. Meischer removed the nuclei from a salmon's white blood cells and isolated an apparently new chemical substance that he called nuclein. In later years, chemists studied the composition of nuclein and identified its major component as deoxyribonucleic acid.

The consistent isolation of DNA from chromosomes and the participation of chromosomes in heredity made DNA a likely candidate for the hereditary substance. Then, in the 1920s, experiments performed at New York's Rockefeller Institute helped forge the link between DNA and heredity. Alfred Mirsky and his colleagues reported that, with only two exceptions, all cells of a complex organism have virtually the same amount of DNA in their nuclei. The exceptions are the reproductive cells (sperm and egg cells). These cells contain half as much DNA as other cells, an observation correlating with the theory that sperm and egg cells are the vehicles for bringing together genetic information from parents to form offspring.

Also in that decade, in 1928, the English bacteriologist Frederick Griffith reported some puzzling results of experiments with bacteria. Griffith was working with pneumococci, the Gram-positive bacteria that cause pneumonia. Griffith found that he could take chemical debris from a pathogenic strain of pneumococci, mix it with a harmless strain, and transform the harmless strain into a pathogenic strain (FIGURE 4.2 illustrates Griffith's work). But Griffith was unable to isolate the transforming factor. That isolation was finally accomplished years later by Oswald Avery and his research

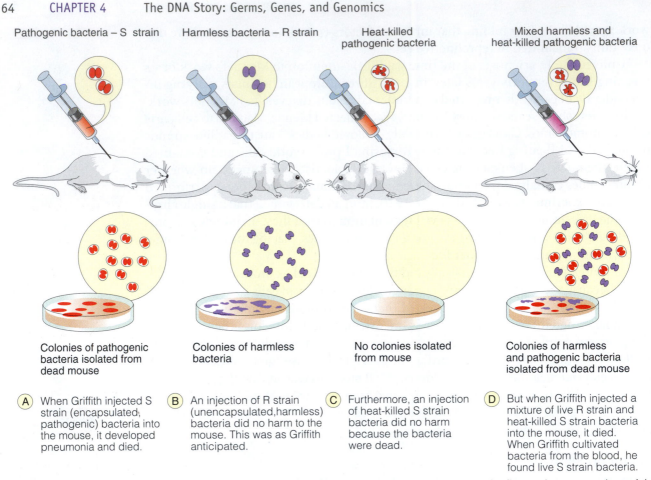

Pathogenic bacteria – S strain   Harmless bacteria – R strain   Heat-killed pathogenic bacteria   Mixed harmless and heat-killed pathogenic bacteria

Colonies of pathogenic bacteria isolated from dead mouse

Colonies of harmless bacteria

No colonies isolated from mouse

Colonies of harmless and pathogenic bacteria isolated from dead mouse

(A) When Griffith injected S strain (encapsulated, pathogenic) bacteria into the mouse, it developed pneumonia and died.

(B) An injection of R strain (unencapsulated, harmless) bacteria did no harm to the mouse. This was as Griffith anticipated.

(C) Furthermore, an injection of heat-killed S strain bacteria did no harm because the bacteria were dead.

(D) But when Griffith injected a mixture of live R strain and heat-killed S strain bacteria into the mouse, it died. When Griffith cultivated bacteria from the blood, he found live S strain bacteria.

FIGURE 4.2   The Transformation Experiments of Griffith. Griffith's experiments were among the first to demonstrate bacterial transformation and the transfer of genetic information.

group at Rockefeller Institute. In 1944, Avery's group identified the transforming substance as DNA. This work led to a clearer understanding of the process of transformation, described in Chapter 8.

It was becoming clear that DNA was a key molecule in heredity. But the developing focus on DNA among some scientists irritated an opposing group of biochemists, who believed that proteins act as genetic materials and provide the chemical codes to form new proteins. Their belief would soon be dealt a severe blow by experiments to be performed at Cold Spring Harbor Laboratory in New York.

At that lab, **Alfred Hershey** and **Martha Chase** worked with the bacterium *Escherichia coli* and the viruses that attack it. Bacterial viruses are called **bacteriophages**, or **phages** for short (Chapter 10 contains a detailed description of these viruses). In 1952, scientists knew that phages penetrate bacteria and use them as chemical factories for producing copies of themselves. They were also aware that phages consist of nothing more than a core of DNA enclosed in a coat of protein. What they did not know was whether the DNA or the protein (or both) control the production of new viruses. The experiments of Hershey and Chase would answer that question, and their results would help establish the role of DNA in heredity.

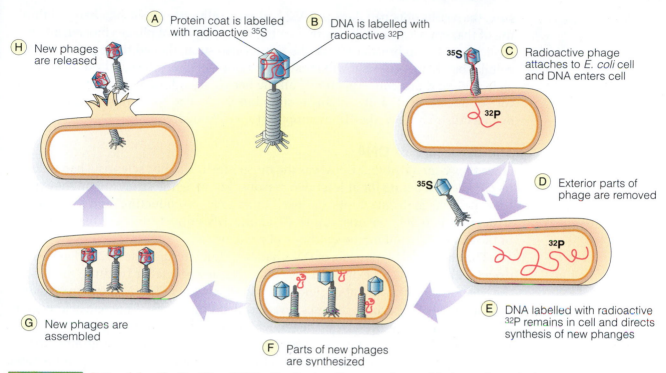

(H) New phages are released

(A) Protein coat is labelled with radioactive 35S

(B) DNA is labelled with radioactive 32P

(C) Radioactive phage attaches to *E. coli* cell and DNA enters cell

35S

32P

(D) Exterior parts of phage are removed

35S

32P

(E) DNA labelled with radioactive 32P remains in cell and directs synthesis of new phanges

(G) New phages are assembled

(F) Parts of new phages are synthesized

FIGURE 4.3   **Determining the Function of DNA.** The Hershey-Chase experiment with viruses (bacteriophages) and their host bacterium *E. coli*. (**a**) Hershey and Chase prepared viruses with two radioactive labels: one in the coat, one in the DNA. (**b**) They combined the viruses with *E. coli* cells and gave the viruses an opportunity to interact. (**c**) Then they agitated the mixture in a blender and thereby removed the viral protein coats. The coats were found to have the radioactive sulfur. (**d**) The researchers found the DNA carrying the radioactive phosphorus within the bacterial cytoplasm. (**e**) Here the DNA directs the synthesis of new viral components, (**f**) which are assembled to new viruses. (**g**) The viruses are then released from the cell cytoplasm. The results indicate that the DNA alone directs the synthesis of both the DNA and the protein in new viruses.

Hershey and Chase made use of the facts that the DNA in the core of the phage contains phosphorus but no sulfur and that protein in the phage coat has sulfur but no phosphorus. They cultivated phages that had radioactive phosphorus (^{32}P) in their DNA and radioactive sulfur (^{35}S) in their coat. Then they mixed the radioactive phages with *E. coli* cells and waited just long enough for phage replication to begin (Chapter 6 details this process). At this point, they sheared away any remaining viruses and debris from the surface of the bacterial cells using an ordinary household blender (now a museum piece). **FIGURE 4.3** shows their experiments.

Now Hershey and Chase tested the *E. coli* cells and the surrounding fluid to find out where the radioactive substances had concentrated. They discovered that most of the radioactive phosphorus was in the cytoplasm of the *E. coli* cells and most of the radioactive sulfur was in the sheared-away phages in the surrounding fluid. These observations indicated that only the DNA core entered the bacterium while the protein coat of the virus remained outside. Hershey and Chase concluded that phage DNA—and only DNA—is the substance responsible for phage replication; protein apparently has no role in the process.

Certain experiments stand out as turning points in scientific history, and the Hershey-Chase experiments at Cold Spring Harbor Laboratory are an example. To be sure, the results obtained there in 1952 had great influence on the biochemical thinking of that era because they clarified an important aspect of phage (indeed, all viral) replication. But in broader terms, the results also strengthened the place of DNA in cellular biochemistry by showing that DNA alone can direct the synthesis of both nucleic acids and proteins. Apparently all the hereditary information for the manufacture of nucleic acids and proteins is stored in DNA. The "protein supporters" were dealt a crushing blow; the "DNA supporters" celebrated.

The Structure of DNA

Once they concluded that DNA was the molecule of heredity, scientists had to know its chemical structure for at least two reasons: knowing the structure might explain how the DNA molecule duplicates itself during cell reproduction; and knowing the structure could help scientists understand how DNA directs the hereditary activities of the cell by controlling the synthesis of proteins.

The chemical components of DNA had been known since the 1920s. During that decade, Pheobus A. T. Levine and his colleagues at Rockefeller Institute established the existence of two types of nucleic acid: Ribonucleic acid (RNA) and deoxyribonucleic acid (DNA). Levine's analyses revealed that both types of nucleic acid contain three components: a five-carbon sugar, which is either ribose (in RNA) or deoxyribose (in DNA); a number of phosphate groups, which are chemical groups derived from phosphoric acid molecules; and a series of compounds called nitrogenous bases (bases, for short), which contain nitrogen and have the chemical properties of bases. The bases in DNA are adenine, thymine, guanine, and cytosine; in RNA, they are adenine, uracil, guanine, and cytosine. Adenine and guanine are double-ring molecules known as purines; cytosine, thymine, and uracil are single-ring molecules called pyrimidines. (Chapter 3 discusses the nucleic acids.)

Levine's group observed that DNA contains roughly equal proportions of phosphate groups, deoxyribose molecules, and bases. Therefore, they correctly concluded that DNA is composed of these three components joined to one another in units. The units are known as nucleotides. A nucleotide of DNA consists of a deoxyribose molecule attached to both a phosphate group and a base, as we shall see presently. The identity of the base distinguishes one nucleotide from another. In later years, scientists found that the identity of the bases is key to DNA action.

In the early 1950s, almost nothing was known about the spatial arrangement of the nucleotides in a DNA molecule. About this time, a new technique called **x-ray diffraction** became available. In x-ray diffraction, crystals of a chemical substance are bombarded with X rays. Electrons in the chemical substance scatter (or diffract) the X rays, and a pattern develops on a photographic plate. The pattern gives strong clues to the three-dimensional structure of the chemical substance.

Among the leading experts on the diffraction patterns of DNA were the British biochemists Maurice H. F. Wilkins and Rosalind Franklin. Wilkins (see A Closer Look, on page 67) had found a way to prepare uniformly oriented DNA molecules, and Franklin was using these preparations to obtain clear diffraction patterns of DNA. About the same time, a young American postdoctural student named James D. Watson was working at Cambridge University with a British graduate student named Francis H. C. Crick. Watson and Crick undertook the task of determining the structure of DNA (FIGURE 4.4). They performed no laboratory benchwork—their great contribution to science was interpreting and combining the data supplied by Wilkins and Franklin to guess DNA's structure.

Watson and Crick began with the basic assumption that the nucleotides in DNA are joined by a chain of phosphate groups bound to deoxyribose molecules in an alternating fashion (P–D–P–D–P . . .). Then, using the x-ray data, they decided that a base was joined to each deoxyribose molecule, hanging out as a side group from the chain, as **FIGURE 4.5** illustrates. Next, they took note of Franklin's x-ray diffraction data, which suggested that the DNA exists as a coil, or helix. But there was a problem: The data indicated that the molecule is not a single helix, but two helices (two coils). Maybe, they suggested, the DNA molecule is a double helix (Figure 4.5). And if the bases were arranged to point inward (not outward, as biochemists were suggesting), the density of DNA would come close to fitting the density revealed by the x-ray diffraction studies.

A CLOSER LOOK

The Third Man

The names James D. Watson and Francis H. C. Crick are familiar to students of biology as those of the scientists who first proposed the structure of DNA. In doing so, they constructed a model of how hereditary material makes replicas of itself and showed how genes could encode the synthesis of protein in a cell. In 1962, Watson and Crick received the Nobel Prize in Physiology or Medicine for their work.

Also cited that year was their coworker, a British scientist named Maurice Hugh Frederick Wilkins. Working with his colleague Rosalind Franklin, Wilkins made available the data that permitted construction of the Watson-Crick model. Wilkins was originally a physicist of New Zealand heritage. Educated in England, he came to the United States during World War II to work on uranium separation for the Manhattan Project, which produced the first atomic bomb. Partly because of the destructive effects of the bomb, his interest turned from the atomic nucleus to the cell nucleus. Returning to England in the late 1940s, he assumed a research position at King's College.

Wilkins' forte was x-ray crystallography. Using a special camera, he placed crystals of DNA before a beam of X rays and charted their patterns of deflection. When substances crystallize, their atoms line up in a lattice work of repeating units. These units deflect X rays in a regular pattern, and by studying the patterns, a scientist can determine the distances between different components of the latticework. (The process is much like determining the structure of monkeybars by analyzing the shadow they cast when hit by the sun at various angles.) Wilkins theorized that DNA molecules exist as a helix, a form resembling a corkscrew wrapped around a cylinder.

In 1951, Watson arrived in Cambridge, England, on a research fellowship to learn about molecular structures. There, at the Cavendish Laboratory, he met Francis Crick. Both were intensely interested in DNA's structure, and they immediately set to work to solve the puzzle. Watson and Crick did no experiments in the usual sense. They worked entirely with other people's material, including the x-ray photographs of Wilkins and Franklin. They assembled jigsaw models of the nucleotide components so that DNA would conform to the helical pattern shown by the X rays. Finally, in the April 25, 1953 issue of *Nature,* a British scientific journal, Watson and Crick made their historic suggestion about the structure of DNA . . . but not before checking with Wilkins.

(a)   (b)

FIGURE 4.4 **Elucidating the Structure of DNA.** James D. Watson and Francis H. C. Crick, the American graduate student and the British biochemist who correctly guessed the structure of DNA. (**a**) The scientists as they appeared in 1952, when the structure of DNA was formulated, and (**b**) more recently.

At this point, the observations made years before by biochemist Erwin Chargaff came into the picture. Chargaff reported that in DNA, the amounts of adenine and thymine are nearly identical, as are the amounts of cytosine and guanine. Watson and Crick thus envisioned that for every adenine molecule on one helix of DNA, there must be a thymine molecule on the opposing helix (and vice versa). This would be the "complementary" base. Similarly, for every guanine molecule on one helix of the DNA molecule, there must be a complementary cytosine molecule on the other (and vice versa). The weak chemical bonds formed between the opposing bases made "chemical sense," and the setup of opposing bases fit the x-ray diffraction data perfectly. The model was complete.

Watson and Crick expressed their hypothesis in a seminal paper published in 1953 in the prominent journal *Nature*. Their suggested structure for DNA fit the available data perfectly, and the scientific community hailed their work as a great achievement. Watson, Crick, and Wilkins shared the 1962 Nobel Prize in Physiology or Medicine; Franklin had previously died of cancer, and the prize was not awarded to her posthumously, although her contribution to deciphering DNA's structure is universally accepted.

The Watson-Crick structure was a great scientific leap forward. It made it possible for biochemists to postulate how DNA replicates and passes duplicates of itself onto the next generation of cells. And knowing DNA's structure gave biochemists an inkling of how DNA provides hereditary information for the synthesis of proteins: They looked at the variable sequence of bases in DNA and envisioned how it might be a code, for encoding a variable sequence of amino acids in a protein. Scientists let

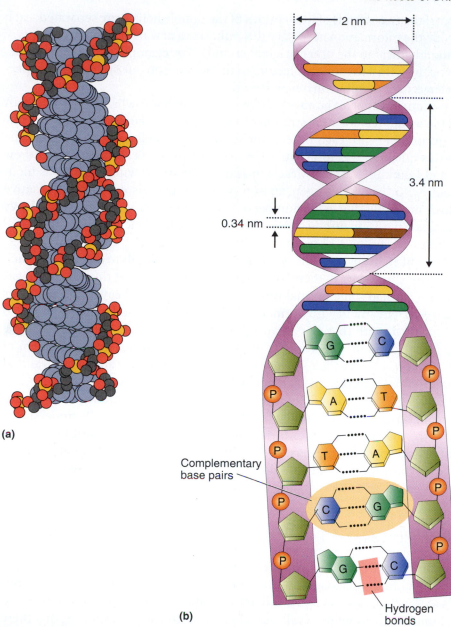

(a)

(b)

Complementary
base pairs

Hydrogen
bonds

FIGURE 4.5   The DNA Molecule. (**a**) A space-filling model of the DNA molecule as determined by x-ray diffraction data. (**b**) A stylized model of DNA. The molecule embodies a double helix of two intertwined strands of deoxyribose molecules and phosphate groups. Complementary nitrogenous bases extend out from the deoxyribose molecules toward one another.

their minds race on the implications of this insight. They could hardly know what lay ahead.

DNA Replication

Toward the end of their article in *Nature* in 1953, Watson and Crick briefly noted that the structure of DNA might provide insight into how the molecule replicates. They implied that if the two strands of nucleotides of the double helix were separated, each could provide the information necessary to synthesize a new partner complementary to itself (and identical to the other original strand). For example, an adenine molecule always pairs with a thymine molecule, so the adenine on one strand would specify a complementary thymine on a new strand.

An important question to be answered by the research was whether the two DNA strands serve as models for new strands of DNA and then combine with the new strands *or* recombine with each other after serving as models for new strands. The answer to this question was found in 1958 in experiments performed by Matthew Meselson and Franklin W. Stahl. Meselson and Stahl worked with *E. coli* cells incubated initially with "heavy" nitrogen, an isotope of nitrogen. (Isotopes are alternative forms of an element differing in the number of neutrons present in their atoms.) Heavy nitrogen atoms have 15 neutrons (^{15}N). The cells were then incubated with normal nitrogen, whose atoms have 14 neutrons (^{14}N).

Meselson and Stahl analyzed the newly forming bacterial cells to determine which form of nitrogen showed up in their DNA. The data indicated that the new DNA was a hybrid; that is, it had some normal nitrogen as well as some heavy nitrogen. This would occur if a DNA strand with normal nitrogen served as a model for a new strand and then combined with that new strand to form a new double helix. FIGURE 4.6 shows the Meselson-Stahl experiments in diagrammatic form. This mechanism of DNA replication has been called semiconservative replication because one strand of the original DNA is conserved in each newly formed double helix.

The work of Meselson and Stahl brought to an end a momentous period in microbiology. The structure and method of replication of DNA were finally understood. Importantly, the insights generated during that period stimulated a series of experiments that would solve the mystery of how DNA acts as the hereditary substance. Understanding this process would provide the first clues as to how the process of protein synthesis could be manipulated to benefit society.

DNA to Protein

Proteins are the working and structural components of all cells. Chemically, they are composed of building blocks called amino acids (much as the building blocks of nucleic acids are nucleotides). Only 20 different amino acids yield the countless combinations found in the proteins of cells. Biochemists estimate, for example, that there are roughly 10,000 different enzymes operating in microbes, which implies that there are at least 10,000 unique combinations of the 20 amino acids.

What makes one protein different from another protein is its sequence of amino acids. For example, two proteins may each consist of 33 amino acids linked together, but if the sequences of the amino acids are different, then the proteins are different. Consider, for a moment, how many words can be composed from our 26-letter alphabet. For proteins, the alphabet is composed of 20 amino acids.

If DNA molecules specify amino acid sequences in proteins, it follows that some message in the DNA molecule must specify an amino acid sequence. The contemporaries of Watson and Crick suggested that the sequence of bases in DNA might constitute a code expressing such a message. The questions were: What was the code?

(a) Possible results when solution containing DNA is centrifuged for 2–3 days

DNA in solution in
centrifuge tubes

Light DNA
Hybrid DNA
Heavy DNA

(b) Results of Meselson and Stahl's experiment

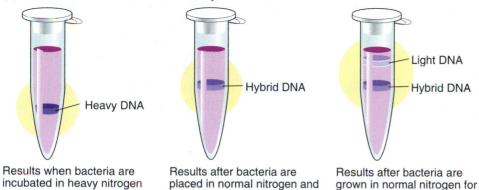

Heavy DNA

Hybrid DNA

Light DNA
Hybrid DNA

Results when bacteria are
incubated in heavy nitrogen

Results after bacteria are
placed in normal nitrogen and
one generation has passed

Results after bacteria are
grown in normal nitrogen for
another generation

FIGURE 4.6   How DNA Replicates. The experiment of Meselson and Stahl. (**a**) DNA is placed in a tube containing the chemical compound cesium chloride (CsCl). The dense solution is centrifuged for 2 to 3 days. During this time, different forms of DNA containing different amounts of various nitrogen isotopes separate according to their weights. (**b**) Meselson and Stahl grew bacteria in heavy nitrogen (^{15}N) and found that the DNA occurred at a certain level in the CsCl. Then they permitted the bacteria to replicate one time and synthesize DNA in a medium with ordinary nitrogen (^{14}N). They discovered that a hybrid DNA containing both ^{14}N and ^{15}N was forming. Thus the DNA contained one old strand (with ^{15}N) and one new strand (with ^{14}N). After another round of replication, light DNA containing only ^{14}N was found. This was also a product of semiconservative DNA replication.

How does the code get translated into an amino acid sequence? We shall explore answers to these questions in this section.

The Intermediary and the Genetic Code

One of the first points that biochemists had to work out was whether the biochemical information in DNA is used directly to make protein or some intermediary is employed. The assembly of amino acids into proteins was known to take place in the cell's cytoplasm, but DNA is largely restricted to the cell's nucleus in eukaryotic organisms and to the nucleoid region of prokaryotic organisms. A direct passage of information from DNA to protein was unlikely.

In the 1940s, biochemists reported that when *E. coli* cells are synthesizing proteins, they contain an unusually large amount of the second type of nucleic acid, ribonucleic acid (RNA). (RNA is compared to DNA in **TABLE 4.1** .) Experiments with

TABLE 4.1 A Comparison of DNA and RNA	
DNA (Deoxyribonucleic Acid)	**RNA (Ribonucleic Acid)**
In prokaryotes, found in the nucleoid and plasmids; in eukaryotes, found in the nucleus and in some extranuclear organelles	In prokaryotes and eukaryotes, found dissolved in the cytoplasm and at ribosomes; in eukaryotes, found in the nucleolus
Always associated with chromosome (genes); each chromosome has a fixed amount of DNA	Found mainly in combinations with proteins in ribosomes in the cytoplasm, as messenger RNA, and as transfer RNA
Contains a pentose (5-carbon) sugar called deoxyribose	Contains a pentose (5-carbon) sugar called ribose
Contains bases adenine, guanine, cytosine, thymine	Contains bases adenine, guanine, cytosine, uracil
Contains phosphorus (in phosphate groups) that connects various sugars with one another	Contains phosphorus (in phosphate groups) that connects various sugars with one another
Functions as the molecule of inheritance	Functions in protein synthesis
Double-stranded	Usually single-stranded
Larger size	Smaller size

radioactive isotopes revealed that RNA molecules could move from the nucleus to the cytoplasm in eukaryotic cells, so attention centered on RNA as the possible intermediary. Moreover, it was easy to see how the sequence of bases in DNA could determine the complementary sequence of bases in RNA; that is, the base sequence of DNA would serve as a model (or template) for synthesizing a molecule of RNA with complementary bases (except that uracil would replace thymine in the RNA molecule). The process would be similar to that taking place during DNA replication, except that single-stranded RNA molecules would result.

As the notion of information flow from DNA to RNA became accepted, remaining questions were: How does the genetic message of DNA transmit to RNA, and what is the genetic code? Answers to these questions came in a series of experiments performed on *E. coli* in 1961 by Francis H. C. Crick and his colleagues. Crick's group reasoned that the genetic code in DNA is in a series of blocks of information, each block composed of three bases. A single block of three bases could specify a single amino acid in the protein to be encoded.

A bit of simple mathematics pointed to the three-base code. Virtually all proteins are made of just 20 amino acids. If a single base encoded a single amino acid, then there would be only enough codes for four amino acids (remember, there are only four bases). If two bases (e.g., AG, TC, and so on) encoded a single amino acid, then only 16 codes would exist and only 16 amino acids could be encoded. However, if *three* bases comprised a code unit, then 64 possible codes could exist in DNA (for example, ATA, GAT, TCG, CAT, and so on). Sixty-four codes were more than enough to specify the 20 amino acids.

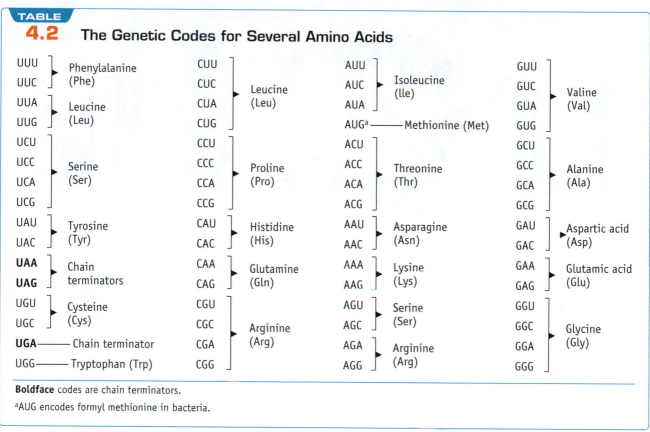

TABLE 4.2   The Genetic Codes for Several Amino Acids

UUU, UUC	Phenylalanine (Phe)	CUU, CUC, CUA, CUG	Leucine (Leu)	AUU, AUC, AUA	Isoleucine (Ile)	GUU, GUC, GUA, GUG — Valine (Val)
UUA, UUG	Leucine (Leu)			AUGª	Methionine (Met)	
UCU, UCC, UCA, UCG	Serine (Ser)	CCU, CCC, CCA, CCG	Proline (Pro)	ACU, ACC, ACA, ACG	Threonine (Thr)	GCU, GCC, GCA, GCG — Alanine (Ala)
UAU, UAC	Tyrosine (Tyr)	CAU, CAC	Histidine (His)	AAU, AAC	Asparagine (Asn)	GAU, GAC — Aspartic acid (Asp)
UAA, **UAG**	Chain terminators	CAA, CAG	Glutamine (Gln)	AAA, AAG	Lysine (Lys)	GAA, GAG — Glutamic acid (Glu)
UGU, UGC	Cysteine (Cys)	CGU, CGC, CGA, CGG	Arginine (Arg)	AGU, AGC	Serine (Ser)	GGU, GGC, GGA, GGG — Glycine (Gly)
UGA	Chain terminator			AGA, AGG	Arginine (Arg)	
UGG	Tryptophan (Trp)					

Boldface codes are chain terminators.

ªAUG encodes formyl methionine in bacteria.

To verify this hypothesis, Crick's group performed experiments with viral DNA molecules, altering their bases to see what would happen. Each time they inserted a single nucleotide or two nucleotides into the viral DNA molecule, the insertion altered the amino acid sequence in the resulting protein. Hence, the insertion was altering the base code. However, a three-base insertion or deletion resulted in the addition or deletion of a single amino acid in the chain; the remainder of the amino acids were untouched. It was apparent that the magic number of bases was three. Thus, the evidence indicated that a block of three bases specifies a single amino acid in a protein.

Scientists next set out to determine the base codes for all 20 amino acids. In a test tube they placed synthetic RNA having a known base sequence along with enzymes, a variety of amino acids, and other essential materials needed to make protein. Their objective was to learn which amino acid would be used to make the protein if the synthetic RNA consisted of a single base. They found, for example, that a synthetic RNA molecule consisting of nothing but uracil (U–U–U–U–U–U–U– . . .) encoded a protein consisting solely of the amino acid phenylalanine. Thus, they reasoned, the RNA code for phenylalanine must be UUU. Working backwards, they concluded that the DNA code for phenylalanine is AAA. In this way, the base codes for all 20 amino acids were determined, as TABLE 4.2 delineates.

In the following years, biochemists determined that the genetic code is nearly universal; that is, the same three-base codes specify the same amino acids regardless of whether the organism is a bacterium, a bee, or a bird. This remarkable discovery

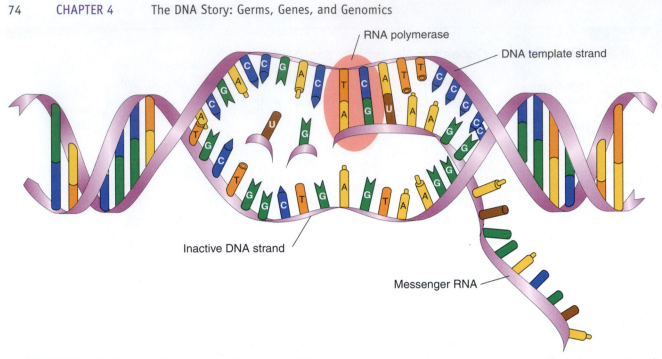

FIGURE 4.7   The Transcription Process. The enzyme RNA polymerase moves along the template strand of the DNA molecule and synthesizes a complementary molecule of RNA using the base code of DNA as a guide. The mRNA will carry the genetic message of DNA into the cytoplasm, where protein synthesis occurs. Note that the inactive strand of DNA is not transcribed.

implies that the essential difference among all living things is the sequence in which the bases occur in DNA. Different sequences of bases in DNA specify different sequences of bases in RNA, and the different sequences of bases in RNA specify different sequences of amino acids in proteins. Thus, the base sequence in an organism's DNA is the so-called central dogma of its heredity and protein synthesis.

Transcription

As evidence about the nature of the genetic code continued to accumulate, a persuasive body of data was being generated on how DNA transfers its message to RNA and what occurs next. As the process unfolded, molecular biologists marveled at its intricacies and tried to absorb the implications of what they were witnessing. It was one of the most awe-inspiring periods in the history of the biological sciences. And microbes were at its core.

Biochemists now know that the process of protein synthesis begins with an uncoiling of the DNA double helix and a separation of the two DNA strands. This separation exposes a gene, a region of the DNA molecule that encodes a protein. Using the enzyme RNA polymerase, cells synthesize a single-stranded molecule of RNA having bases complementary to the bases along that region of one strand of the DNA (as we have discussed above). This production of RNA is called transcription. A base code in part of the DNA molecule is transferred to a base code in the RNA molecule. The RNA fragment so constructed is known as messenger RNA (mRNA), shown in FIGURE 4.7 . Each single-stranded mRNA molecule eventually passes into the cell's cytoplasm carrying the genetic message. Each three-base sequence on an mRNA molecule is called a codon (a codon will specify an amino acid). The mRNA molecule can thus be considered a series of codons.

In the production of the final mRNA molecule, some modification of the RNA may take place. In eukaryotic microbes such as fungi and protozoa and in more complex eukaryotic cells, bits of RNA are snipped out of the preliminary mRNA molecule. These bits of discarded RNA are referred to as introns because they are *intervening* codons between the genes. The sections of remaining mRNA are known as exons, because their code will be *expressed*. In prokaryotic organisms such as bacteria, the preliminary mRNA is the final mRNA molecule—there are no introns.

Two other types of RNA are also encoded by DNA. One type is ribosomal RNA (rRNA), which combines with protein to form ribosomes, the ultramicroscopic bodies existing free in the cytoplasm (in prokaryotes) or associated with membranes (in eukaryotes). Functioning as the cell's "workbenches," the ribosomes are the sites where amino acids are assembled into proteins according to the instructions delivered by mRNA molecules. Ribosomes are composed of two subunits: a smaller subunit where the mRNA molecule binds to the ribosome, and a larger subunit where enzymes that link the amino acids together are located.

The final type of RNA is called transfer RNA (tRNA). Transfer RNA molecules float freely in the cytoplasm. Here they bind to amino acids and deliver them to the ribosomes for use in protein synthesis. A different tRNA molecule exists for each of the 20 amino acids. The amino acid alanine, for example, has its own tRNA; the amino acid glycine has its own tRNA; and each of the other 18 amino acids has a specific tRNA that will bind only to it. The binding requires enzymes and energy obtained from ATP molecules. The amino acid binds at one end of the tRNA molecule. At another location on the tRNA, there is a sequence of three bases that complements a codon on the mRNA molecule; this three-base sequence is known as an anticodon. The matching of complementary codon (on mRNA) and anticodon (on tRNA) is essential for the correct positioning of an amino acid during protein synthesis. This activity takes place in the process of translation, as we shall see next.

Translation

DNA is the starting point for the process in which a protein is synthesized from a collection of amino acids. Essentially, the language of the base code of DNA is translated into the language of the amino acid sequence in a protein. This process is somewhat like visiting a library to write a paper: The library is the DNA—it holds the information; students transcribe the information into notes—the notes are the mRNA; then the notes are translated into a paper—the paper is the protein. We must note, however, that RNA may also serve as a starting point, as A Closer Look on page 76 discusses.

The mechanism of protein synthesis begins at the DNA molecule with transcription of part of its code to a molecule of mRNA. The synthesis of mRNA by RNA polymerase is specified by a segment of DNA bases called the promoter site. For a given gene, the promoter site exists on one DNA strand but not on the other, which is why one strand transcribes its message to the mRNA and the other does not. Moreover, in mRNA production in eukaryotes, a cap is added to the molecule's front end. The cap consists of 7-methylguanosine, a molecule that is essential for binding mRNA to a ribosome. At the end of transcription, special enzymes add a string of adenine-containing nucleotides to the other end of the mRNA molecule, an addition known as a poly-A tail. (These details are important in biotechnology, as we note in Chapter 14).

A CLOSER LOOK

An RNA World

Until the 1980s, one of the bedrock principles of biochemistry was the division of labor in cells: Nucleic acids (DNA and RNA) hold the information for directing the biochemical reaction in the cell; proteins serve as the functional molecules (that is, the enzymes) that catalyze the thousands of chemical reactions taking place. But research in the late 1970s and early 1980s helped modify this principle. Scientists now believe that RNA acting by itself can trigger certain chemical reactions.

The seminal research on RNA was performed independently by Thomas R. Cech of the University of Colorado and Sidney Altman of Yale University. In the late 1970s, Altman found an unusual enzyme in bacteria, an enzyme composed of RNA and protein. Initially, he thought the RNA was a contaminant, but when he separated the RNA from the protein, the enzyme would not function. After several years of exhaustive research, Altman and his colleagues showed that RNA was the enzyme's key component—indeed, under carefully controlled laboratory conditions, it could act alone. At about the same time, Cech discovered that a molecule of RNA from a protozoan could cut internal segments out of itself and splice together the remaining segments.

The findings of Cech and Altman hit the scientific community squarely on the chin because they implied that proteins and nucleic acids do not always depend on each other (as researchers had long assumed). The work also opened the possibility that RNA could have evolved on Earth without protein and that a self-catalyzing form of RNA could have been the first primitive molecule with the ability to reproduce itself. If so, then perhaps the biochemical machinery for translating the DNA-based genetic code in modern cells evolved much later. In essence, scientists conjured a whole new image of how life might have begun on Earth. The Nobel Prize committee was equally impressed. In 1989, it awarded the Nobel Prize in Chemistry to Cech and Altman.

By 1990, the self-reproducing molecule of RNA had a name—ribozyme. Biochemists at Massachusetts General Hospital soon modified a ribozyme by removing its internal segments. Then they showed that the new ribozyme could join separate short nucleotide segments aligned on specially designed, external templates. The research was a step toward designing a completely self-copying RNA molecule. Would such a ribozyme enclosed in a membrane constitute a primitive cell? If so, the cell would be quite different from what most scientists have imagined the first cells to be.

Once formed, the mRNA molecule moves away from the DNA molecule, and translation is ready to begin. In the cytoplasm, the mRNA molecule complexes with one or more ribosomes. Only a portion of the molecule contacts the ribosome at a given time, and several ribosomes may be using the information in a single mRNA simultaneously to form several molecules of a protein. **FIGURE 4.8** shows the interaction.

While the RNA-ribosome complex is developing, amino acids are binding to their specific tRNA molecules in the cytoplasm. Then, the tRNA molecules transport the amino acids to the ribosome where mRNA is stationed. One portion of the mRNA molecule attaches to the smaller subunit of the ribosome. Next, one codon of the mRNA molecule attracts its complementary anticodon on a tRNA molecule, and the pairing brings a certain amino acid into position. Now, the smaller subunit of the ribosome moves along the mRNA molecule to a second codon; immediately, a second tRNA molecule (holding a second amino acid) pairs its anticodon with the second codon. Two tRNA molecules have thus positioned their amino acids next to each

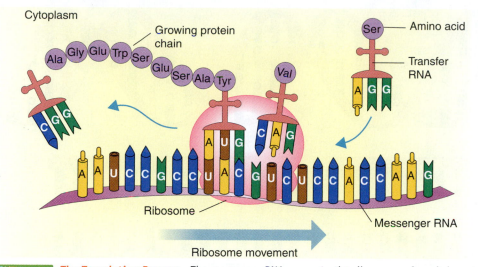

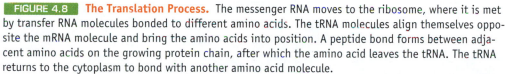

FIGURE 4.8 **The Translation Process.** The messenger RNA moves to the ribosome, where it is met by transfer RNA molecules bonded to different amino acids. The tRNA molecules align themselves opposite the mRNA molecule and bring the amino acids into position. A peptide bond forms between adjacent amino acids on the growing protein chain, after which the amino acid leaves the tRNA. The tRNA returns to the cytoplasm to bond with another amino acid molecule.

other. Next, an enzyme joins the two amino acids to form a dipeptide (two amino acids joined together). The first tRNA molecule then breaks free from its amino acid and moves back into the cytoplasm to bind to another molecule of that amino acid.

Now, the smaller unit of the ribosome moves to a third codon of the mRNA molecule. The codon attracts its complementary anticodon on a tRNA molecule, which brings a third amino acid into position. As before, an enzyme chemically joins the amino acid to the first two amino acids to form a tripeptide (three amino acids in a chain). The second tRNA is then dispatched back into the cytoplasm. And so it goes: Codon after codon is paired with its anticodon, and amino acid after amino acid is positioned and joined to the growing peptide chain. The genetic message of DNA is being translated into an amino acid sequence in a protein. Protein synthesis is in full swing and will continue until hundreds or thousands of amino acids have been added one by one to the growing chain. We are at the crux of one of life's most fundamental processes.

The final codons of the mRNA molecule are chain terminators, or "stop" signals. When the ribosome reaches one of these codons (UAA, UAG, or UGA), no complementary tRNA molecules approach and no amino acids are added to the chain. The stop signals activate release factors that discharge the protein chain from the ribosome, and protein synthesis comes to a conclusion. However, further modification of the protein's three-dimensional structure may be necessary before the protein is able to function in the cell.

With the conclusion of the process, the genetic message of DNA has been expressed as the amino acid sequence in a protein. This genetic message will manifest itself through the activity of an enzyme, or a structural element of the cell, or a signaling protein, or any of numerous other protein-based organic compounds. This extraordinary process, summarized in **FIGURE 4.9** , is one of the key underpinnings of microbial life—indeed, of all life.

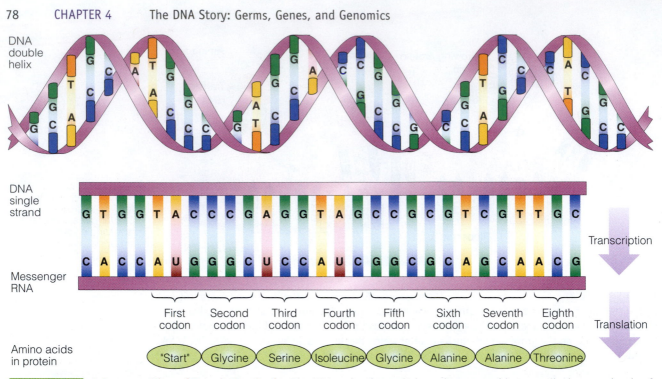

DNA double helix

DNA single strand

Messenger RNA

First codon | Second codon | Third codon | Fourth codon | Fifth codon | Sixth codon | Seventh codon | Eighth codon

Transcription

Translation

Amino acids in protein

"Start" Glycine Serine Isoleucine Glycine Alanine Alanine Threonine

FIGURE 4.9   A Summary View of Protein Synthesis. The DNA molecule unwinds, and one strand is transcribed as a molecule of messenger RNA. The mRNA then operates as a series of codons, each codon having three bases. During translation, a codon specifies a certain amino acid for placement in the protein. Note that codons 6 and 7 are identical; they thus specify the same amino acid, alanine. Similarly, codons 2 and 5 are identical; they encode glycine.

Gene Regulation

As biochemical knowledge increased, scientists began to wonder how microbes regulate the complex machinery of protein synthesis. They reasoned that the continuous synthesis of all enzymes for all possible chemical reactions would probably represent a waste of energy for the microbes (as well as create an enzyme-storage problem). They asked: What prevents a microbe from running at full throttle all the time? How do microbes economize in the synthesis of protein? Their experiments indicated that certain enzymes appear only when their substrates are present and that a delicate and flexible regulation of gene control is the rule rather than the exception.

In 1961, two Pasteur Institute scientists, François Jacob and Jacques Monod, were among the first to propose a scientific mechanism for controlling protein synthesis. Jacob and Monod suggested that genes in *E. coli* and other bacteria fall into several types: structural genes, which provide the base codes for proteins (in the process we just discussed); an adjacent operator gene, which stimulates the expression of structural genes; and a somewhat distant repressor gene, which controls the operator gene. Jacob and Monod gave the name operon to the entire DNA unit for expressing a particular trait. In 1965, they won the Nobel Prize in Physiology or Medicine for explaining the operon model of gene regulation (which we consider next). Scientists later expanded the unit and identified a promoter site next to the operator gene. The enzyme RNA polymerase binds to the promoter site to "promote" transcription.

The *lac* operon model helps us understand how protein synthesis is controlled at the level of the gene. When the nutritious disaccharide lactose is absent from a microbe's environment, there is obviously no need for the microbe to produce lactose-

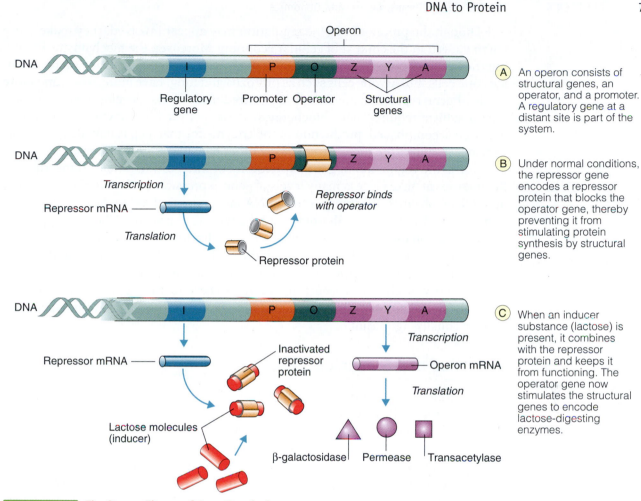

A An operon consists of structural genes, an operator, and a promoter. A regulatory gene at a distant site is part of the system.

B Under normal conditions, the repressor gene encodes a repressor protein that blocks the operator gene, thereby preventing it from stimulating protein synthesis by structural genes.

C When an inducer substance (lactose) is present, it combines with the repressor protein and keeps it from functioning. The operator gene now stimulates the structural genes to encode lactose-digesting enzymes.

FIGURE 4.10   **The Operon Theory of Gene Regulation.**

digesting enzymes. Accordingly, the microbe's repressor gene encodes mRNA to form a repressor protein (**FIGURE 4.10**). The repressor protein binds to the operator gene and overlaps the adjacent promoter site. This action prevents RNA polymerase from reaching the structural genes and bringing about the transcription of the enzymes that normally break down lactose. (The effect is somewhat like placing a log across railroad tracks.) The result is that the microbe does not produce lactose-digesting enzymes (which would be metabolically efficient because there is no lactose to break down).

Then, at some later time, lactose enters the microbe's environment. In this context, lactose is known as an inducer because it will "induce" protein synthesis; the inducer joins with the repressor protein and changes the protein's shape, thus making it unable to bind to the operator gene. With the repressor protein neutralized, the operator gene and promoter site are free to stimulate the structural genes to encode lactose-digesting enzymes. The enzymes soon appear and digestion of lactose to glucose and galactose begins. The *lac* operon was first worked out for *E. coli*. It explains why lactose-digesting enzymes form only when lactose (the inducer) is present in the bacterial cytoplasm. When the lactose has been totally consumed, the repressor protein changes back to its original shape and binds to the operator gene, effectively shutting off enzyme production.

Though the processes of gene regulation may appear involved, they make sense when viewed in the context of cellular economy. Moreover, the mechanisms involving the bacterial operon are relatively simple compared to the control systems in eukaryotic cells. Eukaryotic cells such as protozoa and fungi have more genes, and more levels of gene regulation are therefore required. Indeed, the development of a complex organism requires many biochemical steps catalyzed by enzymes. Once a step has been accomplished, production of the enzyme for that step is halted, and the organism's resources are directed at making other enzymes for other steps.

Nor is the regulation of gene expression confined to the time of transcription. Regulation can take place at many levels of gene expression, such as in the processing of mRNA molecules, the transport of mRNA molecules to the cytoplasm, the binding of mRNA molecules to the ribosome, and the folding of proteins. These multiple levels of regulation must be addressed by scientists who attempt to reproduce the system for gene expression in test tubes. Indeed, the evolution of biotechnology and recombinant DNA research (Chapter 14) is intimately associated with an understanding of the regulatory mechanisms occurring in cells. To be sure, the regulatory mechanisms are probably as important as the processes for protein synthesis, and understanding them is equally significant.

Genes and Genomics

By unlocking the secrets of DNA, James D. Watson and Francis H. C. Crick, together with numerous colleagues and contemporaries, stood at the forefront of what would become the biotechnology revolution. In a more practical sense, they also opened a proverbial "can of worms." Soon after scientists learned how genes work in bacteria, they developed an insatiable desire to know how genes work in other microbes, not to mention plants, animals, and humans. Soon they were focusing on an organism's entire set of genes, its genome. As they developed genomics, the study of genomes, they began asking certain probing questions: How do the genes of a bacterium differ from those of a eukaryote? What is the minimum number of genes necessary for life? How do genes influence the course of infectious disease? And how can genes can be put to work to make medicines, industrial products, and other materials to benefit society? What followed was a crescendo of brilliant research efforts.

To answer these questions, scientists realized that they had to know the exact components of the genome; that is, they had to learn the nature of each base in the DNA. In this arena, the work of the English biochemist Frederick Sanger in the 1970s was ground-breaking because it showed that deciphering a genome was possible. Sanger laboriously worked out the entire sequence of the 5368 bases of virus PhiX174 (and won the 1980 Nobel Prize in Chemistry, his second Nobel Prize), and he showed that the process could be adapted to a larger organism. Now scientists began to dream bigger dreams: Why not decode the entire genome of a bacterium? Or the genomes of yeasts, worms, plants, mice, virtually any organisms? Why not decode the entire genome of a human being? Why not, indeed.

The Beginning

Science historians generally agree that the effort to learn the nature of genomes had its beginning at a 1985 meeting at the University of California. Robert Sinsheimer, a molecular biologist and Chancellor of the University, brought together a group of scientists to discuss a colossal effort to map and sequence the human genome. This human book of life encodes in its pages all the biochemical information to build and maintain the body. Human genes contain about 3 *billion* bases; in comparison, an *E. coli* cell has only about 4.7 million. To appreciate its epic proportions, consider that if the human genome were written out in capital letters (e.g., CTTACGTGG . . .), it would fill 200,000 pages of 8½-by-11 paper; the *E. coli* genome, by contrast, would use a mere 300 pages.

As the discussions went on, scientists expanded the genome project to include a wide spectrum of organisms, including *E. coli,* the yeast *Saccharomyces cerevisiae,* the nematode worm *Caenorhabditis elegans,* the fruit fly *Drosophila melanogaster,* the corn plant, the mouse, and a host of others. The technology had to be developed; the financial resources had to be found; the leadership had to be identified. But the scientists were determined to succeed.

As the years unfolded, the genome projects took shape. Three billion dollars were allocated by Congress to fund the human genome effort, and substantial sums were set aside for other genome projects. James Watson was appointed the first director of the human genome project, and he was succeeded in 1994, by Francis Collins, its current leader. Federal funding for the project was supervised largely through the Department of Energy and the National Institutes of Health. Moreover, several private companies and university laboratories undertook genome research. An interesting aspect of public genome funding is that 3% of the funds have been set aside for ethics studies because genome research will make available much detailed and sensitive information that may affect personal lives. (For example, should an insurance company know that a person carries a gene for an incurable disease?)

The Methods of Genome Research

Determining a genome requires a multistep approach involving assigning genes to particular chromosomes, locating the genes on each chromosome, and determining the sequence of bases in the DNA chain. During the process, scientists must make a series of biochemical maps of each chromosome. The mapping process takes place in steps, with each section of the map describing the order of genes on a particular chromosome and the spacing between them. As the process moves along, maps of finer detail emerge, and the map resolution (its clarity) increases.

Scientists construct two kinds of maps as genome research proceeds: gene linkage maps and physical maps. A gene linkage map is the coarsest type of map (i.e., it has the lowest resolution). This map identifies the relative positions of genes on the chromosome by describing a genetic marker found close to each gene. The marker consists of a series of bases that can be easily identified by biochemical methods, thereby making it unnecessary to identify the more complex gene. Gene linkage maps narrow the hunt for a gene down to a region of the chromosome only a few million base pairs long.

Using a gene linkage map to find one's way along a chromosome is somewhat like knowing landmark cities on a road map, but not the number of miles between them. A physical map, by contrast, describes the biochemical makeup of the chromosome itself. It shows the location of genes and genetic markers on the chromosome, and it denotes the number of bases between them. The map cites the actual distances between genes on a chromosome in terms of the number of bases. Thus, it describes the structural characteristics of the chromosome. Physical maps are created by chemically snipping the chromosomes into pieces, replicating (cloning) the pieces, and reordering them in the way they originally occurred on the chromosome.

Next comes the determination of the name of each base in the genome, the actual genetic code. One of the major methods for determining DNA's base sequence is the Sanger method, first used by Frederick Sanger, the Nobel laureate. Also known as the chain-termination method, this method uses an enzymatic procedure to synthesize DNA chains of varying lengths, stopping the synthesis at one of the four bases. DNA technologists begin with the single-stranded DNA fragment they wish to sequence. Next, they add a mix of free nucleotides and nucleotides tagged with unique dyes, different ones for adenine, guanine, cytosine, and thymine. Then they mix in a piece of primer DNA, which is a short piece of DNA that complements and pairs with the single-stranded DNA to provide a starting point for DNA synthesis. DNA polymerase, the DNA-synthesizing enzyme, is then added.

Next, the mixture is placed into a machine called a thermal recycler. At high temperature, the primer attaches to the single-stranded DNA. Then, as the temperature cools, the DNA polymerase binds to the primer and incorporates free and dye-tagged nucleotides in the growing DNA chain. But wherever a dye-tagged nucleotide is added, chain formation terminates. The process continues over many heating and cooling cycles, resulting in new DNA chains of every possible length. Each of these chains has a dye-tagged nucleotide at one end.

Next is the identification step. The mixture is transferred to a thin, two-foot long, glass tube containing gel that is connected to a high-voltage electrical source. In this process, called electrophoresis, the DNA fragments migrate at different speeds according to their size, the smallest fragments moving the fastest. As the fragments reach the end of the tube, a camera records the color emitted by each dye-tagged nucleotide that passes by. Fed into a computer, the information yields a string of As, Gs, Ts, and Cs, which represent the base sequence of the new DNA fragment and, by inference, the genetic code of the original DNA molecule.

This method for sequencing DNA works for a small piece of DNA but not for an entire genome. Here a somewhat more cumbersome and more "traditional" method must be used: A chromosome from the organism is cut by enzymes into random fragments, each with about 150,000 to 200,000 base pairs of DNA. Each of the fragments is randomly inserted into a plasmid or a bacterial artificial chromosome (BAC) for incorporation into a bacterium. The bacterium multiplies and produces millions of copies of the same fragment. After the copying (or "cloning") step, the fragments are used to create physical maps. This is accomplished by sequencing the ends of the fragments and using computers to look for overlapping regions of DNA. Specified sequences of bases, the genetic markers, are also used to locate individual genes and create a gene linkage map. Once the maps have been established, the DNA fragments are randomly shattered into smaller segments of about 2000 base pairs. The segments are then separated into single-strands of DNA, and the sequencing can begin.

This traditional method has been used in the government-funded effort to sequence the human genome. It is illustrated in **FIGURE 4.11** . J. Craig Venter and his colleagues from the private firm Celera Genomics have taken a slightly different approach. Venter and his colleagues developed sophisticated computational methods to devise a "whole genome shotgun method." In this strategy, a genome is randomly shredded

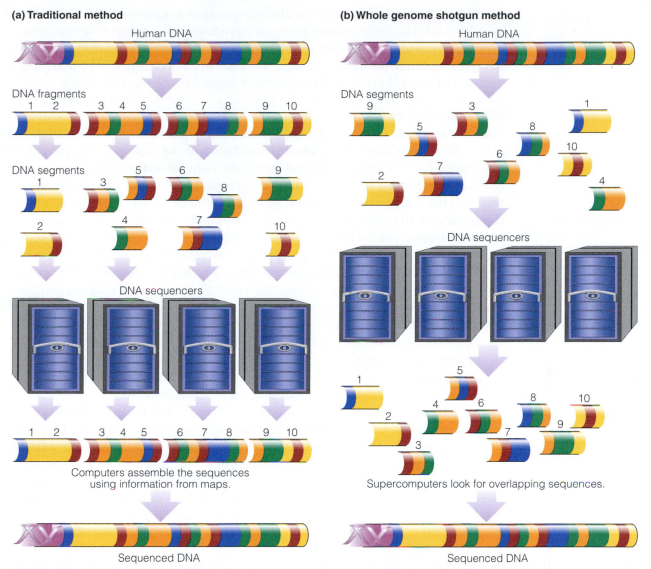

FIGURE 4.11 **Sequencing Methods.** Two methods for determining the base sequence of a molecule of DNA are shown. (**a**) In the traditional method employed by the Human Genome Project, the DNA is broken into fragments of decreasing size; the fragments are copied in bacteria, then put in order by making physical and gene linkage maps. Next, the fragments are broken into still smaller segments, and the order of the bases is determined in an automated sequencer. Computers use the maps to assemble the segments. (**b**) In the whole genome shotgun method used by Celera Genomics, the DNA is fragmented into random segments and copies are made in bacteria. The copies are analyzed, and the order of the bases is determined. Supercomputers are then used to assemble the segments into a complete DNA molecule.

into DNA segments of 2000, 10,000, or 50,000 base pairs by squeezing chromosomes through a pressurized biochemical syringe. Copies of the fragments are made by inserting them into bacteria via plasmids or BACs. Next, the segments are sequenced, as Figure 4.11 shows. The method has been likened to shredding many copies of a book and reassembling one copy, sentence by sentence. The traditional method, by comparison, is like breaking a book into chapters, then shredding and reassembling the chapters before putting the book together.

Although both the traditional and whole genome shotgun methods have their advantages and disadvantages, the methods have been used effectively to uncover the genomes we discuss next. This research has been so extraordinary that one scientist has remarked that one day the biological history of an organism will be neatly classified into the period before and the period after its genome was elucidated.

Microbial Genomes

With the methods in place, molecular biologists began their genomic searches. In May 1995, fifteen years after Sanger's elucidation of the viral genome, history was made when J. Craig Venter, Hamilton Smith, and their associates won the race to specify the first complete genome of a free-living organism. That month, the researchers unveiled the nature and sequence of the 1.8 million base pairs in the 1749 genes of the genome of the rod-shaped bacterium *Haemophilus influenzae*. The Venter-Smith team fragmented the bacterial genome with ultrasonic (extremely high-pitched) vibrations, then sequenced the fragments, and arranged them in order using innovative computer software and sequence overlap information. In a few short months, they were able to decipher the genome for a second microbe, *Mycoplasma genitalium*. This reproductive tract pathogen is among the smallest bacteria known, having only 580,000 base pairs and about 250 genes.

By 1996, Venter's group had completed the sequencing of the genome of the bacterium *Methanococcus jannaschii*. This organism belongs to the domain Archaea (discussed in Chapter 5). The genome research supported Carl Woese's proposal of a third domain by showing that the genes of *M. jannaschii* are fundamentally different from those of traditional bacteria. The genome sequence for *Staphylococcus aureus* was also announced in 1996. But the big news of that year was the completion of the genome sequencing for the yeast *Saccharomyces cerevisiae*. In a field already littered with milestones, the sequencing of *S. cerevisiae* marked the first success with a eukaryotic genome. Sixteen chromosomes were analyzed, 12 million bases were sequenced, and 6000 genes were identified. The sequencing revealed many previously unknown genes. It also showed a high degree of redundancy in the genome; that is, many genes are repeated over and over again, the significance of which remains unexplained.

In 1997, the field of whole genome sequencing took off. Among the bacterial genomes sequenced were those of *Helicobacter pylori*, a cause of gastric ulcers; *Borrelia burgdorferi*, the agent of Lyme disease; and *Streptococcus pneumoniae*, among the most serious causes of bacterial pneumonia. That same year, European and Japanese biochemists determined the sequence of bases in the genome of *Bacillus subtilis*. This organism is familiar to most students of microbiology because it is widely used as a laboratory test organism. It is also familiar to industrial microbiologists as a source of enzymes for the production of vitamins, detergents, and food products.

But possibly the outstanding achievement of 1997 was the sequencing of the genome of *Escherichia coli*. The long-awaited announcement was made by Frederick R. Blattner of the University of Wisconsin-Madison. Studied for decades and a regular on the pages of this book, *E. coli* is the organism of choice for studying how bacteria work. Because there is an enormous biological literature on *E. coli*, its genes and gene products can be fit into what is already known about its biology. Blattner's research group identified 4,638,858 base pairs in the 4288 genes of the *E. coli* genome.

■ *Haemophilus influenzae*
hē-mä′fil-us in-flü-en′zī

■ *Methanococcus jannaschii*
meth-a-nō-kok′kus jan-nä′shē-ē

■ *Borrelia burgdorferi*
bôr-rel′ē-ä burg-dôr′fèr-ē

By 1998, genome sequencing of microorganisms was reaching unprecedented heights. Scientists elucidated the genomes for the bacterial pathogens *Treponema pallidum* (syphilis) and *Mycobacterium tuberculosis* (tuberculosis). Then they announced that all 97 million bases of the microscopic worm *Caenorhabditis elegans* were finally decoded, and soon, they presented the genome of the mustard plant, *Arabidopsis thaliana*. By the end of the century, the genomes of over 50 organisms were known. Indeed, the technology was so advanced that it was not unusual to learn that a genome was decoded in a single day.

The Human Genome

The work with microbial genomes has been dwarfed by the effort to elucidate the human genome. One of the most ambitious projects in the history of molecular biology (indeed, the history of biology), the mapping of the human genome was launched in January 1989. That month, a group of biologists, computer experts, ethicists, industrial scientists, and engineers began the monumental effort to identify all the bases and their sequence in all the human genes. The human genome has been compared to a rope 2 inches in diameter and 32,000 miles long, all neatly arranged in a structure about the size of a domed football stadium.

At the outset, the federally funded work on the genome proceeded in American and British laboratories. Researchers divided up the work, first making physical and gene linkage maps of increasing detail, then improving the technology for sequencing. They tested the technology on model organisms, including the microbes noted above; then, in 1997, they launched into full-scale sequencing of the human genome.

But the work was slow and plodding because the technology was weak. And much of the data were unrewarding because most functional genes (exons) consist of 10,000 to 150,000 bases, and between them are endless stretches of apparently functionless bases (introns). Still the work went on. Then, in 1998, J. Craig Venter broke from the group and formed a new company, Celera Genomics, for the express purpose of sequencing the human genome. His company adopted the whole genome shotgun method and bypassed the mapping steps. Not to be outdone, Francis Collins' team obtained additional public funding and drastically scaled up its effort, creating five sequencing supercenters. And the consortium adopted a new game plan: It would attempt a rough draft of the genome and leave the finishing touches for later.

The race for the human genome was on. Collins' public effort used DNA obtained from six anonymous individuals, and researchers sequenced each fragment multiple times to increase the certainty of the base positions, reducing the error rate to less than 1 base in 100. The Celera team took advantage of the fact that the public data are deposited nightly in a database called GenBank. It fed the sequence data into its own supercomputers for comparison purposes and to fill in the missing pieces. Using the GenBank and Celera data, the supercomputers had to make 500 million trillion comparisons.

History was made on June 26, 2000 when President Bill Clinton and Britain's Prime Minister Tony Blair (by satellite) convened a White House press conference to announce a draft copy of the human genome (FIGURE 4.12). Flanked by Collins and Venter, President Clinton pointed to the "most wondrous map ever produced by humankind." He expressed his pleasure at an "epic-making triumph of science and reason."

The announcement seemed to herald the end of the genome race. But knowledgeable scientists pointed out that this was only a "draft" copy of the genome, a version

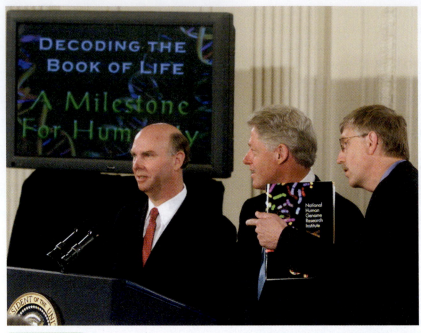

FIGURE 4.12 **An Important Announcement.** President Clinton with J. Craig Venter and Francis Collins at a White House press conference announcing the draft copy of the human genome.

containing the location of about 90% of the more than 3 billion bases. (Venter claimed over 95%.) Then, in February 2001, scientists from the competing groups published in *Science* and *Nature* the long-awaited details of the draft DNA sequence. Although many mysteries remained, the details gave scientists the first panoramic view of the human genetic landscape and revealed a wealth of information, as well as some surprises.

We should caution that details of the human genome did not turn out to be the Holy Grail or Rosetta Stone for understanding all of biology—two early metaphors widely used to describe the coveted prize. Nevertheless, knowing the genome did provide a number of challenging and interesting insights: The number of human genes is in the neighborhood of 35,000 to 50,000, much lower than the 100,000 genes traditionally believed present in the nuclear material; the number of bases in the human genome is 3,164,700,000, close to the estimate; an average gene consists of 3000 bases, with the largest known gene (2.4 million bases) encoding dystrophin, a protein that strengthens muscle tissue; about 99.9% of the bases in the genome are the same in all people; about 50% of the newly discovered genes have no known function; less than 2% of the genome codes for proteins, and "junk" DNA makes up over 50% of the genome; and, finally, chromosome 1 has the most genes (2968) and chromosome Y has the fewest (231).

The map of the human genome provides a magnificent and unprecedented resource that will serve as the basis for research and discovery throughout the twenty-first century and beyond. Indeed, geneticist Norton Zinder has compared revealing the genome to publishing the first book on the human body in the year 1543; today, five hundred years later, scientists are still struggling to understand many of the body's parts.

Beyond the Genome

The publication of genomes for prokaryotic and eukaryotic organisms presented scientists with great volumes of data. For the first time in history, researchers had available the entire genetic content of several living things. But it also created the dilemma of what to do with the data. As studies continued and years passed, the science of genomics emerged, focusing on the mapping, sequencing, and study of genomes. Bioinformatics, an offshoot of genomics, is the discipline in which the genome is carefully scrutinized to determine how it directs biological activity. To understand the volume of work ahead, consider the statement made by geneticist Eric Lander on publication of the human genome: "Just because you have the parts list of a Boeing 727," said Lander, "you don't necessarily know how the airplane flies. The genome is a parts list for a living thing—it does not tell you how a living thing works."

For the microbiologist, the field of medical genomics offers fresh insights about the disease process. From genome analysis, scientists can deduce the importance of a specific function. *Mycoplasma genitalium,* for instance, uses a full 5% of its genes to encode a protein allowing it to adhere to the reproductive tissue. Research directed toward altering these genes might be a useful starting point for therapy. Furthermore, *Haemophilus influenzae* lacks the genes for three Krebs cycle enzymes, a fact that may help microbiologists understand its ability to thrive in the human body. Comparing the genomes of virulent and avirulent strains should open a window onto the pathogenic process.

Some researchers see genes as agents for diagnosing and treating disease and envision snippets of DNA that they can insert in patients' cells and then send the patients home, cured of their illness. In that sense, all the DNA knowledge uncovered in the past half-century is merely a preamble to the startling discoveries waiting to be made in the decades ahead. For many physicians, DNA will be a pharmacological substance of extraordinary potency that can be used to correct the imperfections that make patients susceptible to disease. Knowing the genome will amplify human power, while affecting every nook and cranny of the practice of medicine.

However, there is one point of consternation: the issue of patents. It irks many scientists that the U.S. government awards patents for human genes, yet without patents companies could not afford to invest the resources to develop diagnostic tests to detect mutant genes. One company has already received over 500 patents on human genes, and myriad other companies continue to "mine" the public database of sequences in GenBank to help build their gene catalogs. On finding a gene of interest, company scientists enter the base sequence into a computer program that determines the amino acid sequence of the protein, and by comparing this protein with known proteins, they can determine whether the gene is potentially useful for developing a new drug or diagnostic test. Such a process has led to over 800 new genetic tests so far.

Knowledge of genomes will yield a veritable tsunami of information as the discipline of bioinformatics continues to develop. Analyzing genomes will help drug companies locate better drug targets and increase the efficiency of drug development. For example, in patients with osteoporosis, a crippling bone disease, researchers found that bone cells have genes that overexpress themselves and yield a glut of an enzyme called cathepsin. The scientists learned about this overexpression by sequencing the genes of the bone cells and comparing the base sequence to sequences that encode known proteins. Pharmaceutical scientists then launched an effort to develop a drug to destroy the cathepsin and possibly halt the progression of osteoporosis.

Identifying and studying the proteins encoded by genes is the subject matter of the discipline of proteomics. Proteomics seeks to determine what a protein does in an organism. The discipline is still in its infancy, but has tremendous potential and a mountain of work ahead, for over 75% of proteins in a eukaryotic organism have no known cellular function. However, proteomics is poised to yield remarkable discoveries because this group of unknowns includes new enzymes, new signaling proteins, new regulatory proteins, new structural proteins, and a host of other chemical compounds.

The greatest challenge in the postgenomic era will be sifting through the sea of base sequence data to sort out the useful information. Indeed, at the fundamental level, genomes have changed the face of biological research. Traditionally, researchers began their work with a biological phenomenon that needed explaining or a medical problem that required addressing. To resolve the problem, they often had to identify the gene involved and decipher what it does. Knowledge of genomes radically changes the approach: The researchers can start with the gene and work toward its function. In this regard, a basic shift is now developing in how researchers do biology. There are few parallels in scientific history.

■ A FINAL THOUGHT

For many decades, scientists pondered the nature of the gene. Try as they might, however, they could not imagine how DNA could be involved. They thought it too simple ("too stupid" some said) for the sophisticated function of heredity.

But by the 1950s, compelling experiments were suggesting that DNA was anything but stupid. Then, when Watson and Crick announced their model of DNA's structure, a flurry of brilliant research started things snowballing; soon thereafter, scientists cracked the code that makes DNA so smart. Next, they were cloning genes and examining their structures, while researching the proteins they encode. Then, they were ready to map the genome. What they accomplished made heads spin in the scientific community.

Deciphering the human genome has been compared to landing an astronaut on the moon. However, the milestone of stepping onto the lunar surface was the grand finale to years of space research, while the milestone of sequencing the genome is the stepping-off point for a daunting journey that may take a century to complete. The sequencing of the human genome has vast and still unrecognized implications for people's lives. It is safe to assume that scientists will not be turning off their sequencing machines or reading help-wanted ads for many decades to come.

What, then, can we expect in future years? First, scientists know that an *E. coli* cell has 4288 genes, but they are still not sure how many genes a human cell has (current estimates point to a number in the range from 35,000 to 50,000). Within a few years, the genetic foundations for innumerable diseases will be known, and predictive tests will be available. Gene therapy will be common, and primary care providers will practice genetic medicine. Gene-based designer drugs for several diseases will be marketed, and cancer therapy will target the molecular basis for each tumor type. The genes for aging will be catalogued, and trials to increase the human life span will be ongoing. Individualized preventive medicine will be available, and comprehensive genomics-based health care will be standard. In short, the world will be very unlike what we have grown accustomed to. Indeed, we stand on the threshold of knowledge so powerful that we will wonder how we ever got along without it.

QUESTIONS TO CONSIDER

1. Try to put yourself in Griffith's position in 1928. Genetics is poorly understood, DNA is virtually unknown, and bacterial biochemistry has not been clearly defined. How would you explain the remarkable results of his experiments?

2. The "central dogma" is a supposedly firm principle that explains how genes function in cells. The dogma states in part that RNA is transcribed from DNA. During the AIDS generation, the central dogma has come into question because of new biochemical insights. What is the basis for that question?

3. In one sense, DNA is a relatively simple molecule, having only four different subunits. And yet, DNA can specify extraordinarily complex proteins having at least twenty different amino acids in chains of many thousands or more. How does DNA accomplish this seemingly impossible task?

4. The names of Watson and Crick are indelibly etched in the history of twentieth-century biology as the "discoverers" of the structure of DNA. Knowledgeable students of biology know, however, that Watson and Crick did not "discover" DNA. Which names should be remembered along with theirs, and why?

5. Determining the sequence of bases in the human genome has been equated to landing a human on the moon. Many would dispute that comparison and suggest that elucidating the human genome is far more important. What do you think they have in mind?

6. Tube A of DNA contains the following percentages of bases: 23.1% A, 26.9% C, 26.9% G, and 23.1% T. By contrast, tube B has 32.3% A, 32.3% C, 17.7% G, and 17.7% T. Which tube has the double-stranded DNA and which has the single-stranded DNA. Why?

7. You maintain that enzymes make proteins, but your colleague suggests that proteins make enzymes. Which of you is correct, and what is the basis for your conclusion?

■ KEY TERMS

Informative facts are necessary for the expression of every concept, and the information for a concept is founded in a set of key terms. The following terms form the basis for the concepts of this chapter. On completing the chapter, you should be able to explain and/or define each one:

bioinformatics
chromosome
codon
doexyribonucleic acid (DNA)
exon
gene
gene linkage map
genomics
intron
messenger RNA (mRNA)
nucleotide
operon

physical map
poly-A-tail
promoter site
proteomics
repressor protein
ribonucleic acid
ribosomal RNA (rRNA)
ribosomes
semiconservative replication
transcription
transfer RNA (tRNA)
translation

■ http://microbiology.jbpub.com/book/microbes

The site features **eLearning,** an online review area that provides quizzes and other tools to help you study for your class. You can also follow useful links for in-depth information, read more stories of microbiology, or just find out the latest microbiology news.

Bacteria: The First Microbes

5

Looking Ahead

Bacteria are among the most successful organisms on Earth. Over the billions of years of their existence, they have evolved to occupy every conceivable niche on Earth; indeed, they influence almost everything we experience. In this chapter, we shall get to know the bacteria as we examine their structures and activities.

On completing this chapter, you should be able to . . .

- appreciate the enormous span of time for which bacteria have existed on Earth and understand their contributions to the formation of the world as we know it.
- summarize the various forms of known bacteria and define many of the submicroscopic structures associated with a bacterial cell.
- describe the process by which bacteria reproduce and grasp the significance of the frequency of bacterial replication.
- identify some of the environments in which bacteria thrive and recognize the different types of cultivation techniques available for growing bacteria in the laboratory.

Ancient cyanobacteria built these structures. Called stromatolites, they are our earliest record of life on Earth. These three-billion year old fossils are the built up layers of the cyanobacteria mats, sediment, and minerals. The inset shows a colored scanning electron micrograph (SEM) of a filamentous cyanobacterium.

- outline several important groups of bacteria in order to appreciate their diversity.
- identify the importance of bacteria in the disease process and briefly summarize some of the mechanisms of bacterial disease and body resistance.

The Earth came into being an almost incomprehensibly distant 4.5 billion years ago. For the first several millions of years of its existence, Earth was a ball of molten rock. Then, as millennia passed, a thin crust formed over the hot core, and violent volcanic activity filled the days and nights. Now the Earth was awash with energy: There was radiation from the sun, lightning from intense electrical storms, and heat from radioactive decay and the ever-present volcanic eruptions. In the incessant rain and the tropically warm oceans, organic molecules were forming—amino acids, proteins, nucleotides, and carbohydrates—the stuff that one day would compose living things.

For its first billion years, the Earth was as barren of life as the surface of the moon is today. Then, about 3.5 billion years ago, microscopic cells, the first living things, came into being. Although scientists are uncertain how these cells arose, they are reasonably sure that these first life forms were bacteria. The ancient bacteria (like the modern forms) were tiny, single-celled creatures, with little evidence of internal structure. Organisms like these are called prokaryotes, a name that reflects the absence of a nucleus (*karyo* is Greek for "nut" [or nucleus], and *pro* means "primitive"). Chapter 2 compares prokaryotes and eukaryotes more fully.

Bacteria (or prokaryotes) were the only inhabitants of the Earth for over 2 billion years, nearly half of the planet's existence. Then, about 1.5 billion years ago, the eukaryotes evolved, as FIGURE 5.1 illustrates. These organisms, typified by protozoa, fungi, plants, and animals, have larger cells, with nuclei and complex internal structures. Although the eukaryotes were larger organisms, the bacteria were not threatened because by that time bacteria had the advantage—they were well-established and in full swing.

And what a group they were: Growing wildly in the oxygen-free atmosphere of Earth, the ancient bacteria chemically combined hydrogen with the carbon of carbon dioxide to form methane (not unlike what we find in a swamp today). Then the photosynthetic bacteria evolved. These forms are called cyanobacteria (or blue-green algae in the older literature). Cyanobacteria used their chlorophyll pigments in the process of photosynthesis to capture light energy and produce carbohydrates as energy-storage compounds.

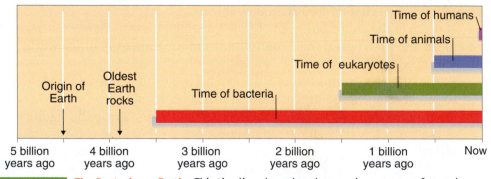

FIGURE 5.1 **The Bacteria on Earth.** This timeline shows how long various groups of organisms have existed on Earth. The bacteria have been in existence for a notably longer period of time than any other group, particularly humans. They have adapted well to Earth's varied conditions simply because they have had the longest opportunity to adapt.

In so doing, they played a decisive role in Earth's development because photosynthesis dramatically increased the atmospheric concentration of oxygen from 1% to the current 21%. More than any other organisms on Earth, cyanobacteria were catalysts for the oxygen-based chemistry that now pervades earthly life.

All the contemporary bacteria developed from the evolutionary experiments that Mother Nature performed on these ancient forms. And the results have been spectacular, for modern bacteria have come to occupy every conceivable niche on Earth. Every crack, every crevice, every cranny—whether at the bottom of a 6-mile deep Pacific trench, at the pinnacle of Mount Everest, or in the blazing desert of the Sahara—contains some sort of bacterial life. A single pinch of rich soil contains over a billion bacteria. Indeed, a handful of that same soil contains more bacteria than all the people who have ever lived on planet Earth.

It should not be surprising that bacteria occupy a critical place in the web of life. Many bacterial species make major contributions to the mineral balance of the world by metabolizing the nutrients in freshwater, marine, and terrestrial environments. The bacteria residing in soil influence the Earth's ecology by breaking down the remains of all that die and recycling the elements. And certain bacterial species do what few other species of organisms on Earth can do—they trap nitrogen from the air and convert it to substances used by plants to make protein, protein that ultimately winds up on our plates as grains, meat, and dairy products.

In industrial and biochemical laboratories, bacteria are both workhorses and lab rats. Growing in enormous numbers in mammoth fermentation tanks, bacteria carry on their day-to-day chemical routines and yield products of substantial value. For example, they manufacture organic compounds; they produce fermented foods; they synthesize antibiotics and vitamins; and they serve as biological factories for genetic engineering. They even have a bearing on politics, as A Closer Look on page 93 explains.

Despite all the good they do, bacteria also pose a threat to humans because a small percentage of species cause infectious disease. Indeed, human history is replete with accounts of ravaging epidemics of cholera, plague, typhoid fever, and syphilis. But even as the great "slate-wipers" of history, bacteria may have served a useful purpose in giving the human population an opportunity to renew and improve itself. Some evolutionary biologists argue that bacteria are the key agents of natural selection for the human species; that is, in a sense, they help improve our species by selecting the fitter individuals through the distasteful task of infectious disease.

As our constant companions, bacteria have impacted our lives in some ways that are clear and in some ways that have not yet been discovered. These remarkable and fascinating microbes are the subject matter of this chapter.

Bacterial Structure and Physiology

Bacteria differ structurally from all other kinds of organisms, which are the more complex eukaryotes. This difference was first noted in the 1950s with the development of the electron microscope and during studies into the biochemical and physiological properties of organisms. In very broad terms, a bacterium differs from a eukaryotic organism because it is much smaller and less complex. But, as we shall see in the pages to follow, the differences go much deeper.

A CLOSER LOOK
Of Powder and Politics

The situation was quickly becoming desperate: It was 1915, Great Britain was at war with Germany, and the British supply of acetone was rapidly dwindling. Acetone is a solvent for making cordite, the powder used to make explosives for naval guns. Before the war, German industrial firms were major suppliers of acetone, but to expect any more shipments was unrealistic.

That year David Lloyd George, then Minister of Munitions, learned of the work of Chaim Weizmann and arranged to meet him. Weizmann, a Russian chemist living in England, was interested in bacterial fermentations. George explained the seriousness of the situation and told Weizmann that the usual wood distillation process was not working well enough to meet the demand for acetone. Weizmann's enthusiasm was fired; he resolved to do what he could.

Weizmann returned to his laboratory and worked relentlessly. Within a few weeks, he isolated an organism that transforms ordinary cornstarch into a mixture of acetone and other biochemicals. The organism was a bacterium belonging to the genus *Clostridium*. Weizmann reported his success to Lloyd George, and with the help of Canadian and American corporations, the process was scaled up to industrial proportions. By 1917, the British had their acetone, and cordite production resumed at full proportions.

Thirty-one years later, the state of Israel was created. Its first president was Chaim Weizmann.

General Morphology

Bacteria generally occur in variations of three major forms: Rod-shaped bacteria are known as bacilli (sing., bacillus); other bacteria are spherical and are called cocci (sing., coccus); and still other bacteria have a long, spiral form and are known as spirochetes if the cells are rigid or spirilla (sing., spirillum) if they are flexible. FIGURE 5.2 summarizes these forms.

Species of bacilli vary in size, with the shortest ones measuring about 0.5 μm in length and the longer ones about 20 μm. However there are extremes: In 1993, scientists from Indiana University discovered a bacterium of extraordinary size—600 μm long—in the gut of a surgeonfish. That record was surpassed in 1999, when a pearl-like bacterium approximately 750 μm in diameter was isolated from sediment samples obtained from the ocean floor off the coast of Namibia in Africa. To place its size in perspective, consider that this bacterium is roughly the size of the period at the end of this sentence. And, because a bacterium by definition is an invisible organism, the discovery of a visible bacterium may necessitate a reworking of the definition. (To be sure, one could argue that the phrase "visible bacterium" is what an English professor would call an oxymoron, two words that contradict each other.)

Cocci measure roughly 0.5 μm in diameter and occur in several configurations; that is, the bacteria may appear as diplococci (two cocci in a group) or as streptococci (cocci in a chain) or as tetrads (cocci in groups of four or eight). When cocci occur in random clusters, the configuration is called a staphylococcus. This term is derived from the Greek word *staphyle*, which means "grape." The configuration can be important to identification of cocci. For instance, the bacterium that causes gonorrhea is a diplococcus, while the cause of "strep throat" is a streptococcus. The common "staph infection" is caused

■ Staphylococcus
staf-i-lō-kok'kus

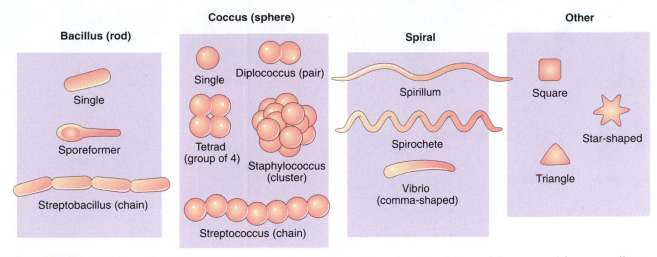

FIGURE 5.2 **Variations in Bacterial Structure.** The forms of bacterial species are variations of three general forms as well as a few miscellaneous forms.

by a species of *Staphylococcus*. In medical labs, technicians can use a microscope to distinguish these configurations and assist physicians in their diagnoses.

Other forms exist as well. For example, cyanobacteria consist of elongated microscopic cells usually occurring in filaments, which may measure a meter or more in length. Although the cells are connected at their outer walls, each cell operates independently. Still other bacterial species we shall study later in this chapter have stalks and grow long, branching cells that develop complex structures at their tips.

Studying the structures of bacteria gives us an appreciation of their complexity and opens a window to their activities. It also helps us to realize that bacteria are much more than dots on a microscope slide. As we survey the structural makeup of bacteria, watch for the various functions that bacteria can perform. These functions give us a sense of their day-to-day existence.

Staining Procedures

Some microbes, such as protozoa and fungi, are usually large enough to be seen without staining, but almost all bacteria are too small. Moreover, bacterial cytoplasm is transparent, which adds to the difficulty of seeing bacteria. As early as the 1880s, scientists such as Robert Koch and Paul Ehrlich used stains to improve their ability to see bacteria. The basic notion underlying staining is that the chemical components of bacterial cytoplasm have an overall negative charge and attract stains such as crystal violet and methylene blue, which have a positive charge. Use of a single stain characterizes the so-called simple stain technique illustrated in FIGURE 5.3.

The Gram stain technique was developed in the 1880s by Christian Gram, a Danish physician. This technique has multiple steps and not only stains bacteria, but also separates them into two groups: the Gram-positive bacteria and the Gram-negative bacteria. The bacteria are first stained with crystal violet (see A Closer Look on page 97) and then with Gram's iodine solution. At this point, all the bacteria are blue-purple. Now the bacteria are washed with ethyl alcohol, and the Gram-positive bacteria remain blue-purple, but the Gram-negative bacteria lose their color and become transparent. Then

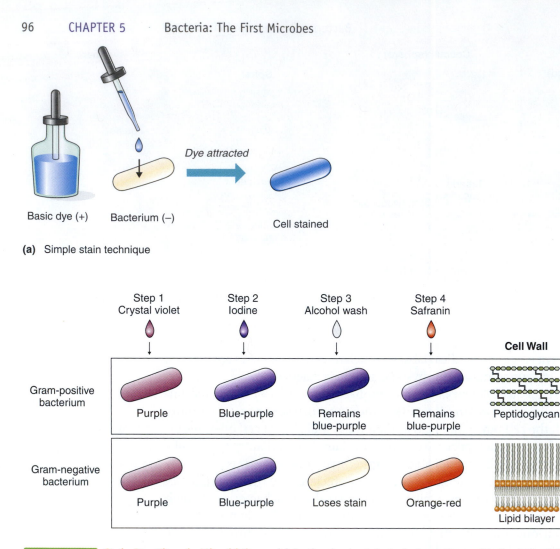

(a) Simple stain technique

FIGURE 5.3 **Stain Reactions in Microbiology.** (**a**) In the simple stain technique, the positively charged dye (+) is attracted to the negatively charged bacteria (−), and staining takes place. (**b**) The Gram stain technique is a differential procedure. All bacteria stain with crystal violet and iodine, but only Gram-negative bacteria lose the blue-purple color in the alcohol wash. Subsequently, these bacteria stain with the safranin dye. Gram-positive bacteria remain blue-purple.

in the final step, the red dye safranin is applied to the bacteria. It stains the Gram-negative bacteria (causing them to become orange-red), but it has no effect on the Gram-positive bacteria (which remain blue-purple). Microscopic observation reveals the color of the bacteria and tells us whether they are Gram-positive or Gram-negative.

The Gram stain technique is called a differential technique, because it separates bacteria into two different groups. Knowing whether a bacterium is Gram-positive or Gram-negative is important because most bacterial species fall into one or the other group. Thus, Gram staining is a key first step in describing the characteristics of a bacterium. Certain antibiotics such as penicillin are effective against Gram-positive bacteria but less so against Gram-negative bacteria, and certain antiseptics and disinfectants affect Gram-positive bacteria but not Gram-negative bacteria (or vice versa). Other characteristics of these two groups are noted in other chapters.

A CLOSER LOOK

"A.O. Means What?"

In modern bacteriology laboratories, the crystal violet dye used for Gram staining is prepared by mixing dye particles with a solution of a salt called ammonium oxalate. This procedure has not changed since 1929, when a graduate student named Thomas Hucker introduced it. How this "Hucker modification" came about is part of the folklore of microbiology.

Hucker was studying bacteriology at Yale University. Early in 1929, his advisor suggested that he contact several hospital and university laboratories to see how they were performing the Gram stain technique. Hucker was to report his findings in a presentation at an upcoming scientific meeting in Philadelphia. He dutifully sent out a series of letters and learned that the standard procedures were being used at all laboratories—all, that is, except Dartmouth's.

The reply from Dartmouth College piqued his interest. At the time, the usual procedure was to dissolve crystal violet in aniline oil. But Dartmouth bacteriologists apparently were using ammonium oxalate. Hucker tried ammonium oxalate and found that the stain improved with age and gave clearer results. He prepared his paper for the Philadelphia meeting and sent a draft to Dartmouth's biology department with a note of thanks. Soon thereafter he received a phone call from Dartmouth—they had never heard of ammonium oxalate for Gram staining. Hucker was perplexed.

In the days that followed, Hucker learned that a chemist had intercepted his survey letter and sent the reply. In writing out the method for preparing crystal violet, the chemist had read "A.O." on the bottle of Dartmouth's stain and assumed it meant that the dye was dissolved in ammonium oxalate. Aniline oil simply did not occur to him. Moreover, he had not bothered to check with the biology department because it was inventory time and other things were on his mind. Thus, a case of badly interpreted bacteriological shorthand led to the Hucker modification. Hucker became famous; the chemist remained anonymous.

Surface Structures

With certain exceptions, all bacteria are encased in a cell wall. The cell wall contains a tough mesh of polysaccharide and protein called peptidoglycan. Peptidoglycan lends rigidity and strength to the cell wall and is found in no other living thing. Gram-positive bacteria have a thick layer of peptidoglycan in their cell walls (which may contribute to their ability to resist the alcohol wash used in Gram staining), while Gram-negative bacteria have a much thinner central peptidoglycan layer sandwiched between outer layers of other complex biochemicals.

The cell wall is the site of the bacterium's vulnerability to certain antibiotics, such as penicillin and its relatives (ampicillin, amoxicillin, methicillin, and numerous others). These antibiotics prevent the bacterium from synthesizing peptidoglycan, and they leave the microbe with only a cell membrane. Internal pressures soon cause the cell to swell and burst, as **FIGURE 5.4** shows strikingly. Because Gram-positive bacteria contain more peptidoglycan, these microbes are more susceptible to the effects of penicillin. Staphylococci and streptococci are examples of Gram-positive bacteria sensitive to penicillin.

The structure of the bacterial cell membrane (also called the plasma membrane) is similar to its counterpart in eukaryotes. The essential feature is a double layer of

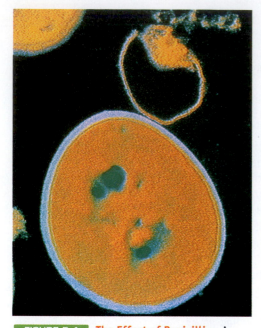

FIGURE 5.4 **The Effect of Penicillin.** A photomicrograph of a *Staphylococcus aureus* cell bursting after treatment with penicillin (×150,000). The antibiotic has prevented construction of the peptidoglycan layer of the cell wall, and internal pressures have led to weakening and disruption of the cell membrane.

phospholipids (a phospholipid spholipid bilayer) with protein molecules suspended in the phospholipids at the surface and spanning the layers. Some of these proteins function as enzymes during chemical reactions, and some transport substances across the cell membrane. Because of its everchanging nature, the membrane is called a fluid mosaic. On the outside of the membrane, small carbohydrate molecules are linked to the proteins. The carbohydrates are believed to be recognition sites, where antibody molecules from the body's immune system unite with the bacteria during infection.

Gram-negative bacteria have a gap between the cell wall and the cell membrane. This gap is called the periplasmic space and is filled with material called periplasm. Research experiments performed in the 1990s indicate that this area is an active and important processing center. For example, organic nutrients too large to pass through the cell membrane are broken down in this space; peptidoglycan synthesis occurs here; and membrane constituents are conveyed through this space to their correct spots in the lipid layer.

Certain species of bacteria have a coating outside the cell wall called the glycocalyx; it is also known as a capsule if tightly bound to the cell, or a slime layer if slimy and flowing. The glycocalyx, shown in FIGURE 5.5, provides protection for the bacterium, shielding it from drying, chemicals, and environmental stresses.

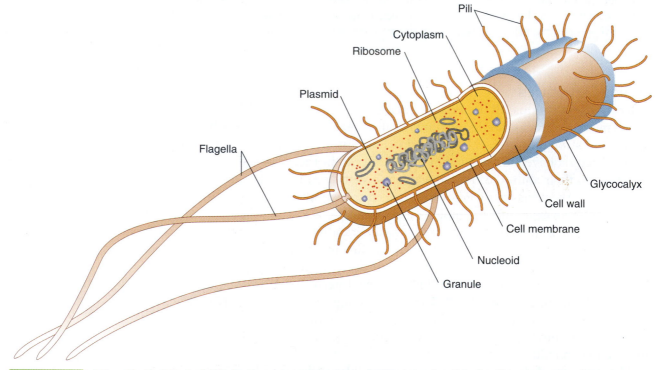

FIGURE 5.5 **A Hypothetical Bacterial Cell.** The structural features of this composite, "idealized" bacterium are drawn in a way that shows their relationships. Such a bacterium probably does not exist.

It should not be surprising that many species of pathogenic (disease-causing) bacteria have a glycocalyx.

Organic materials in the glycocalyx allow a bacterium to cling to surfaces. Glycocalyxes are used, for example, by streptococci and other oral bacteria to bind to pockets between the teeth and gums as a prelude to dental caries. The bacteria break down the sugars and other carbohydrates we consume and produce large amounts of acid. Gradually the acid eats away at the enamel and forms a depression, or cavity. When the bacteria penetrate to the soft tissue below, they continue to multiply and often produce large amounts of gas. The gas presses against nerve endings and causes the throb we call a toothache. (When the dentist drills through to the infection site, the gas is released and the pain ebbs.)

Bacteria may also contain hairlike structures anchored to the cell wall and cell membrane. One such structure, the flagellum, is a rigid filament of protein that rotates and propels a bacterium through its liquid surroundings. Various species of bacteria have a single flagellum, a tuft of flagella at one end of the cell, or flagella that cover their cell surface. A flagellum is too thin to be seen with the light microscope. However, it can be many times the length of the cell, as FIGURE 5.6 shows. Using flagella, bacteria appear to move in pulses along a series of curved lines (a "run") punctuated by periods of tumbling. A bacterium such as E. coli is capable of traveling about 2000 times its body length in an hour. Calculations for a 5 foot 10 inch human walking at about 2.25 miles per hour would yield about the same figure (which means there is at least one way in which humans and bacteria are equal).

Another hairlike structure called the pilus (pl., pili) helps bacteria attach to tissues or other surfaces. Pili are rigid cylindrical rods about 1 µm in length and about 7 nm thick. Each pilus consists of protein units wound in a helical (spiral) coil. When causing infection, bacteria use their pili to hold fast to the surfaces of the host cells. One novel approach to dealing with bacterial diseases is to synthesize drugs that will react with and neutralize the pili. A Closer Look on page 100 relates how research in this field is performed. Pili also help bacteria remain attached while they exchange genetic material, as we explore in Chapter 10.

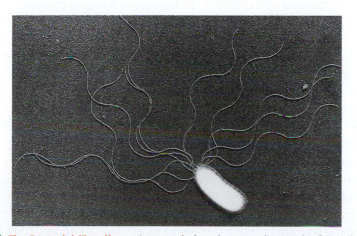

FIGURE 5.6   The Bacterial Flagellum. A transmission electron micrograph of *Pseudomonas marginalis* showing polar flagella (×38,800). Note that the flagella are many times the length of the bacterium and have a characteristic wavy arrangement.

A CLOSER LOOK

Diarrhea Doozies

They gathered at the clinical research center at Stanford University to do their part for the advancement of science (and earn a few dollars as well). They were the "sensational sixty"—sixty young men and women who would spend three days and nights and earn $300 to help determine whether hairlike structures called pili have a significant place in disease.

A number of nurses and doctors were on hand to help them through their ordeal. The students would drink a fruit-flavored cocktail containing a special diarrhea-causing strain of *Escherichia coli (E. coli)*. Thirty cocktails had *E. coli* with normal pili, and thirty had *E. coli* with pili mutated beyond repair. Bacteria with the threadlike pili should latch onto intestinal tissue and cause diarrhea, while those with mutated pili should be swept away by the rush of intestinal movements and not cause intestinal distress—at least that's what the sensational sixty were out to verify or prove false.

On that fateful day in 1997, the experiment began. Neither the students nor the health professionals knew who was drinking the diarrhea cocktail and who was getting the "free pass"; it was a double-blind experiment. Then came the waiting. Some volunteers experienced no symptoms, but others felt the bacterial onslaught and clutched at their last remaining vestiges of dignity. For some, it was three days of hell, with nausea, abdominal cramps, and numerous bathroom trips; for others, luck was on their side (investing in a lottery ticket seemed like a good idea at the time).

When it was all over, the numbers appeared to bear out the theory: The great majority of volunteers with mutated bacteria experienced no diarrhea, while the great majority of those with normal bacteria had attacks of diarrhea, in some cases, real doozies. All appeared to profit from the experience: The scientists had some real-life evidence that pili contribute to infection; the students made their sacrifice to science and pocketed $300 each; and the local supermarket had a surge of profits from sales of toilet paper, Pepto-Bismol, and Immodium.

Cytoplasmic Structures

The genetic information of bacteria is contained in a single chromosome occurring as a closed loop of deoxyribonucleic acid (DNA) in the cytoplasm. The DNA is tightly compacted and has no protein associated with it (in contrast to the DNA in eukaryotic cells). Because they have only one chromosome, bacteria are appealing as research tools in biochemical genetics—scientists can isolate the single chromosome and study its activity without worrying about the other chromosome in the pair, as they must do when working with eukaryotic cells. Also, there is no protein to interfere with their experiments, as happens when working with eukaryotic chromosomes. The cytoplasmic area where the bacterial chromosome concentrates is called the nucleoid. Thousands of genes, the functional units of DNA, make up the chromosome.

In addition to a chromosome, many bacterial species have tiny loops of DNA called plasmids. Suspended in the cytoplasm, the plasmids contain several genes that encode proteins for nonessential cell functions. Plasmids replicate independently of the chromosome and are important tools in modern DNA technology because they can be isolated, opened, and modified with new genes; the modified plasmids can then be inserted into fresh bacteria, where their new genes are activated to encode the protein desired by the biochemist. This imaginative and innovative process underlies much of the current revolution in biotechnology.

Bacteria also have in their cytoplasm the necessary structures for producing proteins and other substances essential to their existence. These include ribosomes (ultramicroscopic bodies consisting of RNA and protein), as well as transfer RNA molecules and a host of organic substances that make up the "body" of the cell. Included here are a variety of proteins, carbohydrates, lipids, and nucleotides. Numerous minerals and growth factors are included in the "living soup" of cytoplasm. The chemical activity going on in the cytoplasm is unceasing.

Certain bacteria, particularly members of the genera *Bacillus* and *Clostridium,* have the ability to produce an extraordinarily resistant structure called the endospore. This structure, commonly known as the spore, is a unique form of the bacterium formed during its life cycle, generally when stress is encountered in the environment. The spore is formed during an involved process, illustrated in FIGURE 5.7 . It contains a chromosome, two cell membranes, a cortex, a spore coat, and a surrounding wall called an exosporium. The spore's remarkable structure gives it the ability to withstand environmental stresses to the extent that it is probably the most resistant form of life known to science.

In 1995, California researchers led by Raul Cano reported the recovery of bacterial spores trapped in the intestine of a bee entombed in amber for 25 million years. Some scientists were skeptical that bacteria could remain alive as spores for 25 million years, but Cano's evidence was persuasive. Then, in 2001, West Chester University researchers reported recovering live *Bacillus* spores from New Mexico salt crystals 250 million years old (these spores were formed before dinosaurs walked the Earth). On a somewhat more practical level, bacterial spores of certain species can be boiled for 2 hours without destroying them, or they can be left in alcohol for 20 years. When placed in a nutritious environment, the spore walls crack open, and the spores revert to bacteria that reproduce and metabolize as if nothing had happened.

The ability of bacteria to form spores is key to certain types of bioterrorism. For example, *Bacillus anthracis,* the bacterium that causes anthrax, is a sporeformer. The organism is easily grown in the laboratory, and its spores can be spread efficiently among large populations (for example, by releasing them in subway tunnels). Falling on the skin and inhaled into the lungs, the spores revert to vegetative (growing) bacteria that multiply furiously and within days cause extensive blood hemorrhaging and almost certain death unless antibiotic treatment is administered. Public health officials are constantly on guard to protect against this sort of biowarfare, and numerous popular novels are built around this topic. One such novel you might enjoy reading is *Rainbow Six* by Tom Clancy.

Bacterial Reproduction

Bacteria reproduce by the relatively straightforward method of binary fission, a process that results in a colony (or clone) of genetically identical cells. During binary fission, a mature bacterium increases in size, and its enzymes replicate its DNA to yield two chromosomes. New cell wall and cell membrane grow inward from the margin of the cell, and as they come together, the two chromosomes separate, one into each compartment. (The effect is somewhat like forming two sausages from a single, long sausage.) The

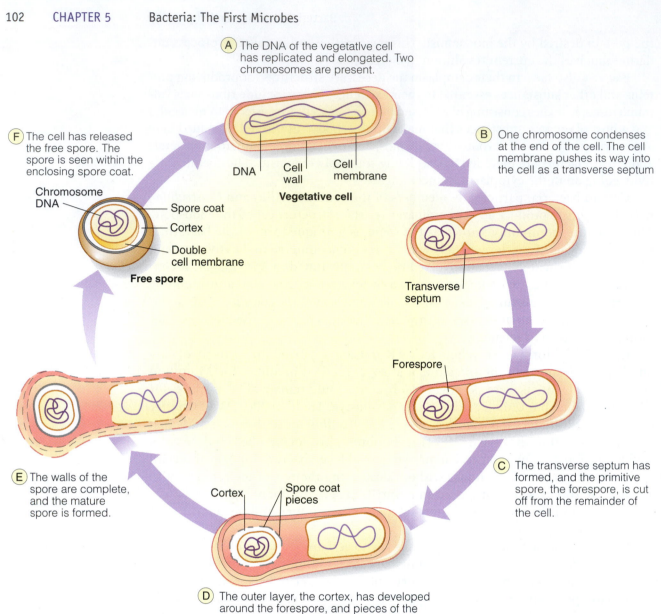

A The DNA of the vegetative cell has replicated and elongated. Two chromosomes are present.

DNA   Cell wall   Cell membrane
Vegetative cell

B One chromosome condenses at the end of the cell. The cell membrane pushes its way into the cell as a transverse septum

Transverse septum

Forespore

C The transverse septum has formed, and the primitive spore, the forespore, is cut off from the remainder of the cell.

F The cell has released the free spore. The spore is seen within the enclosing spore coat.

Chromosome DNA
— Spore coat
— Cortex
— Double cell membrane
Free spore

E The walls of the spore are complete, and the mature spore is formed.

Cortex   Spore coat pieces

D The outer layer, the cortex, has developed around the forespore, and pieces of the spore coat are forming.

FIGURE 5.7 **The Formation of a Bacterial Spore.** The bacterial cell metabolizes nutrients and multiplies for many generations as a vegetative cell. After some time, the vegetative cell enters the sporulation cycle shown here and forms a free spore.

adjoining walls and membranes separate, and two new bacterial cells result. The entire process requires only a few seconds and is shown in **FIGURE 5.8** .

The method of bacterial reproduction is not complex when compared to mitosis, the multistage process of DNA duplication and separation observed in eukaryotic cells. However, the frequency of bacterial reproduction is quite extraordinary. For example, under ideal circumstances, the common intestinal bacterium *Escherichia coli* can undergo binary fission and produce a new generation of bacterial cells every 20 minutes (by comparison, producing a new generation of humans takes about 20 years). Within 1 hour, a batch of 100 of these bacteria can become 800 bacteria; in 2 hours, there will

■ *Escherichia*
esh-ėr-ē′kē-ä

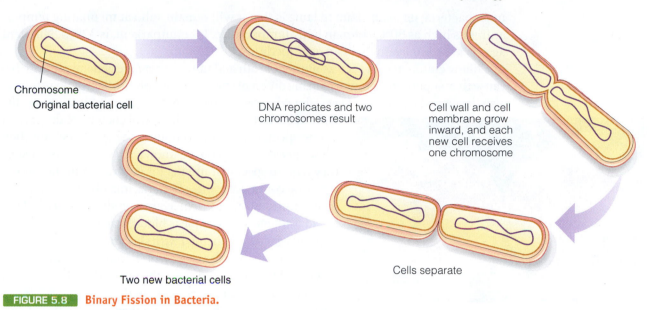

Chromosome
Original bacterial cell

DNA replicates and two chromosomes result

Cell wall and cell membrane grow inward, and each new cell receives one chromosome

Cells separate

Two new bacterial cells

FIGURE 5.8   Binary Fission in Bacteria.

be 6,400 bacteria; and in only 3 hours, the original 100 bacteria will have become 51,200 bacteria. One enterprising mathematician has calculated that a single *E. coli* cell reproducing every 20 minutes could yield in 36 hours enough bacteria to cover the face of the Earth a foot thick!

Then why are we not smothered in bacteria? The answer is that bacteria are susceptible to the same dynamics as any population of plants, animals, or other organisms. Bacteria grow and reproduce wildly for a time, but then the environment catches up to them. Nutrients become scarce, predators become more plentiful, waste products accumulate, water is in short supply, the gaseous environment changes, and from the bacterium's point of view, things get thoroughly out of hand. More bacteria die than reproduce, and the population decreases significantly. We discuss these dynamics and survey some conditions that contribute to microbial reproduction more thoroughly in Chapter 9.

Bacterial Growth

Bacteria are cultivated in the laboratory on or in materials called culture media (sing., medium). A culture medium is a water solution of various nutrients that encourage the growth of a particular species of microorganism. The medium generally contains a source of energy (for example, glucose), plus sources of carbon, nitrogen, and other essential nutrients. Other substances may be added to encourage the growth of a particular bacterial species. The ability to cultivate bacteria in the laboratory is essential to many applications of microbiology. For example, bacteria must be cultivated to obtain the valuable industrial products they synthesize (Chapter 14).

Culture media may be utilized as liquids or gels. In liquid form, the culture medium is generally called a broth. A typical example is nutrient broth, a "beef soup" that contains beef broth, a derivative of yeast cells called yeast extract, and a protein mixture known as peptone. The gel, or semisolid, form of a culture medium is referred to somewhat imprecisely as an agar. This is because it consists of a broth solidified with agar, a complex carbohydrate derived from algae. Agar is not used as a nutrient by most bacteria; its main claim to fame is that it will remain solid at incubating temperatures as high as 80°C (human body temperature, for comparison, is 37°C). Nutrient agar is a typical example of an agar medium.

Many culture media such as nutrient broth and nutrient agar are general purpose because they support the growth of many different species of microbes. Microbiologists have also developed a number of special media, called selective media, for cultivating specific types of bacteria. Such a medium will encourage the growth of one species while discouraging the growth of another. Some species of bacteria are known to be fastidious; that is, they require specific nutrients for cultivation in the laboratory. To accommodate these bacteria, microbiologists add the special nutrients to nutrient agar to produce a so-called enriched medium. An example is blood agar, which consists of nutrient agar supplemented with sheep red blood cells. This medium encourages the growth of streptococci, which break down the blood cells and cause the normally red medium to become clear, a process called hemolysis (FIGURE 5.9). A physician uses this phenomenon when taking a throat culture to help detect the streptococci of strep throat.

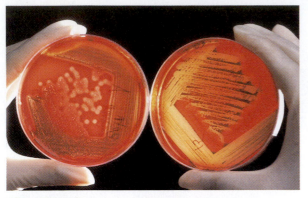

FIGURE 5.9   **Use of an Enriched Medium.**

The extensive research into bacterial culture media has resulted in laboratory protocols for cultivating most human pathogens. Bioterrorists have taken advantage of these advances and have learned to cultivate huge quantities of pathogens in vats capable of holding 10,000 gallons. They add test tube volumes of broth cultures to gallon-sized vessels and incubate the bacteria. When the medium becomes cloudy with growth, they add the contents of the vessel to a larger vessel, one containing a hundred gallons of broth. Following incubation, the larger vessel is emptied into a still larger one, and so on, until the bioterrorists have a vat filled with incalculable amounts of bacteria.

But the bacteria-filled broth cannot be used as a bioweapon, and the bioterrorists must dry up the broth and produce a powder. This is accomplished by any of several means. For example, the broth can be sprayed into the heated barrel cylinder of a spray dryer. As the heat evaporates the broth, the dried bacteria fall to the bottom of the cylinder as a powder. Another method involves a heated drum. In this case, the broth is sprayed onto the surface of a heated drum that resembles a steam roller. The broth evaporates quickly and leaves behind a bacteria-rich powder that can be scraped off.

At this point, the process is still not complete. There are too many clumps of bacteria for the powder to be an effective bioweapon. Therefore, the powder must be further processed to a biodust of fine particles that can reach into the lungs or into cracks in the skin. The sophisticated and patented methods required for producing the biodust ensure that bioterrorism can be carried out only by groups that have the technology and expertise to use them.

The Spectrum of Bacteria

It is a common misconception that a great variety of bacteria have been identified and studied. Despite their importance in medicine, research, ecology, and industry, the truth is that the vast majority of bacterial species remain unknown, with no name or available method for cultivation. Indeed, a gram of ordinary garden soil contains thousands of species that will probably remain obscure for many decades to come.

Yet about 4000 species of bacteria, with varying structures, growth patterns, and biochemistry, have been identified. Those who believe that "once you've seen one bacterium, you've seen them all" are well advised to read this section closely—for indeed, the term "bacteria" embodies a bewildering variety of forms.

Archaebacteria (Archaea)

We open our discussion of bacterial variety by considering a unique type of prokaryote called archaebacteria. These organisms are also known as archaea (sing., archaeon) to distinguish them from the traditional bacteria (in this context, the latter are known as eubacteria, or "true bacteria," as Chapter 2 notes). The reason for the nomenclature shift is the gradually emerging agreement that archaebacteria (archaea) are not bacteria in the traditional sense because their cell walls do not contain peptidoglycan, their cell membranes have unusual lipid compositions, and the RNA in their ribosomes has a unique chemical composition. Largely for these reasons, Carl Woese and his colleagues at the University of Illinois have recommended that these microbes be placed in a domain (or superkingdom) called Archaea.

FIGURE 5.10 **The Habitat of Extremophiles.** This view of the effluent channel of an alkaline spring in Yellowstone Park shows a mat of cyanobacteria in the foreground. The temperature in the channel is about 75°C (37°C is human body temperature). Archaea live in environments such as these.

Another reason for separating archaea from other prokaryotes is the extremely harsh environments in which many species live (FIGURE 5.10). Indeed, the new word extremophiles has been coined for organisms like these. One group of archaea are the thermoacidophiles. These organisms live under extremely acidic and extremely hot conditions. One archaeon, *Sulfolobus acidocaldarius,* grows well at temperatures of 85°C (about 185°F) and in soil with a pH of 1.0 (the acidity of fuming sulfuric acid). Among the most heat-resistant organisms isolated to date is the archaeon *Pyrolobus fumarii,* which is found in hydrothermal vents at the ocean's bottom. In 1995, scientists found dense communities of these microbes in the boiling waters just above the Macdonald Seamount, an active volcano 150 feet below the surface of the Pacific Ocean near Polynesia. The microbes flourish at temperatures up to 113°C (about 260°F); indeed, they will not grow if the temperature drops below 90°C (194°F) because it gets too cold! Waxy molecules in the cell membrane and many strong linkages between sulfur atoms in the proteins help protect these microbes from the heat that exceeds the heat of boiling water.

Another group of archaea are the methanogens. These prokaryotes live solely on carbon dioxide, nitrogen, and water. They produce methane under oxygen-free conditions and thrive in volcanic rock, marshes, lake bottoms, and animal feces (interestingly, they were in animal intestines all along, but remained unidentified until fairly recently).

A third group of archaea are the extreme halophiles. These organisms live in high-salt environments, such as Utah's Great Salt Lake. Some species have red pigments in their cytoplasm; such cellular pigments account for the redness in salt collection ponds near California's San Francisco Bay.

Archaea were so-named because microbiologists believe the organisms existed under primitive Earth conditions. At one time, scientists even thought that archaea

■ *Sulfolobus acidocaldarius* sul′fō-lō-bus as-i-dō-käl-dār′ē-us

■ *Pyrolobus fumarii* pī-rōl′ō-bus fū-mär′ē-ē

predated traditional bacteria, which evolved from them. However, that notion was challenged in recent years by comparative analyses of the DNA in prokaryotes. The current thinking is that an as-yet-unidentified ancestral microbe gave rise to both archaea and traditional bacteria. It is conceivable that traces of the ancestral microbe are waiting to be discovered somewhere in a primordial swamp.

Photosynthetic Bacteria

Among the traditional bacteria (i.e., eubacteria) are the cyanobacteria and other photosynthetic bacteria that have chlorophyll pigments dispersed in their cytoplasm. These bacteria have different colors due to varying forms of the pigments; and because they use sulfur extensively in their chemical reactions, many are grouped as green sulfur bacteria, purple sulfur bacteria, and purple nonsulfur bacteria. These microbes use photosynthesis to produce energy-rich organic matter that other species of bacteria can absorb and use in their metabolism when the photosynthetic microbes die. The photosynthesizers are known as autotrophic microbes because they synthesize their own food (*auto-* refers to "self," and *troph* refers to "feeder"; hence, "self-feeder"). By comparison, microbes that use preformed organic matter for food are known as heterotrophic microbes (*hetero-* refers to "other," meaning from other sources). By providing organic matter at the base of the food chain, the photosynthesizers occupy a key position in the nutritional patterns of nature.

The cyanobacteria have green chlorophyll pigments. When they contaminate aquaria and swimming pools, cyanobacteria cause the water to become an eerie blue-green color with a slimy feel. They also thrive in freshwater ponds, where they form deposits identical to those of cyanobacteria living on Earth over 3 billion years ago. Those massive limestone (calcium carbonate) deposits were abundant in virtually all freshwater and marine environments until about 1.6 billion years ago. Known as stromatolites, such deposits are still being formed, but only where the environment provides much light, acid, and salt. Examples of these environments occur in Western Australia.

In addition to photosynthesis, some species of cyanobacteria carry out nitrogen fixation. In this process, the cyanobacteria take up nitrogen from the atmosphere and use it to synthesize ammonia and other nitrogen-containing substances, which plants can incorporate into organic compounds when the bacteria die. The process occurs within specialized cyanobacterial cells called heterocysts. These cells produce the key enzyme for trapping nitrogen. The nitrogen-containing compounds are also shared with photosynthetic cells of the cyanobacterial colony, which, in turn, produce carbohydrates to supply the energy needs of the heterocysts. Such a mutually beneficial relationship is called symbiosis. Because they carry out photosynthesis as well as nitrogen fixation, cyanobacteria are among the most independent organisms on Earth. Moreover, they were probably the first organisms to introduce atoms of oxygen into the planet's atmosphere.

Heterotrophic Eubacteria

Most species of what we commonly refer to as "bacteria" are the heterotrophic eubacteria. The majority of these bacterial species are key players in the global cycling of nitrogen, sulfur, iron, phosphorus, and other nutrients. Most species are decomposers (organisms that break down chemical compounds in the environment), but many are producers (organisms that synthesize compounds for their own use and for other organisms). For example, heterotrophic bacteria belonging to the genera *Azotobacter* and *Rhizobium* are among the few organisms (other than cyanobacteria) that trap nitrogen from the atmosphere and synthesize useful organic compounds. We discuss these microbes at length in Chapter 16.

Among other important heterotrophic eubacteria are *Escherichia coli* and *Lactobacillus* species. Strains of *E. coli*, shown in **FIGURE 5.11**, live in the human intestine and help newborns digest milk by breaking down its lactose. This microbe is also a highly effective research tool in physiology, biochemistry, and genetics (probably more is known about *E. coli* than any other organism on Earth), and it produces many industrial enzymes, vitamins, and amino acids. *Lactobacillus* species live in the female genital tract and help guard against infection by other microbes. They are also used in the large-scale manufacturing of cheese, sour cream, yogurt, and other fermented milk products, as we note in Chapter 14.

The heterotrophic bacteria include many other species with medical, industrial, and environmental importance. Among the more interesting are members of the genus *Pseudomonas*, One species, *P. aeruginosa*, produces a soluble blue-green pigment that emits a green glow when it infects burnt tissue and wounds. In the soil, other species of the genus produce a large variety of enzymes that contribute to the breakdown of pesticides and similar waste chemicals. Unfortunately, some species also detract from the quality of soil by breaking down nitrates and converting these valuable plant nutrients to nitrogen gas, which is then released to the atmosphere.

Heterotrophic bacteria include many pathogens responsible for human disease as well as several other microbes. Included here is *Serratia marcescens*, a Gram-negative bacterium distinguished by the blood-red pigment it produces when it forms colonies. Although *S. marcescens* can infect the respiratory tract in patients with compromised immune systems, the organism is better known for its influence on the course of history. In 332 B.C., Alexander the Great and his army of Macedonians were laying siege to the city of Tyre in what is now Lebanon. The siege was not going well. Then one morning, blood-red spots appeared on several pieces of bread (actually they were colonies of *S. marcescens*). At first, the "blood" was thought to be an evil omen, but a soothsayer named Aristander pointed out that the red material was coming from inside the bread. This suggested that blood would be spilled within Tyre and that the city would fall. Alexander's troops were buoyed by this interpretation; with renewed confidence, they charged headlong into battle and captured the city. The victory opened the Middle East to Alexander and the Macedonians. Their march did not stop until they reached India.

Because of the bright red pigment of its colonies, *S. marcescens* was also used in the past to test wind currents that might carry bioweapons. Balloons filled with broth cultures of the bacteria were burst at a given point, and open plates of nutrient agar were set out for a specific time period at numerous other points. If bright red colonies appeared on any of the plates after incubation, that was evidence that bacteria released by a bioterrorist could travel from the starting point to that point. Since somewhat pathogenic strains of *S. marcescens* are now known to exist, tests such as these are no longer performed.

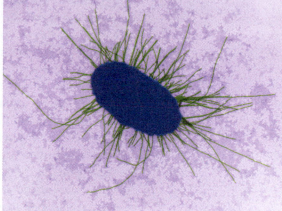

FIGURE 5.11   The Heterotrophic Bacterium *Escherichia coli*. In this transmission electron micrograph, *E. coli* displays short, hairlike pili on its surface. The pili encourage the bacteria to adhere to the host's intestinal tissues (×40,000).

■ *Azotobacter*
ä-zo′to-bak-tėr

■ *Rhizobium*
rī-zō′bē-um

■ *Lactobacillus*
lak-tō-bä-sil′lus

■ *Pseudomonas*
sū-dō-mō′näs

■ *Serratia marcescens*
ser-rä′tē-ä mär-ses′sens

Spiral and Filamentous Bacteria

The spiral bacteria resemble springs that have been partially stretched out. They can be either spirochetes or spirilla. Spirilla have flexible cells and flagella at the ends of the spiral, and spirochetes have rigid cells and long flagella inserted beneath the outer membrane of the cell wall. These flagella rotate and allow the spirochetes to move rapidly. One species of spirochete causes syphilis; another causes Lyme disease (Chapter 19).

Many bacteria occur as long, chainlike branching forms known as actinomycetes (*actino-* means "radiating"; *mycetes* refers to "fungus," since the branching chains of cells are funguslike). Within this group are species of *Streptomyces*, which produce numerous antibiotics including tetracycline, erythromycin, and neomycin. Actinomycetes form very resistant spores at the tips of their filaments, a feature that allows them to remain alive in difficult environments such as soil (to which they give a characteristic mustiness). It should be noted that although actinomycetes form chains of cells, each operates independently.

Gliding and Sheathed Bacteria

Certain species of bacteria produce a slimy substance, then move by gliding along in it. Wavelike contractions of the outer cell membrane propel these bacterial "gliders" through the slime. Two important genera of gliding bacteria, *Beggiatoa* and *Thiothrix*, live in sulfur-rich muds and break down foul-smelling hydrogen sulfide, releasing sulfur that can then be used by other organisms.

Another group of gliding bacteria, the myxobacteria, have a unique developmental cycle involving the formation of reproductive structures called fruiting bodies. When nutrients are exhausted, the bacteria congregate and interact with one another, eventually producing a stalk with a mass of cells at the top. The cells differentiate into sporelike bodies that are highly resistant to environmental stresses and thus permit the myxobacteria to survive until conditions improve.

Sheathed bacteria are a type of filamentous (chainlike) bacteria having cell walls enclosed in a sheath of complex carbohydrates and proteins. The sheath helps protect the bacteria from predators and environmental stresses, and it permits the bacteria to attach to food particles. Members of the genus *Spherotilus* are in this group.

Predatory and Other Bacteria

Among the more intriguing of bacterial forms are bdellovibrios, a group of rod-shaped bacteria that prey on other bacteria. A bdellovibrio bacterium attaches to the surface of a host bacterium, then rotates and bores a hole through the cell wall. Now it takes control of the host cell and grows in the space between the cell wall and cell membrane, killing the host bacterium in the process. *Bdellovibrio bacteriovorus* is the most thoroughly studied species of the group.

The bacteroides are a group of Gram-negative bacterial rods that live in oxygen-free environments. Several species live in the stomach of the cow and digest the cellulose in plant cells, a chemical process accomplished by few other organisms (we explore the implications of this process more completely in Chapter 15.) Human feces contain large amounts of bacteroides, which are probably very helpful in our digestive processes. Indeed, an estimated 30 percent of the bacterial mass isolated from human feces consists of bacteroides.

Chemolithotrophic bacteria are a group of bacteria that derive their energy from chemical reactions and use simple carbon compounds and inorganic materials to syn-

■ *Beggiatoa*
bej′jē-ä-tō-ä

■ *Thiothrix*
thī′ō-thriks

■ bacteroides
bak-tė-roi′dēz

thesize larger molecules. For example, chemolithotrophic bacteria such as *Nitrosomonas* and *Nitrobacter* species use carbon dioxide for their carbon needs and grow in environments containing nitrate or nitrite ions. This type of biochemistry is an essential feature of the cycles of nature.

The acid-fast bacteria include members of the genus *Mycobacterium,* the so-called mycobacteria. These rod-shaped bacteria have large amounts of mycolic acid in their cell walls, a factor that makes them very difficult to stain. However, when heat or other agents are used to force stain into the cytoplasm, the bacteria resist decolorization even though a dilute acid-alcohol solution is used. Therefore, they are said to be acid-fast (or acid-resistant). Many mycobacteria are free-living, but there are two notable pathogens in the group: *M. tuberculosis,* the cause of tuberculosis; and *M. leprae,* which causes leprosy. The acid-fast characteristic is a key factor in their detection.

■ *Nitrosomonas*
nī-trō-sō-mo>näs

■ *Nitrobacter*
ni-trō-bak'tẻr

Submicroscopic Bacteria

Among the very small bacteria are the rickettsiae, chlamydiae, and mycoplasmas. These forms are said to be submicroscopic because they cannot be seen clearly with a light microscope; an electron microscope is required to see them in detail.

Rickettsiae were discovered by and named for Howard Taylor Ricketts, a bacteriologist from the University of Chicago. In 1910, Ricketts was investigating the cause of Rocky Mountain spotted fever, a blood disease accompanied by a skin rash and high fever. Ricketts identified ticks as the mode of transmission for the disease, but in the course of his work he was stricken with a closely related disease and died. Rickettsiae are about 0.45 µm in diameter and are extremely difficult to cultivate in the laboratory; they are transmitted among their hosts almost exclusively by arthropods (such as fleas, ticks, and lice). During the 1990s, a tickborne rickettsial disease called ehrlichiosis occurred sporadically in the northeastern United States.

At 0.25 µm, chlamydiae are a step below rickettsiae in size. One species of chlamydiae called *Chlamydia trachomatis* causes a common sexually transmitted disease known as chlamydia (note that the disease and the organism have the same name). Chlamydia, the disease, is accompanied by burning pain in the reproductive tract and possible complications in the pelvic organs; it is estimated to affect over 4 million Americans annually. One case is presented in FIGURE 5.12 .

Mycoplasmas are the only bacteria that lack a cell wall. A single membranous layer of lipids encloses these tiny cells, which measure about 0.15 µm in diameter and are the smallest known bacteria. Mycoplasmas appear under the electron microscope as irregular blobs, and certain species cause mild pneumonia in humans.

Bacterial Pathogens

The bacteria are probably best known to the public as agents of human disease. The notion that bacteria are involved in infectious disease, the germ theory of disease, was developed by Louis Pasteur during the latter part of the 1800s, as Chapter 1 explores. Verification of the germ theory was offered by Robert Koch, a German physician and a contemporary of Pasteur. Following the lead of Koch and Pasteur, microbiologists from around the world participated in the Golden Age of Microbiology, a 60-year period lasting from about 1860 to 1920, during which a majority of the bacterial pathogens of infectious disease were identified.

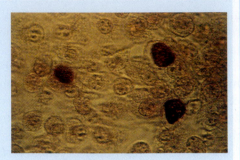

1 An educated, professional woman met a gentleman at a friend's house one evening. She was director of a New York law firm. The man was equally successful in his professional career.

2 The couple hit it off immediately. There were many evenings of quiet candlelit dinners, and soon, they became sexually intimate. Neither one used condoms or other means of protection.

3 Three days after having intercourse, the woman began experiencing fever, vomiting, and severe abdominal pains. She immediately made an appointment to see her doctor.

4 Assuming the illness was an intestinal upset, the physician prescribed appropriate medication. The woman was actually suffering from chlamydia, but the fact that she was sexually active did not come up during the examination.

5 Feeling better, the woman continued her normal routine. But 6 months later, with no apparent warning, she collapsed on a New York City sidewalk.

6 The woman was rushed to a local hospital, where doctors diagnosed a chlamydial infection by observing the dark inclusion bodies typical of chlamydia (see figure). They performed emergency surgery: her uterus and Fallopian tubes were badly scarred. She recovered, but the scarring left her unable to have children.

FIGURE 5.12   A Case of Chlamydia. This case occurred in a 32-year-old professional woman. Tragic complications of the disease resulted because she and her doctor neglected to consider that she could be suffering from a sexually transmitted disease.

Bacteria can cause disease in a number of ways. In some cases, bacteria grow and multiply in extraordinary numbers in body tissues and cause cellular death—the bacteria that cause tuberculosis work in this way. At other times, bacteria produce substances to help them overcome body defenses—staphylococci, for example, produce an enzyme that clots blood and forms a protective layer of clotting material around the invading bacteria; this layer negates the body's attempt to attack the bacteria and destroy them.

In other instances, bacteria synthesize poisonous substances called toxins. Toxins interfere with the chemical activities of cells and tissues. For example, the bacterium that causes tetanus produces a powerful toxin that interferes with the relaxation of muscles after they contract. The muscles undergo severe spasms and "lock" into place ("lockjaw" is one effect). Another toxin producer is *Clostridium botulinum,* the organism that causes botulism. The toxin produced by *C. botulinum* has been developed as a weapon of bioterrorism because it can be added to water and, when ingested, causes nerve paralysis that is highly lethal.

Bacterial disease can also result from an excessive immune response in the body, a response that leads to inflammation and local tissue damage. In the case of endocarditis, for instance, streptococci elicit antibodies from the immune system, and the antibodies unite with bacteria at the heart valves; progressive deterioration of the valves results in heart failure.

The human body responds to disease by defending itself with chemical substances such as interferon (an antiviral substance) and lysozyme (an antibacterial substance that destroys the cell wall). White blood cells attempt to engulf and destroy bacteria through the process of phagocytosis. In addition, the body calls the immune system into action. The production of activated T cells and antibodies is stimulated by chemicals associated with bacteria and other microbes. Ultimately the T cells and antibodies attack the bacteria and destroy them, as we describe in depth in Chapter 17. We also look more closely at specific bacterial diseases in Chapter 19.

Despite these negative images, it is important to remember that of the thousands and thousands of bacterial species, a tiny percentage (less than 1%) are pathogenic. The vast majority of bacterial species are beneficial to humans: They participate in numerous industrial processes, synthesize many vital chemicals, and work their magic in recycling processes taking place in the soils and waters of the Earth. Although bacteria and other microbes have largely earned their reputations from the diseases they cause, they have numerous positive characteristics and functions. We shall see many of those in other chapters.

■ A FINAL THOUGHT

When you pick up this book for the first time, you may experience a moment anticipating your first look at bacteria. Perhaps you leaf through the pages, turn back to see a photograph a second time, then pause and think: "Is that all there is? Little rods and circles?"

Our first encounter with bacteria is often a disappointing (and perhaps exasperating) experience. Since early childhood, we have been taught to loathe and despise bacteria, and we have been schooled in all the dastardly deeds they do. "Wash your hands" we are admonished; "Don't eat it if it falls on the ground." And on and on. We expect bacteria to be grotesque and fearsome monsters. We expect them to rank with death and taxes. But they turn out to be little sticks and rods, not very dangerous-looking at all. So what's the big deal?

Perhaps that's good. Perhaps we need to wipe away any preconceived notions of bacteria and start rebuilding our views—a process that will take us through the complexity of bacterial structures, give us insight into their chemistry, point up their remarkable genetics, and stress their importance in food production, industrial manufacturing, and soil ecology.

It's going to take some time to absorb the importance of bacteria to our lives, but for now try to understand that the bacteria are more than the tiny rods you see on these pages or under the microscope. The electron microscope has yielded a wealth of information about their structure (as this chapter demonstrates) and studying bacterial structure gives us a clue to what they do. This "what they do" part explains their importance to us. Please stay tuned.

■ QUESTIONS TO CONSIDER

1. Suppose a bacterium had the opportunity to form a glycocalyx, a flagellum, a pilus, or a spore. Which do you think it might choose? Why?
2. Extremophiles are of interest to industrial corporations, who see these bacteria as important sources of enzymes that function at temperatures of 100°C and in extremely acidic or basic conditions (the enzymes have been dubbed "extremozymes"). What practical uses can you foresee for these enzymes?
3. In the fall of 1993, public health officials found that the water in a Midwestern town was contaminated with sewage bacteria. The officials suggested that homeowners boil their water for a couple of minutes before drinking it. Would this treatment remove all traces of bacteria from the water? Why?
4. Researchers have estimated that, in broad terms, about one-third of human feces are composed of bacteria. That being the case, about how much bacteria do we "produce" in a week? In a year? How can this be possible?

5. Suppose this chapter on the structure and growth of bacteria had been written in 1940, before the electron microscope became available. Which parts of the chapter would probably be missing?

6. "Bacteria are all the same. Once you've seen one, you've seen 'em all!" What evidence could you present to counter this statement?

7. There are thousands of species of bacteria, yet with few exceptions, all of them have variations of three shapes: the rod, the sphere, and the spiral. Do you find this strange? Why or why not?

■ KEY TERMS

Informative facts are necessary for the expression of every concept, and the information for a concept is founded in a set of key terms. The following terms form the basis for the concepts of this chapter. On completing the chapter, you should be able to explain and/or define each one:

acid-fast bacteria
actinomycetes
archaebacteria
autotrophic microbes
bacilli
bacteroides
bdellovibrios
binary fission
cell wall
chemolithotrophic bacteria
chlamydiae
cocci
culture medium
cyanobacteria
differential technique
enriched medium
eukaryotes
extreme haplophiles
extrêmophiles
flagellum
fluid mosaic
glycocalyx
Gram stain technique

heterotrophic microbes
methanogens
mycoplasmas
myxobacteria
nitrogen fixation
nucleoid
nutrient agar
nutrient broth
peptidoglycan
periplasm
pilus
prokaryotes
rickettsiae
selective medium
Serratia marcescens
simple stain technique
spirilla
spirochetes
spore
staphylococcus
symbiosis
thermoacidophiles

■ http://microbiology.jbpub.com/book/microbes

The site features **eLearning,** an online review area that provides quizzes and other tools to help you study for your class. You can also follow useful links for in-depth information, read more stories of microbiology, or just find out the latest microbiology news.

Viruses: At the Threshold of Life

6

■ Looking Ahead

In the microbial world, viruses stand out for their simplicity, extraordinarily small size, and distinctive method of replication. We highlight these characteristics in this chapter.

On completing this chapter, you should be able to . . .

- explain the events leading to the discovery of viruses and recognize some turning points in the development of virology.
- appreciate the distinctive structure of viruses, including their incredibly small size and relatively simple components.
- describe the stages and details of the replication process exhibited by viruses.
- explain the significance of viral replication with respect to the disease process and the development of resistance to viral disease.
- identify some methods for controlling viruses and for using viruses to benefit society.
- recognize the existence of certain types of subviral particles.
- summarize the development of cancer and understand the involvement of viruses in this dreaded disease.

It was February 2003 and China revealed that over 300 people had died in the previous months from a deadly respiratory illness soon known as SARS (Severe Acute Respiratory Syndrome). The victims suffered from a rapid onset of high fever, muscle pain, chills, sore throat, and shortness of breath, or worse, severe pneumonia. When a man who had traveled in mainland China brought SARS to Hong Kong and spread it to health care workers, the city virtually shut down. Many people on the streets of Hong Kong wore masks. Restaurants closed and business at hotels and the airport were down 80%. Ironically, the Hong Kong tourism bureau had just proudly unveiled its new tourism ad campaign featuring the slogan: *Hong Kong: It will take your breath away.*

Every science has its borderland where known and visible things merge with the unknown and invisible. Startling discoveries often emerge from this uncharted realm of speculation, and certain objects manage to loom large. In the borderland of microbiology, a particularly curious and puzzling object is the virus.

Viruses exist at the threshold of life. Although many individuals speak of "live virus vaccines" and many scientists "*grow* viruses" in their laboratories, it appears that viruses have a level of simplicity that places them somewhere between living objects and chemical compounds. Indeed, one scientist has written that viruses are transitional forms between inert molecules and living organisms. He has whimsically suggested that viruses be called "organules" or "molechisms," depending on one's point of view. Another scientist has suggested that perhaps viruses have evolved beyond organisms—why be mechanically or physically complex, he queries, when you can subjugate other organisms or entities for your own purpose? As we shall see in the pages ahead, viruses truly do not fit into our traditional vision of living things.

The Discovery and Structure of Viruses

In the period between 1880 and 1915, a wealth of discoveries occurred in the field of microbiology. Researchers finally pinpointed the organisms causing tuberculosis, typhoid fever, syphilis, and many other infectious diseases; they improved methods of sanitation and food preservation; they increased their understanding of the importance of microbes in the environment; and they learned to use microbes for industrial processes to benefit society. As Chapter 1 notes, this period is part of the Golden Age of Microbiology.

But there were several researchers whose work was somewhat unproductive. These researchers devoted their energies to discovering the microbes that cause such diseases as measles, chicken pox, polio, and hepatitis. Their labors bore little fruit because the methods available for isolating and studying the microbes causing these diseases were technologically primitive. As it turned out years later, these diseases were not caused by bacteria, but by viruses. And in the period around 1900, neither the microscopes nor the laboratory methodologies were capable of dealing with viruses. We rarely know who these unfortunate researchers were, but we can be certain that they suffered much frustration and depression because their work was largely unrewarded.

At the beginning of the 1900s, the word "virus" was being used in science to refer to any unseen or unknown entity involved with infectious disease. For example, a bacterium was identified as the "virus" of tuberculosis, and a protozoan was found to be the "virus" of malaria. Thus, scientists spoke freely of "viruses," although not in the context in which we currently use the term. (To be sure, their inappropriate use of the word "virus" in their scientific reports makes reading them very confusing.) The true nature of viruses remained a mystery.

The Development of Virology

■ Iwanowski
I-wan-OW-ski

One of the first scientists to study viruses in depth was Dmitri Iwanowski, pictured in FIGURE 6.1 . Iwanowski was interested in tobacco mosaic disease, a disease that causes tobacco leaves to shrivel and die. Attempting to isolate the "bacterial" cause of the disease, Iwanowski crushed diseased tobacco leaves and filtered them to sep-

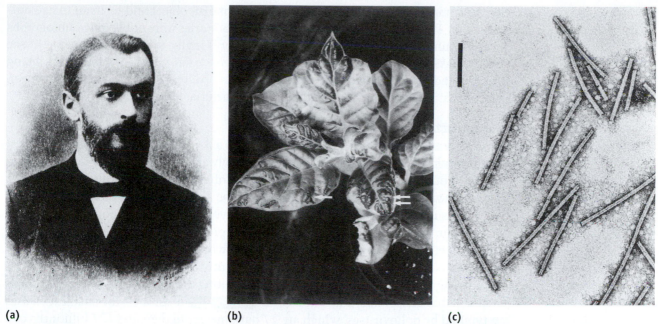

(a)   (b)   (c)

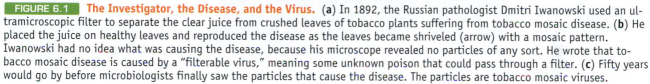

FIGURE 6.1   **The Investigator, the Disease, and the Virus.** (**a**) In 1892, the Russian pathologist Dmitri Iwanowski used an ultramicroscopic filter to separate the clear juice from crushed leaves of tobacco plants suffering from tobacco mosaic disease. (**b**) He placed the juice on healthy leaves and reproduced the disease as the leaves became shriveled (arrow) with a mosaic pattern. Iwanowski had no idea what was causing the disease, because his microscope revealed no particles of any sort. He wrote that tobacco mosaic disease is caused by a "filterable virus," meaning some unknown poison that could pass through a filter. (**c**) Fifty years would go by before microbiologists finally saw the particles that cause the disease. The particles are tobacco mosaic viruses.

arate their particulate debris from their juices. He used a filter that would trap the smallest known bacteria. To his surprise, Iwanowski found that the clear juice coming through the filter contained the infectious agent. He had no idea what was in this juice, and he suggested that a "filterable virus" caused tobacco mosaic disease, meaning that whatever caused the disease could pass through the filter.

Some years later, in 1898, the Dutch investigator Martinus Beijerinck repeated Iwanowski's work and expanded its scope by showing that even though the juice was diluted many times, the infectious agent was still present. Beijerinck called the unknown agent *contagium vivum fluidum*, or "contagious living fluid." How it caused the disease remained a mystery.

Real progress in identifying viruses would not occur until the 1930s. During that decade, scientists led by Wendell M. Stanley found that they could form crystals of tobacco mosaic virus. Crystal formation implied that the viruses were some sort of chemical molecule. However, viral diseases were accompanied by fever, a response by the immune system, and other symptoms associated with microbial infection, and so scientists continued to debate the nature of viruses. While the debate continued, virologists found that they could cultivate viruses in the living cells of fertilized eggs and in tissue cells. This breakthrough allowed them to obtain large quantities of viruses for study purposes.

Then the electron microscope came on the scene. This instrument grew out of an engineering design made in 1933 by the German physicist Ernst Ruska. Ruska used electrons trapped in a sealed vacuum tube to strike an object and form an image out-lining it. By 1941, American engineers at RCA laboratories were publishing photographs with magnifications of 10,000 times and higher. Finally, scientists were able to see viruses, and soon they had an image of the tobacco mosaic virus—it appeared as a long thin rod. Indeed, it seemed that viruses were tiny bacteria. Later research, however, showed that viruses were much different from bacteria. Much different, to be sure.

The Structure of Viruses

If you were to stop 100 people on the street and ask if they recognize the word virus, all 100 would probably nod their heads knowingly and say "yes." They might tell you about the flu they recently had, or they might offer that AIDS is caused by a virus. Were you to ask the same individuals to describe a virus, they would probably reply that a virus is "a tiny microbe," or "something you need a microscope to see," or "a germ." After a couple of moments, they would probably scratch their heads and ad-mit that even though they are familiar with the word " virus," they don't really know what a virus is (unless, of course, they had recently read this chapter).

Virtually everyone knows that viruses are among the smallest microbes that cause disease in humans, plants, and animals. As   FIGURE 6.2   illustrates, the smallest viruses are typified by polioviruses, which are 27 nanometers in diameter (27 billionths of a meter), while the larger ones, such as smallpox viruses, measure about 250 nanome-ters in diameter, about the size of the smallest bacteria. It is difficult for the human mind to comprehend the scale of a nanometer (a billionth of a meter), since we are used to lengths of inches, feet, and yards. Consider, however, that 500 or more viruses could fit inside a single bacterial cell.

Despite their difficulty in seeing viruses, scientists have elucidated the shapes of different viruses. Certain viruses, for example, exist in the form of a helix, similar to a tightly wound coil (they are said to have "helical symmetry"). The tobacco mosaic and rabies viruses are examples of helical viruses. Numerous viruses have the shape of an icosahedron (they are said to have "icosahedral symmetry"). An icosahedron is a geometrical figure with 20 equal-sized triangular faces, 12 edges, and 12 corners. Among the icosahedral viruses are the poliovirus, herpesvirus, chickenpox virus, and infectious mononucleosis virus.

Some viruses have other shapes. The smallpox virus, for example, has the shape of a brick, with a swirling pattern of tubes on its surface. The viruses that attack bac-teria, the bacteriophages, have a head and a tail assembly that makes them resemble miniature tadpoles (bacteriophages are discussed at length in Chapter 10). Still other viruses, such as the Ebola virus, occur as long threadlike structures, often with a hook at the end of the thread. And others, such as the human immunodeficiency virus (HIV), begin as an icosahedron, then change shape and become a cone. Scientists thus classify the viruses in three forms: helical viruses, icosahedral viruses, and complex viruses.   FIGURE 6.3   summarizes these forms.

The Components of Viruses

Plant, animal, and human cells are highly specialized, with various compartments in which different cellular functions are performed. These compartments exist in cyto-plasm, a warm, gel-like organic soup of proteins, lipids, carbohydrates, minerals, and numerous other factors essential to growth. Viruses are extraordinarily different. All

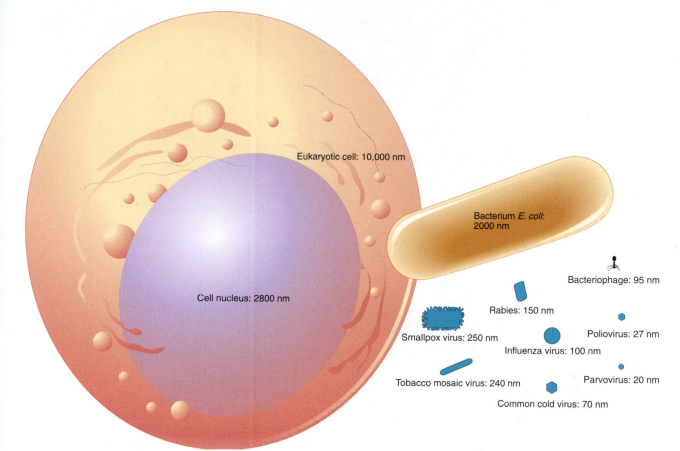

Eukaryotic cell: 10,000 nm

Bacterium *E. coli*: 2000 nm

Cell nucleus: 2800 nm

Bacteriophage: 95 nm

Rabies: 150 nm

Smallpox virus: 250 nm

Poliovirus: 27 nm

Influenza virus: 100 nm

Tobacco mosaic virus: 240 nm

Parvovirus: 20 nm

Common cold virus: 70 nm

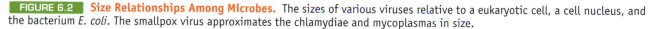

FIGURE 6.2 **Size Relationships Among Microbes.** The sizes of various viruses relative to a eukaryotic cell, a cell nucleus, and the bacterium *E. coli*. The smallpox virus approximates the chlamydiae and mycoplasmas in size.

viruses consist simply of a core of nucleic acid enclosed in a coat of protein. In some types of viruses, a membranelike envelope encloses the nucleic acid–protein combination, but there is nothing else.

The nucleic acid core of the virus, known as the genome, consists of either deoxyribonucleic acid (DNA) or ribonucleic acid (RNA), but not both. Sometimes the nucleic acid molecule is long and helical (as in tobacco mosaic and rabies viruses); but often it is folded, as in polioviruses, herpesviruses, and HIV. Usually the nucleic acid is a single molecule, but occasionally it exists in segments. The influenza virus, for example, consists of eight segments of RNA, each enclosed in a protein coat (the eight coated segments are then enclosed in protein). Although the nucleic acid portion of the virus is called the genome, you should be cautious when using this term because it also refers to the complete set of genes of any organism. (The human genome, for example, is all the genes in the 46 chromosomes of a human cell.) **FIGURE 6.4** shows the components of viruses.

The outer protein coat of the virus is called a capsid. In a helical virus, this protein layer binds to the nucleic acid and maintains the shape of the helix. In icosahedral viruses, the capsid takes the form of 20 equilateral triangles joined together to

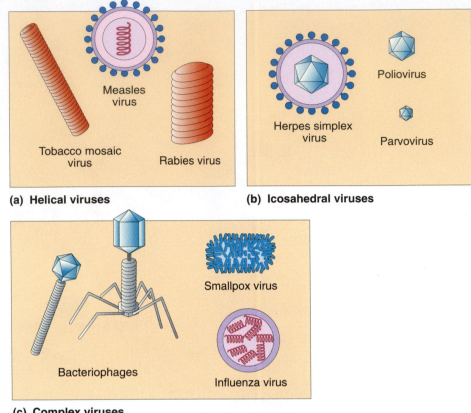

(a) **Helical viruses**

(b) **Icosahedral viruses**

(c) **Complex viruses**

FIGURE 6.3   **Viral Shapes.** Viruses exhibit variations in form. (**a**) The nucleocapsid has helical symmetry in the tobacco mosaic, measles, and rabies viruses. The helix resembles a tightly coiled spiral. (**b**) Certain viruses, such as herpesviruses, polioviruses, and parvoviruses, exhibit icosahedral symmetry in their nucleocapsids. The icosahedron is a polyhedron having 20 triangular faces and 12 points. (**c**) In other viruses, neither helical nor icosahedral symmetry exists exclusively. The bacteriophage, for example, has an extended icosahedral "head" and a helical tail with extended fibers. The smallpox virus has a series of rodlike filaments embedded within the protein at its surface. And the influenza virus consists of a series of helical segments enclosed by an envelope.

give the virus its shape. Smaller protein units called capsomeres are bound together chemically to form the capsid, much as patches are joined to make up a quilt. The combination of genome and capsid is known as the nucleocapsid. Several different viruses are depicted in   FIGURE 6.5 .

Certain viruses have an envelope, an enclosing structure similar to the membrane that encloses a cell. Acquired at the last stage of the virus' replication process, this envelope contains specific viral proteins that identify it as a viral structure. It protects the nucleocapsid and helps the virus penetrate a living cell during the replication process, as we shall see shortly.

In many viruses, the envelope has projections called spikes. Sticking out like the tines of a fork, the spikes help the virus contact its host cell, and then assist the viral nucleocapsid in its penetration into that host cell. HIV, for example, contains in its spikes two proteins that help it "dock" on its host cell. Many scientists are working to understand the physiology and biochemistry of the spikes in the hope of preventing

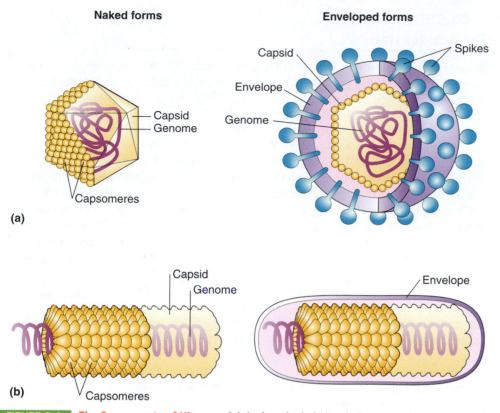

Naked forms

Enveloped forms

Capsid

Spikes

Envelope

Capsid
Genome

Genome

Capsomeres

(a)

Capsid
Genome

Envelope

Capsomeres

(b)

FIGURE 6.4   **The Components of Viruses.** (**a**) An icosahedral virus in both naked and enveloped forms. Capsomere units are shown on one face of the capsid. The genome consists of either DNA or RNA, which is folded and condensed. (**b**) A helical virus in both naked and enveloped forms. The genome winds in a helical fashion. The capsomeres are protein subunits that form the capsid.

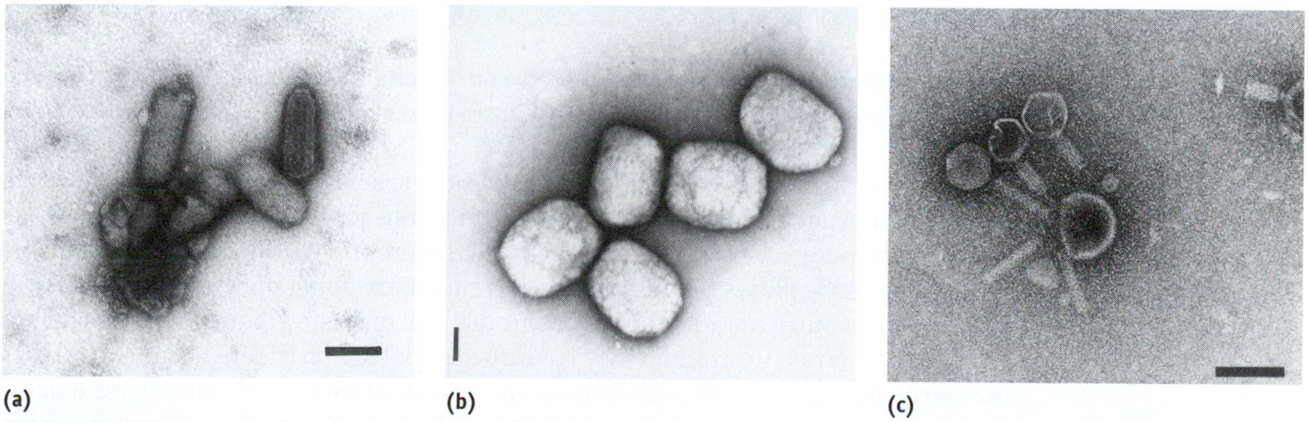

(a)

(b)

(c)

FIGURE 6.5   **Viral Symmetry.** Transmission electron micrographs of viruses displaying various forms of nucleocapsid symmetry. (**a**) The vesicular stomatitis virus, which causes skin sores in bovine animals. This virus has a helical nucleocapsid and appears in the shape of a bullet. It is related to the rabies virus. (Bar = 100 nm.) (**b**) The vaccinia virus, which causes cowpox. This viral nucleocapsid is rectangular with a series of rodlike fibers at its surface. The symmetry is described as complex. (Bar = 100 nm.) (**c**) A bacteriophage with an icosahedral head and an extended tail. This symmetry is also designated complex. (Bar = 100 nm.)

A CLOSER LOOK

The Achilles Heel

Students of ancient history know that the Greek warrior Achilles was cloaked in armor except at one vulnerable spot near his heel. An arrow directed at his heel in battle eventually killed Achilles (whose name has been given to a key tendon connecting three lower leg muscles to the bone).

In their search for antimicrobial agents, scientists are continually on the lookout for an organism's Achilles heel. And, for influenza, such a spot may exist at the enzyme neuraminidase (nur-ah-MIN-i-dase). This enzyme is found in the spike of the virus. It destroys sialic acid, an organic compound remaining in the viral envelope at the conclusion of the replication cycle. This reaction is important because, if left in place, sialic acid would clump influenza viruses together on the cell surface, and they could not enter fresh cells to replicate. But neuraminidase efficiently breaks down the sialic acid, and without this organic "glue," the viruses are free to enter a person's respiratory cells and begin the disease process. It stood to reason that crippling the neuraminidase would cripple the virus by making viral particles clump together.

Enter the scientists. Graeme Laver and his colleagues at the Australian National Laboratory found that neuraminidase is a protein consisting of four units extending out from the viral envelope like four ultramicroscopic balloons. The amino acid chains in the protein units are similar among influenza viruses, so a drug that reacts with the neuraminidase of one influenza virus will react with that of all.

Then the group set to work to synthesize a drug that would "muddy" the site where neuraminidase binds to sialic acid. They hunted for the amino acids where a drug could anchor itself and sought methods to bind a drug tightly there. It was molecular biology at its best, plus a bit of logic and a little luck. Fifteen long years passed as the group slogged on.

Finally, the new drug was ready: Its name is zanamivir (Relenza). A second drug called oseltamivir (Tamiflu) later became available. Both portend a new generation of anti-influenza drugs. Watch for their names this flu season.

viral replication and bringing viral infection to an end. For instance, interrupting the activity of enzymes in the spikes of influenza viruses has proven fruitful, as A Closer Look above discusses.

In summary, a virus consists of a core of nucleic acid, a covering of protein, and, in some cases, an enclosing envelope. You will note that the term "cytoplasm" has not been mentioned; nor has there been discussion of anything other than nucleic acid and protein. This is because there is no chemistry going on within a virus, there is no intake of nutrients, and there is no production of waste products. Viruses do not increase or decrease in size, and they have no metabolism. In the environment, viruses are inert particles. (One researcher has even suggested that a jar of viruses would make a perfect pet—you don't have to feed them or clean up after them.)

There is, however, one thing viruses do, and they do it particularly well—they replicate. During replication, an infecting virus biochemically programs a host cell to produce hundreds, sometimes thousands of copies of the virus. In so doing, viruses usually destroy the cell that has "hosted" their replication. Partly for this reason, a prominent scientist has called viruses "bad news wrapped in protein." We shall see how replication takes place in the next section.

Viral Replication

Outside a host cell, a virus is an inert particle; when it encounters a host cell, however, it becomes a highly efficient replicating machine. The virus invades the host cell, utilizes the metabolism of the cell, and produces multiple copies of itself. The process occurs in several stages, as we shall see next.

The Stages of Replication

The first stage in viral replication, called the attachment stage, usually occurs with a high degree of specificity; that is, most viruses will attach to and replicate within specific cells only. For example, hepatitis viruses replicate only in liver cells, and human immunodeficiency viruses replicate only in cells of the immune system. This specificity derives partly from the presence of protein molecules embedded in the envelope or capsid surface of the virus and in the cell membrane of the host cells. In the host cells, these molecules, called receptor sites, represent the chemical "lock" that molecules of the viral "key" must fit. In many instances, the spikes of the viral envelope contain the necessary molecules that bind the virus to the receptor sites. (It has not escaped the attention of researchers that interfering with this attachment could conceivably bring an end to a viral infection.) In other cases, the virus attaches by its "tail" to receptor sites at the host cell surface, as shown in FIGURE 6.6 .

Many viruses have envelopes that are composed of the same basic material as the cell membrane. Therefore, when such a virus associates with the host cell, the envelope "blends" with the cell membrane, much like bringing olive oil together with corn oil. This blending opens the cell membrane to the virus. In contrast, many nonenveloped viruses enter the host cell through the process of phagocytosis. Here, the host cell engulfs the virus much as white blood cells engulf foreign particles in the body tissues. In still other cases, the capsid of the nonenveloped virus is left outside the host cell, and the genome passes into the cell (see A Closer Look on page 123). This is the penetration stage.

Once within the cellular environment, enzymes of the cell cytoplasm remove the protein capsid from the viral genome. Next, in the uncoating stage, the viral nucleic acid is released. Then the synthesis stage begins. If the nucleic acid consists of DNA, it operates like a set of genes: The DNA encodes messenger RNA (mRNA) molecules. The resulting mRNA molecules carry the genetic code for synthesizing the proteins needed for viral reproduction. On the other hand, if the viral nucleic acid consists of RNA, the RNA acts as an mRNA molecule to encode the necessary proteins. Through complex chemical processes, explored in Chapter 4, proteins soon appear in the cell's cytoplasm. Some of these proteins function as enzymes to hook nucleotides together and synthesize new fragments of viral nucleic acid. Other enzymes are used to stitch together the amino acids for viral capsids. And some enzymes are used to break down cellular parts to provide building blocks for new viruses. Once synthesis of viral parts is complete, they are combined to form new viral particles. This is the assembly stage.

The replication process of a typical DNA virus is summarized in FIGURE 6.7 . Replication takes place over a period of minutes or hours, depending on the virus and its host cell. Note that the virus does not bring along any ribosomes, the biochemical "workbenches" where amino acids are joined to make proteins; instead, the cell's ribosomes must be used. The virus has no energy-rich molecules to fuel the chemical

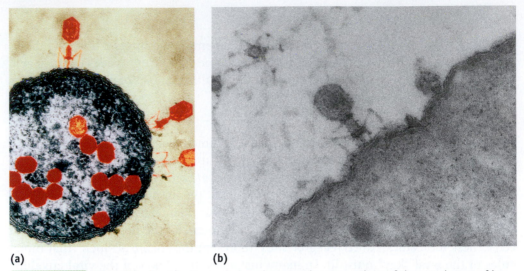

(a)   (b)

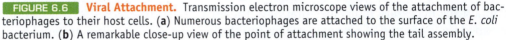

FIGURE 6.6   **Viral Attachment.** Transmission electron microscope views of the attachment of bacteriophages to their host cells. (**a**) Numerous bacteriophages are attached to the surface of the *E. coli* bacterium. (**b**) A remarkable close-up view of the point of attachment showing the tail assembly.

reactions, so it depends on the energy-rich molecules available in the cell's cytoplasm. The virus has no amino acids that it can contribute to protein synthesis, so it must appropriate amino acids in the cell. These processes constitute a substantial drain on the cell, and in many cases it breaks down and dies. As it does, the new viruses emerge in the release stage.

And things are even worse if enveloped viruses are involved, because the viral envelope is built from components of the cell's membrane: After the nucleocapsid forms, it moves toward the perimeter of the cell and forces its way through the membrane during the release stage. Thus, it encloses itself in pieces of cell membrane (which contain viral proteins). This action, which disrupts the cell membrane from within, occurs hundreds or even thousands of times as viral nucleocapsids make their exit. Normally, the cell would be able to repair such holes, but a large number of them are developing while the internal cellular environment is breaking down and while the cell's carefully balanced chemical processes are being disrupted. The widespread puncturing of the cell membrane by the nucleocapsids represents the *coup de grace* for the cell. Indeed, the viral replication cycle is often termed the lytic cycle (from lyse, meaning "to break or disrupt").

Viral Replication and Disease

If we stop for a moment and look back, we can develop a broad picture of how viral disease takes place. An essential feature of viral disease is the destruction of body cells. As the cells are destroyed, the critical functions performed by those cells come to an end. Consider, for example, what happens during the liver disease hepatitis: Viruses replicate within the liver cells, and as the cells die, the liver tissue disintegrates. The

A CLOSER LOOK

Empty Boxes

What do you do after you've opened the box and removed the gift? You probably admire the gift, thank the giver, and think about how you'll use it. Perhaps you try it on or otherwise begin weaving it into your life. The poor box gets shunted aside until someone thinks to put it in the garbage pail (or the recycle bin).

It's not much different with viruses: The DNA or RNA of the viral genome enters the cell's cytoplasm and encodes new viruses, while the capsid is broken down or, in the case of a bacteriophage, is left outside the cell. But things may be different in the future. Researchers are investigating viral capsids as miniature reaction chambers for creating ultramicroscopic wires or synthesizing crystals for microelectronic components. Because of a capsid's uniform size and shape, scientists see it as the ultimate small test tube.

Investigators at Montana State University have put the theory to work. Mark Young and his colleagues have cultivated masses of viruses, separated the capsids from the genomes, and reassembled the capsids. They combined the protein shells with tungsten salts and found that the salts penetrate the capsids when the acidity level is varied to control the pore size. Soon they were dreaming of constructing computer chips inside viral capsids—and perhaps bringing new meaning to the phrase "computer virus."

chemistry of the liver is radically altered: It cannot process amino acids for the body's use, and protein metabolism is blocked; production of bile drops off, and fat digestion is affected; liver cells are not available to process carbohydrates, and carbohydrate chemistry is disrupted; the liver is normally a storehouse for many vitamins, and its degeneration changes the body's vitamin balance; and on and on.

Or consider AIDS. In this disease, the human immunodeficiency virus (HIV) destroys the T cells essential to the activity of the immune system. These cells react to infection by protozoa, fungi, and other microbes. They also hold in check the microbes that are harmless under normal circumstances (the so-called opportunistic microbes). However, as HIV replicates in the T cells, these cells are destroyed, and the patient suffers numerous and varied infections of the lungs, brain, intestine, and blood. Unless the viral replication is interrupted and the T cells are restored in the immune system, the patient will continue to suffer these illnesses.

Still another example occurs with the disease rabies. In this case, the virus replicates in and destroys the nerve cells of the brain. This deadly disease can be contracted from dogs and cats, or any other warm-blooded animal. A case related to bats is reviewed in FIGURE 6.8. Ironically, viruses can also cause disease in bacteria, as A Closer Look on page 125 explores.

Defense Against Viruses

We have painted a rather bleak picture of the interaction between viruses and their host cells. And yet we do not die of measles, chickenpox, herpes simplex, mononucleosis, or numerous other viral diseases (although such diseases as rabies and AIDS

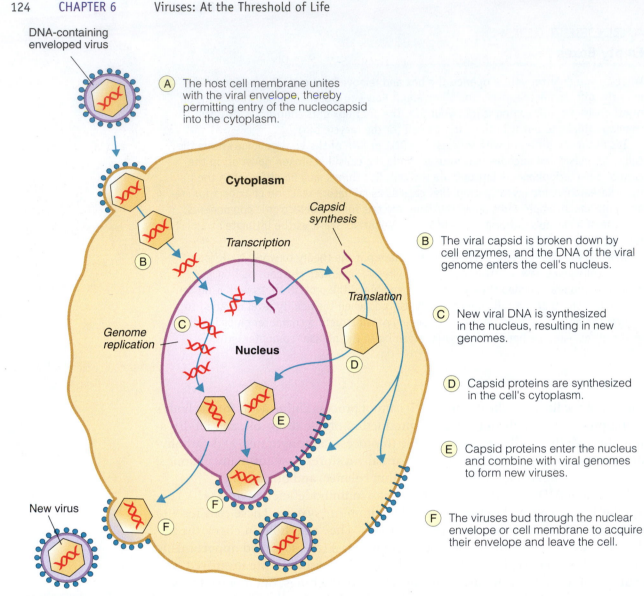

DNA-containing enveloped virus

Cytoplasm

Capsid synthesis

Transcription

Translation

Genome replication

Nucleus

New virus

A   The host cell membrane unites with the viral envelope, thereby permitting entry of the nucleocapsid into the cytoplasm.

B   The viral capsid is broken down by cell enzymes, and the DNA of the viral genome enters the cell's nucleus.

C   New viral DNA is synthesized in the nucleus, resulting in new genomes.

D   Capsid proteins are synthesized in the cell's cytoplasm.

E   Capsid proteins enter the nucleus and combine with viral genomes to form new viruses.

F   The viruses bud through the nuclear envelope or cell membrane to acquire their envelope and leave the cell.

FIGURE 6.7   Replication of a DNA Virus. The process illustrated here is for a herpes virus (such as one that might cause genital herpes).

are life-threatening). Thus, we must be able to defend ourselves from most viral diseases, but the question is how?

A major part of the answer is in the immune system, especially its antibodies. Antibodies are protein molecules synthesized by the immune system in response to the presence of viruses, as we examine in depth in Chapter 17. Molecular components of a virus, especially the capsid proteins, stimulate B cells of the immune sys-

1 During early July, a man spotted a bat in the living room of his apartment in New Jersey. The apartment was on the second floor of an old wooden house in poor repair. Above the apartment was the attic, also in disrepair. After some minutes, the man captured the bat by isolating it in a corner. He grabbed the bat with a cloth and carried it to the doorway, where he released it outside. This was the second time the man had captured a bat in the house.

2 On October 12, the man developed an aching sensation in his right shoulder and neck, and he became restless. The symptoms worsened, and the next day, with chills and a sore throat, he visited a hospital emergency room.

3 The man was given an anesthetic throat spray for the pain and antibiotics by oral administration. He was sent home with other recommendations on how to care for a respiratory ailment.

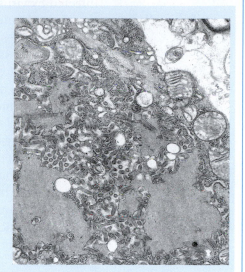

4 The symptoms became more acute, and the man was admitted to the hospital with hallucinations, high fever, and terrible muscle aches. Within days, he developed shock and kidney failure, and he died on October 23.

5 When public health officials visited the man's home, they discovered bat droppings in the attic and numerous openings to the outside. The man's tissues had revealed evidence of bat-associated rabies virus, and the observations confirmed the diagnosis of rabies.

FIGURE 6.8 **A Case of Rabies Occurring in Warren County, New Jersey, in October 1997.** This was the first case of human rabies in New Jersey since 1971.

A CLOSER LOOK
"Big Fleas Have Little Fleas"

For the person suffering the miseries of a bacterial disease, there may be some consolation in knowing that bacteria suffer from viral disease as much as people do—for bacteria are the prey of bacteriophages, ultramicroscopic viruses that replicate within and destroy bacterial cells. Without the host bacterium, the bacteriophage remains as inert as a grain of sand. Possibly this is the culmination of the eternal predator-prey relationship first enunciated by the English mathematician Augustus DeMorgan:

Big fleas have little fleas upon their backs to bite 'em.
And little fleas have lesser fleas, and so on ad infinitum.

tem, and these cells produce huge numbers of highly specific antibody molecules that bind to viral particles. The antibodies prevent the viruses from reaching their host cells by covering the attachment sites on the viral capsid or envelope. Obviously, if the viruses cannot replicate, the infection will come to an end. Another defense mechanism is the activity of T cells. These immune system cells attack cells that have been infected by viruses, because the cells display on their surfaces a set of viral-derived protein molecules. The T cells destroy the infected cells and the viruses within them.

Beyond these natural defenses, there are several drugs that can be used against viruses. One example is acyclovir (known commercially as Zovirax). This drug inhibits the replication of herpes simplex and chickenpox viruses by interfering with the replication of their DNA. Another useful drug is amantadine, a drug believed to prevent attachment of influenza viruses to their host cells in the respiratory tract. Still another drug is azidothymidine (AZT). Physicians use this drug to treat AIDS because it inhibits a key enzyme that functions during HIV's replication cycle.

Another potential antiviral drug still in development is interferon. This protein is actually a group of over 20 substances (interferons) that are produced by host cells when attacked by viruses. The interferons protect neighboring cells against viral penetration, and some interferons are active against cancer cells. The drug has been produced by genetic engineering techniques; scientists have already used it to treat certain viral diseases, and they hope that it will be a valuable antiviral medication for future generations.

Viral Vaccines

As we have noted, the primary defense against viruses is the production of antibodies by the body's immune system. It makes good sense, therefore, to have the antibodies available in the bloodstream and tissues even before the virus enters the body. This is the theory underlying vaccines. A person is exposed to chemically crippled viruses, and the immune system produces antibodies that circulate in the body. Should the viruses show up at a later time, the antibodies will immediately bind to them and prevent the disease from taking hold.

Various kinds of vaccines for viral diseases are available. In one kind, scientists chemically destroy the viral genome but leave the capsids intact. In scientific jargon, such viruses are said to be inactivated, meaning that they are unable to replicate in cells (some people refer to the viruses as "dead"). An example of a vaccine consisting of inactivated viruses is the Salk vaccine used against polio. Such vaccines must be given by injection to ensure a successful immune response.

Another kind of antiviral vaccine contains viruses that infect cells and replicate but at an extremely low rate, so low that disease symptoms do not develop. These viruses are obtained by cultivating viruses for many generations until a low-replicating strain appears. The viruses can be taken orally, and they stimulate the immune system for a long period of time, eliciting a stronger response than inactivated viruses. The viruses in these vaccines are sometimes called "live" viruses because they replicate, but the preferred adjective is attenuated. Indeed, the first vaccine was a live virus known as the vaccinia virus, from which the word vaccine is derived. The vaccinia virus causes a smallpox-like disease in cows and derives its name from Vacca, which is Latin for "cow." In 1796, Edward Jenner used inoculation with the vaccinia virus to protect people from being infected with small pox and thus the first vaccine was born. Unfortunately, vaccines consisting of live attenuated viruses carry an element of risk because the viruses may infect body cells. At this writing, attenuated virus vaccines are used to prevent chickenpox, measles, mumps, and rubella. In addition, the Sabin vaccine used against polio consists of attenuated viruses.

A third kind of viral vaccine is the genetically engineered vaccine. In this case, viral proteins are produced by yeast cells that have been altered to express the genes from a specific virus. The proteins are then concentrated, purified, and used in the vaccine. The highly successful vaccine against hepatitis B is an example. It carries a low risk of side effects because it contains no viruses or viral fragments. Biochemists are seeking to further improve the vaccine by splicing the viral genes into fruits such as bananas so that one can be immunized to hepatitis B simply by eating a banana. Chapter 14 explores still other innovative approaches to enhancing the public health with genetic engineering.

Vaccines can play a key role in preventing bioterrorism. For example, one of the most feared diseases is smallpox. For centuries, it exacted a heavy toll of death among Europeans and was rampant in North and Central America as well. The World Health Organization led the campaign to eradicate smallpox, and it achieved success in 1977. However, the eradication of this disease has left open the possibility that bioterrorists could spread a calamity of epic proportions by releasing smallpox viruses. Fortunately, the existence of an effective vaccine for smallpox is considered a deterrent to such bioterrorism.

When Viruses Don't Replicate

In some situations, a virus enters a host cell but does not replicate immediately. The scientific name for this phenomenon is lysogeny; the process is sometimes called the lysogenic cycle to distinguish it from the lytic cycle. During lysogeny, the virus incorporates its genes into the host cell's genes and becomes part of the cell. It then remains with the host cell and multiplies when the host cell multiplies. Such an indwelling virus is believed to be responsible for some types of cancer: For instance, scientists believe that in certain cases of leukemia, viruses have entered normal T cells of the immune system and transformed them into cancerous T cells. Moreover, liver cancers may be due to lysogenic hepatitis B viruses, and lysogenic papilloma viruses may be involved in cervical cancers. These types of cancer are relatively rare compared to cancers in which viruses are not involved.

A more common example of lysogeny occurs during HIV infection and AIDS. The human immunodeficiency virus has RNA in its genome. After the virus has entered a host T cell, the capsid is stripped away and the RNA is released. Next, a viral enzyme called reverse transcriptase uses the RNA as a template (a model) and synthesizes a complementary strand of DNA, as shown in **FIGURE 6.9** . (Incidentally, this is where AZT interferes with HIV replication.) The DNA then migrates to the cellular nucleus, enters it and integrates itself in to the cell's DNA. Such a fragment of viral-derived DNA is called a provirus. From its site within the nucleus, the provirus encodes all the proteins needed for HIV replication. A person having such an indwelling provirus is said to have HIV infection. The new viruses encoded by the provirus attack and seriously reduce the number of available T cells, and there soon comes a point at which the body cannot mount an effective response against normally harmless microbes. Chapter 18 explains the consequences of HIV infection in more depth.

In many plant cells, the lysogenic cycle predominates over the lytic cycle, and plant viruses do not cause any apparent harm to the cells. Indeed, the viruses are often responsible for remarkably beautiful variations in the plant's flowers. Petunias,

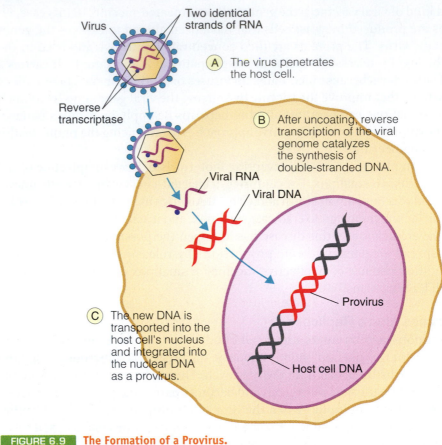

Virus

Two identical
strands of RNA

(A) The virus penetrates
the host cell.

Reverse
transcriptase

(B) After uncoating, reverse
transcription of the viral
genome catalyzes
the synthesis of
double-stranded DNA.

Viral RNA

Viral DNA

(C) The new DNA is
transported into the
host cell's nucleus
and integrated into
the nuclear DNA
as a provirus.

Provirus

Host cell DNA

FIGURE 6.9   **The Formation of a Provirus.**

carnations, and tulips, for example, owe their unusual patterns of pigmentation to proviruses in the cell. These plants usually grow more slowly than uninfected plants, but they show few other symptoms of viral presence. The viruses replicate only in certain cells of the plant and do not destroy these host cells.

Viral Cultivation

As the decades of the early 1900s unfolded, scientists developed relatively easy ways to cultivate bacteria, but viruses remained difficult to work with. Animal inoculations were used to increase the quantity of viruses and study their effects at the organ and tissue level, but observing the effects of viral infection at the cellular level was tedious.

Then, in the 1930s, scientists led by Claud Johnson and Ernest Goodpasture discovered that fertilized chicken eggs would support the replication of viruses. This was a key breakthrough because it made large populations of viruses available for study. In the next decade, biochemists found that live cells could be freed from their surrounding tissues, then washed, dispersed, and dispensed to tubes or flasks, where they remained alive as long as the scientists supplied the proper balance of nutrients for growth. The next step was the successful introduction of viruses to these so-called cell cultures. Vaccine production soon followed—the first vaccine for polio, developed in the 1950s by Jonas Salk and his coworkers, was a direct result of the technology of viral cultivation in cell cultures.

A second type of cell culture is the diploid fibroblast culture. Such a culture consists of immature cells called fibroblasts, which are derived from fetal tissues and retain the capacity for rapid and repeated cell division, a property not retained by mature cells. The third type of cell culture is the continuous cell line culture. The cells in this type of culture have been propagated for many generations and are relatively easy to cultivate because they have adapted to laboratory conditions. The three types of cell cultures have facilitated the development of virology.

We should note that cultivation of viruses is far more difficult than cultivation of bacteria. Viral cultivation requires an expertise that is not easily acquired and equipment that is highly sophisticated and expensive. Thus, most terrorist groups lack the resources to develop viruses as weapons.

Viruses as Research Tools

Although viruses are generally cast in a negative light as agents of cell destruction, these microbes are also sophisticated biochemical tools. Because it is relatively simple, the viral genome can be studied in depth and its activity can be systematically plotted. Molecular biochemists working with viruses as their tools have made significant breakthroughs. For instance, in the 1950s, scientists used bacteriophages and *E. coli* cells to discover the roles of ribosomes and messenger RNA molecules in the synthesis of proteins.

In modern biotechnology, viruses are used to carry genes into cells, where the genes are activated. For instance, contemporary research in cystic fibrosis uses weakened common cold viruses to carry normal genes into the respiratory tract cells of patients who have defective genes. (The viruses are shown in FIGURE 6.10.) The imported genes soon encode the proteins missing in the patient, and the symptoms of cystic fibrosis are diminished. Moreover, vaccinia (cowpox) viruses have been used to produce state-of-the-art vaccines, as we discuss in Chapter 14. These viruses can accommodate large stretches of foreign DNA. When they release their genome in the cytoplasm of host cells, it will express its own genes as well as those it is carrying. Such a vaccine, known as a viral-vector vaccine, has the advantage of reducing the need for multiple inoculations.

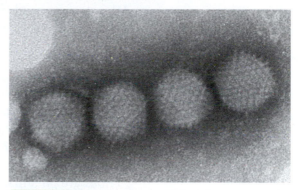

FIGURE 6.10   **Adenoviruses.** Adenoviruses are respiratory tract viruses that can be crippled by altering their genes. Then the viruses are engineered to carry human genes to relieve the symptoms of cystic fibrosis. When introduced to respiratory tract cells of the patient, they encode enzymes that the patient could not otherwise produce.

Agricultural biotechnologists have also used gene-altered viruses to develop disease resistance in plants. Tomato plants and tobacco plants, for example, suffer infection by viruses, and their leaves shrivel and display a mosaic pattern (as in tobacco mosaic disease). Researchers at Washington University have genetically transformed the cells of these plants by inserting the gene that encodes the viral capsid. The plant cells then produce viral capsids, and the plants display unusually high resistance to mosaic disease. Scientists remain uncertain as to how the resistance comes about, but it is possible that the viral protein blocks essential steps in the replication of the virus associated with the disease.

Viruses and Evolution

In the scheme of life on Earth, viruses appear to lie somewhere between the living and the nonliving, and it is tempting to view them as a sort of "missing link." Because viruses require living cells in which to replicate, it is unlikely that viruses preceded their host cells on Earth.

Many contemporary biologists believe that viruses are little more than "renegade genes"—that is, the DNA or RNA of viruses was once part of a cellular chromosome or messenger RNA molecule. These nucleic acid segments may have acquired the genetic information for self-replication using the biochemical machinery of a cell. With the passage of time, new genes for synthesizing capsid proteins may have been added to the viral genome, and the viruses may then have been freed from their host cells to lead a completely independent existence.

Although their own evolution is uncertain, viruses have played a key role in the evolution of other organisms. For instance, viral pathogens can act as selective agents to eliminate individuals or species having little resistance to their effects. In addition, viruses play important roles in gene transfers among bacteria (as we explore in Chapter 10). In this case, certain viruses carry genes between bacterial cells and increase their diversity, which allows for a better chance of survival. An example of this phenomenon occurs when viruses bring into bacterial cells the genes that foster antibiotic resistance. Some scientists also believe that viruses transfer genes among complex animal and plant cells and thereby increase diversity within the species. Increased diversity should eventually lead to a population of individuals better adapted to the environment.

Viroids and Prions

Although they are not viruses, viroids and prions merit brief attention here because these viruslike particles have an influence on society as agents of illness. Viroids are ultramicroscopic, single-stranded molecules of RNA without any protein coat. They infect plants, resulting in stunted growth and abnormal development. And their effects can be devastating: Viroids are responsible for a disease in coconut palms that has nearly destroyed the entire palm population of the Philippines.

Prions are *pro*teinaceous *in*fectious particles thought to cause a number of diseases, including kuru and mad cow disease (Chapter 18). Prions were named in 1982 by Stanley B. Prusiner, who was among the first to postulate their existence and who won the 1997 Nobel Prize in Physiology or Medicine (**FIGURE 6.11**). Researchers have shown that prions can survive the heat, radiation, and chemical treatment that normally inactivate viruses. Moreover, prions appear to be composed solely of protein: They are susceptible to some protein-digesting enzymes but are not altered by enzymes that break down nucleic acids. The absence of nucleic acids raises the question of how prions replicate, because one of the central dogmas in biology is that inheritance operates through DNA and RNA.

In the 1990s, investigators discovered that prions are apparently deviant versions of a harmless protein found on the membrane surfaces of most mammalian cells, particularly brain cells. This protein, called prion protein (PrP), protects the cells by helping rid their environment of dangerous biochemicals. Sometimes a transformation occurs, and the PrP folds into a different form, thereby becoming a prion. The pleated sheets of the misshapen proteins tend to clump together, and the prions block

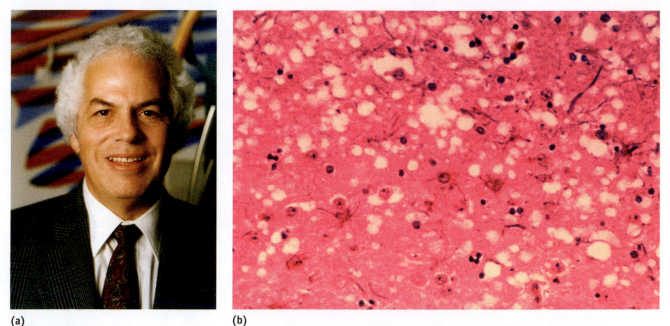

(a)   (b)

FIGURE 6.11 **Prion Research and Disease.** (**a**) Stanley Prusiner, winner of the 1997 Nobel Prize in Physiology or Medicine for his work on prions as agents of infectious disease. The award surprised the scientific community because the Nobel Prize is generally awarded to two or more individuals in each category, and because the work on prions remains somewhat speculative and incomplete. (**b**) A photomicrograph showing the vacuolar degeneration of gray matter characteristic of human and animal prion diseases. Mad cow disease is believed to be a type of spongiform encephalopathy.

the molecular traffic into the cells, eventually killing them. Brain tissues develop a spongelike appearance with empty areas of dead tissue. These are the so-called spongiform encephalopathies. Bovine spongiform encephalopathy, or mad cow disease, is one example of this kind of prion-related condition.

Many questions remain: For example, researchers have not yet confirmed the normal function of PrP; nor have they established how the transformation from PrP to prion takes place, how prions move from host to host, and how they bring about disease symptoms. Furthermore, synthetic prions (free of any possible contaminating substance) have not yet been shown capable of causing disease. More research is needed to clarify these issues.

Viruses and Cancer

Cancer is indiscriminate. It affects humans and animals, young and old, male and female, rich and poor. In the United States, over 450,000 people die of cancer annually, making the disease the second most common cause of death among Americans, after cardiovascular disease. Worldwide, over 2 million people die of cancer each year.

The Development of Cancer

Cancer results from the uncontrolled reproduction of cells through the process of mitosis: The frequency of mitosis is greater for cancer cells than for normal cells. The cells escape controlling factors and, as they continue to multiply, form an enlarging cluster. Eventually, the cluster grows into an abnormal, functionless mass of cells. This mass is called a tumor.

Normally, the body will respond to a tumor by surrounding it with a capsule of connective tissue. Such an encapsulated tumor is designated benign. If, however, the cells multiply too rapidly and break out of the capsule and spread, the tumor is described as malignant. The individual now has cancer, a reference to the radiating spread of cells that resemble a crab (the word "cancer" is derived from the Greek *karkinos*, meaning "crab"). The term oncology, designating the study of cancer, is derived from *onkos*, the Greek word for "tumor."

Cancer cells differ from normal cells in three major ways: They grow and undergo mitosis more frequently than normal cells, they stick together less firmly than normal cells, and they undergo dedifferentiation. Dedifferentiation means that the cells revert to an early stage in their development, often becoming formless cells that divide as rapidly as early embryonic cells. Moreover, cancer cells fail to exhibit contact inhibition; that is, they do not stop growing when they contact one another, as normal cells do. Rather, they overgrow one another to form a tumor. Sometimes, they metastasize, or spread, to other body locations where new tumors begin. Also, they invade and grow in a broad variety of body tissues. There appears to be no boundary limiting the growth of a tumor.

How can a mass of cells bring illness to the body? By their sheer force of numbers, cancer cells invade and erode local tissues, thereby interrupting normal functions and choking organs to death. For example, a tumor in the kidney blocks the tubules and prevents the flow of urine during excretory function, a brain tumor cripples this organ by compressing the nerves and interfering with nerve impulse transmission, and a tumor in the bone marrow disrupts blood cell production. In addition, tumor cells rob the body's normal cells of vital nutrients to satisfy their own metabolic needs. Moreover, some tumor cells are known to produce hormones identical to those normally produced by the body's endocrine glands, thereby overloading the body with chemical regulators. Some tumors block air passageways; others interfere with immune system functions so that microbial diseases take hold. Ultimately, tumor cells weaken the body until it fails.

The Involvement of Viruses

Scientists are uncertain as to what triggers a normal cell to multiply without control. However, they know that certain chemicals are carcinogens, or cancer-causing substances. The World Health Organization (WHO) estimates that carcinogens may be associated with 60 to 90 percent of all human cancers. Among the known carcinogens are the hydrocarbons found in cigarette smoke, as well as asbestos, nickel, certain pesticides, and environmental pollutants in high amounts. Physical agents such as ultraviolet radiation and X rays are also believed to be carcinogenic.

There is considerable evidence that viruses are also carcinogens. Experiments with animals have indicated that some viruses can induce tumors. Although federal law and ethical considerations prohibit these experiments from being repeated with human volunteers, a number of viruses have been isolated from human cancers and transferred to animals and tissue cultures, where an observable transformation of normal cells to tumor cells takes place. Examples of such viruses are the herpesviruses associated with tumors of the human cervix and the Epstein-Barr virus, which is linked to Burkitt's lymphoma, a tumor of the jaw.

One of the clearest virus-cancer links emerged in 1980, when a research team led by Robert T. Gallo of the National Cancer Institute isolated a virus that transforms normal T cells into the malignant T cells found in a rare cancer called T cell leukemia. Evidence has recently demonstrated that the so-called human T cell leukemia virus (HTLV) can also cause neurological pain disorder and a condition marked by destruction of the sheaths that surround the nerve fibers. The virus appears to be bloodborne, and the primary mode of transport is through the intravenous use of illegal drugs.

How Viruses Transform Cells

The mechanism by which viruses and other carcinogens transform normal cells into tumor cells remained obscure until the oncogene theory was developed in the 1970s. This theory suggests that the transforming genes, referred to as oncogenes, normally reside in the chromosomal DNA of cells. Researchers J. Michael Bishop and Harold Varmus located oncogenes in a wide variety of creatures, from fruit flies to humans. Bishop and Varmus also made the astonishing discovery that very similar oncogenes exist in some viruses, and they hypothesized that these genes could have been captured by the viruses. It appeared that the oncogenes were not viral in origin but part of the genetic endowment of every living cell. So far, over 60 different oncogenes have been identified.

In recent years, the theory of oncogene activity has been revised slightly. Researchers now suspect that cells have a set of normal genes, called proto-oncogenes, which are the forerunners of oncogenes. In the cells, proto-oncogenes may have important functions as regulators of growth and mitosis. Scientists now know that proto-oncogenes can be converted to oncogenes by carcinogens, such as viruses, radiation, or chemicals; then tumor formation begins (FIGURE 6.12 illustrates the process). Indeed, the oncogene related to bladder cancer differs from its counterpart proto-oncogene by only 1 nucleotide out of 6000.

But how do viruses convert proto-oncogenes to oncogenes? Virologists have observed that when a virus enters a cell, it may begin a lysogenic cycle within the cell and thereby transform it. If the viral genome is composed of DNA, for example, the viral DNA may integrate itself directly into the cell chromosome and become a provirus (as we noted previously). If the viral genome is composed of RNA, an enzyme synthesizes DNA using the RNA as a template. The DNA is then inserted into the cell chromosome as a provirus.

Once the proto-oncogene becomes an oncogene, it can influence cellular growth and mitosis in several ways. Oncogenes, for example, may provide the genetic codes for growth factors that stimulate uncontrolled cell development and reproduction. Or the oncogenes may be incapable of encoding the substances that turn off cell growth, a function their proto-oncogenes could carry out. During the 1980s, a team led by Michael Wigler conducted a particularly interesting series of studies at Cold Spring Harbor Laboratories and isolated a protein whose production was encoded by an oncogene. The protein was injected into normal cells, and within hours, the normal cells began to multiply rapidly and show unmistakable signs of conversion to cancer cells.

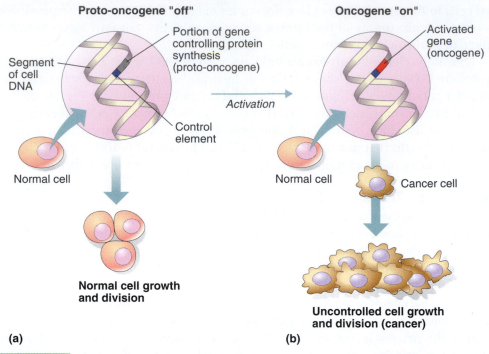

FIGURE 6.12 **The Oncogene Theory.** The oncogene theory helps explain the process of cancer development. (**a**) The normal cell grows and divides without complications. Within its DNA, it contains proto-oncogenes that are "turned off." When the genes are activated by viruses or other factors, they revert to oncogenes, which "turn on." (**b**) An abnormal cancer cell results. The oncogenes encode proteins that regulate the transformation from a normal cell to a cancer cell and the development of a tumor.

In addition to the Epstein-Barr virus and HTLV-1, some strains of the human papilloma viruses cause cervical and other forms of cancer. In the spring of 2006, the Food and Drug Administration approved a vaccine for the oncogenic strains of the human papilloma viruses, which represents the first vaccine for cancer in humans and brings hope that the rate of cervical cancer in women will be drastically diminished in the future.

■ A FINAL THOUGHT

When I was in college, I was fascinated by viruses. During my senior year, I wrote a paper with the ominous title "Viruses—Alive or Dead?" I spent hours in the library hunting for studies by researchers willing to take a stand either way, and I compiled the thoughts of six investigators. I wrote the paper and proudly presented it to Dr. Marruzella, my professor. Some days later, I approached him and asked what he thought about the issue. To my dismay, he shrugged his shoulders and said "Ed, what difference does it make? We treat viral diseases the same way regardless of whether their causes are dead or alive."

But the question has continued to bother me to this day, and now I put it to you. Are viruses alive? At present, the tendency of many biologists is to sidestep the question.

However, I shall make a suggestion. Although this book refers to viruses as microbes for the sake of convenience, I have avoided references to "live" or "dead" viruses. Instead, I prefer to use the words "active" for replicating viruses and "inactive" for viruses unable to replicate. I suggest, therefore, that we consider viruses to be inert chemical molecules with at least one property of living things—the ability to replicate. Thus, viruses are neither totally inert nor totally alive, but somewhere on the threshold in between.

Incidentally, Dr. Marruzella did point out one mistake I made"Your title is wrong," he said. "The alternative is not alive or dead, but living or nonliving. If they are dead, it means they were once alive, no?"

QUESTIONS TO CONSIDER

1. A textbook author referring to viruses once wrote: "Certain organisms seem to exist only to reproduce, and much of their activity and behavior is directed toward the goal of successful reproduction." Would you agree with this statement? Can you think of any creatures other than viruses that fit the description?

2. Oncogenes have been described in the literature as "Jekyll and Hyde genes." What factors may have led to this label, and what does it imply? In your view, is the name justified?

3. Wendell M. Stanley's 1935 announcement that viruses could be crystallized stirred considerable debate about the living nature of viruses. Suppose a virus were discovered that contained both DNA and RNA. Might this stir an equal amount of controversy on the nature of viruses? Why?

4. Researchers studying the bacteria that live in the oceans have long been troubled by the question of why bacteria have not saturated the oceanic environments. What might be a reason?

5. In broad terms, the public health approach to dealing with bacterial diseases is treatment. Can you guess the nature of the general public health approach to viral diseases? What evidence do you have to support your answer?

6. Bacteria can cause disease by using their toxins to interfere with important body processes, by overcoming body defenses (such as phagocytosis), by using their enzymes to digest tissue cells, or by other similar mechanisms. Most viruses, by contrast, do not encode toxins, cannot overcome body defenses, and produce no digestive enzymes. How, then, do viruses cause disease?

7. When Ebola fever broke out in Angola, Africa in 2005, the death toll was high, but the epidemic was short-lived. By comparison, when influenza breaks out at the start of winter, the toll is low, but the epidemic lasts for 6 or more months. From the standpoint of the viruses, what dynamics do you see in these two types of epidemics?

◼ KEY TERMS

Informative facts are necessary for the expression of every concept, and the information for a concept is founded in a set of key terms. The following terms form the basis for the concepts of this chapter. On completing the chapter, you should be able to explain and/or define each one:

acyclovir	interferon
amantadine	lysogenic cycle
antibodies	lysogeny
attenuated virus	lytic cycle
azidothymidine	nucleocapsid
cancer	oncogene
capsid	oncology
capsomeres	prion
carcinogen	proto-oncogene
cell culture	provirus
dedifferentiation	receptor site
genome	spike
helix	viroid
icosahedron	virus
inactivated virus	

◼ http://microbiology.jbpub.com/book/microbes

The site features **eLearning,** an online review area that provides quizzes and other tools to help you study for your class. You can also follow useful links for in-depth information, read more stories of microbiology, or just find out the latest microbiology news.

Protists: A Microbial Grab Bag

7

Looking Ahead

The protists are a group of eukaryotic microbes that include the protozoa, certain types of algae, and other organisms. This chapter surveys the significance of these microbes and describes some of the complex structures they possess.

On completing this chapter, you should be able to . . .

- explain some general features of protists, including their structures and physiological processes.
- appreciate the characteristics of four groups of protozoa, and explain how these microbes influence society.
- name and briefly describe several protozoal diseases that affect humans.
- describe the complex patterns of reproduction displayed by various protists and compare them to patterns observed in other microbes.
- list the key features of several groups of single-celled (unicellular) algae.
- conceptualize the important roles played by unicellular algae in the oceans of the world.
- make some generalizations about the two major types of slime molds and understand why they are important as research tools.

While we can point to several historical events that mark the beginning of Microbiology as a science, the old saying: "Seeing is Believing" perhaps best describes the most significant. In 1674, Anton Van Leeuwenhoek was the first to observe microbial life when he examined a drop of pond water through his microscope. Among several forms of life he documented the first description of protists. The inset shows the protists *Euglena* and *Phacus*, from a drop of pond water.

April 12, 1993 should have been a festive day in Milwaukee, Wisconsin. The first home game of the baseball season was scheduled, and fans were eager to see the Brewers play the California Angels. But the scoreboard contained an ominous message: or "For your safety, no city of Milwaukee water is being used in any concession item." The city was in the throes of an epidemic, and a protist was to blame.

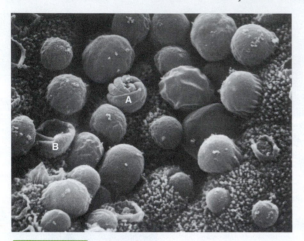

FIGURE 7.1   A Waterborne Protozoan. A scanning electron micrograph of *Cryptosporidium coccidi* at the surface of intestinal tissue. The parasites are globular saclike bodies 2 to 6 µm in diameter. Within the globes are numerous long, thin forms of the parasite, called merozoites. After release from the globes, the merozoites will infect nearby cells. One parasite (A) has lost its membrane, and the merozoites can be seen crowded together. The craterlike structures (B) are parasites from which the merozoites have already been released.

The protist in this instance was the protozoan *Cryptosporidium coccidi*, pictured in **FIGURE 7.1**. *C. coccidi* is an intestinal parasite that causes mild to serious diarrhea, especially in infants and elders. As the protozoa attach themselves to the intestinal lining, they mature, reproduce, and encourage the body to release large volumes of fluid. The infection is accompanied by abdominal cramps, extensive water loss, and, in many cases, vomiting and fever.

Even as the first ball was being thrown out at the stadium, health inspectors were analyzing Milwaukee's water purification plants to see how a protozoan could have reached the city's water supply. *Cryptosporidium* is a waterborne parasite commonly found in the intestines of cows and other animals. Perhaps, they guessed, the heavy rain and spring thaw had washed the protozoan from farm pastures and barns into the Milwaukee River. The river might have brought *Cryptosporidium* into Lake Michigan, from which the city drew its water. Indeed, the inspectors noted how close the mouth of the river was to the intake pipe from the lake. Moreover, *Cryptosporidium* can resist the chlorine treatment used to control bacteria in water, and the tests for bacterial contamination do not detect protozoa, such as *Cryptosporidium*.

As researchers worked to unravel the mystery, the baseball game went on. Soda was available, but only from bottles. Drinking fountains were turned off. Two huge U.S. Army tank trucks stood by to provide reserve water for the 50,000 fans in attendance. And in the city, tens of thousands of Milwaukeeans made the mildly embarrassing trip to the drugstore to stock up on toilet paper and antidiarrheal medications. (A large window sign at a local Walgreen's proudly proclaimed: "We have Imodium A-D.") Back at the ballgame, things were not going much better for the locals—the Brewers lost to the Angels 12 to 5.

■ *Cryptosporidium coccidi*
Krip′tō-spô-ri-dē-um
kok′sid-ē

Cryptosporidium coccidi is one of the protozoa and other protists we shall encounter as we continue to explore how microbes affect society. The protists are a mixed group of microbes sometimes considered "taxonomic misfits." Although they are quite distinct from bacteria and fungi, many protists resemble members of other kingdoms of eukaryotes. More or less by default, three major microbial groups have been assigned to Whittaker's Protista kingdom even though they do not share broad taxonomic similarities. The three groups we shall consider are the protozoa, the unicellular (single-celled) algae, and the slime molds.

Over 100,000 species of protists are currently known to exist and another 35,000 species are recognized by their fossils. The protists are exceeded only by the bacteria in the number of environments to which they have adapted. They are a grab bag of disparate creatures lying in the classification gap between prokaryotes and the smallest animals and plants. Indeed, the word "protist" comes from Greek stems meaning "very first," and the protists are considered to be the first eukaryotes to appear on Earth. It is reasonably certain that modern protists evolved from the earliest eukaryotes, which apparently emerged from one or more lines of prokaryotes. Evolutionary biologists estimate that eukaryotic organisms arose approximately 2.5 billion years ago, approximately 1 billion years after the first prokaryotes appeared on Earth.

Although all protists are unicellular organisms, some exist as colonies of cells and are referred to as colonial organisms. While most species of protists are microscopic, some species have diameters close to a millimeter. As we shall see, protists practice all known modes of feeding: Some capture their food by phagocytosis, while others absorb simple organic nutrients across their plasma membranes. Still others are parasites that cause human and animal disease. Many protists are photosynthetic—they use their chlorophyll pigments to trap the sun's energy and form carbohydrates such as glucose. A number of species display movement mediated by cilia, flagella, or cytoplasmic extensions called pseudopodia. Certain protists are funguslike, others are plantlike, and still others are animal-like. Indeed, there are certain species of protists that combine the characteristics of plants and animals, and other species that combine the characteristics of fungi and animals. **FIGURE 7.2** displays the broad variety in this group of microbes.

Asexual reproduction by mitosis is the most common means of population increase among the protists. But sexual reproduction also occurs—it is among the evolutionary traits first appearing in the protists. Sexual exchanges vary greatly: Some

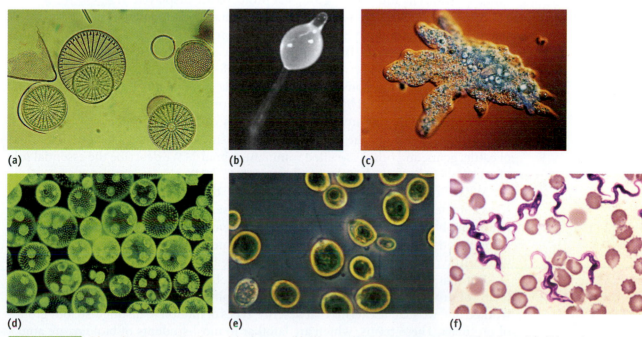

(a) (b) (c)
(d) (e) (f)

FIGURE 7.2 **Diversity Among Protists.** (**a**) Diatoms. (**b**) Slime mold. (**c**) *Amoeba proteus*. (**d**) Volvox. (**e**) *Chlamydomonas*. (**f**) *Trypanosoma*, among red blood cells.

species fuse their entire cells, and other species come together to exchange only their nuclei. Some species of protists show the pattern of sex cell (gamete) formation followed by fertilization that occurs in animals.

A major evolutionary advance of the protists is the development of eukaryotic cellular features, such as nuclei, mitochondria, chloroplasts, and other membranous organelles discussed in Chapter 2. Most researchers believe that the three kingdoms of complex eukaryotes (Fungi, Plantae, and Animalia) evolved from ancestral protists having these cellular features. The Whittaker system of classification is based on this supposition.

The Protozoa

The protozoa are so named because they were once believed to be the first animals (*proto* means "first," and *zoa* is derived from the word *zoan*, for "animal"). The protozoa lack cell walls, ingest food particles, generally move about freely, and produce no sporebearing structures. Traits such as these distinguish the protozoa from other protists such as the unicellular algae and the slime molds.

Approximately 40,000 species of protozoa have been named; most are found in aquatic environments, in moist soil, or as parasites. Under ideal conditions, the protozoa exist as active feeding forms called trophozoites. As conditions become difficult, however, the cells of some species transform into protective bodies known as cysts and withstand adverse environmental conditions in this form. Because of this high resistance, cysts have the potential to be used as weapons of bioterrorism (although their resistance is inferior to that displayed by bacterial spores such as anthrax spores). The four major groups of protozoa are displayed in FIGURE 7.3 .

As essential participants in the cycles of elements taking place in the soil, protozoa act with other microbes to decompose the remains of dead animals and plants and recycle the nutrients. In addition, many species of protozoa are part of the zooplankton, the animal-like component of aquatic food chains. Feeding on microscopic algae, the zooplankton convert the algal cellular components into nutrients, which are absorbed and digested by other "consumer" organisms such as sponges, jellyfish, worms, and tiny marine invertebrates (animals without backbones). And on land, other protozoa perform the same nutrient-releasing functions in the digestive tracts of cattle, goats, and other so-called ruminant animals. (We describe the essential role of microbes in beef production in Chapter 15.)

When Robert Whittaker assigned protozoa to the kingdom Protista in 1969, he did so as a matter of convenience. However, new research by Carl Woese and other contemporary biologists has shed light on the biochemical and evolutionary relationships among protists, and it is conceivable that the protozoa may eventually be placed elsewhere. In Woese's three-domain system, protozoa are considered Eukarya (Chapter 2).

But that does not resolve the immediate problem of how to classify the various forms of protozoa. Therefore, while the fine points of classification continue to be worked out, we shall use terms that describe types of motion to distinguish four groups of protozoa. These terms, which are familiar to most students of biology, are amoebas (protozoa that move by pseudopodia), flagellates (those that move by flagella), ciliates (those that move by cilia), and sporozoa (those that are nonmotile in the adult form). Discussions of each of these groups follow.

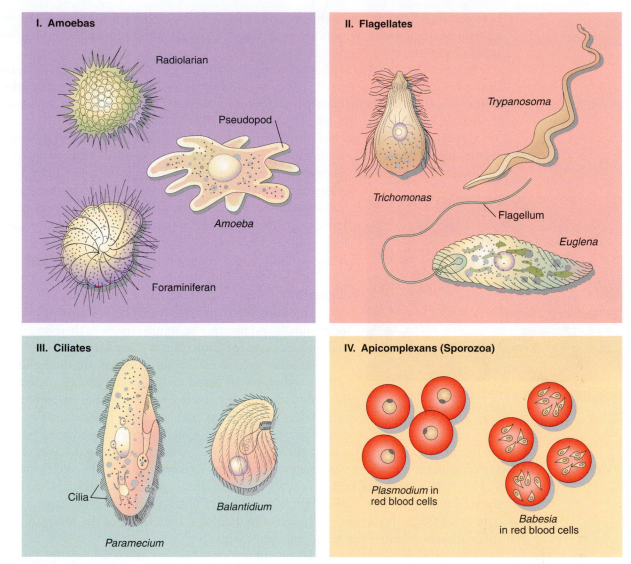

FIGURE 7.3 **The Four Major Groups of Protozoa.** The organelles of motion are shown for members of each of the four groups.

Amoebas

Amoebas are renowned for their plasticity: They have no definite form and are constantly changing shape by sending out pseudopodia (sing., pseudopodium, which translates to "false foot"). Pseudopodia are temporary cellular extensions used to move and feed in any direction. The movement of the pseudopodia occurs as alternating liquid and gel states form within the cell. Amoebas are also known as sarcodines (from the Greek *sarkodes* for "fleshlike substance," a reference to their lack of definite form).

Amoebas have often been portrayed in popular writing as blobs of cytoplasm, the simplest form of "animal" life imaginable. Despite this impression, the amoeba is probably not a primal organism. Compelling evidence shows that the amoeba's simplicity is a product of evolution. Members of the group probably originated from ancestral protozoal forms, probably flagellates, because certain forms intermediate between flagellates and amoebas continue to exist. Notable in this group is the organism *Mastigamoeba*, which has both flagella and pseudopodia and exhibits two types of motion.

None of the amoebas is photosynthetic, so all species must obtain their nutrients from preformed organic matter, the heterotrophic mode of nutrition. Reproduction in the amoebas occurs primarily through mitosis. There is also evidence for sexual reproduction in certain species.

Certain species of amoebas form hardened, shell-like casings called tests. For example, the protozoa called foraminifera produce elaborate chalky shells of calcium carbonate having numerous pores through which the pseudopodia move in and out as the organism feeds (*foramen* is a Latin term meaning a "passageway"). The remains of foraminifera build up as dense deposits on the ocean floor, and in some areas, geologic upthrust has brought these deposits to the surface. The White Cliffs of Dover in England, shown in FIGURE 7.4 , are a well-known example. Moreover, foraminifera flourished during the Paleozoic era, about 225 million years ago, when many oil fields were forming from the remains of prehistoric animals and plants. Foraminiferan shells therefore serve as depth markers for geologists drilling for oil.

Two other groups of shelled amoebas are the heliozoa and the radiolaria. Heliozoa are freshwater amoebas encased in hardened capsules of silicon dioxide that resemble the sun (*helios* is Greek for "sun"). From these capsules, the amoebas extend their long, slender pseudopodia. Radiolaria are found exclusively in the sea, where they live within finely sculptured glassy skeletons of silicon dioxide reminiscent of Christmas tree ornaments. The skeletons are as varied as snowflakes and have elaborate radiating, geometrical designs, making them a favorite subject of photographers. Some radiolaria are among the largest protists, with skeletons several millimeters in diameter. Like those of the foraminifera, these skeletons contribute to oceanic sediments.

FIGURE 7.4   **The White Cliffs of Dover, England.**
These cliffs are composed of the remains of foraminifera that thrived in the oceans millions of years ago.

Few pathogenic species are found among amoebas, but one acknowledged infectious agent is *Entamoeba histolytica,* the cause of amoebiasis. Transmitted by contaminated food and water, this organism is responsible for intestinal ulcers and sharp appendicitislike pain. Another health problem is caused by a species of *Acanthamoeba*. These protozoa infect the cornea of the eye in people who wear contact lenses; therefore, recommended care procedures should be followed closely. Evidence is also accumulating that amoebas are found in home humidifiers. When aerated and breathed into the respiratory tract, they may cause an allergic reaction called humidifier fever.

■ *Entamoeba histolytica*
en-tä-mē′bä
his-to-li′ti-kä

■ *Acanthamoeba*
a-kan-thä-mē′bä

Flagellates

All flagellates have flagella, the long, hairlike organelles that permit independent motion. As in other eukaryotic cells (such as sperm cells), flagella of protozoa have the 9 + 2 arrangement of ultramicroscopic protein fibers called microtubules. Nine pairs of microtubules are arranged in a ring, and two microtubules lie at the center of the ring. This similarity in arrangement provides further evidence for the evolutionary relationship between unicellular and multicellular organisms. The flagella whip about (the Latin *flagellum* translates to "whip"), and the protozoan moves.

In one sense, the flagellates can be regarded as the most fundamental protists: If all the other organisms on Earth were to disappear, many flagellate species could survive as long as a supply of nitrogen existed in some organic form (there are no nitrogen-fixing microbes in the group). Protozoal flagellates have the combination of chemical talents and diversity of adaptive types that could ensure survival to serve as the basic stock for future evolution.

How flagellates and other eukaryotes came into being is a puzzle that has received much attention in contemporary biology. The prevailing wisdom points to a theory referred to as the endosymbiosis hypothesis. Espoused by Lynn Margulis and her colleagues at the University of Massachusetts, the hypothesis states that primitive eukaryotic cells descended from prokaryotic cells (i.e., bacteria) that had established mutually beneficial relationships with other, smaller prokaryotic cells. Such a relationship may have begun when a primitive prokaryote developed the ability to move its plasma membrane inward and engulf food particles, thereby forming food vacuoles and other internal membranous structures. The membrane may have eventually surrounded the hereditary material to form a nucleus and result in the first eukaryote. This eukaryote may also have been the first predator because it could bring materials into its cytoplasm.

But the process was not yet complete. It is conceivable that the primitive eukaryote may have engulfed a bacterium. Rather than breaking down the bacterium into bits, the eukaryote may have retained the bacterium in its cytoplasm and ceded to this microbe the responsibility for energy metabolism. Thus, the engulfed bacterium may have eventually become a mitochondrion. Evidence for this hypothesis derives from the discoveries that the mitochondria of eukaryotic cells have ribosomes that are similar to bacterial ribosomes, that mitochondria have their own set of transfer RNA molecules used in protein synthesis (Chapter 12), and that mitochondria have a closed loop chromosome similar to that found in a bacterium. Of course, this is all speculation, but the biochemical and physiological evidence lends credence to the hypothesis. **FIGURE 7.5** illustrates the hypothesis.

Carrying the endosymbiosis hypothesis further, the flagella of protists may have evolved from spiral bacteria, which are long, thin, and highly motile. It is possible that a spirochete (a spiral bacterium) attached to the outer surface of the primitive eukaryote, and, over great expanses of time, its descendents evolved to a flagellum; later, the flagellar proteins may have been used to synthesize cilia and other cellular fibers. Moreover, chloroplasts may have once been cyanobacteria that were engulfed and remained in the cytoplasm and permitted the eukaryotic cells to participate in photosynthesis.

An impressive array of flagellates live within animals. In the gut of a wood-eating termite, for example, flagellates of the genus *Trichonympha* break down the cellulose in wood and release the glucose for use by the termite. This symbiotic relationship benefits both the termite and protozoan (but is of scant interest to the homeowner who must repair the structural damage caused by termites). The large intestine of humans is another dwelling place for extensive populations of flagellates.

Flagellates also span the gap between single-celled and colonial organisms. For example, colonies of pigmented cells in microbes belonging to the genus *Volvox* dis-

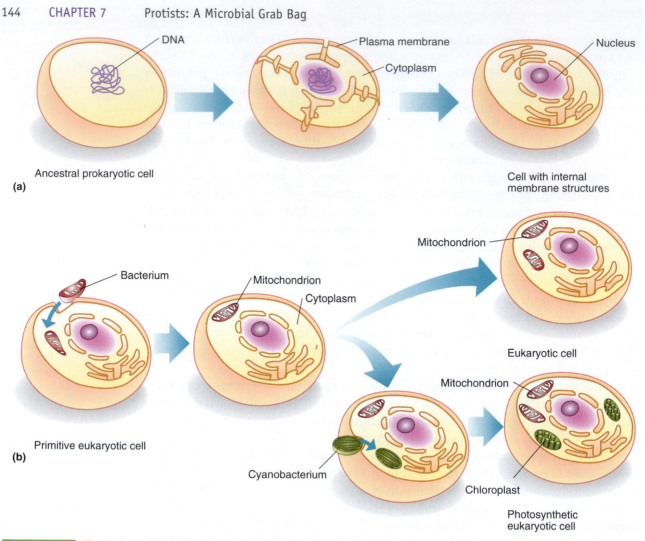

FIGURE 7.5   **The Endosymbiosis Hypothesis.** The endosymbiosis hypothesis presents a possible mechanism for the development of several structures seen in eukaryotic cells. **(a)** Internal membranes and membranous structures may have evolved when an ancestral prokaryote developed the ability to move its plasma membrane inward. **(b)** The mitochondrion and chloroplast may have originally been a bacterium and a cyanobacterium, respectively.

play a preliminary step in the development of multicellular organisms. Scientists have found that certain cells in the colony are adapted for feeding, while others specialize in reproduction, as they might in a multicellular species of organism. Other colonial forms are found in the genera *Gonium* and *Pandorina*; these microbes are also regarded as algae because of their photosynthetic pigments.

■ *Trypanosoma cruzi*
tri-pa′nō-sō-mä krüz′ē

Among the flagellated pathogens are *Trypanosoma brucei* and *Trypanosoma cruzi*, the causes of African and South American sleeping sickness, respectively. Trypanosomes are transmitted by tsetse flies (for *T. brucei*) or cockroachlike triatomid bugs (for *T. cruzi*). When introduced into the bloodstream of a human, the protozoa invade the brain tissue, where they cause a comalike condition. South American sleeping sickness is also accompanied by severe heart infection. Charles Darwin, renowned for his theory of evolution, is believed to have acquired this disease during his stopoff in South America on his voyage on the H.M.S. Beagle. A Closer Look (on page 145) explains a modern approach to dealing with the disease.

A CLOSER LOOK
Proud to Serve

Dealing with South American sleeping sickness has always been difficult: Health-care workers have used a variety of drugs to kill the protozoal parasite; they have plastered the walls of native homes to try and eliminate the triatomid bugs that spread the disease; and they have used insecticides of various formulations to kill the bugs. Still the epidemic continues to rage.

Now, genetic engineers have entered the picture. Yale University researchers have isolated *Rhodococcus rhodnii*, a bacterium living in the gut of the triatomid bug. Within the bug, the bacterial rod lives a peaceful, unassuming existence, deriving nutrients from the insect and contributing a natural defense against invading bacteria. The symbiotic relationship is an excellent example of mutualism in nature.

Researchers have cultivated a strain of *R. rhodnii* and have successfully altered its genome so that it produces a powerful drug called cecropin A. This protein compound is extremely active against *Trypanosoma cruzi*, the cause of South American sleeping sickness. When produced in the insect's gut, the drug kills the parasite and thereby interrupts its transmission to humans. In the next step, researchers hope to introduce to the environment a number of triatomid bugs with this built-in medical arsenal. In the best-case scenario, the new bugs will replace the natural ones and bring the drug to where it is needed most: deep inside the insect's gut. The bugs will survive, the humans will survive, and the most dangerous member of the triumvirate, the protozoal parasites, will be eliminated.

Giardia lamblia, the bane of campers, hikers, and backpackers, is also a flagellate. This protozoal flagellate is contracted from contaminated water, especially in mountain streams and lakes. It causes nausea, gastric cramps, and a foul-smelling watery diarrhea. The disease affects wild animals, and they may be the source of water contamination. Anton van Leeuwenhoek, among the first to visualize the microbial world, is believed to have seen and described Giardia from his own stools.

■ *Giardia lamblia*
jē-är′dē-ä lam′lē-ä

Trichomonas vaginalis is the protozoan that causes trichomoniasis ("trich"). Trichomoniasis is a sexually transmitted disease that affects over 2 million Americans annually. Patients suffer intense itching, burning pain, and a frothy discharge from the reproductive tract. The disease is more common in women than men, and it can be treated with antibiotics such as metronidazole and miconazole. The final flagellate we consider is *Leishmania tropica*, which is transmitted by the sandfly. This parasite caused illness in military personnel during the Persian Gulf War, as **FIGURE 7.6** chronicles.

■ *Trichomonas vaginalis*
trik-o-MONE-as
vaj-in-AL-is

■ *Leishmania*
leesh-MA-nee-ah

Ciliates

The ciliates are an extremely diverse group of heterotrophic protozoa ranging in size from a microscopic 10 μm to a huge 3 mm (about the same relative difference as between a football and a football field). Their cells have a greater variety of specialized compartments than most other protists, and they have hairlike cilia with the same 9 + 2 arrangement as the flagella of flagellates.

The movement of the cilia is coordinated by a primitive nerve network running beneath the surface of the cell. The cilia sway in a synchronized fashion much as a field of wheat bends in the breeze or the teeth of a comb bend as you pass your finger down

1 During 1991, approximately 500,000 military personnel took part in Operation Desert Storm in Saudi Arabia, Kuwait, and other countries of the Persian Gulf region.

2 While stationed in the Middle East during and after the fighting, many individuals were subjected to the bites of sandflies, the arthropods that transfer the protozoan *Leishmania tropica*.

3 On returning to the United States after several months' duty, seven men displayed the symptoms of leishmaniasis, including high fever, chills, malaise, liver and spleen involvement, and gastrointestinal distress. Several had low-volume watery stools and abdominal pain.

4 When bone marrow samples from the seven patients were examined, the tissue yielded evidence of *L. tropica,* the agent of leishmaniasis. The men were treated over a period of weeks with an antimony compound called sodium stibogluconate. All recovered.

FIGURE 7.6   An Outbreak of Leishmaniasis. This outbreak occurred among military personnel who fought in Operation Desert Storm during 1991.

the row. This ciliary movement is coordinated so precisely that the microbe can move forward or backward or turn. Researchers believe that certain cilia serve a sensory function, since they appear to transmit stimuli and coordinate rapid movements in the organism. For example, when *Paramecium*, a common genus of ciliate, encounters an unpleasant stimulus, it backs off rapidly.

A second unique characteristic of the ciliates is the presence of two types of nuclei: a large macronucleus that controls cellular metabolism and growth; and one or more micronuclei. Both function in reproduction. During cell division, the micronuclei divide by mitosis, while the macronucleus simply pinches apart to yield two daughter macronuclei.

Members of the genus *Paramecium* display astonishing complexity in structure and behavior. This slipper-shaped microbe is enclosed by a pellicle, a composite covering consisting of an outer membrane and an inner layer of closely packed structures including defensive organelles called trichocysts. Trichocysts are expelled from the pellicle as sharp, dartlike fibers (*tricho* means "hair") that are driven forward at the tip of a long expanding shaft and are used to catch prey. *Paramecium* also has in its cytoplasm a set of contractile vacuoles. These bubblelike organelles are used to "bail out" excess water.

Paramecia exhibit an elaborate form of sexual behavior called conjugation (**FIGURE 7.7**). In this process, two paramecia line up alongside one another and fuse at the oral region. An extensive reorganization and exchange of nuclear material occurs during the next several hours, and much biochemical activity occurs in both micronuclei and macronuclei. The exchange of DNA is reciprocal—genetic material passes from each cell to the other. At the conclusion of conjugation, the organisms

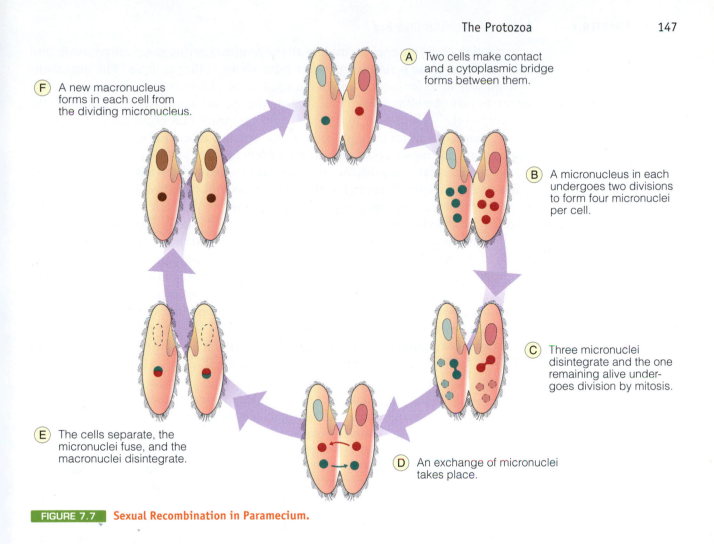

F A new macronucleus forms in each cell from the dividing micronucleus.

A Two cells make contact and a cytoplasmic bridge forms between them.

B A micronucleus in each undergoes two divisions to form four micronuclei per cell.

C Three micronuclei disintegrate and the one remaining alive undergoes division by mitosis.

E The cells separate, the micronuclei fuse, and the macronuclei disintegrate.

D An exchange of micronuclei takes place.

FIGURE 7.7 **Sexual Recombination in Paramecium.**

separate, each now having altered (recombined) DNA. This remarkable exchange of genetic material does not occur in complex plants or animals, and it gives new genetic characteristics to each *Paramecium*, possibly allowing the individuals to deal better with their environment. Chapter 8 explores microbial genetics in more detail, including a form of conjugation that takes place among bacteria.

Not all ciliates are motile. Some forms have stalks (for example, the trumpetlike *Vorticella*), while others use their cilia to set up currents in the surrounding water to bring food close (species of the genus *Stentor* are examples). In still other ciliates, the cilia fuse to form leglike organelles called cirri (sing., cirrus), which the organism uses to "walk" over surfaces. Complexities like these are fairly rare among microbes.

Sporozoa

The fourth and final group of protozoa we shall consider are the sporozoa, so named because at some stage in their life cycles, they develop a sporelike body. These microbes are also known as apicomplexans because they contain a "complex" of organelles at one end of the cell (the apex). Virtually all the protozoa in this group are parasitic, and many of them cause serious diseases in humans and other animals. The organisms have no structures for locomotion in the adult stage, although some of the intermediate stages may exhibit motility. In most cases, the sporozoa move by flexing their cells.

Among the well-known sporozoa are the microbes that cause malaria, a blood disease that accounts for over 300 million cases of illness per year worldwide (malaria is considered the most important health problem in today's world). Several species of *Plasmodium* can cause the disease, and mosquitoes are involved in transmission of all of these. The mosquito bites an infected person, becomes infected, then transmits the parasite to the next unsuspecting individual it bites. In the patient, the parasites destroy the red blood cells, causing a wave of intense cold followed by intense fever, the so-called malaria attack. Extensive anemia develops, and the hemoglobin from ruptured red blood cells makes the urine dark (giving rise to the name "blackwater fever").

Since its discovery in about 1640, quinine has been the mainstay treatment for malaria, but as resistance to it has been observed more frequently in the parasites, scientists have been researching new forms of the drug. An interesting slant on malaria is presented by evolutionary biologists, who point out that the malarial parasite has encouraged adaptation in the red blood cells of individuals living in the Earth's tropical zones (where malaria is prevalent). Distorted red blood cells arising from the adaptation are able to resist penetration by the parasite. Outside the malarial zones, however, this adaptation is a liability—it is called sickle cell anemia.

Another sporozoan of contemporary concern is *Toxoplasma gondii*, the agent of toxoplasmosis ("toxo"). Toxoplasmosis is a blood disease often acquired through contact with a cat, such as when changing the litterbox. The protozoa, pictured in **FIGURE 7.8** along with its complex life cycle, can also be consumed in improperly cooked beef or from the soil while working in the garden. Adults experience a mild mononucleosislike illness, but a pregnant woman can transmit the protozoa across the placenta, and a child may be born with serious nerve damage. Persons with compromised immune systems, such as AIDS patients, experience seizures and brain lesions.

One of the most serious illnesses in AIDS patients is *Pneumocystis carinii* pneumonia, a lung disease that causes over 50 percent of all deaths associated with AIDS. *P. carinii* commonly lives in the human lung without consequence, but when the immune system is suppressed, as during AIDS, the parasites multiply quickly and take up all the air spaces. Progressive deterioration of the lung tissue leads to filling of the lungs with fluid, followed by suffocation. It should be noted that analysis of its ribosomal RNA indicates that *P. carinii* may be a fungus, but the evidence is not conclusive.

Still another sporozoan parasite is *Cryptosporidium coccidi,* the microbe we encountered at the opening of this chapter. The final one we shall note is *Cyclospora cayetanensis,* a cause of intestinal illness. During the 1990s, this microbe was linked to imported raspberries, as A Closer Look on page 150 relates.

- *Plasmodium*
plaz-mō′dē-um

- *Pneumocystis carinii*
nü-mō-sis′tis
kär-nē-ī (or kär-i′nē-ē)

- *Cyclospora cayetanensis*
sī′klō-spô-rä
kī′ē-tan-en-sis

■ Other Protists

A gentle breeze, a sunny seashore, the soft lapping of the water . . . what vacationer could ask for more? Unfortunately, the same scenario appeals to red-pigmented dinoflagellates, a group of organisms we study in this section. When the ocean is warm

(a)

A Birds and rodents acquire the parasites from the soil.

B The cat is infected when it consumes an infected bird or rodent.

C The child is infected by contact with the cat or by the sand in the sandbox.

D A woman is infected by contact with contaminated cat litter.

E The fetus is infected by passage across the placenta.

F Consumers are infected by contaminated beef.

(b)

FIGURE 7.8   *Toxoplasma gondii*, the Cause of Toxoplasmosis. (a) A scanning electron micrograph of numerous parasites in the crescent-shaped trophozoite stage (×31,000). (b) The cycle of toxoplasmosis in nature.

A CLOSER LOOK
The Best News Since Shortcake

The strawberry growers and distributors of California were about to take a hit. A serious outbreak of diarrheal illness was being blamed on the source of their livelihood, and they were shaking in their boots. The outbreak was developing in Houston, Texas, where 62 cases of sickness were confirmed. All indications pointed to the strawberries. Already the government memos were being prepared: People should avoid California strawberries. Clearly, 1996 was not shaping up as a banner year.

Then came news of another outbreak, this time in Charleston, South Carolina. Intestinal illness had developed after a luncheon attended by 64 people, and 37 were sick with abdominal cramping, vomiting, and diarrhea. For dessert, the management had served strawberries.

But wait—raspberries were also served at the Charleston luncheon.

And there was another luncheon at the same restaurant on the same day, and none of 95 attendees became ill—the only fruit the management served there was strawberries. Maybe it wasn't the strawberries after all. Maybe it was the raspberries?

The ensuing weeks would be considerably brighter for strawberry growers, but nightmarish for raspberry producers. Intestinal illness related to the protozoan *Cyclospora* broke out in New York City, New Jersey, Toronto, and a host of other locales. In each case, raspberries were implicated, but strawberries were nowhere to be seen. The strawberry industry had dodged the bullet. Whew! Bring on the shortcake.

and nutrients are plentiful, dinoflagellates experience a burst of reproductive activity and fill the water with their trillions of descendants. So many are present that the water appears to turn a bloody or rusty color. These so-called red tides occur periodically on the Atlantic and Pacific coasts of the United States, virtually anywhere conditions are favorable.

Red tides are hazardous to health because certain species of dinoflagellates produce a toxin that is poisonous to humans and animals. The toxin concentrates in molluscs such as mussels, clams, and scallops. When ingested by humans, the toxin may cause transient neuromuscular disturbances, such as tingling and numbing of the lips, tongue, and fingertips, followed by uncertain balance, lack of muscular coordination, slurred speech, and difficulty in swallowing. There is no known antidote to the toxin.

Red tides are also dangerous to those who swim in the water because the toxins can be ingested directly from the water. Moreover, the depletion of oxygen in the water contributes to the death of plants and animals and causes the water to develop a stench due to the decaying organic matter. Beachgoers often wish they had opted for a vacation in the mountains.

We shall look more closely at the dinoflagellates presently, as we survey the unicellular algae. These protists benefit society in numerous ways, but as we have seen, they can also be dangerous to health.

Unicellular Algae

Biologists have always had a difficult time attaching a definition to algae. The term *algae* is a Latin word that translates literally to "seaweeds." To be sure, seaweeds are considered algae, but the term also encompasses a multitude of simple plants including many considered to be microbes. For our purposes, the key word is "simple" in comparison to a "complex" plant, which is a more highly evolved photosynthetic organism such as a moss, fern, or seedbearing plant. Biologists recognize two general types of algae: unicellular algae, consisting of independent cells; and multicellular algae (such as seaweeds) that have linked cells benefiting one another. Both algal types are photosynthetic eukaryotes. Unicellular algae are considered microbes.

The importance of unicellular algae to society cannot be overstated. As the major type of simple life in the seas, unicellular algae comprise the microbial population known as phytoplankton (*phyto* refers to plants, and *planktos* is Greek for "wander," thus a "wandering community of plant life"). The phytoplankton live near the ocean surface and use the sun's energy during photosynthesis to generate most of the molecular oxygen available in the atmosphere. Virtually every animal on land or in the sea depends directly or indirectly on phytoplankton for food, and every animal on land as well as in the sea depends on the oxygen that these algae produce to carry out their metabolic activities (Chapter 9). Phytoplankton are among the most important members of the oceanic food chains. Indeed, about half the world's organic matter is produced by phytoplankton.

Consistent with concepts set down by botanists, the algae are classified into six groups, known technically as divisions. Three of the divisions contain only unicellular algae, and one contains both unicellular and multicellular forms. We shall discuss members of these four divisions and leave the other two divisions of multicellular algae for botany courses.

Pyrrophyta. Microbes of the division Pyrrophyta are the so-called fire-algae. The adjective "fire" refers to the bright red and orange pigments most of the species possess. All members of the group are dinoflagellates having cells encased in rigid walls composed of cellulose coated with silicon. Because of their pigments, dinoflagellates have the plantlike capacity to perform photosynthesis, and, like protozoa, they move by using their two flagella. The first flagellum moves the cell forward, while the second whirls the cell on its axis, a property that gives dinoflagellates their name (*dinos* is Greek for "whirling").

Dinoflagellates are key microbes among the phytoplankton. Using their photosynthetic chlorophyll and carotenoid pigments, they produce carbohydrates by incorporating the energy from the sun, a process we study in detail in Chapter 9. Some species are bioluminescent; that is, chemical reactions occurring in their cytoplasm yield energy in the form of light, allowing the organisms to give off a greenish glow called bioluminescence. On a clear night, the sea can literally light up with a bioluminescent glow. Unfortunately, dinoflagellates sometimes experience a "bloom," or population explosion, that leads to a red tide, as mentioned earlier.

Chrysophyta. Microbes of the division Chrysophyta are golden-brown and yellow-green algae that include the diatoms (*chrysos* is Greek for "golden"). Diatoms are distinguished by their exquisitely beautiful intricate shells of silicon dioxide that overlap much like the two halves of a shoe box. These delicate glasslike shells are perforated to allow contact between the cell and its external environment. The shells form excellent fossils, and scientists can trace blooms of diatoms to the Cenozoic era, 65 million years ago.

Diatoms (FIGURE 7.9) are probably the major component of the phytoplankton. The 10,000 species of diatoms share with the dinoflagellates foundation roles in the oceanic food chains. In addition, their shells have economic importance because they

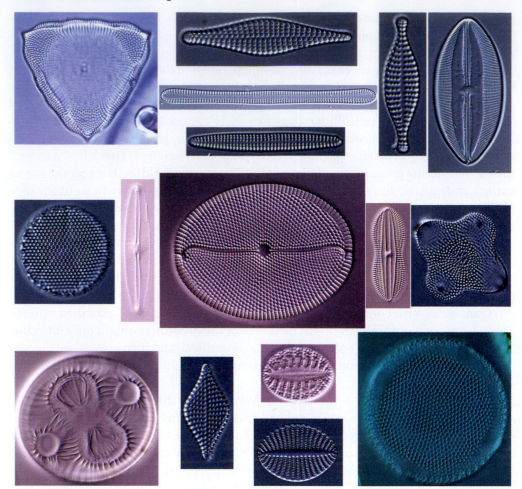

FIGURE 7.9   **Diatoms.** A collection of light micrographs of diatoms. Note the broad variety of their shapes and sizes. Diatoms trap the sun's energy in photosynthesis and use it to form carbohydrates that are passed on to other marine organisms as food.

are used as polishing and insulating materials. Furthermore, those who have swimming pools or aquaria have probably used diatomaceous earth, a filtering material that removes cyanobacteria and other contaminants in the water. The diatomaceous earth, resulting from millions of years of accumulation of diatom shells, is gathered from geological deposits as a fine, white powder. Over 300,000 metric tons are quarried annually from one deposit in California.

It is interesting to note that some animals use the pigments of chrysophytes to manufacture vitamins A and D. For example, researchers have discovered that when a fish consumes chrysophytes, the algal pigments gather in its liver and undergo several chemical conversions resulting in these two vitamins. Indeed, for many decades, humans have used the oils from fish livers as important sources of vitamins A and D (before the advent of vitamin tablets, many children received a daily teaspoon of cod liver oil).

Unfortunately, some forms of diatoms are toxic to animals. In the year 2000, for example, California scientists linked species of the genus *Pseudonitzschia* to the death of hundreds of sea lions. The diatoms concentrated in the tissues of anchovies, a favored food of the sea mammals, and they produced a neurotoxin that caused brain lesions and seizures before death. Public health officials raised the concern that similar diatom concentrations within oceanic food chains could affect humans.

Euglenophyta.   The division Euglenophyta includes over 800 known species, the most notable of which are in the genus *Euglena*. This common type of freshwater microbe has a nucleus for its hereditary material, two flagella for propulsion, and very flexible nutrient requirements: In sunlight, it is fully autotrophic; that is, it synthesizes its own nutrients using its chloroplasts to manufacture organic compounds by photosynthesis (Chapter 9). In the dark, the *Euglena* loses its photosynthetic pigments and feeds exclusively on organic matter, thereby displaying a heterotrophic mode of nutrition. When returned to light, the *Euglena* resynthesizes its photosynthetic pigments and once again becomes autotrophic. These microbes have plantlike as well as animallike characteristics, as noted earlier in this chapter.

Euglena species possess a characteristic light-detecting eye spot that allows them to detect sunlight and swim toward it. The name of the genus is derived from this structure (*eu* is the Greek stem for "true," and *glene* is Greek for "eyeball"). So far as scientists know, reproduction occurs only by asexual means involving mitosis. Overabundant reproduction by the microbe causes water to become a pea soup green.

Chlorophyta.   Microbes of the division Chlorophyta include both unicellular and multicellular algae. An important difference between these algae and those of the other three divisions is that chlorophytes have carotenoid pigments as well as unique variants of the "green grass" chlorophyll molecules called chlorophylls A and B. There is widespread belief among biochemists that multicellular algae and complex plants evolved from a unicellular chlorophyte because both plants and green algae (and no other algal group) have both carotenoids and chlorophylls A and B. Moreover, the chlorophytes synthesize starch and store it in their chloroplasts, a characteristic shared only with plants.

A well-studied member of the Chlorophyta is the flagellated organism *Chlamydomonas*. This microbe is often studied as the prototypical chlorophyte. It has a complex life cycle that includes an alternation of generations, a characteristic found in multicellular green algae and in complex plants (another bit of evidence for the ancestral relationship mentioned above). During alternation of generations, two forms of the organism take part: a diploid form having two sets of chromosomes (*diploos* is Greek for "double"), and a haploid form having a single set of chromosomes (from the Greek *haploos* for "single"). The mature *Chlamydomonas* is a single haploid cell, as shown in FIGURE 7.10 . When it reproduces asexually, the microbe undergoes several rounds of mitosis to form many new cells. Sexual reproduction occurs when environmental stress is present. Haploid gametes (sex cells) form within the parent cell and are released. They unite with gametes from opposite mating types (designated + and −) and fuse. This fusion forms a diploid zygote surrounded by a durable coat to protect against harsh environmental conditions. After the stress has disappeared, the zygote undergoes meiosis, a form of cell reproduction that results in a halving of the chromosome number. The process results in four haploid cells that emerge as new *Chlamydomonas* cells. Alternation of generations is not at all common among microbes, but it is notable in the chlorophytes.

Among the fascinating members of the Chlorophyta are the snow algae. Over 350 species of these microbes are known to thrive in the near-freezing, sun-blasted

■ *Chlamydomonas*
klam-i-dō-mō′äs

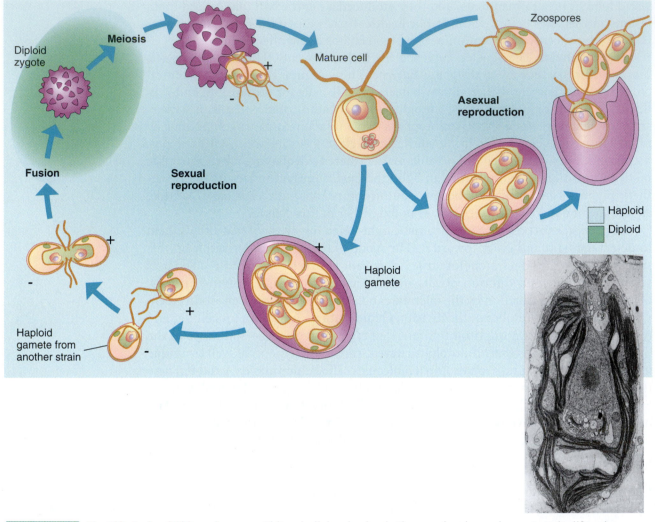

Meiosis

Diploid zygote

Mature cell

Zoospores

Asexual reproduction

Fusion

Sexual reproduction

Haploid

Diploid

Haploid gamete

Haploid gamete from another strain

FIGURE 7.10 **The Life Cycle of *Chlamydomonas*.** This unicellular alga has both asexual and sexual aspects to its lifecycle. Haploid and diploid forms are present at various points in the cycle.

slush of melting snowfields around the world. In high snowfields of the western United States, they blush red in footstep-sized patches (where the phenomenon is called "watermelon snow"); in the northern New England mountains (for example, the White and Berkshire Mountains), they give salmon-orange streaks to the last snows of winter; and in high-elevation shady forests, they lend a lime-green glow to snow several feet below the surface. Species of *Chlamydomonas* and *Chloromonas* are among the most common snow algae; their pigments allow these algae to capture energy from the sun and form the basis for food chains in the snow. Scientists who study snow algae are said to be among the "coolest" microbiologists.

Another well-known member of the Chlorophyta is the colonial alga *Volvox*. This microbe (also considered a protozoan) consists of a community of cells living independently but deriving some benefit from each other. Although not directly related to any multicellular organism, *Volvox* suggests how multicellularity might have evolved. Scientists have traced the origin of multicellular land plants to about 400 million years ago. It is conceivable that one or more kinds of green algae were among the first to try this experiment in community living. From this humble beginning, all of today's land plants may have come into being.

Slime Molds

As you have probably noted, some protists—such as protozoa—bear a striking resemblance to animals, while other protists—such as unicellular algae—are similar to plants. It is appropriate, therefore, to include a discussion of slime molds here. These are a collection of microbes that are both animal-like and plantlike. Slime molds resemble plants because their cells are enclosed in cell walls composed largely of cellulose (a plant feature), and they resemble animals because at some point during their life cycles they have flagellated cells that move. They are called "slime molds" because they are slimy and have a threadlike structure similar to that of fungi (Chapter 8). Indeed, the slime molds are a prime reason why the beginning of this chapter referred to the protists as a "grab bag of disparate creatures."

Scientists recognize two groups of slime molds: the acellular slime molds and the cellular slime molds. Acellular slime molds are classified in a division called Myxomycota (*myxo* refers to mucus or slime, while *mycota* means "funguslike"). Also called plasmodial slime molds, these microbes consist of a huge, multinucleate "cell" that resembles a giant amoeba and is called a plasmodium. A single plasmodium can grow large enough to cover an entire log, although it will be only a millimeter thick. The plasmodium feeds on dead organic matter and thus recycles numerous nutrients back into the environment. When the habitat dries out, the plasmodium produces fruiting bodies, spore-producing stalks that extend upward. The spores germinate into sex cells (gametes), which unite to form a zygote. The zygote develops into a multinucleated plasmodium, completing the life cycle.

Cellular slime molds are classified in the division Acrasiomycota (*acrasia* means "confusing," a reference to the confusing nature of their life cycle). These slime molds differ from the acellular slime molds in that their small, amoeboid forms retain their individuality when they congregate. When food and water are plentiful, the individual slime mold cells move about, engulfing bacteria, organic debris, and other available food sources. When food is in short supply, however, the cells come together within a cellulose sheath and congregate to form a large, many-celled stage called a "slug." (Scientists studying this formation of a community of cells hope to understand how multicellular organisms arose from unicellular organisms.) The entire mass of cells migrates together for a time, a behavior that helps to disperse the species.

But the cycle is not yet finished. At a later time, the slug transforms itself into a threadlike structure resembling a fungus. Next, structures extend upward as spore-producing stalks. Within the saclike sporangia at the tips of these stalks, spores are produced by mitosis and are released for wide dispersal. Each spore germinates and forms another amoebalike cell to begin the process anew. The interesting life cycle of the cellular slime mold *Dictyostelium discoideum* is shown in FIGURE 7.11 .

Despite its somewhat distasteful name, the slime mold is rather interesting to watch under a microscope. It contains fine channels through which cytoplasm streams

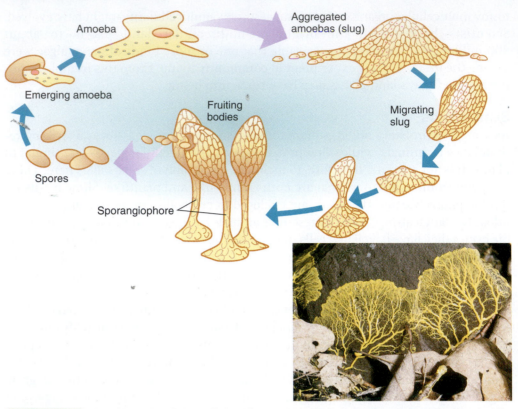

Amoeba

Emerging amoeba

Aggregated amoebas (slug)

Migrating slug

Fruiting bodies

Spores

Sporangiophore

FIGURE 7.11 **The Life Cycle of the Cellular Slime Mold.** *Dictyostelium discoideum*. The amoeboid form of this organism resembles a protozoan, while the fruiting body stage resembles a fungus.

in pulsing flows—first one way, then another. Biochemists are interested in its thousands of nuclei, all of which undergo mitosis independently. It is not unusual, therefore, to see the various phases of mitosis happening at the same time in the cell. Partly for this reason, slime molds have come to occupy an important place in research laboratories where cell division is studied. Indeed, scientists hope to learn how and why a cancer cell undergoes such rapid division by studying the process in the slime mold.

Water Molds

Water molds include a number of protists that live in water and damp soil and cause the unsightly and furry growths that plague fish in home and commercial aquaria. The organisms have threadlike bodies reminiscent of some fungi and are sometimes classified in the division Oomycota of the fungi (and then called oomycetes). However, the water molds are not true fungi because they exhibit few or no cross walls separating the cells in their threads. Furthermore, they have more cellulose than chitin in their cell walls, a characteristic of plants, not fungi; histone proteins occur in their chromosomes (and do not in fungi); and their sexual phase of reproduction differs markedly from that observed in fungi.

But perhaps the most notable characteristic of the group is the presence of flagellated spores, or zoospores, that develop through mitosis in a structure called a sporangium. The spores have two flagella—one directed forward, the other backward. The motility of the zoospores is frustrating to mycologists wishing to include the microbes as fungi because no other motile spores are known to exist among fungi. For these reasons, the water molds are generally classified with the protists.

Several species of water molds have significant economic importance. For example, one species causes downy mildew of grapes, and another has destroyed millions of avocado trees in California. Still another species, *Phytophthora infestans*, is responsible for the late blight of potatoes, a disease that destroyed the Irish potato crop during the extremely damp years of the mid-1800s and led to mass Irish immigration to North America.

■ *Phytophthora infestants*
fi-tof'thô-rä in-fes'tans

Ireland of the 1840s was an impoverished country of 8 million people, mostly tenant farmers paying rent to English landlords. The sole crop of these farmers was potatoes, grown season after season on small tracts of land. But early in the 1840s, heavy rains and dampness portended calamity. The potatoes had suffered before, but not as badly as with this new disease. It struck down the plants like frost in the summer. Beginning as black spots, it decayed the leaves and stems and left the potatoes a rotten, mushy mass with a peculiar and offensive odor. Farmers left the potatoes to rot in the fields, and the disease spread. After two years, the potato rot seemed to slacken, but in 1847 ("Black '47") it returned with a vengeance. Despite limited relief efforts by the English, over 2 million Irish people died from starvation. Eventually, over a million starving, demoralized survivors set off for Canada and the United States, and many Americans are descended from those great waves of immigrants. The great Irish potato famine changed the face of two continents—few microbes have had such a dramatic effect on society.

■ A FINAL THOUGHT

In 1997, *Newsweek* magazine labeled it the "cell from hell." It was a protozoan known as *Pfiesteria piscicida*, and it was destroying the fish population along the southeastern coast of the United States. Silvery menhaden (a type of herring) were turning up with quarter-sized lesions on their bodies, and *Pfiesteria* hysteria was setting in. During August of that year, tens of thousands of fish died in a 4.5 mile stretch of Maryland's Pocomoke River.

■ *Pfiesteria piscicida*
fes-ter'ē-ä pis-si-sē'dä

Although we have been highlighting the positive aspects of microbes in society, occasionally we are jarred into the realization that microbes can also make life very difficult. Microbes cause massive kills of plants, animals, and humans, and public health officials must be continually vigilant, for much is not yet known about microbes. Despite all our efforts to understand their physiology, biochemistry, genetics, evolution, and ecology, microbes will continue to pop into our midst and leave a trail of destruction. A friend of mine, a chemist, once made this analogy: "If you dip a tennis ball in the ocean, the water dripping from the tennis ball represents all that is known; the ocean represents all that is waiting to be learned."

■ QUESTIONS TO CONSIDER

1. Suppose a research scientist discovered that a toxic chemical was wiping out the populations of diatoms and dinoflagellates in the oceans of the world. What horror story could occur if this were true?

2. You and a friend who is three months pregnant stop at a hamburger stand for lunch. Based on your knowledge of toxoplasmosis, what helpful advice can you give your friend? On returning home, you notice that she has two cats. What additional information might you be inclined to share with her?

3. The flagellated protozoan *Giardia lamblia* is named for Alfred Giard, a French biologist of the late 1800s, and Vilem Dusan Lambl, a Bohemian physician of the same period. Unfortunately, this information does not tell us much about the organism (except who did much of the descriptive work). In contrast, the names of other protozoa in this chapter tell us much. What are some examples of more informative names?

4. In the movie *Star Wars*, there is a famous bar scene where Han Solo is seeking transport for himself and his traveling companions. The scene is memorable for the grotesque diversity of the extraterrestrial creatures. Suppose you were to illustrate this scene using protists as models for your creatures. What might you include in your drawing?

5. Members of the genus *Euglena* illustrate the difficulty in assigning the designation animal-like or plantlike to protists. Why is this the case? Why could *Euglena* or a similar organism be considered the basic stock of evolution?

6. How does studying slime molds help one to understand the developmental cycle of animals? Other than their rapid division, what other characteristics of slime mold cells might provide researchers with insight into cancer cells?

7. Microbes of the genus *Chlamydomonas* are among the few microbes that display an alternation of generations. What is the importance of this characteristic in the life of the microbe?

■ KEY TERMS

Informative facts are necessary for the expression of every concept, and the information for a concept is founded in a set of key terms. The following terms form the basis for the concepts of this chapter. On completing the chapter, you should be able to explain and/or define each one:

alternation of generations	contractile vacuole
amoeba	diatomaceous earth
apicomplexan	diatoms
bioluminescence	dinoflagellates
chlorophyte	endosymbiosis hypothesis
chrysophyte	euglenophyte
ciliate	flagellate
conjugation	foraminifera

fruiting body
hypothesis
heliozoan
macronucleus
micronucleus
microtubules
pellicle
phytoplankton
protozoa
pseudopodia
pyrrophyte

radiolarian
red tide
sarcodine
slime mold
sporangia
sporozoan
test
trichocyst
trophozoite
zooplankton
zoospore

http://microbiology.jbpub.com/book/microbes

The site features **eLearning**, an online review area that provides quizzes and other tools to help you study for your class. You can also follow useful links for in-depth information, read more stories of microbiology, or just find out the latest microbiology news.

8

Fungi: Yeasts and Warm Fuzzies

Ringworm, or Tinea is caused by several different species of fungus commonly known as dermatophytes. It is strange, but true, that fungi not only cause disease, but are also the original source of penicillin, our first antibiotic. The inset shows a picture of a penicillin producing fungus. Penicillin, a derivative of mold, is used to cure many common infections.

◼ Looking Ahead

Before microbiology was established as a separate discipline of biology, fungi were studied in botany courses because they were thought to be plants. In contemporary biology, however, fungi are considered microbes, as we shall see in the pages ahead.

On completing this chapter, you should be able to . . .

- appreciate some of the novel features of fungi that encourage scientists to place them in their own kingdom.
- understand the sexual and asexual reproductive cycles of fungi and compare these cycles to those of other microbes.
- explain how fungi benefit society in the production of various foods, the synthesis of key industrial products, and the cycling of elements in the environment.
- describe several important fungal diseases that affect plants and humans.
- understand the nature of yeasts and the biochemistry of their metabolism as it relates to the bread and fermentation industries.
- name and describe four groups of fungi and identify key members of each group.
- recognize the important roles assumed by fungi in lichens and in mycorrhizae.

About 250 million years ago, at the close of the Permian period, a catastrophe of epic proportions hit the Earth. Scientists estimate that over 90 percent of animal species in the seas vanished virtually overnight during this unexplained cataclysm. The great extinction also wreaked havoc on land animals and cleared the way for dinosaurs to dominate the planet.

But land plants apparently managed to survive the catastrophe, and before the dinosaurs emerged, they spread and carpeted the world—at least, that is what scientists long believed. In recent years, however, many of them have revised their views and proposed that fungi filled a significant niche in this part of the Earth's history. In 1996, Dutch scientists from Utrecht University presented evidence that most land plants were also destroyed by the Permian extinction, and for a brief geologic span, dead wood covered the planet. During this period, they suggest, the fungi emerged, and wood-rotting species experienced a huge spike in their population. Numerous findings of fossil fungi from the post-Permian period offer support for this theory. These plentiful fossils have been found in all corners of the globe. Significantly, the fossils contain the active feeding forms of the fungi rather than the dormant spores they produce.

If the evidence holds up, it would appear that fungi proliferated wildly and entered a period of feeding frenzy in which they were the dominant form of life on Earth. Even today, fungi are literally everywhere: from the food we eat, to the air we breathe, to the soil we walk on. Two types of microbes, pictured in FIGURE 8.1, make up the bountiful group of fungi: the single-celled yeasts and the filamentous "fuzzy" molds. Yeasts are known for their fermentation abilities and as causes of some human diseases; they are also a regular part of our diet, as A Closer Look on page 162 explains. Molds are long, tangled filaments of cells that form hairlike masses, especially in acidic environments such as bread, cheese, and sour dairy products. Sometimes the filamentous mass of a mold clings together tightly and forms a compact structure we call a mushroom.

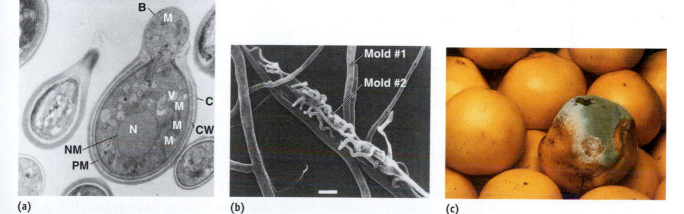

(a)   (b)   (c)

FIGURE 8.1   Two Types of Fungi.   (a) A transmission electron micrograph of a yeast cell showing typical eukaryotic features. Note the cell wall (CW), capsule (C), and plasma membrane (PM) on the cell surface. In the cytoplasm, clear bodies called vacuoles (V) and mitochondria (M) may be seen. The nucleus (N) is surrounded by a nuclear membrane (NM). In the upper portion of the photograph, a reproductive structure, the bud (B), is growing out from the parent cell. (b) Two molds, one wrapped around the other, in a parasitic relationship. The long filamentous form of each mold and the many branches are visible. (Bar = 10 µm.) (c) A mold growing on a grapefruit.

A CLOSER LOOK

The Drier the Better

Care to do some microbiology at home? Simply go to the grocery store and purchase a package of "active dry yeast." Open the package, pour the contents into a bit of warm water, and you're on your way—instant microbes. You can study the yeasts with the microscope, investigate their physiology, and if you're really with it, rearrange their genes. All this from a package.

But it wasn't always that way. For decades, scientists could not figure out how to keep yeasts alive and in a dry state. If people wanted to make bread, for example, they had to get "starter yeast" from the "mother" dough where the yeast was growing; if the objective was wine fermentation, a bit of old wine was necessary.

Then, during World War II, German prisoners were found in possession of a curious brown powder—it was the elusive dry yeast. They had been given the yeast for use as a nutritious food or for making bread by adding it to a bit of dough (or "Battlefield Red," if added to crushed grapes). The prisoners refused to reveal the secret of the dried yeast, probably because they did not know what it was.

The postwar period was a different matter. In the euphoria of sharing, the world learned the secret of trehalose. Trehalose is a disaccharide, a simple two-molecule sugar. Added to yeast cells, it stabilizes the cell membrane, prevents cell damage due to drying, and keeps the yeast alive and active. Now everyone knew the answer, including an entrepreneur named Arthur Fleischmann—founder and owner of Fleischmann's Active Dry Yeast.

Fungi are an unheralded and often overlooked group of microbes. They are, however, a critical link in the web of life on Earth, serving as important decomposers of organic matter. Indeed, they make incalculable contributions to ecosystems by liberating nutrients from organic materials and making those nutrients available to insects, worms, bacteria, and myriad other organisms within the environment (Chapter 16). Without fungi (and bacteria), the nutrients in organic materials would be locked up; the cycles of elements would grind to a halt; the fertility of soil would decline precipitously; and ecosystems would collapse.

Fungi are eukaryotic microbes, and the fungal cell is usually several times larger than the bacterial cell. With its great size and complexity, the fungal cell can have broad diversification in structure and mode of life. A key property of a large cell such as a fungal cell is its capacity to carry large amounts of genetic information and transmit that information reliably to the next generation. A large cell can also specialize different cellular functions in various organelles. This specialization and division of labor lead to great efficiency in cellular activities, such as an enhanced ability to acquire food (which results in enhanced growth). It also encourages the eukaryotic organism to adapt more readily to life-threatening changes in its environment.

The size of a eukaryotic cell is limited by several factors. One of these factors is the surface-to-volume ratio. A larger cell requires a proportionally larger cell surface so that a greater quantity of materials can move into and out of the cell. Fungi have adapted to avoid this potential problem by developing flat cells, thereby increasing the surface-to-volume ratio. Another limiting factor is whether the nucleus of a larger cell can provide genetic instructions to regulate the immense number of chemical reactions occurring in the large volume of cytoplasm. Many species of fungi have adapted by evolving multiple nuclei in a single cell.

Classification, Structure, and Growth of Fungi

About 100,000 species of fungi have been named and identified, and scientists estimate that another 200,000 are waiting to be discovered. After the 1660s, when fungi were first described by Robert Hooke, these microbes were collected and studied by botanists who considered them simple plants. In fact, until the mid-1900s, those wishing to study the fungi would typically have to enroll in a course in botany (even though it was clear that fungi have no photosynthetic pigments and therefore stand apart from traditional plants).

In modern biology, the fungi are classified in their own kingdom in the Whittaker system, a kingdom appropriately named Fungi. Although some fungi (such as yeasts) are single-celled microbes, most species of fungi are composed of strands of cells. A single fungal strand is a microscopically thin, threadlike unit called a hypha (pl., hyphae). All the hyphae of a single fungus collectively form an interwoven mass called a mycelium (pl., mycelia), as shown in FIGURE 8.2 . Incidentally, you will note that throughout microbiology the prefixes *myco-* and *myce-* generally apply to a fungus or something funguslike. For example, a mycologist is one who studies fungi; a mycosis is a fungal disease; an actinomycete is a funguslike bacterium; and *Mycobacterium* is a genus of bacteria that appear in cultures as funguslike threads.

Structure of Fungi

When a certain author was a student in college (many moons ago), he memorized the following definition of a fungus: "A fungus is a nucleated, sporebearing, achlorophylous organism that generally reproduces by both sexual and asexual means, and whose usually filamentous, branched, threadlike structures are surrounded by cell walls of chitin." Most of this definition is still valid today: Fungal cells have nuclei ("nucleated"); most species form spores ("sporebearing"); fungal cells have no chlorophyll ("achlorophylous"); most species reproduce by both sexual and asexual means; except for yeasts, fungi have "filamentous, threadlike strands"; and their cell walls contain a unique polysaccharide called chitin.

In addition to these characteristics, many species of fungi have partitions between cells in each hypha. These partitions are called septa (sing., septum). The partitions are not complete, however; they have pores that permit the cytoplasm from adjacent

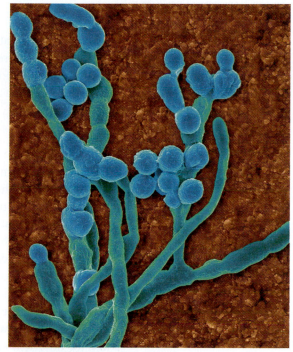

FIGURE 8.2   **A Typical Fungus.** A false-color scanning electron micrograph of *Cladosporium cladosporioides*, one of the most common fungi isolated from air samples. Outdoors, this fungus is commonly found on decaying vegetation. Indoors, it may be isolated from refrigerator moldings, tile grout in showers, and vinyl shower curtains. In the mycelium, the conidiophores and conidia of the fungus can be seen ($\times$ 2970).

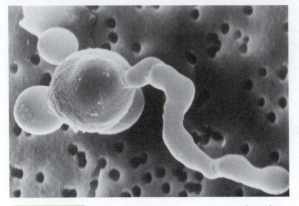

FIGURE 8.3 **A Germinating Spore.** A scanning electron micrograph of a spore of the fungus *Cephalosporium* germinating to form a hypha (× 6000). Note the septa, the partitions between cells of the hypha.

cells to mingle. In some fungal species, the septa between the cells break down, and the cytoplasm from adjacent cells mixes freely in a single, enormously long tubular cell. These fungi are said to be coenocytic (from the Greek word *koinos*, meaning "shared in common," a reference to the shared cytoplasm). As we noted previously, the cell walls of the fungal hypha are composed primarily of chitin, a polysaccharide not found in plants (although it is a component of the outer skeleton of arthropods such as insects, crabs, and lobsters, where it gives structural integrity). The presence of chitin in their cell walls is one reason why fungi are no longer classified with plants.

The reproduction of a fungus depends heavily on its production of highly resistant spores, as we shall discuss in depth presently. A hypha normally arises by growth from a single spore, as pictured in **FIGURE 8.3** . As the hypha continues to elongate, new cell formation takes place primarily at its tip, where proteins are concentrated. Although fungi are nonmotile, continued growth of the hypha brings the microbe in contact with new food sources. This growth can be so rapid that a single hypha can produce enough cells in a 24-hour period to span over half a mile if placed end-to-end.

Nutrition in Fungi

With very few exceptions, all fungal species are heterotrophic, meaning that they live on preformed organic matter. Fungi digest organic matter by excreting enzymes into the environment and then absorbing the simple products of digestion. This process is necessary because fungal cells cannot engulf food particles by phagocytosis. Some products of the enzymatic digestion remain outside the fungus and provide nutrients for other organisms in the environment. For example, certain fungal species are among the few organisms that can produce the enzyme cellulase and use it to decompose cellulose (the principal polysaccharide in wood). When cellulose is digested, it yields glucose molecules, which are extremely useful sources of energy for a broad variety of organisms. Some species of fungi also have the ability to synthesize ligninase, an enzyme that breaks down the tough lignin-containing fibers of plants.

Another biochemical feature that distinguishes fungi from plants is that fungi store carbohydrates as the polysaccharide glycogen, while plants store carbohydrates as a different polysaccharide, starch. Although most fungal species are saprobes (i.e., they live on dead organic matter), some are parasites (they thrive on living things and cause disease, especially in agricultural products). As we shall see, fungi cause numerous diseases that have brought untold human misery.

Many fungal species live under conditions that are acidic (pH between 5 and 6). For this reason, fungal contamination is common in acidic foods such as sour cream, yogurt, citrus fruits, and most vegetables. Moreover, the acidity in bread and cheese favor fungal contamination. In some cases, this "contamination" can be helpful: Blue cheese (**FIGURE 8.4**) consists of milk curds supporting the fungus *Penicillium roqueforti* (the blue streaks in the cheese are the fungus). Most people would prefer to remain blissfully ignorant of this bit of microbiological trivia.

(a)

(b)

FIGURE 8.4 **Products of Fungal Activity.** When fungi are cultivated under industrial conditions, they yield a variety of chemical substances in their metabolism and an equally broad variety of useful products. (**a**) Carbon dioxide gas in soda pop. (**b**) Saki. (**c**) Blue cheese. (**d**) Bread.

(c)

(d)

Reproduction in Fungi

Most known species of fungi reproduce by both asexual and sexual methods. In some species, asexual reproduction occurs when the hypha undergoes fragmentation and each fragment separates to become a so-called arthrospore (*arthron* is Greek for "joint," a reference to spore formation at the joint). Each arthrospore then develops into a new hypha and, eventually, a new mycelium. Asexual reproduction can also occur when chlamydospores form. These are thick-walled spores that form along the margin of a hypha (*chlamys* is Greek for "cloak," a reference to the spores' wall). When they break free and germinate, new hyphae begin to form.

Asexual reproduction may also take place through the process of mitosis, in which a single cell reproduces to yield two genetically identical daughter cells. The daughter cells then develop into highly resistant spores. Normally the spores form on or in a specialized structure. This structure may be a sac called a sporangium, in which the spores are stored; or it may be the tip of a specialized hypha called a conidiophore. Spores produced free and unprotected at the tip of a condiophore are called conidia (sing., *conidium*, the Latin term for "dust").

Like bacterial spores, fungal spores have tough walls that resist environmental extremes. Millions of spores produced by a single mycelium are blown about by the wind. A single spore falling into a cup of yogurt will soon cause green fuzzy mold to appear on the yogurt's surface. The black, red, yellow, green, and other vibrant colors associated with molds are due to the spores and their sporangia. Unfortunately, spores bring misery to people with mold-spore allergies. Furthermore, one might think that fungal spores would make good bioweapons, but they are not developed for this purpose primarily because fungal diseases are not renowned killers and do not occur in broadscale epidemics.

The process of sexual reproduction involves cell fusion and spore formation, as displayed in FIGURE 8.5 . The nuclei of fungal cells are haploid; that is, each nucleus contains a single set of chromosomes (the Greek word *haploos* means "single"). Spore formation by the sexual process begins with the fusion of cells from compatible mating types of fungi (the cells are known as gametes, the same term used for all eukaryotic sex cells). Then the two haploid nuclei fuse. The fusion results in a diploid cell, one in which two sets of chromosomes are present (*diploos* is the Greek word for "double"). This cell, called a zygote, then undergoes meiosis, a form of cell reproduction resulting in daughter cells that each have a single set of chromosomes. The process results in sexually produced haploid spores. The spores are dispersed, and some eventually come to rest in a nutritious environment. They germinate (much as a seed germinates) and divide by mitosis to form the cells of a hypha, and then a mycelium. The nuclei of all the cells in the mycelium are haploid.

Asexual reproduction by fungi is advantageous because it provides huge numbers of spores, each of which can become a new mycelium. However, the spores are genetically identical; thus, if an environmental agent is able to destroy one, it will destroy all of them. Sexual reproduction generally results in many fewer spores, but they have genetic variability. This variability may permit a spore to survive an environmental agent that might destroy its sister spores, and it will develop into a better adapted fungal mycelium. Having both means of reproduction available is a survival edge for fungi.

Divisions of Fungi

Within the general theme of sexual reproduction in fungi, there are many variations and complexities. These variations have been traditionally used to classify fungi into categories. As we have noted, the Whittaker classification places fungi into their own kingdom, and the newer system developed by Woese and his colleagues (Chapter 2) retains the kingdom Fungi in the domain Eukarya. Within that kingdom, the term "division" is used for individual categories, as it is used in the plant kingdom. However, to distance themselves from botanists, mycologists are beginning to use the term "phylum." Because this transition is not yet established, we shall stick with "division" in this discussion.

It should be noted that mycologists are not in complete agreement regarding fungal categories, and it is often difficult for those writing about fungi to decide which system is best. This book will follow a generally accepted pattern and separate the fungi into four divisions as follows: Zygomycota, the division in which the fungi form sexually produced zygospores; Ascomycota, the division in which sexually produced ascospores form within a saclike ascus; Basidiomycota, the division in which sexually

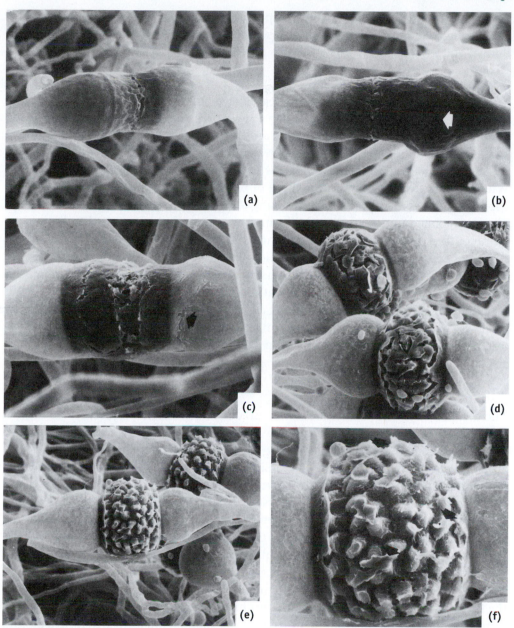

FIGURE 8.5 **Sexual Reproduction in Fungi.** A sequence of scanning electron micrographs showing sexual reproduction in the mold *Gilbertella persicaria*. (**a**) The hyphae of compatible mating types fuse and form a fusion septum. The haploid nuclei within the compatible hyphae have fused at this point to form a diploid zygote. (**b**) Cells at the septum begin to swell and show early signs of a zygosporangium, the structure that will enclose the zygote. (**c**) The outer primary wall begins to rupture. (**d**) The rupturing continues, and (**e**) the zygosporangium is revealed. The zygosporangium continues to mature as the primary wall separates away. (**f**) A magnified view of the zygosporangium, showing its surface characteristics and the remnants of the primary wall. Within the zygosporangium, the zygote is undergoing meiosis to yield one or more haploid spores that will be released to propagate the fungus.

produced basidiospores form on a supportive structure called a basidium; and Deuteromycota, the division in which a sexual cycle of reproduction is currently unknown and no sexually produced spores are recognized.

Zygomycota

Fungi of the division Zygomycota are called zygomycetes. They are also known as "zygote fungi" for their ability to produce zygospores, the only diploid cells in the life history of the zygomycetes. Zygospores are thick-walled, environmentally resistant spores that develop from the fusion of two sexually opposite cells.

■ *Rhizopus stolonifer*
rī'zo-pus stō-lon-i-fẽr

Among the zygomycetes is *Rhizopus stolonifer*, the common black bread mold. When a *Rhizopus* spore lands on a piece of bread, fruit, or other organic matter, the spore germinates and spreads by producing feeding hyphae as well as specialized hyphae that extend up into the air and form sporangia at their tips. The black pigmentation of the sporangia gives the mold (and the bread) its dark color.

The occasional *Rhizopus* contamination we see in food is counterbalanced by the beneficial roles this fungus plays in industry. One species, for example, is the microbial "factory" that facilitates the fermentation of rice to sake, as we note in Chapter 12. (Incidentally, although sake is spoken of as "rice wine," it is more properly classed as a beer because beers are products of grain fermentation, while wines result from fruit fermentation.) The industrial production of cortisone, an anti-inflammatory agent, is also dependent on the chemistry of *Rhizopus*, as Chapter 14 explains.

Ascomycota

The division Ascomycota includes about 30,000 species of ascomycetes, also known as sac fungi. The name *ascomycete* is derived from the presence of an ascus (pl., asci), a tiny spore-containing sac formed during the life cycle of the fungi. Most ascomycetes are saprobes, but many important plant parasites are found in this division, including the fungi that cause powdery mildew, Dutch elm disease, chestnut blight disease, and peach leaf curl disease. Most species of ascomycetes are composed of filaments, but a few species such as the yeasts used in fermentation and baking (discussed below) are single-celled, or unicellular. The septa in the hyphae of ascomycetes have perforations that allow the cytoplasm of one cell to mingle with that of neighboring cells.

The sexual stages of most ascomycetes are less conspicuous than the asexual stages. First, the hyphae of different mating types fuse, a process that brings together the haploid nuclei. Then, the sexually produced ascospores, shown in FIGURE 8.6, form within the ascus. To form the ascospores, the haploid nuclei fuse to form a diploid cell that undergoes meiosis and gives rise to eight haploid ascospores per ascus.

In most ascomycetes, the asci are formed in complex structures called ascocarps, which can be flask-shaped or cup-shaped. In the "cup fungi," the ascocarp is composed of millions of hyphae tightly packed together to form the body of the cuplike container, and asci are exposed on the upper surface. In several species, such as morels and truffles, which are prized as edible fungi, the ascocarp is crowned by bell-shaped tissue that contains the asci.

Ascomycetes also include various *Penicillium* species, the fungi that produce penicillin and related antibiotics such as ampicillin, amoxicillin, methicillin, and many others. Moreover, some species of *Penicillium* are used to ripen and flavor blue cheese,

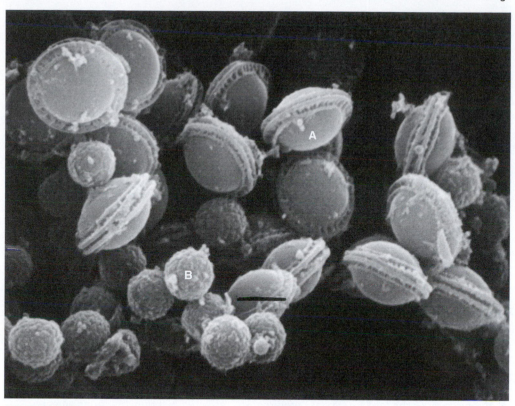

FIGURE 8.6 **Two Types of Fungal Spores.** Ascospores (A) and conidia (B) of the fungus *Aspergillus quadrilineatus*. This fungus was cultivated from the nasal sinuses of an ill patient. The sexually produced ascospores display a series of so-called equatorial crests at their midlines. The asexually produced conidia are rounder, with tightly folded surfaces. (Bar = 1 µm.)

Roquefort cheese, and Camembert cheese. Also in the division Ascomycota are species of the common black mold *Aspergillus*. Certain *Aspergillus* species contaminate house dust and cause allergies and respiratory illness. *Aspergillus* species can also produce dangerous toxins (called aflatoxins) in foods. These toxins poison the nervous system, and they have been investigated as agents of bioterrorism because they can be produced as bioweapons and added to municipal water supplies. On the positive side, *Aspergillus* species synthesize citric acid, a major component of beverages. They are also used in the production of soy sauce and vinegar (Chapter 12). One species, *A. niger*, is used to produce Beano, the trade name for an enzyme preparation that breaks down galactose, a sugar in cabbage, broccoli, Brussels sprouts, and other "strong" vegetables (Chapter 2).

Ascomycetes also number among their species a number of plant and human pathogens. For example, *Claviceps purpurae* infects rye grains, and when humans consume the contaminated grain or bread baked from it, the fungus causes a nervous system disorder called ergot disease. There is an intriguing body of evidence indicating that ergot disease may have influenced the outbreak of the French Revolution (see A Closer Look on page 170). Ironically, *C. purpurae* is now cultivated for medical purposes: In small quantities, its chemical products can be used to treat migraine

■ *Aspergillus*
a-spĕr-jil′lus

■ *Claviceps purpurae*
kla′vi-seps pür-pü-rē′ä

A CLOSER LOOK

A Fungus and the French Revolution

In the early summer of 1789, a great wave of panic spread over France. Rumors circulated that brigands were everywhere, and many townsfolk fled to the woods to hide. Peasants stockpiled weapons, and soon they turned their hostility on landowners, burning homes and destroying records of their debts.

The incident came to be called the Great Fear *(la Peur Grande)*. After it subsided, the rich remained apprehensive. They gradually realized that the peasants had enough power to seize property and commit acts of violence. The momentum for reform soon built to a fever pitch, and on the night of August 4, 1789, the French Assembly met and voted to abolish many ancient rights of the nobility. The French Revolution was under way.

Historians have often wondered what roused the peasants and precipitated the events of 1789. Essentially the fears were groundless—no more brigands than usual were about; and no evidence of a conspiracy existed among the peasants (the panic broke out in widely scattered communities, some separated by mountains). Nor did the timing seem to fit any political, economic, or sociological pattern. The episode seemed to be one of sheer wildness.

In the 1980s, Mary Kilbourne Matossian, from the University of Maryland, proposed a solution to the mystery of the Great Fear. She blamed the episode on ergot disease. Her studies of provincial records revealed a deterioration in public health in sections of France during mid-1789, as well as instances of nervous attacks and manic behavior. Bad flour was thought to be the cause, a factor that would tie in with the ergot disease theory. Another important clue had surfaced in 1974, when a historian reported that the rye crop of the late 1700s was "prodigiously" affected by ergot. He reported evidence of *Claviceps purpurea* in one-twelfth of all the rye. (In modern times, if one three-hundredth of the rye is infected, it cannot be sold.)

But why would so many peasants eat bad bread in 1789? Apparently it was a bad year for rye, and the cold winter and wet spring contributed to widespread ergot disease. Coincidentally, the Great Fear broke out just after the rye harvest. In retrospect, it is clear that the peasants resembled victims of ergot poisoning—hallucinations are common and delirium often sets in, together with seizures, jaundice, numbness, and a belief that ants are crawling under the skin. During the Middle Ages, the disease was called the "holy fire."

It would be overly simplistic to suggest that a fungus precipitated the French Revolution. Nevertheless, the evidence is substantial that ergot disease was a contributing cause. Certainly, the wild displays of the peasants must have been a terrifying sight to landowners. No doubt the fever pitch of the panic and the far-reaching consequences are better explained by the political and cultural climate of the times. Still, if the ergot disease had not happened. . .

headaches or to induce childbirth. The microbe also has potential as a bioweapon because it can be used to poison food or water supplies.

Basidiomycota

The best-known members of the division Basidiomycota are the mushrooms. Mushroom is the common name for the spore-producing body of the fungus, a structure called the basidiocarp; the basidiocarp (i.e., the mushroom) is composed of densely packed

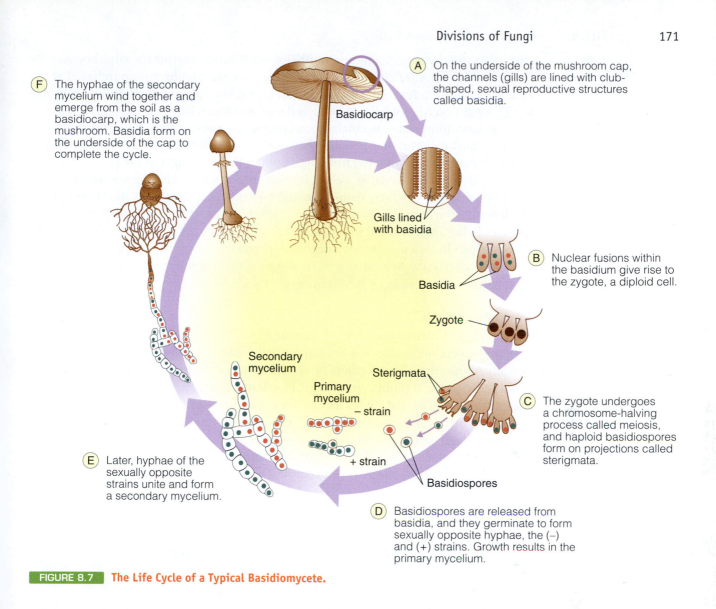

F. The hyphae of the secondary mycelium wind together and emerge from the soil as a basidiocarp, which is the mushroom. Basidia form on the underside of the cap to complete the cycle.

A. On the underside of the mushroom cap, the channels (gills) are lined with club-shaped, sexual reproductive structures called basidia.

Basidiocarp

Gills lined with basidia

Basidia

B. Nuclear fusions within the basidium give rise to the zygote, a diploid cell.

Zygote

Sterigmata

Secondary mycelium

Primary mycelium
− strain

C. The zygote undergoes a chromosome-halving process called meiosis, and haploid basidiospores form on projections called sterigmata.

E. Later, hyphae of the sexually opposite strains unite and form a secondary mycelium.

+ strain

Basidiospores

D. Basidiospores are released from basidia, and they germinate to form sexually opposite hyphae, the (−) and (+) strains. Growth results in the primary mycelium.

FIGURE 8.7 The Life Cycle of a Typical Basidiomycete.

hyphae. In its formation, the mycelium of the fungus forms underground, and the mushrooms develop at the outer edge of the mycelium where growth is most active. FIGURE 8.7 illustrates this life cycle. Sometimes mushrooms appear in rings, the "fairy rings" that seem to sprout overnight in meadows after heavy rains.

Like the hyphae of a scomycetes, those of basidiomycetes are divided by perforated septa. The basidiomycetes, however, have a more lengthy dikaryotic stage in their life cycle, a stage during which they have cells with two haploid nuclei—the basidiocarp is a dense mass of dikaryotic hyphae. Eventually, some of the nuclei fuse to form diploid nuclei; the cells then undergo meiosis and form sexually produced basidiospores. The latter form on a specialized hypha called a basidium (the Greek *basis* means "base"). Each basidiospore is haploid and develops at the tip of a spikelike process. It has been estimated that the average mushroom can contain 15 billion spores.

Members of the division Basidiomycota are called basidiomycetes, as well as "club fungi" because of the club-shaped basidia they form. There are about 25,000 named species in the division, including puffballs, shelf fungi, earthstars, stinkhorns, jelly fungi, and the familiar gill fungi, among which are the mushrooms purchased at the local supermarket. It should be noted that certain mushrooms are highly poisonous. **FIGURE 8.8** describes what can happen if one is not careful.

The basidiomycetes also include certain plant parasites that cause rust and smut diseases. Rust diseases are so named because of the distinct orange-red color on the infected plant, a color derived from pigments in the sporangia of the parasitic fungi. Wheat, oat, and rye plants, as well as lumber trees such as white pines, are susceptible. Smut diseases get their name from dark-pigmented fungi that infect and give a black sooty appearance to plants such as corn, blackberry, and various grains. These diseases bring about untold millions of dollars of damage each year. Another basidiomycete produces a unique enzyme that destroys the tough polymers of wood, causing white rot disease. Scientists are investigating this enzyme for its ability to destroy such toxic substances as DDT, dioxin, and other pollutants.

Deuteromycota

The division Deuteromycota includes about 25,000 fungal species for which a method of sexual reproduction has not been identified. For this reason, these fungi do not fit into any of the other three divisions, and the members are described as deuteromycetes, or Fungi Imperfecti ("imperfect fungi" in scientific jargon). In some cases, a sexual stage has been observed under laboratory conditions, but not in nature. If a sexual stage is identified, the species is usually classified in one of the other divisions.

Many fungi pathogenic to humans are found in the division Deuteromycota. These fungi usually reproduce by forming conidia (unprotected spores) or fragments of hyphae, which cling to surfaces. For example, fragments of the athlete's foot fungus are

1 In early January 1997, a man decided to play a round of golf at a course near his home in northern California. At the ninth hole, the man drove his ball into the woods. As he searched for the ball, he noticed a cluster of wild mushrooms and decided to pick some for dinner.

2 He cooked the mushrooms that evening, along with a steak. As he lived alone, he did not share the mushrooms with anyone else.

3 Three days later, the man experienced severe diarrhea and became very weak. He went to the emergency room, where liver tests revealed abnormal liver function.

4 His symptoms worsened, and within 2 days, his liver and kidneys were deteriorating. Doctors placed him on a kidney dialysis machine but it was too late. He died soon thereafter.

5 Investigators retraced the man's recent activities. At the golf course they found *Amanita phalloides*, an extremely poisonous mushroom. Lab tests confirmed the man's death from mushroom poisoning.

FIGURE 8.8 **A Fatal Case of *Amanita* Poisoning.** This case was related to wild mushrooms. Warm, heavy rainfall in the preceding weeks may have contributed to their unanticipated appearance in the environment.

picked up from towels used by infected individuals or from shower room floors, as we discuss in the next section.

One particularly interesting deuteromycete was found in 1999. This microbe, known as *Pseudomassaria*, apparently produces a chemical compound that mimics insulin; that is, the compound facilitates the passage of glucose into human cells. Researchers hope to develop the compound as a replacement for the insulin currently used by diabetics. One advantage of using this compound is that it may be taken by mouth, while insulin must be injected.

Beneficial and Harmful Fungi

Many fungal species live in a mutually beneficial relationship with other species in nature, an association called mutualism. In the southwestern Rocky Mountains, for instance, a fungus of the genus *Acremonium* thrives on the blades of species grass called *Stipa robusta* ("robust grass"). The fungus produces a powerful poison that can put horses and other animals to sleep for about a week (the grass is called "sleepy grass" by the locals). The animals soon learn to avoid the grass, and it survives where other grasses are nibbled to the ground. The results reflect the mutually beneficial interaction between plant and fungus.

Other benefits arising from fungi are equally apparent—as we shall see in this section, without fungi there would be no beer, wine, or other alcoholic products; lichens could not exist; and plants would not be able to grow as well as they do. Of course, without fungi, there would also be no fungal disease.

Yeasts

In microbiology, there are many uses of the word "yeast." A species of yeast, for example, causes the "yeast infection" that often occurs in women. This disease brings misery and pain (as we shall discuss presently) and is usually due to the yeast *Candida albicans*. By contrast, the yeasts we consider in this section are nonpathogenic and are found within the genus *Saccharomyces*. Species of *Saccharomyces*, pictured in FIGURE 8.9 , are used extensively in baking and brewing. Indeed, the word *Saccharomyces* translates as "sugar-fungus," a reference to the ability of the microbe to ferment sugars. Two species

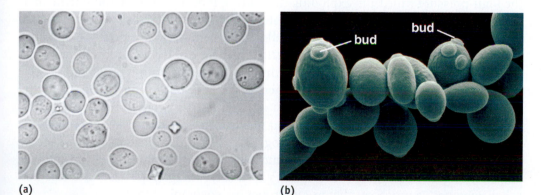

(a)   (b)

FIGURE 8.9   **Yeasts.** Two views of *Saccharomyces*, the common baking and brewing yeast.
(**a**) A photomicrograph of yeast cells used in beer fermentation (× 1000). (**b**) A colored scanning electron micrograph of *S. cerevisiae*. Several cells are budding.

of *Saccharomyces* called *S. cerevisiae* and *S. ellipsoideus* are of interest to us here: The former is commonly used for bread baking, and the latter is employed for making alcohol.

Yeast cells of the genus *Saccharomyces* are single-celled oval microbes about 8 μm in length and 5 μm in diameter. Yeasts usually reproduce by an asexual process called budding, in which a new cell grows at the periphery of the parent cell and then pinches off. Sexual reproduction also occurs in yeasts when two sexually opposite cells unite to form an enlarged cell equivalent to an ascus. Within the ascus, eight smaller cells similar to ascospores form. When these cells are released, each develops into a mature yeast cell. Because of this reproductive process, *Saccharomyces* is classified in the division Ascomycota.

Humans are particularly interested in the chemistry that yeast cells use to metabolize sugars because of the end-products that are formed. The metabolism of sugar begins with glycolysis, a process that most living things use to break down the simple sugar glucose into a smaller molecule called pyruvic acid. This process yields small amounts of energy trapped in a molecule called adenosine triphosphate (ATP). ATP is a type of chemical battery that can be used anywhere in the cell when energy needs arise.

Next comes a cyclic series of chemical reactions called the Krebs cycle. During the Krebs cycle, the pyruvic acid is further broken down, and when oxygen is present, many ATP molecules are produced. At the same time, six molecules of carbon dioxide are also produced for each glucose molecule metabolized. This chemistry has practical implications: When *S. cerevisiae* grows within fresh dough, the carbon dioxide accumulates, and the dough rises, resulting in the soft and spongy texture of bread. The yeast enzymes also break down some of the gluten proteins in the flour, thereby adding to the spongy texture. This very interesting process is explored in more detail in Chapter 12.

The chemistry in yeast cells is also a key element in the alcohol industry. When the oxygen in the yeasts' environment is depleted, the Krebs cycle grinds to a halt, and the yeast cells, especially those of the species *S. ellipsoideus*, shift their metabolism to the process of fermentation. Here, the yeasts change the pyruvic acid into carbon dioxide and ethyl alcohol—consumable ethyl alcohol. During controlled fermentations, *S. ellipsoideus* cultures are added to grape juice, and wine results. If there is some leftover fruit sugar after fermentation, a sweet wine results; if there is little or no leftover sugar, the wine is dry. Champagne contains extra carbon dioxide and is made by adding a bit of sugar to the wine to allow a second fermentation in the bottle (the thick glass bottle and the wire cage for the cork are necessary to prevent the high-pressure wine from exploding the bottle). If beer is desired, the starting material is barley grain, and flowers of a plant called hops are added to the grain to give beer its distinctive flavor. Indeed, yeast cells can ferment virtually anything that contains simple sugars. Chapter 9 discusses the chemistry of fermentation in more detail, and Chapter 12 explores some other interesting fermentation products.

In addition to their practical importance, yeasts have become increasingly important in biotechnology. They were the first eukaryotes subjected to manipulation by the techniques of genetic engineering, and they continue to play a leading role as organisms in which products of biotechnology are produced. For example, the vaccine for hepatitis B is currently produced by modifying yeast chromosomes through the addition of genes obtained from hepatitis B viruses. Chapter 14 describes this and other imaginative processes that use microbes as biochemical tools.

Moreover, yeasts have been used as a source of genes to produce yeast artificial chromosomes (YACs). These artificial nucleic acid molecules are essential tools for discovering the nature and sequence of the units of the hereditary material of humans, the so-called human genome project (Chapter 4). Yeasts are being used as the eukaryotic cells of choice for many experiments in microbiology because they are relatively harmless, multiply rapidly, and have yielded a wide array of biochemical information. In this regard, yeasts such as *Saccharomyces cerevisiae* have become the eukaryotic equivalent of *Escherichia coli*, a key workhorse of biotechnology.

Lichens

Lichens are organisms resulting from associations between a fungus and a photosynthetic partner, usually a cyanobacterium or a unicellular green alga (the fungus is a non-photosynthetic partner in the relationship). Such a combination is very different from either the fungus or the photosynthetic partner growing alone. Scientists point out that lichens provide an outstanding example of mutualism, the type of association in which both partners benefit. In the approximately 15,000 species of lichens, most of the fungal partners belong to the division Ascomycota. The most frequently encountered photosynthetic partners are species of the green algae *Trebouxia* and *Trentephila* and the cyanobacterium *Nostoc*. Species of one of these three genera are found in almost 90 percent of all lichens. About 90 percent of a lichen mass is fungal and about 10 percent is the photosynthetic partner. FIGURE 8.10 shows the relationship.

Lichens are extremely common in nature. They grow in such diverse environments as arid desert regions and Arctic zones, as well as on bare soil and tree trunks. Lichens are often the first organisms to occur in rocky areas, and their biochemical activities begin the process of rock breakdown and soil formation, a process that yields an environment in which mosses and other plants can gain a foothold.

Part of the reason lichens can grow in harsh environments is their minimal requirements for light, air, and nutrients. They do not require an organic food source because the photosynthetic partner provides sugars and other carbohydrates, while the fungal partner lends protection from harsh external environments plus nutrients for building the biochemical products of photosynthesis. Lichens rapidly absorb nutrients and

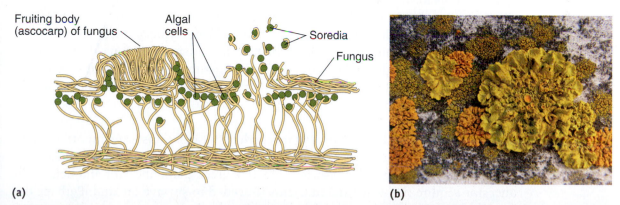

(a)   (b)

FIGURE 8.10 **Lichens.** (a) A cross section of a lichen, showing the upper and lower surfaces where tightly coiled fungal strands enclose photosynthetic algal cells. On the upper surface, a fruiting body, or ascocarp, has formed. Airborne clumps of alga and fungus called soredia are dispersed from the ascocarp to propagate the lichen. Loosely woven fungi at the center of the lichen permit the passage of nutrients, fluids, and gases. (b) A typical "crusty" lichen growing on the surface of a rock. Lichens are rugged organisms that can tolerate environments where there are few nutrients and extreme conditions. Their organic matter often forms the foundation of a local food chain.

other substances from rainwater, and because they are very susceptible to toxic compounds, they provide a sensitive index of air and water pollution.

In a lichen, specialized fungal hyphae either penetrate or envelop the photosynthetic cells. The photosynthetic partner is generally not exposed to the light directly, but enough light penetrates the layers of fungal hyphae to make photosynthesis possible. In addition, certain lichens include cyanobacteria, which trap atmospheric nitrogen gas and incorporate it into organic molecules that can be used for protein production. Lichens can survive with as little as 2 percent water by weight (compared to about 90 percent by weight for other organisms). When water is lacking, the lichens "rest," a condition that changes rapidly as water arrives during a rainstorm.

Reproduction in lichens is not well understood, but it may take place by a combination of normal sexual processes involving spore formation in the fungal component and mitosis of cells in the photosynthetic component. To establish a new population, a lichen breaks into fragments, which give rise to new populations (so long as both photosynthetic and nonphotosynthetic partners are present).

Lichens grow very slowly (the growth of a lichen's diameter may be less than 10 millimeters in a year). This very low but relatively constant rate of growth has encouraged scientists to use lichens to date rocks and human artifacts. For example, the giant carved statues found on Easter Island in the Pacific have been estimated to be 400 years old based on nearby lichen size and growth rate.

Mycorrhizae

Mycorrhizae are mutualistic associations between fungi and vascular plants (trees, flowers, vegetables, and other complex plants). The term *mycorrhizae* literally means "fungus-roots." Over 5000 species of fungi participate in mycorrhizal associations, and the roots of over 80 percent of all vascular plants are believed to have these fungi (most species of trees, for example, have fungal partners). Mycorrhizal associations benefit both the fungus and the plant: The fungal cells supply the plant with more nutrients (especially phosphorus) than it can absorb through its roots alone; meanwhile, the plant supplies the fungus with products of photosynthesis that provide the raw materials for its metabolism. The fungus also absorbs water and passes it to the plant, a great advantage to the plant when the soil is dry and sandy.

All common garden plants also form mycorrhizae, typically with species of zygomycetes. An intriguing observation is that, under certain conditions, the fungi function as bridges through which phosphorus, carbohydrates, and other substances pass from one host plant to another. Plants that participate in mycorrhizal associations tend to grow larger and more vigorously than plants that lack a fungus, especially in soils that are poor.

The study of fossils indicates that mycorrhizal associations were important in the plant invasion of the land over 400 million years ago. It is conceivable that an aquatic fungus may have formed an association with a green plant ancestor and helped the ancestor acquire the water and nutrients it needed to survive on land. Perhaps this step was the critical turning point in the evolution of plants.

Fungi and Human Diseases

Fungal diseases affect many body regions and result in mild problems such as athlete's foot as well as extremely dangerous diseases such as spinal meningitis. Moreover, fungal diseases may affect numerous body regions at the same time, and many of these diseases are related to a weakened immune system.

Among contemporary health problems is the yeast infection caused by *Candida albicans*. This organism is an oval yeast when it infects body tissues, but it grows as filamentous hyphae when cultivated in the laboratory. An organism having this variability in form is described as dimorphic. *C. albicans*, shown in FIGURE 8.11, is often found in a woman's vaginal tract, where it is held in check by the lactobacilli and other types of bacteria. Many of these bacterial species produce lactic acid and other acids in their metabolism, and the acids create an inhospitable environment for *C. albicans*. However, when a woman takes an excessive amount of antibacterial antibiotic, the lactobacilli are killed, the acid soon disappears, and the *C. albicans* flourishes. Other predisposing factors are corticosteroid treatment, pregnancy, diabetes, and tight-fitting garments (which increase the local temperature and humidity). Transmission of the yeast can occur during sexual intercourse (yeast infection is considered a sexually transmitted disease).

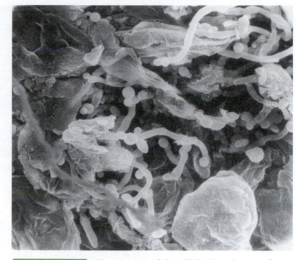

FIGURE 8.11 **The Agent of Candidiasis.** A scanning electron micrograph of *Candida albicans* associated with the tissues of an animal. The oval structure of the cells and the tendency to form hyphae are apparent.

The symptoms of yeast infection include itching sensations, burning internal pain, and a white "cheesy" discharge. Treatment is usually successful when women use antifungal antibiotics, including miconazole, ketoconazole and clotrimazole. Women are also encouraged to eat yogurt, since this dairy product contains large amounts of lactobacilli. The lactobacilli work their way from the intestinal tract to the vaginal tract and reestablish themselves to restore its acidity. Suppositories containing lactobacilli are also available.

Candida albicans is also an inhabitant of the mouth. The microbe proliferates when the immune system is compromised (as in an AIDS patient) or when other controlling microbes disappear. The proliferating yeasts form small white flecks that grow together to become soft, crumbly, milklike curds on the tongue and mucous membranes of the oral cavity. This disease is usually called thrush (because of the appearance of the tissues), although the more correct term is candidiasis. Oral suspensions of antibiotics will bring it under control. An unexplained and severe case of candidiasis is regarded as an early sign of AIDS.

Cryptococcosis is a dangerous disease of the human lungs and meninges, the three membranous coverings of the spinal cord. It is caused by *Cryptococcus neoformans*, a fungus that is found in soil and grows in pigeon droppings (FIGURE 8.12). Gusts of wind bring the fungus to the human respiratory tract, and from there it passes into the bloodstream and meninges. Piercing headaches, neck stiffness, and paralysis usually follow. Resistance to the disease depends on a healthy immune system. Indeed, when the T cells of the immune system are destroyed (as in AIDS patients), the fungus causes serious illness and death.

■ *Cryptococcus neoformans*
krip′tō-kok-kus
nē-ō-fôr′manz

Skin infections due to fungi are referred to as athlete's foot, barber's itch, jock itch, and numerous other names, including ringworm. A general name applied to the infections is tinea, meaning "worm." In premodern times, people thought they saw worms in the scaly ring of blisters occurring on the head—hence, "ringworm." Physicians now

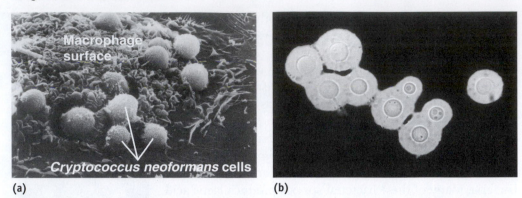

(a)   (b)

FIGURE 8.12 *Cryptococcus neoformans.* (**a**) A scanning electron micrograph of *Cryptococcus neoformans* clinging to the surface of a human macrophage (a type of phagocytic white blood cell). Phagocytosis of the fungal cells is retarded by the capsules they possess. In this view, the cells have been attached to the macrophage surface for 30 minutes, and still have not been phagocytized. (**b**) A photomicrograph of negatively stained *Cryptococcus neoformans*. A distinct capsule surrounds each oval, yeastlike cell.

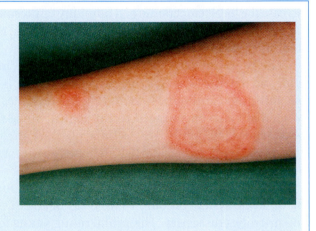

1 In the fall of 1992, a number of wrestlers from the United States and abroad attended a meet in Schaumberg, Illinois.

2 The wrestlers competed in several divisions until the divisional champions were decided. Skin contact between the wrestlers is routine during the bouts.

3 On returning home, a number of participants noticed scaly, pink blotches on their shoulders, neck, or face.

4 Mycologists took skin samples and scrapings from those affected and cultivated *Trichophyton tonsurans*. The diagnosis was ringworm (tinea corporis).

FIGURE 8.13 **An Outbreak of Ringworm.** This outbreak occurred among participants at an international wrestling meet. The incident happened in Schaumberg, Illinois, in 1992. It was believed to be one of the first epidemics of transmissible ringworm reported in the United States.

recognize that a variety of fungi cause the conditions characterized by blisters on the skin, along the surface of the nails, or in the webs between the fingers and toes. Unfortunately the fungi live for long periods of time on shower room floors or mats, as well as on combs, hats, towels, and numerous other objects. Individuals can also acquire the tinea diseases by contact with household pets or infected persons (a serious outbreak of ringworm occurred after a wrestling meet in Illinois in 1992, as **FIGURE 8.13** relates).

■ *Coccidioides immitis*
kok-sid-ē-oi-dēz
im′mi-tis

Travelers to the dry regions of the southwestern United States should be on the alert for respiratory symptoms associated with infection by *Coccidioides immitis*. This fungus exists in the soil and is carried with air into the lungs, where it causes an influenzalike illness that may progress to meningitis. Residents of the San Joaquin Valley of California have been cautioned to watch for "valley fever," especially after dust storms or earthquakes. *C. immitis* has also attracted the attention of bioterrorists because it is easily inhaled with air. Although the illness it causes is not lethal, it is nevertheless incapacitating, and the fear and hysteria generated in a population by an attack with this fungus would be extremely disruptive.

Other fungal lung diseases are caused by *Histoplasma capsulatum* and *Blastomyces dermititidis*. Both microbes are associated with bird droppings, particularly near barns and sheds, and they are probably transmitted in breezes and wind gusts. They cause mild respiratory illnesses, unless the immune system has been compromised; in that case, the illness can involve multiple body organs and become quite serious. During 1995, singer Bob Dylan suffered illness due to *H. capsulatum*. He was successfully treated for infection of the heart valves and resumed his singing tour. When his doctor was asked about how Dylan may have acquired the infection, the physician whimsically made reference to one of the singer's more famous compositions "The fungus," he said, "was probably 'blowing in the wind.'"

■ *Histoplasma capsulatum*
his-tō-plaz′mä
kap-sulä′tum

■ *Blastomyces dermititidis*
blas-tō-mī′sez
dėr-mä-tit′i-dis

A FINAL THOUGHT

In ancient Rome, mushrooms were the food of the gods, and only the emperors were permitted to partake of their pleasures. Today, exotic mushrooms enjoy an equally high reputation among the world's gourmets. Some experts know how to spot them in the wild, but for amateurs, the key word is "caution"—in mushroom hunting, ignorance is disaster.

Mushrooms come in a huge variety of shapes, colors, and sizes. Among the interesting wild mushrooms are the jack-o-lantern fungus, known for its luminous gills; the beefsteak fungus, whose cap resembles a piece of raw beef; and the bird's nest fungus, in which the fruiting body and its spores look like a bird's nest with eggs. On the debit side, about 100 of the 2000 known species can cause fatal mushroom poisoning and death. High on the list of dangerous mushrooms are *Amanita verna*, the destroying angel, and *Amanita phalloides*, the deathcap.

■ *Amanita phalloides*
am-an-ī′tä fal-loi′dez

Botanists urge that mushrooms be hunted with a camera rather than a fork and plate. They point out that the colors and settings encourage prize-winning photography, and they urge that mushroom consumption be limited to those species cultivated for use as food. After all, they reason, birdwatchers do not eat birds, so why should mushroom watchers eat mushrooms?

For those who insist on eating wild mushrooms, mycologists recommend joining a society, reading extensively, and treading lightly into this hobby. As one sage has written:

There are old mushroom hunters,
And there are bold mushroom hunters,
But there are no old, bold mushroom hunters.

QUESTIONS TO CONSIDER

1. A homemaker decides to make bread. She lets the dough rise overnight in a warm corner of the room. The next morning she notices a distinct beerlike aroma in the air. What is she smelling, and where did the aroma come from?
2. Fungi are extremely prevalent in the soil, yet we rarely contract fungal disease by consuming fruits and vegetables. Why do you think this is so?

3. A student of microbiology proposes a scheme to develop a strain of bacteria that could be used as a fungicide. Her idea is to collect the chitin-containing shells of lobsters and shrimp, grind them up, and add them to the soil. This, she suggests, will build up the level of chitin-digesting bacteria. The bacteria would then be isolated and used to kill fungi by digesting the chitin in fungal cell walls. Do you think her scheme will work? Why?

4. Mr. A and Mr. B live in an area of town where the soil is acidic. Oak trees are common, and azaleas and rhododendrons thrive in the soil. In the spring, Mr. A spreads lime on his lawn, but Mr. B prefers to save the money. Both use fertilizer, and both have magnificent lawns. Come June, however, Mr. B notices that mushrooms are popping up in his lawn and that brown spots are beginning to appear. By July, his lawn has virtually disappeared. What is happening in Mr. B's lawn, and what can Mr. B learn from Mr. A?

5. A mushroom walks into a bar and orders a beer. "Sorry," says the bartender, "we don't serve mushrooms." The mushroom thinks for a moment and replies, "But I'm a fun-guy." When you have recovered from this dreadful attempt at humor, you might like to try your hand at another "fun-guy," or "fungus" joke.

6. In 1991, the U.S. Food and Drug Administration approved for over-the-counter sale a number of antifungal drugs, including clotrimazole and miconazole (Gyne-Lotrimin and Monistat, respectively). It thus became possible for a woman to diagnose and treat herself for a vaginal yeast infection. Should she do it?

7. In 1992, residents of a New York community, unhappy about the smells from a nearby composting facility and concerned about the health hazard the facility posed, had the air at a local school tested for the presence of fungal spores. Investigators from the testing laboratory found abnormally high levels of *Aspergillus* spores on many inside building surfaces. Is there any connection between the high spore count and the composting facility? Is there any health hazard involved?

■ KEY TERMS

Informative facts are necessary for the expression of every concept, and the information for a concept is founded in a set of key terms. The following terms form the basis for the concepts of this chapter. On completing the chapter, you should be able to explain and/or define each one:

arthrospore	cellulase
ascocarp	chlamydospore
ascomycete	chitin
ascus	conidia
ascospore	deuteromycete
basidiocarp	dikaryotic
basidiomycete	diploid
basidiospore	fermentation
basidium	fungi

haploid
hypha
lichen
ligninase
mycelium
mycorrhizae
mycosis
rust disease

septa
smut disease
yeast infection
yeasts
zygote
zygomycete
zygospore

http://microbiology.jbpub.com/book/microbes

The site features **eLearning**, an online review area that provides quizzes and other tools to help you study for your class. You can also follow useful links for in-depth information, read more stories of microbiology, or just find out the latest microbiology news.

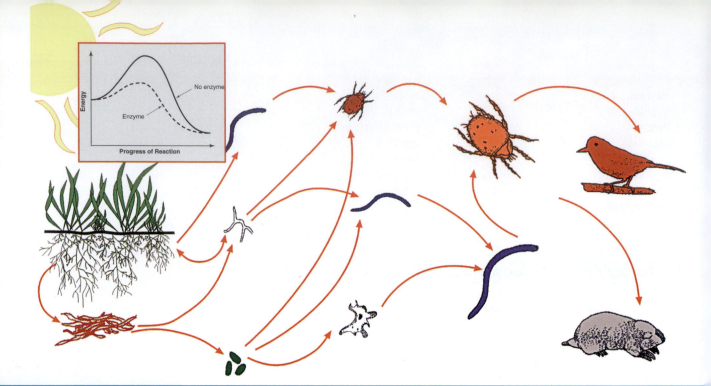

9

Growth and Metabolism: Running the Microbial Machine

The food chain shows that all food is originated by photosynthesis and that otherwise, all food comes from the growth of living things. If one can find magic in life, it is in the regulation of energy flow; a process known as metabolism. While machines and engines can burn fuel, release energy and do work, only living things burn fuel (food) and use the energy released to make self (growth). Metabolism and growth require enzyme activity to control the step-wise release of energy. The inset shows that enzymes facilitate and regulate the flow of energy from food to the making of self.

Looking Ahead

Microbes share with humans many aspects of growth and metabolism. However, there are numerous aspects that set microbes apart and lend them uniqueness, as we shall see in this chapter.

On completing this chapter, you should be able to . . .

- understand the dynamics of a growth curve for a microbial population and specify some of the influences affecting the shape of the curve.
- recall some factors such as variations in water, temperature, and oxygen level that encourage or limit the growth of microbes.
- discuss the broad diversity of microbes in the Earth's environment and understand why that diversity is important to survival.
- recognize the roles of enzymes and energy in microbial metabolism.
- describe how carbohydrate molecules are metabolized through a series of pathways to supply the energy needed for microbial growth.
- appreciate the chemistry, significance, and application of the fermentation process used to produce wines and spirits.
- explain how energy from the sun is used by microbes during photosynthesis to construct the carbohydrates used by myriad life forms on Earth.

Among the more engaging stories of 1996 was the hypothesis that life may now exist or once did exist on Mars. In August of that year, David E. McKay and his colleagues at NASA raised eyebrows with their announcement that a 4.3-pound, potato-sized meteorite from space apparently contained fossilized microbes.

To be sure, scientists agreed that the meteorite (named ALH84001) was from Mars because, on heating, it gave off a mixture of gases unique to the Martian atmosphere; they agreed that crystals in the meteorite looked like crystals produced by Earth's bacteria; and they agreed that the meteorite contained polycyclic aromatic hydrocarbons (a type of compound containing rings of carbon atoms bonded to hydrogens). The latter are found on Earth and are often associated with living things. But they sharply disagreed about whether the meteorite really had "microbial fossils."

Nevertheless, the electron microscope views of the "fossils," shown in **FIGURE 9.1A**, grabbed the attention of researchers throughout the world. The wormlike, tubular shapes with rounded edges were about 20 nm long and resembled microbes from Earth, but in miniature. Wishing to avoid the term "breakthrough," McKay and his colleagues expressed cautious optimism about their discovery.

And well they did, for within fifteen months, scientists had answered most of McKay's observations with reasonable alternatives. New images of the Martian meteorite (**FIGURE 9.1B**) indicated that the microfossils could be narrow ledges of mineral protruding from the underlying rock. Furthermore, scientists concluded that 20 nm of space is simply too small to house even the most basic chemical machinery of life. Moreover, the hydrocarbons might have come from inorganic substances just as plausibly as from life-associated substances (the possibility of earthly contamination was raised). Even if the hydrocarbons were of Martian origin, they could have come from a primordial soup that never gave rise to a life form.

Is the controversy over? Not as of this writing (although the McKay team has withdrawn some of its hypothesis). A NASA probe to Earth's distant neighbor is scheduled to return in 2008 with soil samples for more careful study. For the moment, "the hypothesis has not fared well," as one Harvard scientist understated in 1999.

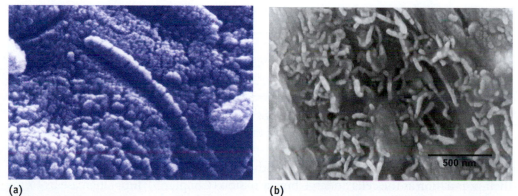

(a)   (b)

FIGURE 9.1   Life on Mars? (**a**) In 1996, investigators presented electron microscope views of fragments from a meteorite originating from Mars. Although the study is controversial, the investigators suggested that the photographs provide evidence for possible microbial life on Mars. This photograph supposedly depicts a wormlike microfossil observed in meteorite ALH84001. (**b**) Investigators disputing the claim took photographs of narrow mineral ledges from the meteorite and showed how they resemble bacteria.

Many people are intrigued by the possibility of extraterrestrial life, and one day microbes may be found in outer space. For the time being, it is astonishing to note the out-of-the-way places on Earth where microbes have been found. For example, one study in 1995 reported bustling microbial communities living at about 115°C (about 265°F) within an undersea volcano near Polynesia. Another study in 1997 reported microbial populations found at an almost unbelievable 33,000 feet below the surface of the Earth, in the mud at the bottom of the Marianas Trench—the lowest spot on Earth. Many of the species had never been seen or classified before.

When such microbes are brought to the laboratory for study, microbiologists are faced with the task of duplicating the conditions under which the organisms grew. There follow many hours and weeks of trial and error before the correct nutritional patterns are established (sometimes they are never established). Temperature and pressure are important considerations, but scientists must also locate sources of carbon and energy for the microbes while developing methods for their cultivation. As we shall see in this chapter, varied microbes require equally varied conditions for growth.

◼ Microbial Growth

Microbes are metabolic machines with the potential for explosive growth and multiplication. Sometimes we suffer the consequences of microbial growth, such as when microbes grow in our bodies, but other times scientists are able to take advantage of microbial growth by harvesting the valuable products of their metabolism. For example, sewage plant operators stimulate bacteria, algae, and other microbes to grow and break down sewage and return the by-products to the soil. And industrial microbiologists grow microbes in building-sized vats to obtain vitamins, amino acids, and other growth factors valuable for human nutrition (Chapter 14 explores these industrial applications in depth).

A major objective of microbiologists is to study the vagaries of microbial growth and manipulate it so that they can better use microbes both to enhance the quality of human life and to minimize the microbes' harmful effects. In addition, the study of microbial growth gives researchers a glimpse into the characteristics of microbes and helps them understand their life processes. In doing so, they gain a better appreciation for all life processes.

Population Growth

As a microbial population develops under relatively stable environmental conditions, its numbers undergo a predictable series of events that result in a mathematical graphic display known as a growth curve. Growth curves are drawn for protozoa and algae as the population develops in liquid media, while such curves are drawn for fungi and bacteria based on their development in either solid or liquid media. Bacteria increase their population as the cells undergo binary fission (Chapter 5); fungi increase by forming new cells at the tips of their hyphae through the process of mitosis (Chapter 8). Protozoa and algae also undergo mitosis to increase their numbers (Chapter 7).

The growth curve is plotted with logarithmic numbers (powers of 10); it shows the growth of a microbial population over a period of time. (Logarithmic numbers are used because they are more manageable—normal ordinal numbers would go off

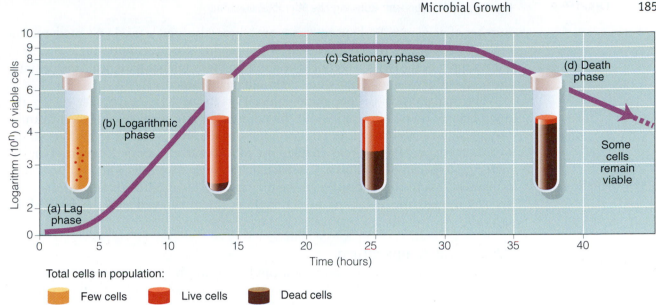

Total cells in population:

Few cells   Live cells   Dead cells

FIGURE 9.2   The Growth Curve for a Bacterial Population. (**a**) During the lag phase, the population numbers remain stable as bacteria prepare for division. (**b**) During the logarithmic phase, the numbers double with each generation time. Environmental factors later lead to cell death, and (**c**) the stationary phase shows a stabilizing population. (**d**) The decline (death) phase is the period during which cell death becomes substantial.

the scale very rapidly.) Plotting the numbers of microbes versus time results in the curve shown in FIGURE 9.2 . Various phases in the curve can be identified, as we discuss next. The phases show the dynamics of the population as time passes.

The first phase of growth is the lag phase. During this "tooling up" phase, no population increase is observed. The microbes have arrived at their new location (a piece of food, for example), and they are synthesizing cell parts and enzymes for digesting the nutrients in the food sample. In addition, the food may have been placed in the refrigerator after microbes entered it (from an errant sneeze, for example), and the microbes may need to adapt to the colder environment during this lag phase. Reproduction and formation of new cells is balanced by the death of others.

The next phase is referred to as the logarithmic (log) phase. Here the microbes grow and divide at the maximum possible rate. A bacterial population, for instance, doubles at regular intervals called the generation time. The growth curve rises smoothly in an upward direction when the logarithms of the actual numbers of bacteria are plotted. In food, for instance, the bacteria find that nutrients are plentiful, the oxygen supply is adequate, and others factors such as acidity and moisture are suitable. In the laboratory, the population growth of bacteria may be so vigorous that visible masses called colonies appear on solid media, each colony consisting of millions of bacterial cells; and broth media may become cloudy (turbid) as bacterial numbers proliferate. Because the population is at its biochemical optimum, research experiments are generally performed during the log phase. The vigorous growth of this phase is attractive to bioterrorists because it means they can cultivate enormous quantities of bioweapons in a relatively short period of time.

After a period of time, the microbial population enters a stationary phase. During this phase, many microbes are dying, and the number of dead microbes balances the number of new ones. Perhaps nutrients are scarce (the food is used up), predator microbes have entered the environment, waste products are accumulating, water is in short supply, or the oxygen is gone. The curve on the graph has flattened. (Incidentally, the food will now smell "ripe," and it should be discarded.)

The final phase of the population's history is the decline (death) phase. Here the microbes are dying off rapidly, and the population size is decreasing significantly. Having a glycocalyx may forestall death for a bacterium by acting as a buffer to the environment, and flagella may enable a microbe to move to a new location. If the organism is a species of *Bacillus* or *Clostridium*, the vegetative cells will revert to spores (Chapter 5), and the stationary phase may extend for months or years. For many species, though, the history of the population comes to an end with the death of the last cell.

■ *Clostridium*
klôs-tri′dē-um

Water and Temperature

The growth of a microbial population can be significantly influenced by aspects of the chemical and physical environment. Microbial cytoplasm is water-based, for instance, and a liquid environment is absolutely required if life is to continue. (Notable exceptions are bacterial and fungal spores, which survive arid environments).

Another significant aspect is temperature: It influences the rate of enzyme activity, and enzymes are protein molecules responsible for all chemical reactions taking place in microbes (as we discuss presently). If the temperature is too low, the rate of the chemical reactions is reduced; if the temperature is too high, enzymes may be broken down by the heat and the reactions will cease.

Microbes have adapted to most temperatures found on Earth (FIGURE 9.3). Those microbes growing best at temperatures between 0°C and 20°C are said to be psychrophilic (the organisms are called psychrophiles); they are found in Arctic and Antarctic environments as well as deep below the ocean surface. One example is an alga called *Chlamydomonas nivalis*. It grows in snowfields, and its spores cause the snow to appear bright red (as we discuss in Chapter 7). It is somewhat disconcerting to see red snow, but after a while you learn to expect the unexpected from microbes.

Psychrophilic bacteria also grow well in the refrigerator (at 5°C) and spoil foods such as milk. Streptococci grow in milk and deposit acid, causing it to become sour. Although these microbes do not represent a threat to health, their presence makes the milk undesirable to the eye, nose, and taste buds. On the other hand, some dangerous bacteria such as species of staphylococci also grow in refrigerated foods and deposit their diarrhea-inducing toxins. Consumption of the food usually leads to a very unpleasant experience (unless the food is thoroughly heated at a high temperature to destroy the toxins).

Another group of microorganisms are described as mesophilic (or mesophiles). These grow best at temperatures between 20°C and 45°C. At one time, scientists believed that most microbial species were mesophiles, but the growing catalog of microbes living at extreme temperatures is causing them to rethink that notion. Because the body temperature of warm-blooded animals, including humans, is about 37°C, mesophiles grow well in the body—most human pathogens are mesophiles.

Microorganisms that grow at high temperatures are said to be thermophilic (or thermophiles). These microbes grow at temperatures of 45°C and higher. One species

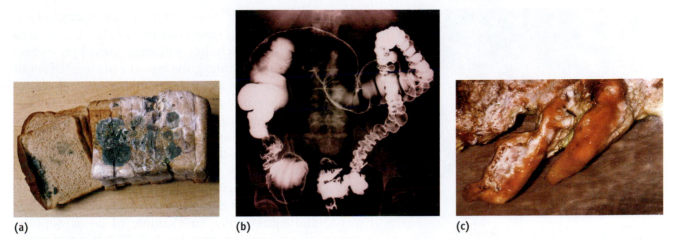

(a)   (b)   (c)

FIGURE 9.3 **Environments of Various Temperatures and the Microbes That Grow within Them.** (**a**) Food spoilage caused by various cold-tolerant microbes in a refrigerator. (**b**) Human large intestines harbor *E. coli*. (**c**) Thermophilic algae.

living at 75°C (167°F) below the Virginia landscape has been named *Bacillus infernus* for its hellish habitat. Some species of thermophilic microbes can grow at temperatures as high as that of boiling water (100°C or 212°F), and some grow at an astounding 113°C (about 260°F); examples of the latter are species of Archaea discussed in Chapter 5. Thermophilic microbes also thrive in such varied environments as compost heaps, hot springs, and thermal vents in the oceans.

Oxygen and Acidity

Another factor governing microbial growth is the presence or absence of oxygen. Most known species of microbes are aerobic; that is, they require oxygen for their metabolism (as do humans). By comparison, certain species of microbes are anaerobic; that is, they can live only in the absence of oxygen and will die when it is present. Such environments as landfills (tightly packed with garbage) and dense, muddy swamps provide breeding grounds for these microbes. Some species of pathogens are included in the anaerobic group: The bacterium that causes tetanus, for instance, grows in the anaerobic, dead tissue of a wound. Here it produces powerful toxins that bring about uncontrolled spasms of the muscles. For this reason, deep puncture wounds necessitate quick attention.

Some species of microbes can grow in the presence or the absence of oxygen. These microbes, described as facultative, are among the most interesting organisms known because they can adapt quickly to and live in either aerobic or anaerobic environments. Scientists believe that a majority of bacterial species may be facultative. Unfortunately, humans cannot make the same claim.

In addition to the aerobic and anaerobic microbes, several species are microaerophilic. These microbes grow best in a low-oxygen environment. In the human body, they can be found in the urinary and digestive tracts, where oxygen is scarcer than it is outside the body. Microaerophiles are also found in deep soil layers.

Another environmental factor of importance is the acidity or alkalinity of the surroundings. The acidity of a medium is expressed as pH, a logarithmic measurement

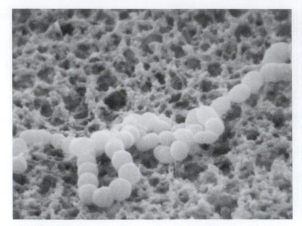

FIGURE 9.4   A Species of Acid-Loving Streptococcus. This species of Streptococcus is *S. lactis*. It thrives under acidic conditions and is used industrially to ferment milk and produce lactic acid in dairy products.

■ *Helicobacter pylori*
    hē'lik-ō-bak-tėr pī'lō-rē

of the hydrogen ion (H⁺) concentration. Most species of microbes have an optimal (most desired) pH level, as well as a pH range within which they will grow. Many known bacterial species, for example, grow best at a pH level of about 7.0, with a range as low as 5.0 and as high as 8.0. Other bacterial species grow best at the very acidic pH level of 3.0—the bacteria that deposit acid in cabbage and convert the cabbage to sauerkraut are an example.

Species of microbes that thrive in acid conditions, such as those that produce sauerkraut, are said to be acidophiles or (acidophilic). Other acidophilic bacteria are used in dairy plants to produce sour dairy products such as yogurt, sour cream, and buttermilk. An example of an acidophilic *Streptococcus* species is pictured in FIGURE 9.4 . Another acidophilic bacterium called *Helicobacter pylori* is known to cause gastric ulcers when it multiplies in the lining of the stomach. Most known species of fungi are acidophilic, tolerating pH levels of about 5.0—this is why fungi contaminate acidic fruits such as oranges, lemons, and limes as well as acidic vegetables such as tomatoes. Fungi are also commonly found in cheeses, which tend to have an acidic pH.

Other Factors

In addition to the factors we have discussed, others such as salt and pressure may influence the growth of a microbial population. Microbes that normally inhabit high-salt environments are said to be halophilic (the organisms are called halophiles, or "salt lovers"). Examples are the halobacteria, a group of archaebacteria that require a minimum salt concentration of 9 percent for growth (as compared to 1 percent for many eubacterial species)—for some halobacteria, an incredibly high salt concentration of 27 percent is most desirable. Although many people are inclined to believe that halobacteria are inhabitants of the oceans, the salt concentration of the ocean is only about 3.5 percent, and true halobacteria are more commonly found in Utah's Great Salt Lake (FIGURE 9.5).

Microbes that tolerate high-pressure environments are referred to as barophilic (or barophiles, meaning "pressure lovers"). Barophiles are found at the bottoms of the oceans, some species living an astounding six miles beneath the surface. Cultivating these microbes in the laboratory presents a considerable challenge because the pressure, oxygen, pH, and nutrient levels of the ocean depths must be matched.

Still another environmental factor is radiation, including X rays, gamma rays, and ultraviolet radiation. Ultraviolet radiation is a component in sunlight and is often deleterious to microbes inhabiting the upper layers of the soil. However, there are microbes that can resist the effects of radiation. One notable species is *Deinococcus radiodurans*, a bacterium found in radioactive wastes. Scientists hope to harness the metabolic powers of this microbe to help dispose of radioactive waste materials (Chapter 16), which are a continuing concern as pollutants of the environment. If the scientists are successful, they will once again show that studying the characteristics of microbes can result in considerable benefit to society.

FIGURE 9.5   **Home to Halophilic Bacteria.** The Great Salt Lake in Utah provides the high-salt environment favored by halophilic bacteria.

Microbial Metabolism

The term "metabolism" is derived from the Greek word *metaballein*, which translates as "change." For our purposes, metabolism refers to all the chemical changes occurring in a microbe during its growth and development. These chemical changes maintain the stability of the microbial cell, while providing a dynamic pool of chemical building blocks for the synthesis of new cellular materials, especially when microbes produce exact copies of themselves.

Metabolism also refers to the chemical changes going on in the complex cells of plants and animals (including humans). Thus, many of the concepts in this section apply equally well to more complex living things. Indeed, much of the metabolism of human cells was first discovered in microbial cells, and scientists were quite surprised to find that the two metabolisms are parallel.

Although there are thousands and thousands of different chemical reactions going on in cells at any one time, the great majority of these reactions fall into either of two general categories: biosynthesis reactions (also known as anabolism), and digestive reactions (also referred to as catabolism). Biosynthesis is a broad term that applies to any chemical process resulting in the formation of cellular structures and molecules. It is essentially a building process in which larger molecules are formed from smaller ones, a process that generally requires an input of energy. Digestion, by comparison, is the general chemical process in which large molecules are broken down into smaller ones. In this process, chemical linkages are usually broken, and energy is often liberated.

The chemical reactions of metabolism are the fundamental underpinnings for such activities as movement, growth, synthesis, and use of foodstuffs. The reactions are highly organized and responsive to cellular controls. For example, a cell will produce the chemical substances necessary for the breakdown of certain carbohydrates only when those carbohydrates appear in its local environment. At other times, production

A CLOSER LOOK

"Hans, Du Wirst Dwas Nicht Glauben!"

Louis Pasteur's discovery of yeast's role in fermentation heralded the beginnings of microbiology because it showed that tiny microbes could bring about important chemical changes. It also opened debate on exactly how yeasts accomplish fermentation. Within months, a lively controversy had evolved among scientists. Some thought that grape sugars enter yeast cells to be fermented, while others believed that fermentation occurs outside the cells. The debate would not be resolved until a fortunate accident happened in the late 1890s.

In 1897, two German chemists, Eduard and Hans Buchner, were preparing yeast as a nutritional supplement for medicinal purposes. According to well-established procedures, they ground yeast cells with sand and collected the cell-free juice. To preserve the juice, they added a large quantity of sugar (as was commonly done at that time) and set the mixture aside. Several days later they noticed an alcoholic aroma coming from the mixture. One taste confirmed their suspicion: The sugar had fermented to alcohol.

The discovery of the Buchner brothers was momentous because it showed that fermentation can occur without living yeast cells. Moreover, it was obvious that a chemical substance from inside the cells was required for fermentation. The scientists named the elusive substance "enzyme," meaning "in yeast."

In 1905, the English chemist Arthur Haden added to the Buchner study by showing that the "enzyme" is really a multitude of chemical compounds; he suggested that a better term would be "enzymes." In doing so, he added to the belief that fermentation is a chemical process dependent on living things. Before long, many chemists began studying life processes in earnest, and biochemistry gradually emerged as a new scientific discipline.

of the substances is suppressed. This form of cellular economy ensures that the reactions of metabolism are used efficiently. We shall begin our discussion by examining the molecules that catalyze the chemical changes of metabolism. As you shall see, many of the descriptions to follow are somewhat involved, but rest assured that they have been simplified as much as possible without skipping key elements of the processes.

Enzymes

The reactions of metabolism are brought about by a special class of molecules known as enzymes (this term translates literally as "in yeast," as A Closer Look above points out). Enzymes are organic molecules that act as catalysts for the reactions of microbial metabolism. They increase the rate of a metabolic reaction without themselves changing. The reaction would probably take place if the substances involved were left to themselves, but it might take days, months, or years to occur; enzymes speed up the rate of the reaction to a few milliseconds. Certain enzymes are composed of protein alone, while others are composed of protein plus a nonprotein portion. Moreover, some enzymes depend for their action on key additions called coenzymes. (We shall encounter several coenzymes presently.) Finally, most enzymes are folded into a three-dimensional structure, and unfolding, which can be caused by excessive heat, can result in their inactivation. This is one reason why extreme heat kills microbes.

An enzyme promotes a reaction by providing a site where reactant molecules can be positioned while they interact. The reactant molecules are known as substrates,

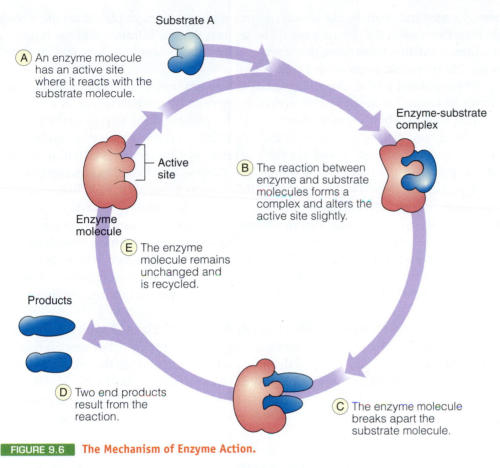

A An enzyme molecule has an active site where it reacts with the substrate molecule.

Substrate A

Enzyme-substrate complex

Active site

B The reaction between enzyme and substrate molecules forms a complex and alters the active site slightly.

Enzyme molecule

E The enzyme molecule remains unchanged and is recycled.

Products

D Two end products result from the reaction.

C The enzyme molecule breaks apart the substrate molecule.

FIGURE 9.6   **The Mechanism of Enzyme Action.**

and the products of the reaction are known as end products. During the reaction, the enzyme is not used up; it functions over and over again. The reaction involving the substrate molecules takes place at a specific location on the enzyme molecule known as the active site. When the correct fit occurs, the enzyme-substrate complex enters a transition state, and the enzyme facilitates the breaking or construction of new chemical bonds.

Over 2000 enzymes have been discovered and named, and many thousands more probably await discovery. The international system for naming enzymes stipulates that each name has two parts: a prefix that usually reflects the substrate on which the enzyme acts and the suffix *-ase*. Thus, lactase is the enzyme that breaks down lactose, and lipase breaks down lipids. In many cases, a common name is used for the enzyme; an example is catalase for the enzyme that breaks down hydrogen peroxide into water and oxygen. Enzyme activity is illustrated in   **FIGURE 9.6** .

Energy

One of the major goals of metabolic reactions is to establish a readily accessible supply of a substance that stores energy and releases it when needed. A steady supply of energy is necessary not only to perform the processes of life but also to maintain order. For example, consider what might take place if the librarians at a community library were to quit or retire all at once. The library might remain open, but as the days passed, more and more books would be strewn about tables or placed on the wrong shelves. Eventually, the disorder would be so great that the library could not function as a library. Without librarians, the "energy" of the library would be lost, and the "life" of the library would come to an end.

Physicists tell us that the universe has a fixed amount of energy and that energy cannot be created or destroyed. However, energy can be converted from one form to another, such as when logs are burned in the fireplace and the energy in their molecules is released as heat and light. In the web of life, certain species of microbes and plants trap the sun's energy and use this energy to synthesize energy-rich molecules such as glucose and other carbohydrates. The process by which this is accomplished is called photosynthesis (we shall consider this process presently). Nonphotosynthetic organisms acquire the energy-rich molecules from the photosynthetic organisms. Then they transfer the energy from the carbohydrates to a usable form known as adenosine triphosphate (ATP). The step-by-step process by which carbohydrates are broken down and their energy is released to produce ATP is known as respiration (because it usually requires oxygen). Photosynthesis and respiration are basic to the metabolism taking place in all living things. They are key aspects of the metabolic activity we discuss in this section.

As noted, a key element of cellular metabolism is ATP. The ATP molecule is somewhat like a portable battery: It can move to any part of the cell where energy is needed and supply the energy. ATP can fuel the transport of nutrients into the cell or the elimination of waste products taking place at the cell membrane; it can be used in the synthesis of protein; it assists the movements of bacterial flagella and protozoal cilia; and it energizes the reproduction process, regardless of whether it is taking place in microbial, plant, or animal cells.

ATP has been referred to as the "universal energy currency." It is a nucleotide whose structure is illustrated in FIGURE 9.7 . The molecule consists of adenine and ribose molecules (called adenosine when chemically linked); the adenosine is linked to three phosphate groups (phosphorus bonded to oxygens). Much of ATP's energy is released when an enzyme breaks the high-energy bond connecting the third phosphate group to the remainder of the molecule. Most people are unaware of the importance of ATP in their lives—but without ATP, life as we know it could not continue. Chapter 6 contains a brief discussion of ATP relative to the fermentation process; more detailed descriptions of ATP's production follow.

Respiration and Glycolysis

Among the most important molecules organisms obtain from the environment are the energy-rich carbohydrates. Formed in photosynthesis by photosynthetic organisms, carbohydrates are the major source of chemical energy for microbes. Among the carbohydrates, glucose is probably the one that is most widely metabolized, and so we shall focus our attention on this six-carbon carbohydrate.

The break down of glucose occurs by the process of glycolysis. This term is derived from the stems *glyco-* referring to glucose or another carbohydrate and *-lysis* meaning "to break down." During glycolysis, glucose molecules are converted into smaller molecules and eventually into the three-carbon compound pyruvic acid. The pathway takes place in the cytoplasm of microbes and does not require oxygen. Each of the multiple chemical reactions of the pathway is catalyzed by an enzyme. The entire pathway is dis-

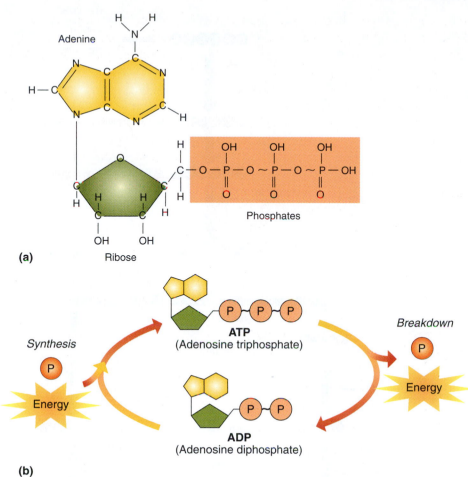

FIGURE 9.7   **Adenosine Triphosphate.** Adenosine triphosphate (ATP) is a key energy source for bacteria and other living things. (**a**) The ATP molecule is composed of adenine and ribose bonded to one another and to three phosphate groups, as shown. (**b**) When the molecule breaks down (right), it releases a phosphate group and much energy to do the cell's work; it then becomes adenosine diphosphate (ADP). For the synthesis of ATP (left), energy and a phosphate group must be supplied to an ADP molecule.

played in   **FIGURE 9.8** . Try not to be dismayed by the complexity of the diagram. Rather, note that one chemical substance is changing to another and that ATP is being invested and, later, gained back. In the next paragraphs, we shall go through the key steps in the chemistry. It is the stuff of microbial life—indeed, of all life.

As Figure 9.8 shows, the pathway of glycolysis begins with a molecule of glucose. (We should pause and note, however, that many other carbohydrates as well as various amino acids and numerous lipids can enter the pathway and be metabolized to get their energy. The pathway is a type of main highway into which pickup trucks, sports cars, SUVs, and eighteen-wheelers converge at the entry point.) In the initial reactions (Figure 9.8), enzymes bring about a number of changes in the glucose molecule. In the first step, for example, ATP is "invested" in the reaction to energize the glucose molecules. For the same reason, ATP is also supplied in the third reaction. Thus, two molecules of ATP must be available in order to initiate the pathway.

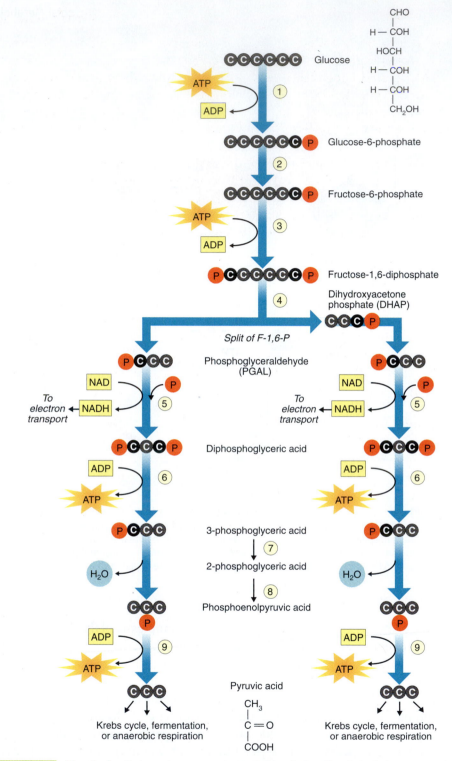

FIGURE 9.8 **Glycolysis.** Carbon atoms are represented by circles. The dark circles represent carbon atoms bonded to phosphate groups. ATP is supplied to glucose in reaction (1) and to fructose-6-phosphate in reaction (3). The splitting of F-1,6-diphosphate in reaction (4) yields PGAL and DHAP. DHAP then converts to another PGAL molecule. Both PGAL molecules proceed through reactions (5) to (9) and yield two molecules of pyruvic acid. Two ATP molecules are generated in reaction (6) and two more in reaction (9). In reaction (5), high-energy electrons and protons are captured by an NAD molecule.

Further along the pathway an important split takes place (note Step 4). Here, the six-carbon molecule is broken into two three-carbon molecules, whose name is abbreviated PGAL (for phosphoglyceraldehyde). Another significant chemical change occurs in Step 5: An enzyme strips two electrons from PGAL and deposits them in a coenzyme known as NAD (for nicotinamide adenine dinucleotide). A reaction such as this in which electrons are removed from one molecule and deposited in a second molecule is called an oxidation-reduction reaction. The NAD molecule acquires one proton (or hydrogen nucleus) as well and becomes NADH. This NADH molecule is used later in the electron transport system (note "to electron transport" in the figure). You will note that two NADH molecules have been formed because the reaction has occurred twice.

In the next series of chemical reactions (Steps 6 to 9), enzymes bring about additional conversions of the three-carbon molecule (we shall skip many of the details), and pyruvic acid eventually forms at the end of the pathway. Two notable events have happened along the way: In Steps 6 and 8, enough energy has been released to encourage the formation of ATP molecules from ADP (adenosine diphosphate) and phosphate. But you will note that a total of four ATP molecules have formed (two on the left side and two on the right side of the pathway). Thus, the two ATP molecules invested in the early reactions have been returned to the cell, and the cell has made out pretty well on its investment—the two invested ATP molecules have returned four ATP molecules, for a net gain of two molecules, a 100 percent return. In metaphorical terms, the cell has acquired four chemical batteries to run its chemical machinery, two more than it had at the start of glycolysis. But the best is yet to come—stay tuned.

We should stop at this juncture and note that we are discussing some rather involved and detailed biochemistry. Indeed, a graduate student in biochemistry will probably be expected to learn the structures and names of all the chemical compounds, the names and characteristics of all the enzymes, and numerous other chemical factors we are omitting here. But a general appreciation of this biochemistry is relevant to the undergraduate student because this (and what is to follow) is the chemistry by which most microbes and all other forms of life get their energy. And, as we have emphasized, where there is no energy, there is no life.

Fermentation

As we shall see presently, the remaining pathways of energy metabolism involve oxygen. There is, however, a vital "side step" of glycolysis, a process called fermentation that takes place in the absence of oxygen. Fermentation is an anaerobic process in which the pyruvic acid formed in glycolysis is transformed into other organic products such as alcohol, acid, and carbon dioxide gas. Fermentation permits certain species of microbes, such as yeast cells, to flourish in the absence of oxygen or when oxygen levels are very low; this ability is an evolutionary advantage to the organism possessing it. Industrial microbiologists use the term "fermentation" in a broad sense to refer to the production of such products as lactic acid, acetic acid, and butyric acid. Chapter 14 explores many of these industrial fermentations in detail.

Fermentation by yeast cells has a special significance because the products include the alcohol in wine, beer, and spirits (as is described in Chapter 8). When yeast cells (*Saccharomyces* species) metabolize glucose by glycolysis, they produce pyruvic acid. In an oxygen-free environment, they convert this pyruvic acid to ethyl alcohol. From the yeast's standpoint, this conversion makes "good chemical sense": The

■ *Saccharomyces*
sak-ä-rō-mī′sēs

yeast cells need to regenerate NAD for reuse in Step 5 of glycolysis; as we shall see, when oxygen is present, the electron transport system regenerates NAD, but in the absence of oxygen, the fermentation process is a valuable alternative. Enzymes in the yeast cells remove the two electrons and the proton from the NADH molecule. They attach the electrons and proton to a molecule of acetaldehyde (which another enzyme has produced from the pyruvic acid molecule). The process regenerates NAD for reuse in glycolysis, and gives us consumable ethyl alcohol.

The production of ethyl alcohol by yeast cells is central to the alcohol industry. Depending on the material present at the beginning of glycolysis, the beverage may be beer, wine, or other spirits. For example, if grapes are the starting material, grape alcohol (wine) is the result; if barley is used, barley alcohol (beer) results; if potatoes are used, the end product is potato wine, which is then distilled to produce vodka. The products of fermentation are as varied as the mind can imagine. In addition, carbon dioxide is liberated during the production of acetaldehyde from pyruvic acid (FIGURE 9.9), and this carbon dioxide accounts for the bubbles that appear in champagne and beer. It is also the gas that makes dough rise when bread is made (as Chapters 8 and 12 explain).

But yeasts are not the only microbes that can perform fermentation; innumerable other microbes can produce myriad fermentation products. For example, species of *Streptococcus* and *Lactobacillus* use the fermentation process to produce large amounts of lactic acid, and this acid converts condensed milk into yogurt. Cheeses obtain their taste from the mixture of acids produced by microbial fermentation during the ripen-

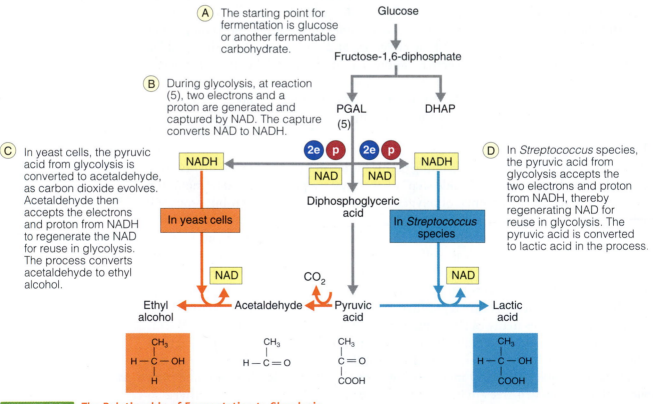

FIGURE 9.9   **The Relationship of Fermentation to Glycolysis.**

ing process (Chapter 15). And such products as acetone, butyl alcohol, vitamins, many antibiotics, and numerous amino acids are also produced by fermentation brought about by microbes. Chapter 14 details the processes behind these products and their uses. For the time being, we return from this "side step" of metabolism and resume our discussion of the main pathway.

The Krebs Cycle

Following glycolysis, the energy metabolism in aerobic organisms continues through a cyclic series of chemical reactions called the Krebs cycle (one enterprising merchant in the author's community calls his bicycle shop "The Krebs Cycle"). In protozoa, fungi, unicellular algae, and other microbes having mitochondria, the chemistry of the Krebs cycle occurs in the membranes of these organelles, since the necessary enzymes are located there. In bacteria and other microbes lacking mitochondria, the reactions take place at the cell membrane.

The Krebs cycle is named for Hans A. Krebs who identified many of the key intermediaries and was a Nobel laureate in 1953. The process is presented in FIGURE 9.10 ; it is termed a *cycle* because the substance formed at the end of the series of events serves as the starting point for another round. (The Krebs cycle is also known as the citric acid cycle because the first compound formed in the series of reactions is citric acid.)

In some oversimplified presentations, the end product of glycolysis—pyruvic acid—is considered the starting point for the Krebs cycle. However, the pyruvic acid molecule does not enter the cycle. Instead, it undergoes a transformation in which an enzyme removes a carbon atom from the molecule (and releases it as carbon dioxide gas), then combines the remainder of the molecule with a coenzyme called coenzyme A. This transformation results in acetyl-coenzyme A (acetyl-CoA). There are two reasons to mention this conversion: During the formation of acetyl-CoA, an NAD molecule accepts two electrons and a proton to become NADH, as Figure 9.10 shows (Step A). This NADH molecule is later used for electron transport, and substantial energy is gained for the cell. Also, note the loss of a carbon atom as carbon dioxide gas.

We are now ready to proceed with the Krebs cycle. In Step B of the process, an enzyme combines the acetyl-CoA molecule with a four-carbon acid called oxaloacetic acid to form a six-carbon acid called citric acid. In the next series of steps, enzymes convert the citric acid to a set of five- and four-carbon molecules. As you follow around the cycle, you will note that in the final step (Step G of Figure 9.10), oxaloacetic acid is once again produced. Along the way, two more carbon atoms have been lost as carbon dioxide gas—a total of three from the original pyruvic acid molecule. (Incidentally, this carbon dioxide is a large part of the gas we breathe out when we exhale.) In the end, the two pyruvic acid molecules are metabolized through the cycle, and six carbon dioxide molecules are produced, thereby accounting for the loss of all six carbon atoms of glucose.

Several key points during the Krebs cycle relate to our discussion of energy metabolism. For example, during the conversion of alpha-ketoglutaric acid to succinic acid (Step D of Figure 9.10), enough energy is released to synthesize a molecule of ATP (via a molecule called guanosine triphosphate, or GTP). This reaction is somewhat similar to one that takes place toward the end of the glycolysis pathway. Because the reaction occurs twice here (i.e., two pyruvic acid molecules enter the Krebs cycle), the net gain to the cell is two more molecules of ATP.

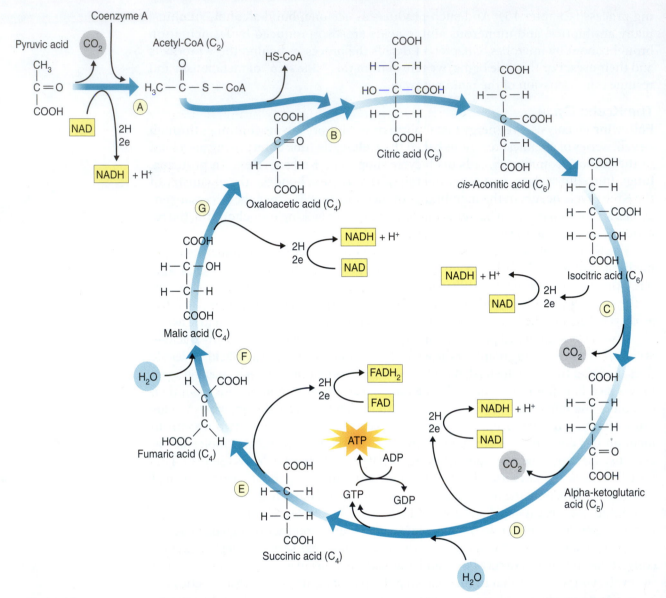

FIGURE 9.10   The Krebs Cycle. Pyruvic acid from glycolysis combines with coenzyme A to form acetyl-CoA in reaction (A). This molecule then condenses with oxaloacetic acid to form citric acid (B). In reactions (C) through (F), citric acid is converted to alpha-ketoglutaric acid and then to other acids in succession. Malic acid regenerates oxaloacetic acid in the last reaction, (G). During the process, the three carbons of pyruvic acid are liberated as three molecules of carbon dioxide. ATP is formed in reaction (D) via GTP, and high-energy electrons and protons are liberated during certain of the reactions. These are captured by NAD or FAD molecules, and the electrons are utilized for electron transport.

Another noteworthy chemical reaction occurs in Step E (Figure 9.10), when electrons are liberated and protons are given off. Two electron pairs and two protons are taken up by the coenzyme FAD (flavin adenine dinucleotide). The coenzyme thus becomes $FADH_2$. Like the NADH we discussed previously, the $FADH_2$ is used in the electron transport system.

A vital part of the Krebs cycle is the formation of several molecules of NADH. For example, NADH is produced as acetyl-CoA is formed (Step A). It also results from the formation of alpha-ketoglutaric acid (Step C), in the synthesis of succinic acid (Step D), and in the conversion to oxaloacetic acid (Step G). These NADH molecules, together with the NADH formed in glycolysis and the $FADH_2$ molecules, comprise the biochemical fuel for the electron transport system.

The Electron Transport System and Chemiosmosis

The electron transport system occurs in the mitochondria (in cells having these organelles) or along the cell membrane (in cells lacking mitochondria). It is the third major aspect of the process of respiration. In this system, a series of losses and gains of electrons take place, beginning with the NADH and $FADH_2$ molecules produced earlier. During the reactions of the electron transport system, a flow of electrons occurs through a chain of transport molecules, and substantial amounts of energy are liberated. Essentially all we have been discussing has set the stage for this process.

The participants in the electron transport system include various molecules: Among these are NAD, FAD, a coenzyme known as coenzyme Q, and a series of complex compounds known as cytochromes. Cytochromes are iron-containing cell proteins that receive and give up electron pairs much like a biochemical bucket brigade. The coenzymes and cytochromes are the vehicles through which the energy of electron pairs is accessed.

The NADH molecules produced in the Krebs cycle and in glycolysis are the starting points for electron transport, shown in FIGURE 9.11 . First, an NADH molecule passes its electron pair to an FAD molecule. (In this way, the NAD molecule is regenerated and can accept the next electron pair and proton in the Krebs cycle.) The FAD molecule then takes on two protons and becomes $FADH_2$. The biochemical bucket brigade continues as the $FADH_2$ molecule passes its electron pair to coenzyme Q, then on to cytochrome B, cytochrome C_1, cytochrome C, and cytochrome A_3, each passing releasing some energy. The last cytochrome, cytochrome A_3, passes the electron pair to an oxygen atom. The oxygen atom now takes on two protons from the surrounding environment and becomes a molecule of water. Thus, the electron pair, once in NADH, has passed through the electron transport system, ultimately winding up in a water molecule.

The $FADH_2$ formed in the Krebs cycle is another starting point. This molecule passes its electrons through the transport system of coenzymes and cytochromes and on to an oxygen atom. The oxygen atom then takes on protons to form more water.

Let us pause for a moment and note the importance of oxygen in this process. In aerobic microbes, the oxygen atom is the only substance that can accept electrons at the end of the transport chain. If oxygen atoms were not present, the cytochromes could not give up their electrons, and the series of electron transfers would soon grind to a halt. Without electron transport, the synthesis of ATP would also cease (as we shall now see), and death would ensue. It is interesting to note that oxygen was not always used for metabolism on Earth, as A Closer Look (on page 201) explains.

During the electron transport process, considerable amounts of energy are released. This energy is utilized to form the ATP molecules in a process known as chemiosmosis. As we have noted above, the enzymes of the Krebs cycle and the components of the electron transport system are located along a membrane (either mitochondrial or cell). As the electrons pass among the various components of the electron transport system, the energy released by the electrons pumps protons (also called hydrogen

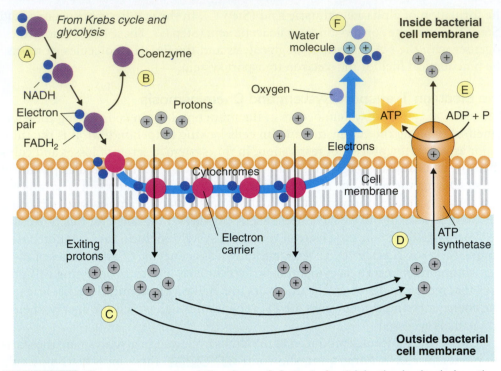

FIGURE 9.11 **Electron Transport and Chemiosmosis in Bacteria.** Originating in glycolysis or the Krebs cycle, (A) a coenzyme (e.g., NADH or FADH2) transports electron pairs to cytochromes in the cell membrane. (B) The NAD or FAD coenzyme is regenerated for reuse. (C) Each time the series of cytochromes pass along an electron pair, they release enough energy to fuel the transport of protons across the cell membrane at three points. (D) Each set of protons then reenters the cytoplasm of the cell through a protein channel lined with ATP synthetase. (E) Three ADP molecules join with three phosphate groups each time a set of protons moves through the channel, thereby accounting for three ATP molecules produced per each electron pair. (F) The electrons combine with other protons and then with an oxygen atom to form water molecules.

ions, or H^+) across the membrane, as illustrated in Figure 9.11. This proton pumping results in a high concentration of protons on one side of the membrane and a low concentration on the other side. Thus, one side of the membrane carries a positive electrical charge due to an accumulation of protons, while the other side of the membrane becomes negative due to a lack of protons.

Suddenly a critical point is reached where no more protons can be handled on one side. A reverse flow of protons begins, and the protons rush through the membrane back to the other side. As they flow, the protons pass through special pores in the membrane containing the enzyme ATP synthetase. During their rush, the protons supply enough energy to allow ATP synthetase to hook together ADP molecules and phosphate groups to form ATP molecules. The proton flow gives the process its name: *chemiosmosis* because the process consists of a flow (*-osmosis*) and is concerned primarily with chemicals (*chemi-*). Figure 9.11 shows the process.

The yield of life-sustaining ATP molecules from chemiosmosis is considerable. For each NADH molecule that delivers its electrons to begin the process, enough energy is released to synthesize three ATP molecules. Biochemists have calculated that a total of 10 NADH molecules result from all the reactions of glycolysis and the Krebs

A CLOSER LOOK
"It's Not Toxic to Us!"

It's hard to think of oxygen as a poisonous gas, but several billion years ago, oxygen was as toxic as cyanide. One whiff by a cellular organism, and a cascade of chemical reactions was set into motion in its cytoplasm. Death followed quickly.

Difficult to believe? Not if you realize that ancient organisms relied on fermentation and anaerobic chemistry for their energy needs. They absorbed organic materials from the environment and digested them to release the available energy. The atmosphere was full of methane, hydrogen, ammonia, carbon dioxide, and other gases. But no oxygen. And it was that way for hundreds of millions of years.

Then the cyanobacteria came on the scene and brought with them the ability to perform photosynthesis. Chlorophyll pigments evolved, and the cyanobacteria could trap energy from the sun and convert it to chemical energy in carbohydrates. But there was a downside—oxygen was a necessary waste product of the process; and the oxygen was deadly.

But not deadly to those organisms that could adapt. As millions of species of organisms died off in the now-toxic oceans and atmosphere, a few species were selected by Mother Nature to live on because they had the enzymes to tuck oxygen atoms safely away in nontoxic chemical compounds. And then, surprise, some organisms even evolved the ability to use oxygen in an election transport system and tap carbohydrates for large amounts of energy (Mother Nature really had to work overtime on that one). And so the Krebs cycle and the electron transport system came into existence.

Also coming into existence were millions of new species, some merely surviving, others thriving in the newly expanding oxygen-rich environment. The face of planet Earth was changing rapidly as anaerobic and fermenting species declined and aerobic species evolved. A couple of billion years would pass, and then, finally, a particularly well-known species of oxygen-breathing creature evolved. *Homo sapiens*.

---

cycle (as the figures note), so 30 molecules of ATP can be produced. In addition, two molecules of $FADH_2$ are formed during the Krebs cycle (Step E in Figure 9.10), and two ATP molecules result from electron transport starting with $FADH_2$, thus a total of four ATP molecules will result, thereby bringing the total to 34 molecules. When we add the two molecules of ATP produced at the end of glycolysis and the two synthesized during the Krebs cycle (Step D), the ATP total becomes considerable— 38 molecules of ATP for cell use. (Not bad for an investment of two ATP molecules.)

Although we have concentrated on glucose in this chapter (it is believed to be the richest source of chemical energy on Earth), numerous other carbohydrates, such as lactose, sucrose, maltose, and starch, are used in microbial metabolism as sources of energy. These carbohydrates do not have separate metabolic pathways. Rather, they utilize the basic pathways we have explored and enter them at various points, thus lending efficiency and economy to the metabolism.

The biochemistry of respiration is how microbes and other living things obtain their energy for life. A single glucose molecule yields over three dozen ATP molecules that the cell can use as portable batteries. This energy runs the microbial machine— indeed, the chemical machinery of all living things.

Anaerobic Metabolism

As we discovered earlier in this chapter, many species of microbes are able to live under anaerobic conditions. These microbes use alternatives to oxygen as an electron acceptor: For example, when it lives anaerobically, *Escherichia coli*, the colon-dwelling bacillus, uses a nitrate ion (NO_3^-) as an electron acceptor in place of oxygen at the end of electron transport. After acquiring electrons, the nitrate ion releases an oxygen atom to become a nitrite ion (NO_2^-). This chemistry is an essential feature of the nitrogen cycle occurring in soils. We discuss the cycle in depth in Chapter 16.

Other electron acceptors used by microbes include sulfate ions (SO_4^-) and carbon dioxide molecules (CO_2). When microbes use sulfate ions as electron acceptors, they convert the ions to molecules of hydrogen sulfide (H_2S). Hydrogen sulfide has the odor of rotten eggs, and it gives a horrid smell to foods or soils where the microbes are living. When carbon dioxide is used as an electron acceptor, it is converted into methane (CH_4).

Anaerobic metabolism was a useful way of obtaining energy in the eons before oxygen filled the atmosphere. Indeed, the anaerobes that practice this type of chemistry remind us that life can exist in an oxygen-free environment. Thousands of anaerobic species are classified as Archaea and Eubacteria, and they continue to exist deep in the soil, in marshes and swamps, and in landfills. The ability to obtain energy is as vital to their continued existence as it is to the aerobic species.

Photosynthesis

In this final section of the chapter, we switch gears and examine the second major aspect of microbial metabolism: biosynthesis, or anabolism as it is often called. Microbes use biosynthetic reactions to construct all the compounds they need to grow and reproduce. We shall continue our focus on energy by emphasizing photosynthesis, the biosynthetic process in which energy is captured and used to construct molecules of carbohydrates. Although this process gives us a glimpse of what is taking place in the microbes' constructive mode, it touches on only a few of the biosynthetic processes taking place. We considered the synthesis of proteins in Chapter 4, when we explored that process in the context of the genetic code in DNA.

Photosynthesis takes place in organisms that can use the sun's energy. This type of synthesis contrasts with chemosynthesis, in which chemical reactions in the cell supply the energy. For photosynthesis, certain microbes, such as cyanobacteria and unicellular algae, have light-absorbing pigments that trap the energy in sunlight and convert it to the energy in ATP molelcules.

Photosynthesis is conveniently divided into the energy-trapping reactions and the carbon-trapping reactions. These reactions are often called the "light" reactions and the "dark" reactions, referring to the fact that the energy-trapping reactions require light and the carbon-trapping reactions can proceed in the dark. However, the labels are misleading because the carbon-trapping reactions can occur in both light and dark. The process of photosynthesis is displayed in FIGURE 9.12 .

In the energy-trapping reactions of photosynthesis, the energy in sunlight is harvested by chlorophyll pigments that absorb light. The light "energizes" electron pairs, and they jump out of the chlorophyll molecules. The electrons move through an electron transport system (much as they do in respiration), jumping from one to the next

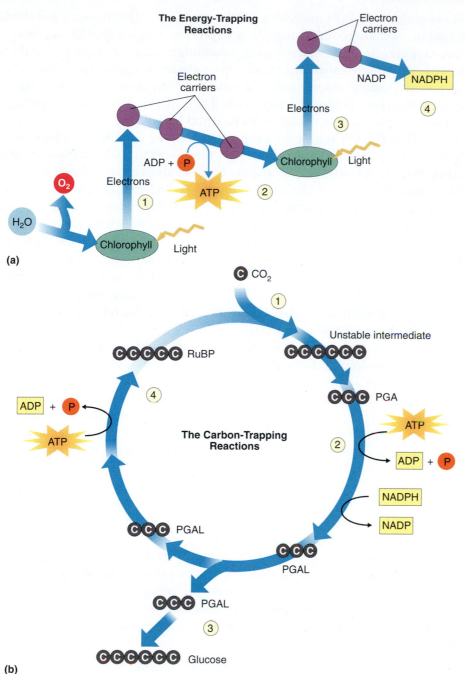

FIGURE 9.12 **Photosynthesis in Microbes.** (a) The energy-trapping reactions that occur along the membranes of a cyanobacterium. (1) Electrons in chlorophyll receive a boost in energy from light, and (2) ATP is synthesized as the electrons pass among electron carriers. (3) In noncyclic photosynthesis, the electrons receive a second boost, and (4) the energy is used to form high-energy NADPH. The ATP and NADPH are used in the carbon-trapping reactions. (b) The carbon-trapping reactions: (1) Carbon dioxide unites with ribulose biphosphate (RuBP) to form an unstable 6-carbon molecule. (2) The latter splits to form two molecules of phosphoglyceric acid (PGA) then phosphoglyceraldehyde (PGAL). ATP and NADPH from the energy-trapping reactions are used in the conversion. (3) Combination of two 3-carbon PGAL molecules eventually yields glucose, and (4) the remainder is used with ATP to form RuBP to continue the process.

in a series of electron carrier molecules. ATP is manufactured by the process of chemiosmosis (as in respiration), involving a flow of protons across a membrane, except the direction is opposite to that in respiration. In addition, once they have passed through a series of cytochrome molecules, the electrons are taken up by a molecule of NADP (nicotinamide adenine dinucleotide phosphate), rather than NAD. The NADP molecule then takes on a proton and is converted to NADPH. Thus, the two main products of the energy-trapping reactions are ATP and NADPH.

After ATP and NADPH have been generated, they are used in the carbon-trapping reactions. In these reactions, an enzyme uses the energy in ATP and traps a molecule of carbon dioxide. It binds the carbon dioxide molecule to a five-carbon carbohydrate called ribulose-biphosphate (RuBP in Figure 9.12). The result is a six-carbon unstable intermediate. The latter immediately splits into two molecules of phosphoglyceric acid (PGA), then phosphoglyceraldehyde (PGAL), which you may recall from glycolysis. The conversion from PGA to PGAL uses the energy in ATP and the NADPH. Now, the two molecules of PGAL undergo a series of enzyme-catalyzed conversions and eventually condense (join) to form a molecule of glucose (having six carbon atoms). The NADPH molecule formed in the energy-trapping reactions is an essential participant in this process; also, NADP is regenerated for reuse during additional reactions. The biosynthesis is complete.

In the process of photosynthesis, the energy of light has been trapped and used to form glucose molecules. These glucose molecules can be stored and used later for glycolysis reactions and ATP production. However, the chlorophyll molecules have been left with two fewer electrons than they previously had, and these electrons must be replaced for photosynthesis to occur again. Cyanobacteria and unicellular algae obtain replacement electrons by breaking down water molecules. Once the electrons have been incorporated, the remaining oxygen atoms from the water molecules are released to the atmosphere. Over billions of years of history, this oxygen has accumulated, and it now accounts for almost 20 percent of the gaseous content of the atmosphere. It is the oxygen we breathe to keep our metabolism going. To be sure, green plants contribute mightily to the atmosphere's oxygen content, but the major contribution comes from the microbes. Thanks, guys.

And so we come to the end of our biochemical journey. It has been long and difficult, but if you have stuck it out, you have a sense of the chemical machinery of microbial life. It is not too difficult to know microbiology, but it is rather difficult to *know* what makes microbes tick. And that's what this chapter has been all about—the chemical machines that lie beneath the images we can see under a microscope or in a photograph.

A FINAL THOUGHT

This chapter contains some of the most complicated concepts you will encounter in microbiology. The concepts are also among the most fundamental because the metabolism applies to all life—bacterial, plant, or human. I hope you can step back and see the forest as well as the trees. Virtually all living things use enzymes for their chemistry and ATP for their energy; a major percentage of species obtain their energy from glucose metabolism; and most depend ultimately on the carbohydrates produced in photosynthesis.

Along with these observations comes the realization that certain mechanisms and molecules have been preserved in microbes, plants, and animals virtually intact through the billions years of evolution. This was one of the great revelations of the twentieth century; it revealed the kinship of all living things. Indeed, one of the corollary benefits of studying the metabolism of microbes is that you come away with a better understanding of the biochemistry of living things in general. Nowhere is this more apparent than in this chapter. There are certainly variations, but there is also an underlying kinship among all forms of life. To understand one form is to understand them all.

QUESTIONS TO CONSIDER

1. An organism is described as a microaerophilic, heterotrophic, mesophilic, chemosynthetic streptococcus. How might you translate this complex bacteriological jargon into a description of the organism?

2. Every state has an official animal, bird, flower, and tree, but only Oregon has a bacterium named in its honor: *Methanohalophilus oregonese*. The species modifier *oregonese* is obvious, but can you decipher the meaning of the genus name?

3. Citrase is the enzyme that converts citric acid to alpha-ketoglutaric acid in the Krebs cycle. A chemical company has located a mutant microbe that cannot produce this enzyme and proposes to use the microbe to manufacture a particular product. What do you suppose the product is? How might this product be useful?

4. A student observes that during the process of respiration, a bacterium exhales before it inhales. What is he thinking?

5. One of the most important steps in the evolution of life on Earth was the appearance of certain organisms in which photosynthesis takes place. Why was this critical?

6. A microbiology professor from Virginia writes: "In the fall, when the apples were ripe in our farm orchard, we had a cow who could not be stopped by any fence (shades of the old nursery rhyme). Topsy would stagger home to the barn every night doing a very good imitation of a drunken sailor. . . ." What two possibilities might you offer for this observation?

7. ATP is an important energy source in all organisms, and yet it is never added to a microbial growth medium or consumed in vitamin pills or other growth supplements. Why do you think this is so?

■ KEY TERMS

Informative facts are necessary for the expression of every concept, and the information for a concept is founded in a set of key terms. The following terms form the basis for the concepts of this chapter. On completing the chapter, you should be able to explain and/or define each one:

acidophile
adenosine triphosphate
aerobic
anabolism
anaerobic
barophile
catabolism
chemiosmosis
cytochromes
decline (death) phase
electron transport
enzyme
facultative
fermentation
glycolysis

growth curve
halophile
Krebs cycle
lag phase
logarithmic log phase
mesophile
metabolism
microaerophile
photosynthesis
psychrophile
respiration
stationary phase
substrate
thermophile

■ http://microbiology.jbpub.com/book/microbes

The site features **eLearning**, an online review area that provides quizzes and other tools to help you study for your class. You can also follow useful links for in-depth information, read more stories of microbiology, or just find out the latest microbiology news.

Microbial Genetics: New Genes for the Germs

10

Looking Ahead

Among the unique characteristics of bacteria is their ability to acquire genes from other organisms. This characteristic is extraordinarily rare in the world of living things. Together with genetic changes resulting from mutation, the acquisition of new genes can have profound practical consequences, as we shall discuss in this chapter.

On completing this chapter, you should be able to . . .

- understand the structure and complexity of the bacterial chromosome and the significance of the plasmids found in bacterial cytoplasm.
- visualize how genetic changes can come about in the chromosome through spontaneous and induced mutations.
- name and explain the mode of action of some agents of mutational change.
- appreciate the remarkable ability of bacteria to acquire genes from their surrounding environment and incorporate those genes into their genomes.
- discuss the process of conjugation in bacteria and understand how it can result in multiple antibiotic resistances.

A toxin produced by the *Bacillus Thuringiensis* (inset) is a natural pesticide. Geneticists can transfer the gene responsible for producing the toxin into corn, cotton, and other crops. Such genetically-modified crops become resistant to insect pests, increasing yield and reducing the need for expensive chemical pesticides.

- describe the role of viruses in transduction and explain two methods by which transduction can occur.
- explain how studies in microbial genetics gave rise to the science of genetic engineering.

In 1968, an extremely serious form of bacterial dysentery broke out in the Central American country of Guatemala. Bacterial dysentery affects the human intestinal tract, resulting in the loss of large volumes of fluid, together with waves of intense abdominal cramps and frequent passage of bloody, mucoid stools. The disease is caused by various species of *Shigella*, a Gram-negative rod-shaped bacterium; the disease is often called shigellosis.

During the Guatemala outbreak, patients were given salt tablets or oral salt solutions to help with rehydration. In addition, various patients also received one of four different antibiotics to kill the bacteria. As the days passed, however, physicians became increasingly frustrated as they attempted to treat the disease. They discovered that none of the four antibiotics was effective—neither tetracycline nor chloramphenicol nor sulfanilamide nor streptomycin could eliminate the *Shigella*. Without useful antibiotics, physicians lost a major weapon in their fight to stop the epidemic. Within a 3-year period, 100,000 people were affected, and 12,000 died.

The outbreak of shigellosis illustrates what can happen when antibiotic-resistant bacteria emerge in society. At the time of the Guatemala outbreak, it was unusual to find bacteria with resistance to one drug, much less four. In this chapter, we shall learn how the *Shigella* cells might have acquired this resistance, as we explore the concept of microbial genetics. The topic is highly relevant because outbreaks of disease due to antibiotic-resistant bacteria have been occurring more and more frequently. Numerous strains of antibiotic-resistant *Staphylococcus aureus*, for example, started to show up in human populations in the 1990s (one strain has been dubbed "super staph"), and headlines have regularly trumpeted the "bugs that won't die."

■ *Shigella*
shi-gel′lä

■ *Staphylococcus*
staf-i-lō-kok′kus

We shall begin our study of microbial genetics by exploring the nature of the bacterial chromosome. This chromosome is different from those of eukaryotic cells, and it can be altered by a change in the genes or through the acquisition of new genes. Either of these processes could lead to drug resistance in a bacterium and, eventually, in a bacterial population. We shall also study the genetics of viruses because viruses are intimately associated with at least one type of bacterial alteration. (The chapters on protozoa, fungi, and algae cover concepts of genetics that apply to those microbes.)

Included under the umbrella of microbial genetics is the topic of genetic engineering, one of the most extraordinary technological advances of all time. Genetic engineering enables molecular biologists to treat genes almost as playthings—isolating them, altering them, inserting them into fresh organisms, and watching to see what they will do. The fruits of this technology are awe-inspiring, and we have only begun to see what is possible. Later in the chapter, we shall explore how studies in microbial genetics led to the science of genetic engineering.

Bacterial DNA

Modern bacteria enjoy the fruits of all the genetic changes their ancestors have undergone over the 3.5 billion years of their existence. Because of their diverse genes, bacteria can thrive in the snows of the Arctic or the boiling hot vents at the bottoms of the oceans. No other organism can compete with bacteria for sheer numbers—a pinch of rich soil has more bacteria than all the people living in the United States today. And considering the fantastic multiplication rate of bacteria (a new generation every half hour), one can easily see how a useful genetic change (such as drug resistance) can be propagated quickly in a stressful environment (such as one containing a drug).

Later in this chapter, we shall study mutation, one of the processes responsible for the genetic changes that have resulted in the myriad bacterial forms we observe on Earth today. A mutation is a permanent alteration in the bacterial chromosome via a change in its DNA. To understand mutation, we must first consider the bacterial chromosome.

The Bacterial Chromosome

Most of the genetic information in a bacterium is located within a single chromosome. The chromosome consists of DNA existing in a double helix in a closed loop. This loop structure is unlike the linear (shoestring) form of chromosomes of eukaryotic organisms. Moreover, the bacterial chromosome exists freely in the cytoplasm without any protein support or surrounding membrane, both of which are found in eukaryotic organisms' chromosomes. The chromosome occupies about half of the total volume of the bacterial cell and, extended its full length, measures about 1.5 mm long, about 1500 times the length of the bacterium that contains it. The tight packing accounts for the explosive release of the DNA when the cell membrane and wall are broken.

How a 1.5-mm-long chromosome could fit into a 1.0-μm *E. coli* cell was poorly understood until 1998, when scientists reported that the chromosome takes on a structure that involves a number of loops. To do this, chromosomal regions attach to one another at anchorage points at intervals of about 50,000 bases. An overall "flower" form results, as shown in FIGURE 10.1 . At present, the elements that form the anchors of the loops remain unknown.

The chromosome of the colon-dwelling bacillus *E. coli* is one of the most intensely studied. Distributed along the chromosome are sites to which genetic activity can be traced. Each site, called a locus (pl., loci), consists of one or more genes. The chromosome of *E. coli* has over 4000 genes. Some viruses, by contrast, have as few as 7 genes, and the 22 pairs of human chromosomes have a total of approximately 50,000 genes. As of 1997, the names of all the nucleotides in the *E. coli* chromosome had been elucidated, as we explore in Chapter 4. (A quick review of Chapter 2 on nucleic acids and their components would be helpful to understanding the chemistry of DNA as we proceed through this chapter.)

The bacterial chromosome replicates in a process called binary fission. The DNA anchors to a point on the cell membrane; then the double helix unwinds. Next, enzymes belonging to a complex called DNA polymerase synthesize a new strand of

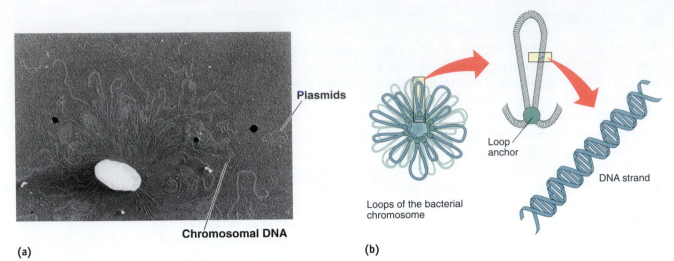

Plasmids

Chromosomal DNA

Loops of the bacterial chromosome

Loop anchor

DNA strand

(a)

(b)

FIGURE 10.1   Bacterial DNA. (**a**) An electron micrograph of an *E. coli* cell immediately after disruption. The tangled mass is the organism's DNA. Changes in DNA occurring through mutation and recombination have helped bacteria adapt to most environments on Earth. (**b**) The loops in the structure chromosome, viewed head-on. The loops in the DNA help account for the compacting of a large amount of DNA in a relatively small bacterial cell.

DNA using nucleotides with bases complementary to those on the original strands (adenine is complementary to thymine, and guanine to cytosine, as Chapter 2 discusses). Each new strand then unites with its complementary "parent" strand and twists to form a new double helix. This method of replication was first worked out in 1958 by Matthew J. Meselson and Franklin W. Stahl. It is known as the semiconservative method of replication because one strand of parent DNA is conserved in the new DNA, while one strand is newly synthesized. Chapter 4 describes the experimental procedures leading to this conclusion.

Because it is a closed loop, a unique type of replication occurs in bacterial DNA. The DNA unwinds at the fixed point, and an enzyme nicks the closed loop at a site known as the origin of replication (**FIGURE 10.2**). The two strands separate, or "unzip," at this point, and a V-shaped replication fork is established. Along one side of the fork, the DNA is synthesized by continuous assembly of nucleotides beginning at the origin of replication. However, along the other side, the synthesis proceeds in discontinuous fashion, and the DNA is synthesized in a series of segments. The segments later join with the help of an enzyme called DNA ligase. The segments involved in this process are known as Okazaki fragments, after Reiji Okazaki, who discovered their presence in 1968. The inner and outer chromosomes then separate, as Figure 10.2 displays.

Much of what characterizes DNA replication in bacteria is not found elsewhere in the world of living things. But bacteria have lived on Earth longer than we can imagine, and the environmental challenges bacteria faced over eons of time resulted in their distinctive genetic patterns. In the next paragraphs, we shall encounter another feature that sets bacteria apart.

Plasmids

Many species of bacteria contain in their cytoplasm a number of closed loops of DNA called plasmids. Plasmids exist as independent units apart from the bacterial chromo-

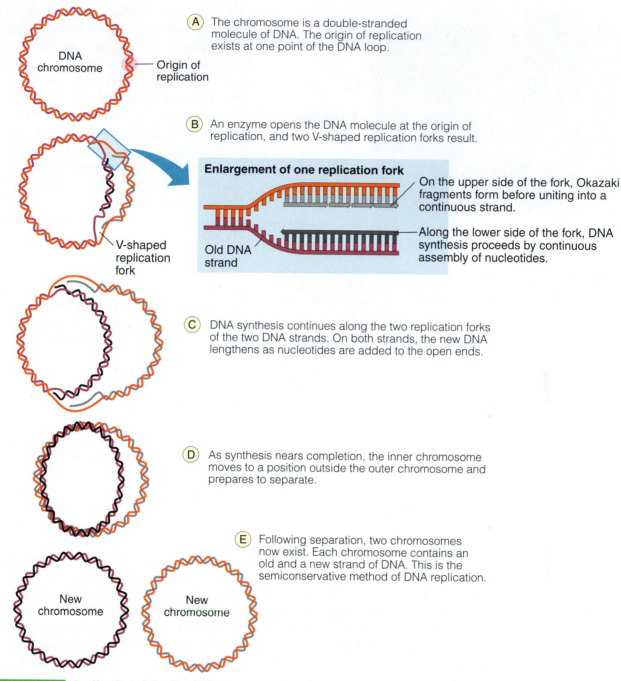

A The chromosome is a double-stranded molecule of DNA. The origin of replication exists at one point of the DNA loop.

DNA chromosome — Origin of replication

B An enzyme opens the DNA molecule at the origin of replication, and two V-shaped replication forks result.

Enlargement of one replication fork

On the upper side of the fork, Okazaki fragments form before uniting into a continuous strand.

Along the lower side of the fork, DNA synthesis proceeds by continuous assembly of nucleotides.

V-shaped replication fork   Old DNA strand

C DNA synthesis continues along the two replication forks of the two DNA strands. On both strands, the new DNA lengthens as nucleotides are added to the open ends.

D As synthesis nears completion, the inner chromosome moves to a position outside the outer chromosome and prepares to separate.

E Following separation, two chromosomes now exist. Each chromosome contains an old and a new strand of DNA. This is the semiconservative method of DNA replication.

New chromosome   New chromosome

FIGURE 10.2   Replication of the *E. coli* Chromosome.

some. They contain about 2% of the total genetic information of the bacterium, and they multiply independently of the chromosome. Numerous species of bacteria, especially Gram-negative ones, are known to have plasmids, but eukaryotic forms of life apparently lack them.

Plasmids do not appear essential to a bacterium's life. However, they may confer a selective advantage to the microbes. For example, certain plasmids known as

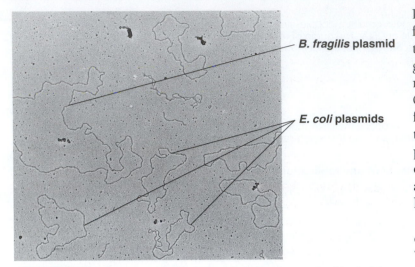

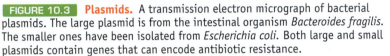

FIGURE 10.3 **Plasmids.** A transmission electron micrograph of bacterial plasmids. The large plasmid is from the intestinal organism *Bacteroides fragilis*. The smaller ones have been isolated from *Escherichia coli*. Both large and small plasmids contain genes that can encode antibiotic resistance.

R ("resistance") factors carry genes that facilitate resistance to antibiotics (scientists believe that the presence of these genes in the *Shigella* species in Guatemala may have encouraged their resistance). Other plasmids permit bacteria to transfer their genetic material to another bacterium, as we shall see. In these cases, the plasmids themselves move from cell to cell. These "traveling plasmids" may have allowed the *Shigella* species to accumulate multiple resistances.

Still other plasmids contain genes that encode the production of toxic proteins. The proteins, called bacteriocins, have a deleterious effect on other bacteria and give the microbes that synthesize them an advantage in the struggle for survival. Toxins that affect human cells may also be encoded by genes in the plasmids. **FIGURE 10.3** shows an electron microscopic view of some plasmids.

Gene Mutations

With rare exceptions, humans go through their lives with virtually the same genes; but bacteria do not necessarily do likewise. This is because the genetic information in the bacterial plasmids and chromosome is subject to changes. Such changes can occur by two major methods: mutation and recombination. In both cases, the changes can have substantial impact not only on the bacterium but on society as well.

The antibiotic tetracycline is a major weapon in the effort to treat disease. It kills a bacterium by passing through the cell wall and membrane and binding to the bacterial ribosome. The ribosome is an ultramicroscopic cytoplasmic body where the genetic message directs the union of amino acids to form a protein, as Chapter 4 explains in detail. The genetic message is carried to the ribosome by a molecule called messenger RNA (mRNA), which has received the message from the DNA in the chromosome or plasmid. However, when tetracycline binds to the ribosome, it prohibits protein synthesis by preventing the mRNA molecule from delivering its message to the ribosome. Scientists now know that some bacteria in a population resist the effects of tetracycline because their ribosomes have undergone a structural change. And this change may come about through the process of mutation.

Mutation is a permanent change in an organism's DNA. Such a change can involve disruption of the nucleotide sequence in a gene or loss of significant parts of the gene. This action can result in the placement of the wrong amino acids in a protein molecule, as we explore in Chapter 4. If the affected protein is slated for construction of a ribosome, then such a change might result in a ribosome with a mangled structure. The altered ribosome is unable to bind with the tetracycline molecule. Thus, as a re-

A CLOSER LOOK

Three Genes

Could the Black Death of the 1300s have resulted from three defective genes? Could 25 million Europeans have succumbed to plague because of three genes? Could the entire course of Western civilization have turned on three genes?

Possibly so, maintain researchers from the federally funded Rocky Mountain laboratory in Montana. In 1996, a research group led by Joseph Hinnebush reported that three genes in the plague bacillus are absent in a harmless form of the organism. And it is possible that the entire story of plague's pathogenicity revolves around these three genes.

The scenario goes like this: Bubonic, septicemic, and pneumonic plague are caused by *Yersinia pestis*, a rod-shaped bacterium transmitted by the rat flea. When the flea is infected, the bacteria amass in its foregut and obstruct its gastrointestinal tract. Soon the flea is starving, and it uncontrollably starts biting humans and rodents and feeding on their blood. During the bite, the flea regurgitates the mass of bacteria in the victim's bloodstream and spreads the plague.

The three genes enter the picture at the very beginning. It appears that harmless plague bacilli have genes encouraging the microbes to remain in the midgut of the flea (although scientists are not sure why this happens). Pathogenic plague bacilli, by contrast, do not have the genes, and they migrate to the foregut of the flea and form a plug of packed bacilli; these are the organisms that pass on to the next plague victim.

Sometimes it is dangerous to oversimplify matters, and this may be one of those times. Still, scientists are inclined to reduce concepts to their least common denominators. And if the tragic Black Death reduces to three genes, then so be it.

sult of the mutation, the bacterium has become resistant to the antibiotic. Another possible result of mutation is discussed in A Closer Look above.

Causes of Mutation

In the environment, mutations occur regularly in bacteria from spontaneous changes in the DNA. It has been estimated, for example, that at least one mutation occurs for every billion replications of a bacterium. Because a mass of bacteria the size of pinpoint has over a billion cells, at least one mutant cell probably exists in this mass. If this mutant is the one with the altered ribosome structure, it will survive exposure to tetracycline when the other 999,999,999 are killed. The surviving mutant then multiplies, and a new mass of bacteria will emerge, all of which have the new ribosome structure. This entire population of bacteria is now resistant to tetracycline. When we consider that some bacteria can replicate (and double their population) in as little as 20 minutes, we can understand how an antibiotic-resistant population can emerge in a relatively brief period of time.

Mutations can also occur as a result of identifiable factors known as mutagens. Ultraviolet (UV) radiation, a component of sunlight, is a known mutagen. When soil-dwelling bacteria absorb UV radiation, thymine molecules next to one another in their DNA are bound together (thymine is one of the four bases of DNA). With its thymine molecules joined in unnatural linkages, the genetic code of the DNA changes,

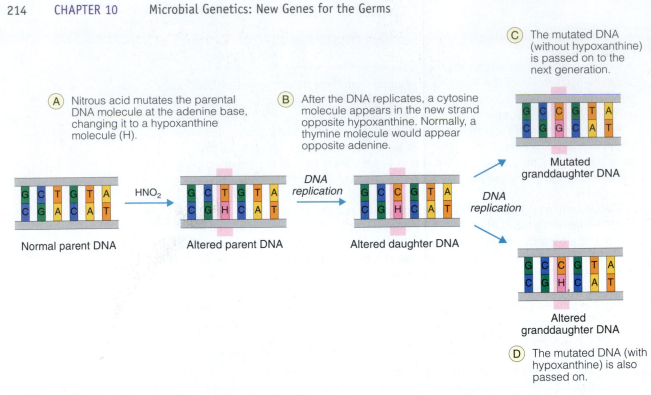

Ⓐ Nitrous acid mutates the parental DNA molecule at the adenine base, changing it to a hypoxanthine molecule (H).

Ⓑ After the DNA replicates, a cytosine molecule appears in the new strand opposite hypoxanthine. Normally, a thymine molecule would appear opposite adenine.

Ⓒ The mutated DNA (without hypoxanthine) is passed on to the next generation.

Ⓓ The mutated DNA (with hypoxanthine) is also passed on.

FIGURE 10.4   How Nitrous Acid Causes Bacterial Mutations.

and an erroneous protein results. This protein could be the one that yields the mangled ribosome. Incidentally, UV radiation is often used in a germicidal lamp for disinfection purposes because it quickly kills bacteria by altering their DNA.

Chemicals in the environment may also act as mutagens. One example is urea, a component of animal urine. Benzopyrene, a component of industrial smoke, is another possible mutagen. Both of these chemicals cause the deletion or insertion of extra nucleotides in the DNA during replication. With a new or missing nucleotide, the entire structure of the DNA changes, and a new genetic message results. Because the genetic message encodes the amino acid sequence in the protein, the protein changes. In most cases, the mutation goes unnoticed or is hidden, but in other cases, the mutation confers a new characteristic on the bacterium (such as antibiotic resistance). Nitrous acid is another chemical that can cause mutations, as FIGURE 10.4 illustrates.

Another way in which a mutation can bring on antibiotic resistance is by changing the DNA so that it encodes a useless enzyme molecule. During the Guatemala epidemic, physicians observed that *Shigella* species were resistant to the antibiotic sulfanilamide. Sulfanilamide acts by preventing the formation of folic acid, a growth factor in many species of bacteria. To prevent folic acid formation, the antibiotic binds with the enzyme the bacterium uses to synthesize the acid. However, if the structure of the enzyme has changed, a union with sulfanilamide is not possible, and the sulfanilamide cannot interfere with folic acid production. The bacterium is now resistant to the antibiotic.

But how can the bacterium fill its need for folic acid once the enzyme has been altered? To do so, the bacterium must evolve other adaptations: Perhaps it gains the capacity to absorb folic acid from the environment, or it may evolve a different pathway for synthesizing folic acid, one that does not need the now-useless enzyme. It has survived the antibiotic through a mutation.

Transposons

Another possible source of mutation is a movable genetic element called a transposon. Transposons are small segments of DNA that have the ability to move from one position to another in the bacterial chromosome. The segments carry no genetic information other than for the ability to insert onto the chromosome. Here they may interrupt the genetic coding sequence of the DNA, thereby causing the DNA to encode an incorrect protein or, in some cases, no protein at all.

Transposons are the "jumping genes" studied by Barbara McClintock, the 1983 winner of the Nobel Prize in Physiology or Medicine. McClintock studied corn plants at Cold Spring Harbor Laboratory on Long Island, New York. In 1951, she unveiled research showing that certain genes move from one chromosome to another in plants. Although her work was ignored at first, other scientists eventually confirmed the existence of transposons in corn plants and discovered the same segments of DNA in bacteria. Thirty years after her first report, McClintock received the Nobel Prize, well-deserved recognition for her discovery. When informed of the Nobel award, she commented that it ". . . seemed unfair to reward a person for having so much pleasure from asking [corn] plants to solve specific problems and then watching their response."

The insertion of a transposon into the DNA of the gene that encodes ribosome protein may have yielded the mutation leading to an altered ribosome structure. Perhaps the bacterium could then resist tetracycline. Researchers have also found that a transposon can transport the genes for antibiotic resistance. Thus, if the transposon carries these genes to a plasmid and the plasmid moves to a new cell (as we shall see momentarily), the genes for antibiotic resistance may join other resistance genes in one bacterium. We shall explore this possibility further in the next section.

Gene Recombinations

One of the horrific features of the Guatemala epidemic was that the *Shigella* species were resistant to all four antibiotics normally used to treat bacterial dysentery. How bacteria might have accumulated these numerous resistances was first shown in an experiment performed in 1955 by Japanese investigators.

The investigators, led by Tomoichiro Akiba, began with a species of *Shigella* isolated from patients and observed to have resistance to the same quartet of drugs as the species in Guatemala (tetracycline, sulfanilamide, chloramphenicol, and streptomycin). The investigators also noted a surprising coincidence: Patients with the drug-resistant *Shigella* also had in their intestine a strain of *Escherichia coli* with resistance to the same four drugs. Because simultaneous mutations resulting in the same four resistances were highly unlikely, the researchers wondered whether the resistances were transferred between *Shigella* and *E. coli*.

Akiba and his colleagues devised a series of experiments to test this hypothesis. They obtained laboratory cultures of drug-resistant *E. coli* and of drug-sensitive *Shigella* species. Then they prepared liquid suspensions of both bacteria and mixed

■ *Escherichia*
esh-ėr-ē′kē-ä

the suspensions together for a short period of time. Next, they carefully isolated the *Shigella* species and tested them. The results were startling: The *Shigella* cells, once sensitive to the four antibiotics, were now resistant to the same four drugs as the *E. coli* cells. It appeared that a transfer of drug resistance had taken place.

In the decades that followed, the concept of transferable drug resistance was studied intensely in laboratories throughout the world, and investigators learned that gene transfers can occur between bacteria by three methods: conjugation, transduction, and transformation. These methods are referred to as gene recombinations. They imply genetic alterations by the acquisition of DNA from another organism.

How important are gene recombinations? Analyses of the chromosomes of bacteria indicate that gene transfers are far more common than scientists previously imagined. In a study reported in 2000, biochemists calculated that *E. coli*, the common bacterium of the human gut, had acquired nearly 20 percent of its DNA (755 of its 4288 genes) from other microbes. The surprisingly large extent of gene swapping and gene acquisition accounts in large measure for the problem of antibiotic resistance, as we shall see in the following paragraphs.

Conjugation

In the recombination process called conjugation, two live bacterial cells come together, and a donor cell transfers some of its genetic material to a recipient cell. The donor is known as an F^+ cell (signifying it has fertility), while the recipient is known as an F^- cell (because it lacks fertility). Conjugation was first postulated in 1946 by Joshua Lederberg and Edward Tatum of New York's Rockefeller University. The two researchers received the 1958 Nobel Prize in Physiology or Medicine for their work. Unknown to Lederberg and Tatum at that time, an important element in the transfer is the bacterial plasmid.

The plasmid of the donor cell is called an F factor, meaning "fertility factor." It contains about 20 genes, most of which are associated with conjugation. The genes encode enzymes that replicate the DNA and move it from donor cell to recipient cell during the conjugation. The genes also encode enzymes and structural proteins used to synthesize special pili known as F pili, or sex pili. These hairlike fibers contact recipient bacteria, then retract so that the surfaces of the donor and recipient cells are very close or touching. A channel referred to as a conjugation bridge then forms between the two cells.

Once the contact has been made, an enzyme nicks one strand of the DNA of the F factor (plasmid), and the single strand of the factor passes through the channel to the recipient cell (FIGURE 10.5). When it arrives in the recipient cell, enzymes synthesize a complementary strand of DNA and a double helix forms. The double helix bends to form a loop and becomes a plasmid, thereby completing the conversion of the recipient cell to a donor cell, or F^+ cell. Meanwhile, back in the donor cell, a complementary new strand of DNA forms and unites with the leftover strand of the original F factor. This double-stranded DNA molecule remains as the new F factor.

Conjugation has been observed in numerous genera of bacteria, including *Shigella*, *Escherichia*, *Salmonella*, and others. If a recipient bacterium already had a gene for resistance to tetracycline, it could through conjugation acquire a gene for resistance to sulfanilamide. If this microbe then engaged in conjugation with a cell having genes for resistance to chloramphenicol and streptomycin, it could acquire resistance genes for those antibiotics as well. Multiple resistances could be acquired in this way.

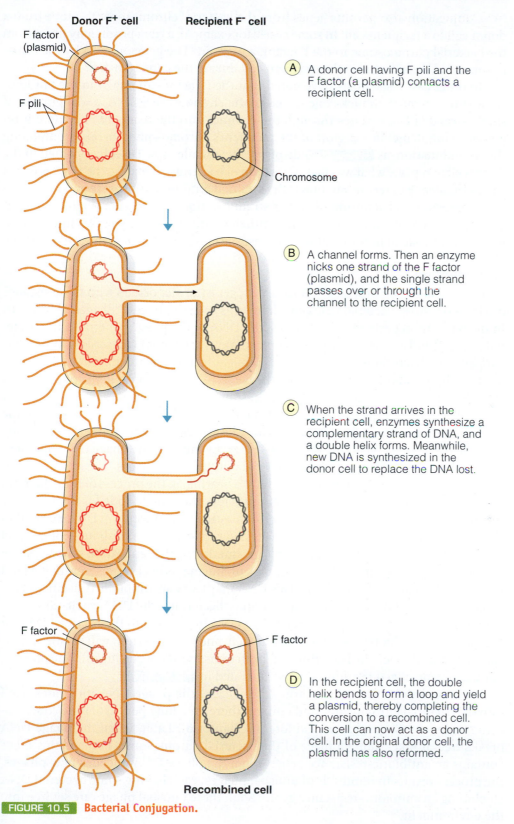

Donor F⁺ cell   **Recipient F⁻ cell**

F factor (plasmid)

F pili

Chromosome

(A) A donor cell having F pili and the F factor (a plasmid) contacts a recipient cell.

(B) A channel forms. Then an enzyme nicks one strand of the F factor (plasmid), and the single strand passes over or through the channel to the recipient cell.

(C) When the strand arrives in the recipient cell, enzymes synthesize a complementary strand of DNA, and a double helix forms. Meanwhile, new DNA is synthesized in the donor cell to replace the DNA lost.

F factor

F factor

(D) In the recipient cell, the double helix bends to form a loop and yield a plasmid, thereby completing the conversion to a recombined cell. This cell can now act as a donor cell. In the original donor cell, the plasmid has also reformed.

Recombined cell

FIGURE 10.5 **Bacterial Conjugation.**

Conjugation also permits genes from the bacterial chromosome to move from a donor cell to a recipient cell. In some cases, for example, a transposon may jump from the bacterial chromosome to the F factor, a plasmid. The genes carried by the transposon could then pass into the recipient cell during the gene transfer.

In other cases, scientists have noted that the F factor can attach to the chromosome. An enzyme then nicks one strand of the chromosomal DNA, and a portion of single-stranded DNA passes through the channel into the recipient cell. Here it replaces a complementary region of the recipient's chromosome, thereby completing the recombination, as **FIGURE 10.6** displays. Meanwhile, new DNA has formed in the donor cell to replace what was lost. The strain of bacterium donating the DNA is called a high frequency of recombination (Hfr) strain. If antibiotic resistance genes exist on its chromosome, a bacterium of an Hfr strain can transfer those genes into a recipient bacterium. Note, however, that the entire chromosome rarely passes into a recipient. Instead, a small portion of the chromosome passes across; then the cells separate.

Transduction

In the recombination process of transduction, gene transfer occurs with the assistance of a bacterial virus. Bacterial viruses are known as bacteriophages ("phages," for short). In the replication cycle of many phages, the DNA of the phage penetrates the bacterial host cell and encodes new phages immediately, as Chapter 6 describes. When this happens, the bacterium serves as a biochemical factory for the production of hundreds or thousands of new phages, and the bacterium is destroyed in the process (this is the lytic cycle explored in Chapter 6). In some cases, however, the phage does not replicate in the bacterial cytoplasm; instead, the phage DNA integrates itself into the bacterial chromosome, where it remains for a long period of time and replicates along with the bacterium (this is the lysogenic cycle). Eventually, the phage DNA may separate and stimulate phage replication to yield multiple new phages. The lytic and lysogenic cycles are important because they have a bearing on the process of transduction.

During the lytic cycle, a phage's DNA provides the genetic code for the synthesis of hundreds or thousands of phages, each consisting of a molecule of DNA encased in protein. Normally the DNA of new phages is synthesized from fresh nucleotides and is the same as the DNA of the starting phage. But during the replication process, some newly forming phages may erroneously pick up fragments of chromosomal or plasmid DNA from the bacterial cytoplasm. This DNA may carry an antibiotic resistance gene. Later, when one of these phages enters another bacterium, the DNA carries along the antibiotic resistance gene. Should the phage DNA integrate into the bacterial DNA, it will insert the antibiotic resistance gene as well, and the bacterium will acquire resistance to the antibiotic. The bacterium has been transduced (i.e., changed). The process, known as generalized transduction, is illustrated in **FIGURE 10.7** .

Now consider a second scenario. In this case, the phage DNA integrates itself into the bacterial chromosome and remains there for a long period of time (the lysogenic cycle, or lysogeny, as we explore in Chapter 6). Later, when the phage DNA breaks free, it takes along a piece of the bacterial chromosome. Perhaps this piece contains an antibiotic-resistance gene. When the phage DNA encodes new phages, the process results in hundreds of unusual phages, each having some bacterial DNA (including an antibiotic-resistance gene). Soon these unusual phages are set free into the environment.

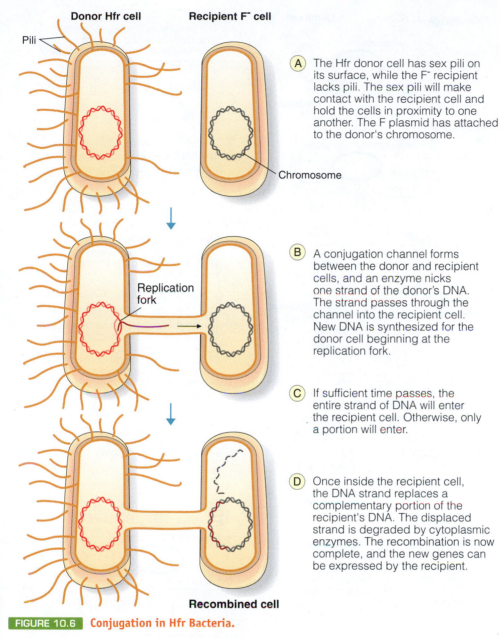

Donor Hfr cell   **Recipient F⁻ cell**

Pili

Chromosome

(A) The Hfr donor cell has sex pili on its surface, while the F⁻ recipient lacks pili. The sex pili will make contact with the recipient cell and hold the cells in proximity to one another. The F plasmid has attached to the donor's chromosome.

Replication fork

(B) A conjugation channel forms between the donor and recipient cells, and an enzyme nicks one strand of the donor's DNA. The strand passes through the channel into the recipient cell. New DNA is synthesized for the donor cell beginning at the replication fork.

(C) If sufficient time passes, the entire strand of DNA will enter the recipient cell. Otherwise, only a portion will enter.

(D) Once inside the recipient cell, the DNA strand replaces a complementary portion of the recipient's DNA. The displaced strand is degraded by cytoplasmic enzymes. The recombination is now complete, and the new genes can be expressed by the recipient.

Recombined cell

FIGURE 10.6   Conjugation in Hfr Bacteria.

Now the process continues. Should an unusual phage enter a fresh bacterium, it may integrate its DNA into the bacterial chromosome. In doing so, it will also integrate the antibiotic-resistance gene and transduce (change) the bacterial cell. A genetic recombination has taken place through specialized transduction, pictured in **FIGURE 10.8**. Through the process, an antibiotic-resistance gene has moved from one bacterium to another. Bacterial reproduction (at its rapid pace) then quickly propagates the antibiotic-resistance genes. Perhaps something like this happened in the years preceding the Guatemala outbreak of shigellosis.

Scientists now recognize that transduction is an infrequent event. This is because genes do not easily break free from bacterial chromosomes, nor do phages always pick

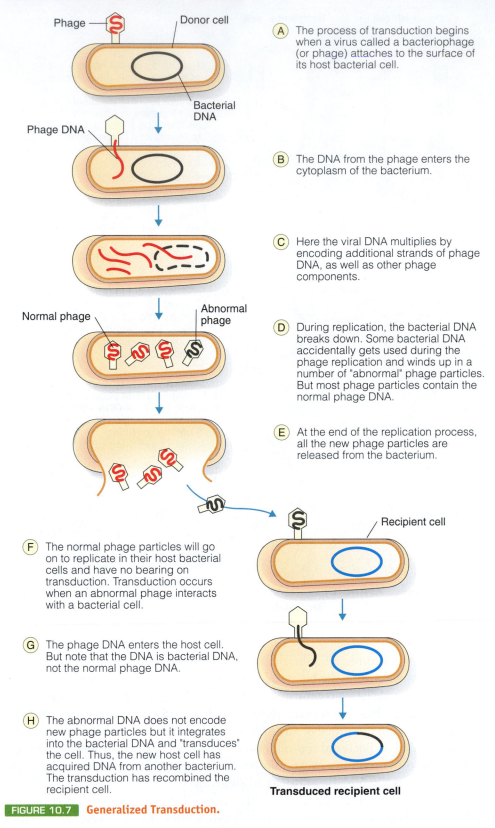

Phage

Donor cell

Bacterial DNA

Phage DNA

Normal phage

Abnormal phage

Recipient cell

Transduced recipient cell

A The process of transduction begins when a virus called a bacteriophage (or phage) attaches to the surface of its host bacterial cell.

B The DNA from the phage enters the cytoplasm of the bacterium.

C Here the viral DNA multiplies by encoding additional strands of phage DNA, as well as other phage components.

D During replication, the bacterial DNA breaks down. Some bacterial DNA accidentally gets used during the phage replication and winds up in a number of "abnormal" phage particles. But most phage particles contain the normal phage DNA.

E At the end of the replication process, all the new phage particles are released from the bacterium.

F The normal phage particles will go on to replicate in their host bacterial cells and have no bearing on transduction. Transduction occurs when an abnormal phage interacts with a bacterial cell.

G The phage DNA enters the host cell. But note that the DNA is bacterial DNA, not the normal phage DNA.

H The abnormal DNA does not encode new phage particles but it integrates into the bacterial DNA and "transduces" the cell. Thus, the new host cell has acquired DNA from another bacterium. The transduction has recombined the recipient cell.

FIGURE 10.7   Generalized Transduction.

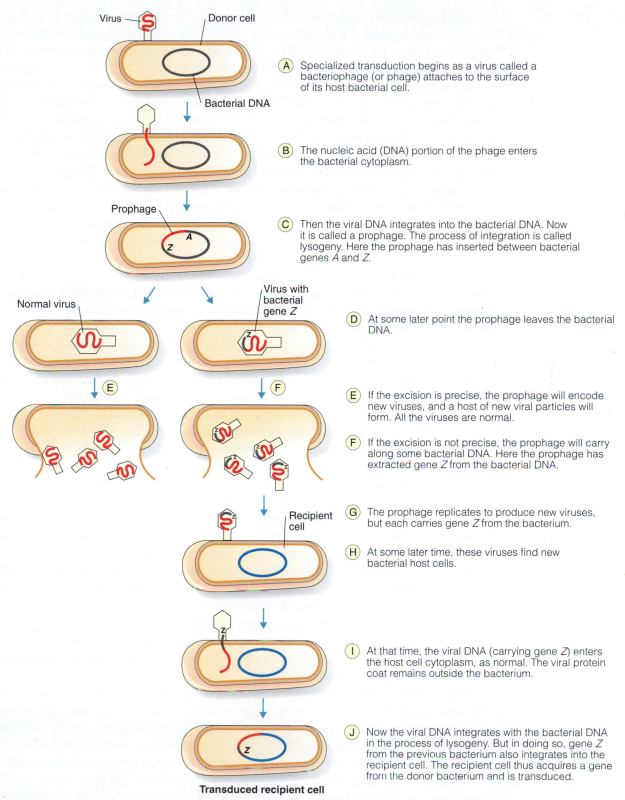

A Specialized transduction begins as a virus called a bacteriophage (or phage) attaches to the surface of its host bacterial cell.

B The nucleic acid (DNA) portion of the phage enters the bacterial cytoplasm.

C Then the viral DNA integrates into the bacterial DNA. Now it is called a prophage. The process of integration is called lysogeny. Here the prophage has inserted between bacterial genes *A* and *Z*.

D At some later point the prophage leaves the bacterial DNA.

E If the excision is precise, the prophage will encode new viruses, and a host of new viral particles will form. All the viruses are normal.

F If the excision is not precise, the prophage will carry along some bacterial DNA. Here the prophage has extracted gene *Z* from the bacterial DNA.

G The prophage replicates to produce new viruses, but each carries gene *Z* from the bacterium.

H At some later time, these viruses find new bacterial host cells.

I At that time, the viral DNA (carrying gene *Z*) enters the host cell cytoplasm, as normal. The viral protein coat remains outside the bacterium.

J Now the viral DNA integrates with the bacterial DNA in the process of lysogeny. But in doing so, gene *Z* from the previous bacterium also integrates into the recipient cell. The recipient cell thus acquires a gene from the donor bacterium and is transduced.

FIGURE 10.8 Specialized Transduction.

up bacterial DNA when they reproduce. However, investigators also recognize that if transduction happens, the consequences can be great. *Escherichia coli*, for example, has always been a harmless inhabitant of human and animal intestines (indeed, this people-friendly microbe is very helpful because its enzymes break down many otherwise indigestible materials while keeping harmful microbes in check). But in 1982, a pathogenic strain called *E. coli* O157:H7 emerged and triggered a rash of violent illnesses linked to rare hamburger meat, fresh fruits, and other uncooked or poorly cooked foods. Researchers now believe that the bacterium turned foe when a virus carried pieces of DNA into an *E. coli* cell, allowing the microbe to encode the dangerous toxins associated with the disease. In 2001, scientists from the University of Wisconsin compared the genomes of the harmless and pathogenic strains of *E. coli* and found that large swatches of their DNA (about 20 percent) did not match; moreover, they found traces of viral DNA, lending credence to the possibility that a virus had brought the troublesome genes to the pathogenic strain via transduction.

Transformation

Like transduction, transformation occurs in a very limited percentage of bacteria in a population. During transformation, a bacterium acquires genes from its surrounding environment; that is, transformation involves the direct uptakes of fragments of DNA by a recipient cell and the acquisition of new genetic characteristics.

Let us consider an example of how transformation can occur. A patient has been treated for an infection of the large intestine caused by a pathogenic strain of *Escherichia coli* (e.g., *E. coli* O157:H7). The physician used an antibiotic called gentamicin, and the bacteria were killed by the drug. Let us suppose that this strain of *E. coli* had a gene enabling the microbes to resist the antibiotic chloramphenicol (the gene could have protected the microbes against chloramphenicol if the physician had prescribed that antibiotic). But the physician chose gentamicin, instead, and the bacteria were killed. Fragments of the bacteria are now strewn about in the intestinal tract of the patient.

Now, let us imagine that, shortly thereafter, a few *Shigella* cells enter the patient's intestine (perhaps in contaminated water). The *Shigella* cells possess the competence factors that allow them to acquire fragments of DNA present in the nearby environment. Competence factors are protein molecules that are believed to allow DNA fragments to pass through the bacterial cell membrane. Let us assume that the *Shigella* cells use the competence factors and pick up the gene for resistance to chloramphenicol. They then incorporate the new DNA into their own DNA (either plasmid or chromosomal) and become transformed, as **FIGURE 10.9** illustrates. They are now resistant to chloramphenicol. Next, they pass out of the intestine in the feces and accumulate in soil or water, where they reproduce to a sizable population of bacteria, all of which have resistance to chloramphenicol.

During other similar transformations, cells of this species of *Shigella* may acquire resistance genes for other antibiotics. Then some cells may undergo conjugation with a bacterium and acquire the gene giving resistance to sulfanilamide. Some cells may next undergo a mutation and acquire resistance to tetracycline. Cells may be infiltrated by a bacteriophage that carries the streptomycin-resistance gene. Over the course of years, decades, and generations, cells of this *Shigella* species could conceivably acquire numerous resistances to numerous antibiotics and become a horrific sleeping giant in the environment. Perhaps, during that fateful period of 1968, such a sleeping giant found its way to thousands of patients in Guatemala, where it caused misery and death.

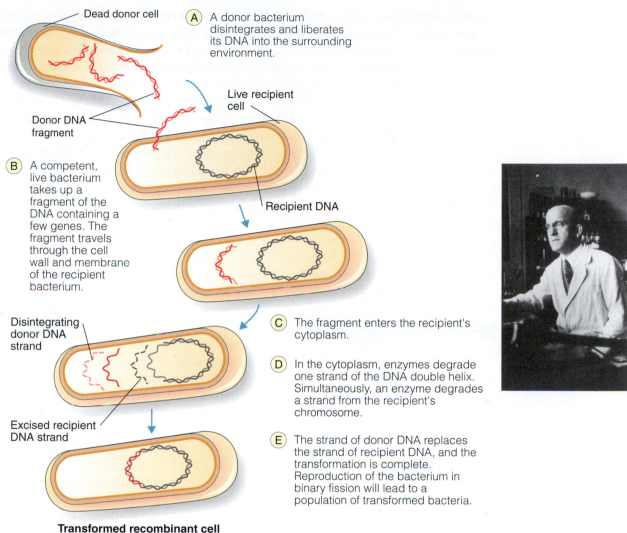

Dead donor cell

A A donor bacterium disintegrates and liberates its DNA into the surrounding environment.

Donor DNA fragment

Live recipient cell

B A competent, live bacterium takes up a fragment of the DNA containing a few genes. The fragment travels through the cell wall and membrane of the recipient bacterium.

Recipient DNA

Disintegrating donor DNA strand

C The fragment enters the recipient's cytoplasm.

D In the cytoplasm, enzymes degrade one strand of the DNA double helix. Simultaneously, an enzyme degrades a strand from the recipient's chromosome.

Excised recipient DNA strand

E The strand of donor DNA replaces the strand of recipient DNA, and the transformation is complete. Reproduction of the bacterium in binary fission will lead to a population of transformed bacteria.

Transformed recombinant cell

FIGURE 10.9 **Bacterial Transformation.** Transformation is the process in which alive bacterium acquires DNA fragments from the environment. Pictured is Oswald Avery, the Rockefeller Institute investigator who in the 1940s led the research effort to identify DNA as the agent of transformation.

In Today's World

Now we fast-forward to today's world. Antibiotic resistance has become one of the most serious problems confronting modern scientists. An alarming number of bacterial species have evolved resistance to one or more antibiotics. High on the list of concerns is *Staphylococcus aureus*.

Staphylococci are normal inhabitants of the human skin, mouth, nose, and throat. Although they generally live in these areas without causing harm, they can initiate disease when they penetrate the skin barrier or the mucous membranes. Penetration is assisted by such occurrences as open wounds, damaged hair follicles, ear piercing, dental extractions, and irritation of the skin by scratching. *Staphylococcus aureus*, the grapelike cluster of Gram-positive cocci, is the species usually involved in disease.

The hallmark of staphylococcal skin disease is the abscess, a circumscribed pus-filled lesion. (A boil is a skin abscess usually beginning as a pimple.) More widespread staphylococcal diseases are staphylococcal septicemia (blood infection), staphylococcal pneumonia (lung infection), staphylococcal endocarditis (heart valve infection), and staphylococcal meningitis (infections of the nerve cord coverings). *S. aureus* is involved in over 250,000 infections per year, primarily in hospitals and nursing homes.

Staphylococcal diseases are usually treated with penicillin, but over the years, strains of *S. aureus* have developed resistance to penicillin and numerous other antibiotics, and the term multidrug-resistant *Staphylococcus aureus* (MRSA) has entered the lexicon of medicine. Through those years, vancomycin remained a viable alternative for treating MRSA infections, even though it is a very expensive and somewhat toxic antibiotic. Then, in 1997, an MRSA strain evolved with partial vancomycin resistance; scientists named it VISA (short for vancomycin intermediately resistant *Staphylococcus aureus*). Although researchers have found useful treatment alternatives in drug combinations, they are grappling with the possibility that one day nothing will be left in the antibiotic arsenal to treat patients infected by this strain of staphylococci. The concern is acute because vancomycin resistant enterococci exist in the human intestine, and gene transfers to *S. aureus* strains are possible. Heightening the concern is the observation that antibiotic-resistance genes are frequently found on easily transmitted plasmids.

Genetic Engineering

If you have read the previous section, you are aware of the gloomy implications of microbial genetics. Through the processes of mutation and genetic recombinations, microbes can modify their genetic material, presenting very real concerns to physicians who must deal with infectious disease. The problem of multiple drug resistances in *Shigella* and *S. aureus* is but one of the many effects of genetic alterations.

However, you should also look on the positive side. Knowledge of microbial genetics helps scientists understand antibiotic resistance, but it also makes available a technology that few could have imagined a generation ago. This is genetic engineering, a technology that grew out of the study of microbial genetics. By applying the principles of microbial genetics in the laboratory, scientists found they could control mutation and genetic recombination almost at will. During the 1970s, researchers discovered that they could alter bacterial DNA and mimic the processes of nature. Soon, they were cutting and splicing DNA, removing and inserting genes, and opening new vistas of pure and applied research. In this section, we shall explore how genetic engineering emerged from studies in microbial genetics. Chapters 4 and 14 explain in detail the mechanics of genetic engineering and the fabulous fruits of this technology.

The Beginnings of Genetic Engineering

As scientists gradually comprehended the implications of gene recombinations in bacteria, they wondered whether they could control those recombinations. But they lacked the biochemical tools until the 1960s, when they discovered and isolated a group of bacterial enzymes called endonucleases (enzymes that react with nucleic acids within the cell). Endonucleases are also called restriction enzymes because they help bacteria "restrict" bacteriophage replication by destroying the phage's nucleic acid within the bacterial cytoplasm (they also act at "restricted" locations in the nucleic acid). The enzymes work by cleaving the phosphate-sugar bonds in the backbone of the nucleic acid (Chapter 2). Importantly, scientists found that these "biochemical scissors" could be directed to snip a bacterial chromosome at a specific point.

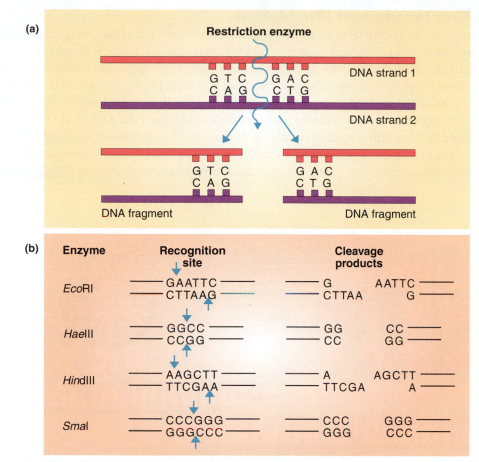

FIGURE 10.10 **Restriction Enzymes.** (**a**) A restriction enzyme cuts through two strands of a DNA molecule to produce two fragments. (**b**) The recognition sites of several restriction enzymes. Note that *Eco*RI and *Hind*III leave dangling ends in the cleavage products, while the other two enzymes do not. Dangling ends ("sticky ends") are desirable because fragments that have them attach to other fragments more easily.

The existence of restriction enzymes was first postulated in the 1960s by Werner Arber when he noted bacterial enzymes cleaving bacteriophage DNA at selected spots (**FIGURE 10.10**). Hamilton Smith subsequently isolated a restriction enzyme from the Gram-negative rod *Haemophilus influenzae*. In 1971, Daniel Nathans used Smith's bacterial enzyme to split the DNA of simian virus 40 (SV40), a cause of tumors in monkeys. It was remarkable that Nathans could use the enzyme from a bacterium to split the DNA of a virus at the same spot. In 1978, the Nobel Prize in Physiology or Medicine was awarded to these three scientists. By that time, biochemists had isolated, purified, and characterized hundreds of different restriction enzymes.

As the 1970s unfolded, scientists were using restriction enzymes to split DNA molecules on demand, regardless of the source of the DNA. It was clear that restriction enzyme X would cleave a DNA molecule at point X, regardless of whether the DNA was from a plant, animal, bacterium, or virus. Experiments with recombination processes such as conjugation showed that bacteria could thrive even though they had foreign DNA, and scientists were soon dreaming of the day when they might use a restriction enzyme to alter DNA in a test tube. Their dreams would soon be realized.

■ *Haemophilus*
hē-mä′fil-us

The First Recombinant DNA Molecule

Among the first scientists to attempt a genetic manipulation was Paul Berg of Stanford University. In 1971, Berg and his coworkers used a restriction enzyme to split the DNA molecule from the SV40 virus, as Nathans had done. Then they went further: Despite numerous biochemical obstacles, they spliced the viral DNA to a bacterial chromosome, that of *E. coli*. In doing so, they constructed the first recombinant DNA molecule. The process was particularly tedious because the bacterial and viral DNAs had blunt ends and would not attach easily (it was much like attempting to attach a brick to the middle of a brick wall). Berg therefore had to utilize exhaustive enzyme chemistry to form overlapping, "staggered" ends that would combine easily **FIGURE 10.11** .

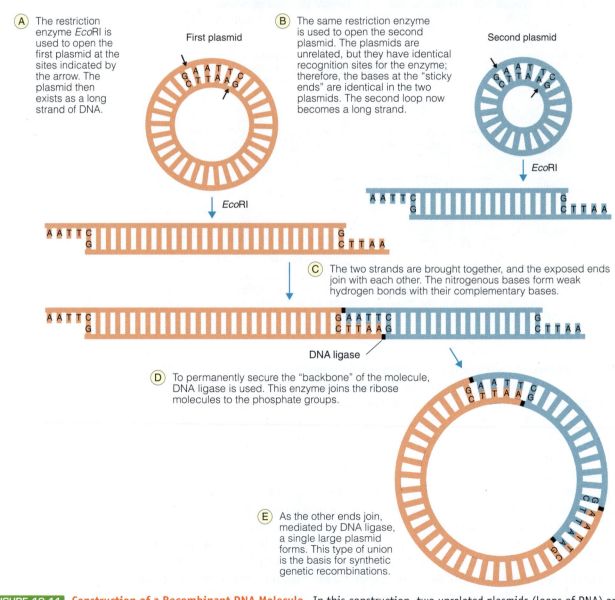

A The restriction enzyme *Eco*RI is used to open the first plasmid at the sites indicated by the arrow. The plasmid then exists as a long strand of DNA.

B The same restriction enzyme is used to open the second plasmid. The plasmids are unrelated, but they have identical recognition sites for the enzyme; therefore, the bases at the "sticky ends" are identical in the two plasmids. The second loop now becomes a long strand.

C The two strands are brought together, and the exposed ends join with each other. The nitrogenous bases form weak hydrogen bonds with their complementary bases.

D To permanently secure the "backbone" of the molecule, DNA ligase is used. This enzyme joins the ribose molecules to the phosphate groups.

E As the other ends join, mediated by DNA ligase, a single large plasmid forms. This type of union is the basis for synthetic genetic recombinations.

FIGURE 10.11 **Construction of a Recombinant DNA Molecule.** In this construction, two unrelated plasmids (loops of DNA) are united to form a single plasmid.

In doing so, he achieved a momentous first and was honored as a corecipient (with Frederick Sanger and Walter Gilbert) of the 1980 Nobel Prize in Chemistry.

While Berg was performing his experiments in 1971, an important discovery was made by Herbert Boyer and his colleagues at the University of California. Boyer isolated a restriction enzyme that nicks a chromosome and leaves it with staggered ends, as Figure 10.11 illustrates. The bits of single-stranded DNA extending out from the chromosome attach easily to other fragments of DNA during recombination experiments. Scientists quickly dubbed these extensions "sticky ends." Boyer's enzyme was named *Eco*RI because it was derived from *Escherichia coli* strain RY13 and was the first restriction enzyme (I) isolated from this strain.

Biochemists were quick to realize that when the complementary ends of DNA fragments are brought together, weak chemical bonds form between the nucleotides on the dangling ends. However, these bonds are not strong enough to hold the ends together indefinitely. To forge a permanent link between the ends, an enzyme called DNA ligase must be used. DNA ligase joins (or ligates) the backbones of DNA strands by forming chemical bonds between the phosphate groups of one fragment and the deoxyribose molecule of the adjacent fragment, as Figure 10.11 demonstrates. This bond, called a phosphodiester bond, exists between all nucleotides in the DNA strands. Its formation by DNA ligase activity seals the fragments together and completes the recombinant DNA molecule.

Engineered Chimeras

Although bacterial chromosomes were widely used in the earliest genetic engineering experiments, Stanley Cohen and his group at Stanford University found that they could perform experiments more efficiently with bacterial plasmids. As described earlier in this chapter, plasmids are small loops of DNA found in bacterial cytoplasm. Plasmids may contain as few as 12 genes or as many as several hundred. Experiments in genetic recombination showed that they could pass out of and into bacterial cells during conjugation.

One of the first plasmid-derived recombinant DNA molecules was constructed in 1972 by Cohen and his colleagues. The biochemists used *E. coli* plasmids and opened the loops at a specified recognition site using the restriction enzyme *Eco*RI. The enzyme was supplied by Herbert Boyer (A Closer Look on page 228), and the inserted DNA was obtained from cells of an African clawed toad. It encoded a protein found in the ribosomes of toad cells. The dangling ends of the toad DNA paired with the complementary ends of the plasmid DNA, and DNA ligase was added to seal the fragments together. The researchers called the recombined plasmid a chimera, after the mythical lion-goat-serpent of Greek mythology.

Cohen had discovered that plasmids could be inserted into fresh bacteria if the bacteria are suspended in cold calcium chloride, then rapidly heated to 42°C. This treatment causes pores in the cell wall and cell membrane to open and permit passage of the plasmids, a type of bacterial transformation. Carefully, he mixed the recombined plasmids with *E. coli* cells and subjected the cell-plasmid mixture to the alternating cold-heat treatment. The research indicated that the plasmids were replicating inside the bacteria within minutes. As the time passed and the excitement mounted, the scientists watched to see what would happen. When they did a biochemical analysis of the proteins in the growth mixture, they found a new protein

A CLOSER LOOK

Of Corned Beef and Plasmids

In 1972, they met at a scientific conference in Hawaii—Stanley Cohen and Herbert Boyer. Cohen was there to lecture about his work with plasmids, the submicroscopic loops of DNA in the bacterial cytoplasm. Boyer was an expert on a restriction enzyme that could cut DNA—any DNA—at a precise point. As he sat in the audience and listened to Cohen, Boyer's mind stirred. Could his enzyme cut Cohen's plasmid and allow a foreign piece of DNA to attach?

Scientific conferences are the last place to talk about science, so Boyer invited Cohen to lunch at a local delicatessen in Waikiki. The corned beef was good that day, and the sandwiches hit the spot. The deli mustard was biting hot, and the beer was ice cold. The time was ripe to talk history—and talk history they did. They would collaborate on a set of genetic engineering experiments, the ones that, in retrospect, revolutionized the science of molecular genetics. As the afternoon wore on, the ideas flowed and the friendship took root. Only one thing about that historic lunch has remained a mystery: Who picked up the tab?

among them; it was the protein of the toad cell ribosome. A successful transplant of vertebrate genes to a bacterium had been accomplished—a bacterium was producing toad proteins. It was the beginning of the era of genetic engineering.

The Implications

Molecular biologists were quick to see the implications of genetic engineering, and they began performing their own gene manipulation experiments (FIGURE 10.12). For example, they soon transferred *Staphylococcus aureus* genes to *E. coli* cells and induced the latter to produce *S. aureus* proteins. Still other biochemists attempted to insert human genes into plasmids, and others speculated about the long-term prospects of genetic engineering. The safety issues were addressed by several conferences, and under the auspices of the U.S. government, a set of guidelines was established to protect the safety of the public and address public concerns about the new technology.

The new field of biotechnology gradually emerged from the experiments in genetic engineering. It offered possible solutions to problems of food production, synthesis of new medicines, pollution abatement, and other human concerns. Scientists envisioned new enzymes to dissolve oil spills, new vaccines against emerging diseases, and new drugs and medicines. The only limits to what could be accomplished lay in the scope of their imaginations.

As with most scientific advances, however, the great achievements of some were put to use by others for deleterious purposes. Among those immoral individuals are the bioterrorists who have attempted to take the genes from highly infectious viruses such as the Ebola virus and splice those genes into already pathogenic viruses such as the smallpox virus to create a fearsome bioweapon. Further, they have taken antibiotic-resistance genes from one bacterial species and attempted to insert them into pathogens such as anthrax and plague bacilli to create strains that could not be treated with antibiotics. Such efforts cast a pall on the great achievements of biotechnology.

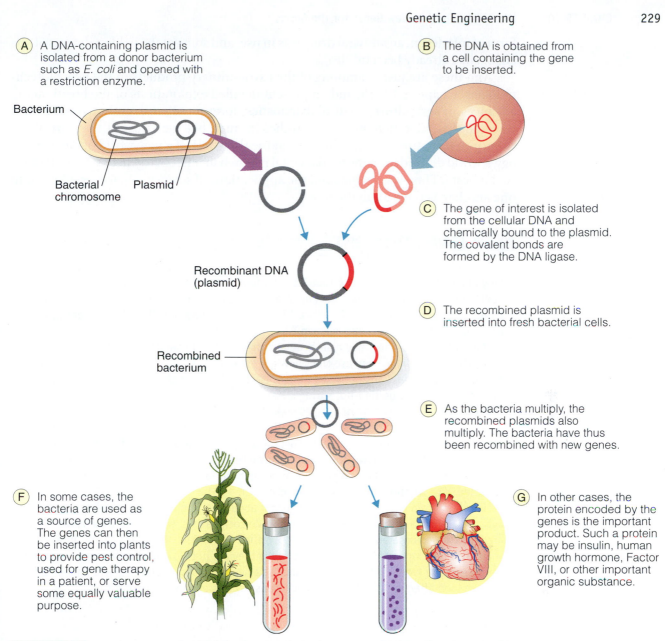

A A DNA-containing plasmid is isolated from a donor bacterium such as *E. coli* and opened with a restriction enzyme.

Bacterium

Bacterial chromosome   Plasmid

B The DNA is obtained from a cell containing the gene to be inserted.

C The gene of interest is isolated from the cellular DNA and chemically bound to the plasmid. The covalent bonds are formed by the DNA ligase.

Recombinant DNA (plasmid)

D The recombined plasmid is inserted into fresh bacterial cells.

Recombined bacterium

E As the bacteria multiply, the recombined plasmids also multiply. The bacteria have thus been recombined with new genes.

F In some cases, the bacteria are used as a source of genes. The genes can then be inserted into plants to provide pest control, used for gene therapy in a patient, or serve some equally valuable purpose.

G In other cases, the protein encoded by the genes is the important product. Such a protein may be insulin, human growth hormone, Factor VIII, or other important organic substance.

FIGURE 10.12 **Developing New Products Using Genetic Engineering.** Genetic engineering is a method for inserting foreign genes into a bacterium and obtaining chemically useful products.

By the year 2000, thousands of companies worldwide were working on the industrial applications of genetic engineering and biotechnology. Some were research companies with special units for these efforts, while others were established solely to pursue and develop new products by gene-splicing techniques. For example, the Eli Lilly company was marketing Humulin, a form of human insulin produced, not by pancreas cells, but by genetically engineered bacterial cells. Hemophiliac patients had access to bacteria-derived Factor VIII to encourage blood clotting, and children of short stature were being treated with human growth hormone obtained not from cadaver tissue, but from bacteria genetically engineered to carry human genes. A genetically engineered form of interferon, an antiviral drug, was in use, and some of the wildest dreams of scientists had already been fulfilled.

And these are just a sampling of the extraordinary products of the new biotechnologies. Chapters 14, 15, and 16 present detailed explorations of the breathtaking advances in agriculture, medical diagnostics, forensic science, and pharmaceutical research that have stemmed from studies in microbial genetics. At the start of the twenty-first century, we find ourselves at the threshold of a new world of biotechnology, a world derived from the fundamental research in microbial genetics. Louis Pasteur said it best: "There are science and the applications of science, separate yet bound to one another as the fruit to the tree."

A FINAL THOUGHT

So you decide you'd rather be 6 foot 2 than 5 foot 1. You decide to go out and get some "tall" genes—maybe you pick them up at a used-gene shop, or order them from a catalogue, or buy them at a local Wal-Mart. "That's absurd," you say. "People can't get new genes or change their genetic makeup."

Ah, but bacteria can. They collect genes; they exchange genes; they store genes; and when they need them, they use their new genes. And that's just one of the extraordinary features of bacteria. Consider some others: Bacterial species thrive in environments ranging from ice to boiling-hot springs. Many species can live either with or without oxygen. A large percentage of species make their own foods from chemicals in the soil. Bacteria have no built-in death age, and they double in number every hour or so. We humans, by contrast, must maintain a constant body temperature; we suffocate without oxygen; we must hunt down complex foods; we reach a certain age, then die; on average, we take a full 25 years to produce a new generation. And, of course, we must be satisfied with whatever genes we were born with.

It is fairly common to read in biology books about the "higher" and "lower" forms of life. Typically, humans are cast as higher forms, while cockroaches, worms, bacteria, and similar creatures are considered "lower" forms. It is difficult to believe that bacteria are "lower" than any other forms of living things. Certainly they are not "lower" than humans. Bacteria were here long (very long) before we humans came on the scene, and they will undoubtedly be here long after we "higher" forms have vanished. It's extraordinary, but true.

QUESTIONS TO CONSIDER

1. In hospitals, it is common practice to clear the air bubble from a syringe by expelling a small amount of the syringe contents into the air. One microbiologist estimates that this practice results in the release of up to 30 liters of antibiotic into a typical hospital's environment annually. How might this lead to the appearance of antibiotic-resistant mutants in hospitals?
2. Some geneticists maintain that the movement of transposons in a bacterial cell is a form of recombination—specifically, "illegitimate recombination." What arguments can be made for and against calling this movement a recombination, and why do you suppose it is labeled "illegitimate"?

3. Which of the recombination processes (transformation, conjugation, or transduction) would be most likely to occur in the natural environment? What factors would encourage or discourage your choice from taking place?

4. In 1976, an outbreak of pulmonary infections among participants at an American Legion convention in Philadelphia led to the identification of a new disease, Legionnaires' disease. The bacterium responsible for the disease had never before been known to be pathogenic. From your knowledge of bacterial genetics, can you postulate how it might have acquired the ability to cause disease?

5. The development of genetic engineering has been hailed as the beginning of a second Industrial Revolution. Do you believe this label is justified? How many products or applications of genetic engineering can you think of?

6. In 1994, the CDC reported that the percentage of antibiotic-resistant strains of *Haemophilus influenzae* had risen from 4.5% to 28% over the previous 5 years. What factors might have accounted for this change?

7. It is not uncommon for students of microbiology to confuse the terms *reproduction* and *recombination*. How do the terms differ?

KEY TERMS

Informative facts are necessary for the expression of every concept, and the information for a concept is founded in a set of key terms. The following terms form the basis for the concepts of this chapter. On completing the chapter, you should be able to explain and/or define each one:

bacteriocins	mutagen
binary fission	mutation
biotechnology	plasmid
competence factor	sex pili
conjugation	transduction
endonuclease	transformation
F factor	transposon
lysogenic cycle	

http://microbiology.jbpub.com/book/microbes

The site features **eLearning**, an online review area that provides quizzes and other tools to help you study for your class. You can also follow useful links for in-depth information, read more stories of microbiology, or just find out the latest microbiology news.

11

Controlling Microbes: Not Too Hot to Handle

Antibiotics and antimicrobial medicine can save a person's life from a bad infection. Because many infected people go to the hospital for treatment, it is important to keep the hospital setting clean to limit the spread of infection, and yet in the United States, roughly 6% of all patients become infected in the hospital. The inset is a picture of *Clostridium Difficile*, which is the cause of one of the most common hospital-associated infections.

■ Looking Ahead

Microbes have the ability to grow and multiply at extremely high rates. Often, this is desirable, as, for example, in industrial situations; on occasion, however, it can be a problem, as when pathogenic microbes grow in our bodies. In this case, microbial growth represents a hazard to good health, and it must be controlled. Several methods of control from both outside and inside the body are discussed in this chapter.

On completing this chapter, you should be able to . . .

- summarize factors that influence the effectiveness of agents used for microbial control.
- explain some of the physical methods of control used to achieve sterilization and destroy all forms of microbes.
- compare the chemical methods of microbial control to the physical methods with respect to the anticipated objectives.
- identify some of the important chemical agents used to retard the growth of microbes on the skin surface and on lifeless objects.
- explore the advantages and disadvantages of using antibiotics to control microbes in the body.

- identify some of the important antibiotics used to treat disease and indicate how these drugs achieve their antimicrobial activity.
- discuss the problem of antibiotic resistance with reference to its origins and implications.

In August, 1961, the *Bacteriological News*, a publication of the American Society for Microbiology, carried this story:

In the 1870s, the son of a British nobleman became dangerously mired in a bog; then a Scotsman happened along. The Scotsman waded into the bog, and with great difficulty, he pulled the boy free. When the nobleman learned of the Scotsman's deed, he offered a monetary reward, but the Scotsman politely refused. Instead, he told the nobleman that he too had a son and perhaps the nobleman would consent to educate the boy. The nobleman agreed, and the bargain was struck.

The Scotsman's son was Alexander Fleming. Fleming attended St. Mary's Hospital School of Medicine, and in the course of his research he discovered the antibiotic penicillin. Meanwhile, the nobleman's son was also gaining fame. He rose to a prominent position in British politics. Then in 1943 he was stricken with pneumonia. Fortunately, he was treated with penicillin and survived. His name was Winston Churchill.

This story describes just one of innumerable instances in which controlling microbes has been of paramount importance. On a grand scale, the development of human civilization has depended in large measure on the ability to manage the microbial world. Although we humans have put microbes to work producing our foods and disposing of our wastes and have relied on them to cycle the essential elements of life, we have also needed to control microbes so as to prevent the epidemics that regularly course through our populations. Such techniques as water chlorination, milk pasteurization, and antibiotic therapy are among the microbial control techniques that have allowed the human species to thrive and prosper.

In this chapter, we shall examine various methods available for controlling populations of microbes. The discussion will focus on three general types of control methods: physical, chemical, and antibiotic. The physical control methods are used on objects distant from the body, such as surgical instruments and microbiology lab apparatus; the chemical control methods include antiseptics used on the skin and disinfectants used on surfaces that contact the skin; and the antibiotic control methods are used within the body. Applied on a broad scale, these control methods constitute a major deterrent to infection and disease.

Physical Methods of Control

The Citadel, a novel by A. J. Cronin, follows the life of a young British physician named Andrew Manson. In the 1920s, Manson arrives at a small coal-mining town in Wales and almost immediately encounters an epidemic of typhoid fever. When his first patient dies of the disease, Manson becomes distraught. For a while, he believes there is nothing he can do to alter the course of the epidemic. But then, in a moment of insight, he realizes that microbes succumb to the effects of intense heat, and before long, he has built a huge bonfire. Into the fire go all his patients' bedsheets, clothing, and personal effects. To his delight, the epidemic subsides shortly thereafter.

Heat has long been known as a fast, inexpensive, and reliable way of controlling microbes. Heat causes biochemical changes in microbes' organic molecules (such as enzymes and structural proteins), and it drives off water, creating a lethally arid environment.

Heat is among the most useful physical methods that we shall study in this section. In most cases, the physical methods are designed to achieve sterilization, a term that implies the destruction or removal of all forms of life. Sterilization is an absolute term that cannot be qualified: An object cannot be "partially sterilized"; either it is sterilized or it remains contaminated with some form of microbial life.

A notable consideration when using physical control methods is the cellular structures that lend resistance to these methods. Certain bacterial species, for example, produce endospores (Chapter 5), whose multiple protein layers prevent heat and other physical agents from reaching the sensitive interior. Members of the genera *Bacillus* and *Clostridium* are notable sporeformers and among the most heat- and chemical-resistant microbes known to science. The fungi, too, produce thick-walled spores in their reproductive cycle (Chapter 8), and many protozoal species have the ability to form cysts (Chapter 7).

Then there is the presumptive record-holder for resistance to radiation, *Deinococcus radiodurans*. This microbe has an extraordinarily high ability to repair any damage occurring within its cell after exposure to radiation. Indeed, the bacterium survives 1000 times the radiation lethal to a human and many times the amount that kills spores. In 1998, the complete genome of *D. radiodurans* was worked out, and researchers began a study to locate clues relating to its extreme resistance. That same year the *Guinness Book of World Records* labeled the microbe the "world's toughest bacterium." **FIGURE 11.1** portrays this microbe.

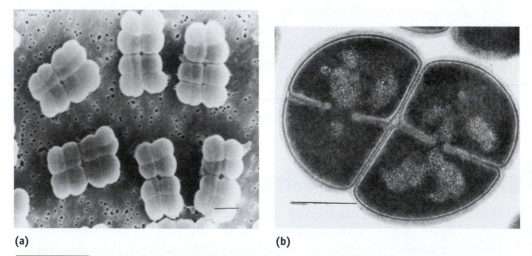

(a)   (b)

FIGURE 11.1 **Extreme Radiation Resistance.** Electron micrographs of *Deinococcus radiodurans*, a bacterium that can withstand a thousand times the radiation lethal to humans. **(a)** This organism is a tetracoccus, as shown in the scanning electron micrograph. (Bar = 2 μm.) **(b)** The transmission electron micrograph shows the complex cell wall of the four cells in the tetracoccus. (Bar = 1 μm.).

Heat Methods

Most microbes live within a specific range of temperatures, and although they survive at the limits of these ranges, they cannot live beyond those points. At high temperatures, for instance, microbial proteins are inactivated as their three-dimensional folded structures change to two-dimensional forms, a process called denaturation. In the denatured form, proteins such as enzymes are inoperative, and the microbe dies as its chemical reactions grind to a halt.

Moist heat methods are among the most widely used techniques for controlling microbes. For instance, the intense heat of the steam generated by boiling water at 100°C kills most microbes in a few seconds, the notable exception being bacterial spores (2 hours or more may be necessary to destroy these, which is why they are such effective bioweapons). Moist heat denatures proteins, as noted above, and it disrupts the integrity of cell membranes by affecting their proteins or causing chemical changes in their lipids.

Pressure is used to efficiently raise the temperature of steam above 100°C. As the pressure increases, the temperature rises and the destruction of microbes increases proportionally. An instrument called the autoclave (see A Closer Look below) increases steam pressure from 0 pounds per square inch (normal sea level pressure) to 15 pounds per square inch. Under these conditions, the temperature of the steam rises from 100°C to 121°C. At this temperature and pressure, endospores die within a few minutes. Technologists use the autoclave to sterilize instruments, glassware, microbial media, hospital and laboratory equipment, and virtually anything else that

A CLOSER LOOK

A Heated Controversy

Among the last defenders of spontaneous generation was the British physician Harry Carleton Bastian. Louis Pasteur had stated that boiled urine failed to support bacterial growth, but in 1876, Bastian claimed that if the urine were alkaline, microbes would occasionally appear. Pasteur repeated Bastian's work and found it correct. This led Pasteur to conclude that certain microbes could resist death by boiling. The spores of *Bacillus subtilis*, discovered coincidentally in 1876 by Ferdinand Cohn, were an example.

Pasteur soon realized that he would have to heat his broths at a temperature higher than 100°C to achieve sterilization. He therefore put his pupil and collaborator Charles Chamberland in charge of developing a new sterilizer. Chamberland responded by constructing a pressure steam apparatus patterned after a steam "digester" invented in 1680 by the French physician Denys Papin. The sterilizer resembled a modern pressure cooker. It attained temperatures of 120°C and higher and became the basis for the modern autoclave. Chamberland would also achieve fame in later years for his work with porcelain filters.

But Chamberland's invention was not universally accepted. A German group of investigators, led by Robert Koch, criticized the pressurized steam sterilizer because they believed its higher temperatures would destroy critical laboratory media. Instead, they preferred an unpressurized steam sterilizer, and in 1881, they developed a free-flowing steam sterilizer. In time, however, they came to appreciate the benefits of pressurized steam as a sterilizing agent, so much so that they modified Chamberland's device to an upright model. Ironically that instrument became known as the Koch autoclave.

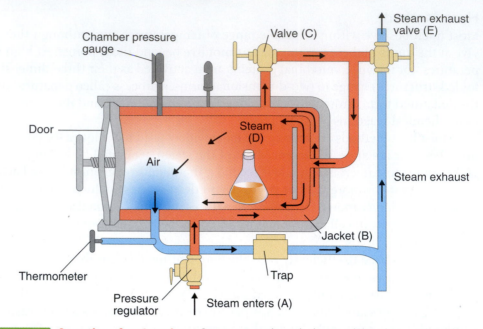

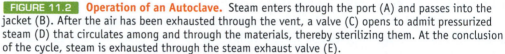

FIGURE 11.2 **Operation of an Autoclave.** Steam enters through the port (A) and passes into the jacket (B). After the air has been exhausted through the vent, a valve (C) opens to admit pressurized steam (D) that circulates among and through the materials, thereby sterilizing them. At the conclusion of the cycle, steam is exhausted through the steam exhaust valve (E).

can withstand the high temperature and pressure. Important exceptions are plastics and certain delicate chemical compounds that must be sterilized by alternative methods such as using gaseous ethylene oxide (as we shall discuss presently). FIGURE 11.2 shows the parts of an autoclave.

A widely used heat method for controlling microorganisms is pasteurization. Pasteurization employs hot water at 62.9°C for 30 minutes (the holding method), or at 71.6°C for 15 to 30 seconds (the flash method), or at 82°C for 3 seconds (the ultraflash method). Depending on the available facilities and the material to be pasteurized (for example milk, wine, or fruit juice), any of these three methods is useful for destroying the most heat-resistant microbial pathogens. The latter include *Mycobacterium tuberculosis*, the agent of tuberculosis, and *Coxiella burnetii*, the cause of Q fever (Chapter 19). It should be emphasized, however, that pasteurization is not a sterilization method since it does not affect bacterial spores.

Another heat method employs dry heat rather than moist heat. Dry heat is used in a hot air oven. The dry heat from the hot air penetrates less rapidly than moist heat, and the sterilizing temperatures tend to be more extreme and the times longer. To achieve sterilization, a temperature of 160°C to 170°C must be applied to a population of microbes for a period of 2 or more hours. In this case, the means of destruction is oxidation of cellular compounds rather than denaturation of proteins. Oxidation involves changes in the chemical nature of molecules, and the process usually requires more substantial energy input than for denaturation. Nevertheless, for sterilizing such things as powders, oily materials, and dry instruments, an oven's dry heat is efficient. Temperature implications for microbial control are presented in FIGURE 11.3 .

■ *Coxiella burnetii*
käks′ē-el-ä bėr-ne′tē-ē

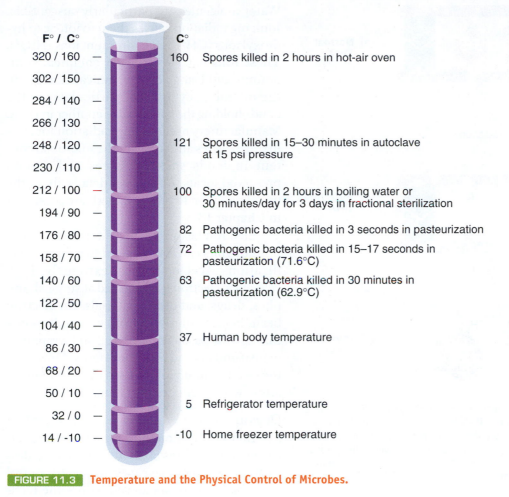

F° / C°		C°	
320 / 160	—	160	Spores killed in 2 hours in hot-air oven
302 / 150	—		
284 / 140	—		
266 / 130	—		
248 / 120	—	121	Spores killed in 15–30 minutes in autoclave at 15 psi pressure
230 / 110	—		
212 / 100	—	100	Spores killed in 2 hours in boiling water or 30 minutes/day for 3 days in fractional sterilization
194 / 90	—		
176 / 80	—	82	Pathogenic bacteria killed in 3 seconds in pasteurization
158 / 70	—	72	Pathogenic bacteria killed in 15–17 seconds in pasteurization (71.6°C)
140 / 60	—	63	Pathogenic bacteria killed in 30 minutes in pasteurization (62.9°C)
122 / 50	—		
104 / 40	—		
86 / 30	—	37	Human body temperature
68 / 20	—		
50 / 10	—		
32 / 0	—	5	Refrigerator temperature
14 / -10	—	-10	Home freezer temperature

FIGURE 11.3   **Temperature and the Physical Control of Microbes.**

Radiation

Various kinds of radiation exert destructive effects on microbes by disrupting the nucleic acid components of their cytoplasm. Ultraviolet radiation, for example, interacts with the DNA of microbial chromosomes and plasmids, and it links together adjacent molecules of thymine (or of cytosine), as we note in FIGURE 11.4 . Ultraviolet radiation also excites the electrons of other molecules in microbes and brings about biochemical changes that lead to death. It is a useful sterilizing agent for dry surfaces such as a tabletop or flat instrument, and it can be used in a closed environment such as an operating room to lower the microbial population of the air. Ultraviolet radiation is particularly effective against bacterial spores, and since it is a major component of sunlight, the release of bacterial spores into the atmosphere by a bioterrorist would result in lower spore survival in the day than at night.

Other kinds of radiation, including X rays and gamma rays, are also useful for sterilization. X rays and gamma rays are about 10,000 times more energetic than ultraviolet radiation, and they induce electrons and protons to jump out of the molecules they strike. This process creates ions, which are atoms or molecules lacking one or more electrons (the high-energy radiation is therefore called ionizing radiation).

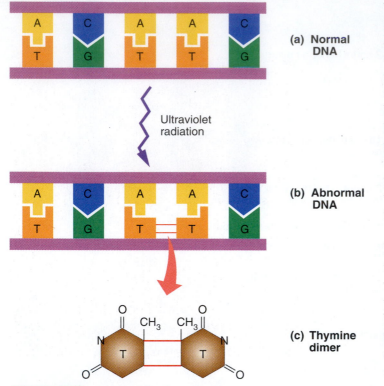

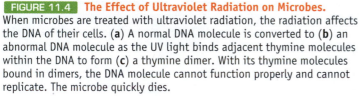

FIGURE 11.4   **The Effect of Ultraviolet Radiation on Microbes.**
When microbes are treated with ultraviolet radiation, the radiation affects the DNA of their cells. (**a**) A normal DNA molecule is converted to (**b**) an abnormal DNA molecule as the UV light binds adjacent thymine molecules within the DNA to form (**c**) a thymine dimer. With its thymine molecules bound in dimers, the DNA molecule cannot function properly and cannot replicate. The microbe quickly dies.

Water molecules are particularly susceptible: Ionizing radiation causes them to ionize to hydroxyl ions (OH^-) and hydronium ions (H_3O^+). These ions react with other chemical compounds and bring about chemical changes in the microbial cytoplasm, particularly in the bonds holding the bases to one another in DNA. Manufacturers use ionizing radiation for sterilizing plastics and to lower the microbial content in foods such as spices and certain preserved meats. Additional details about the use of radiation to preserve foods are presented in Chapter 13.

Electron beams having intensities of millions of volts can also be used to achieve sterilization. Accelerated to extremely high velocities, electrons are directed at surgical supplies, drugs, and other materials. An electron beam is particularly useful because it passes through the packaging material and wrappings of instruments and other apparatus. Moreover, the electron beam can be used at room temperature, which reduces heat damage.

Drying

Another nonheat method for controlling microbial populations is drying (also called desiccation). Humans have used drying to control microbial growth since well before the technology of heat or radiation methods was developed. For example, tradition has it that Peruvian Incas of the Andes Mountains preserved potatoes and other foodstuffs by placing them for several weeks on high mountainsides, where they dried in the open air.

Drying is still used today to prepare foods such as cereals, grains, and numerous other products for storage in the home pantry. Since water is required for most chemical reactions in microbes, it follows that microbes cannot grow where water is very limited or absent. However, many types of microbes, especially bacterial spores, remain alive under these conditions, and if water is introduced to the environment, they will begin multiplying. Bioterrorists could take advantage of this characteristic if they tried to use anthrax spores as bioweapons. The spores would remain alive in the biodust and would revert to vegetative bacilli after being inhaled into people's lungs or entering their tissues via a crack or wound in the skin.

Hikers, campers, and backpackers often carry foods prepared by freeze-drying, or more technically, lyophilization. The food is frozen; then the water is drawn off with a vacuum pump. Lyophilization takes water from the solid phase (ice) to the gaseous phase (vapor) without passing through the liquid phase (water). A freeze-dried product is extremely light and dry. However, it should be noted that microbes

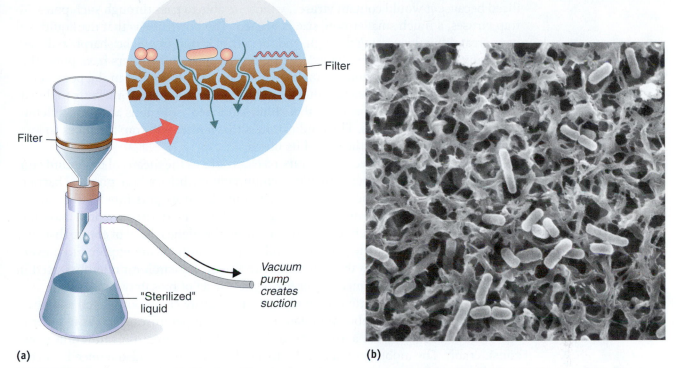

FIGURE 11.5 **The Principle of Filtration.** Filtration is used to remove microbes from a liquid. The effectiveness of the filter is proportional to the size of its pores. (**a**) Bacteria-laden liquid is poured into a filter, and a vacuum pump helps pull the liquid through and into the flask below. But the bacteria are larger than the pores of the filter, and they become trapped. The liquid dripping into the flask is sterilized if all microbial forms, including viruses, are caught. Otherwise, the liquid will remain contaminated. (**b**) A view of *Escherichia coli* cells trapped in the pores of a 0.45-μm nylon membrane filter.

have been preserved as well as the food. Indeed, microbiologists often preserve their microbial cultures by freeze-drying them.

An alternative method for removing water from the microbial environment is by taking advantage of the process of osmosis. When microbes are exposed to a high-salt external environment, for instance, water flows from their cytoplasm, through the cell membrane, and out to the environment. This flow of water, known as osmosis, occurs because the concentration of water is higher inside the microbial cell than outside the cell. In the salty environment, the microbes shrivel and die quickly. Food processors use salt for preserving meats, fish, and numerous other types of foods. Sugar and spices can be used as alternatives to salt.

Filtration and Refrigeration

Microbes can be removed from a liquid solution by the process of filtration. During filtration, liquid passes through a porous material that traps microbes in its submicroscopic pores, as shown in **FIGURE 11.5**. For heat-sensitive solutions such as research materials, filtration is a useful alternative.

Because filters function by mechanical entrapment, they are manufactured in a range of sizes to fit the microbes that are to be removed. For example, if bacteria are to be trapped, then a filter with a pore size of 0.2 μm to 0.5 μm might be used (since the smallest bacteria are in this size range). However, the filtered liquid would not be sterilized because it would contain viruses, which are able to pass through such pores. To trap viruses, a much smaller pore size is needed. The downside is that the liquid will have great difficulty passing through, and the rate of filtration will be sharply reduced.

Many filters have pores shaped like rigid cylinders while others have pores that are like winding and tortuous tunnels. In addition to entrapment in the pores, electrostatic attractions between the filter material and the microbes contribute to the filtration process. It should be noted that "filter-sterilized" solutions are quite different from "sterilized" solutions. The implication is that a filter will trap microbes so long as their size is larger than the size of its pores.

Filters are valuable in contemporary research labs as the means of biohazard control. Biohazard control is performed in containment facilities where physical barriers such as filters prevent pathogens from contacting laboratory personnel or the outside environment. Depending on the type of research performed, there are four biosafety levels. Research in DNA technology often requires the highest level of control because new genetic forms are being generated and extra precautions are required to prevent release of these forms into the natural environment. Research on microbes used in bioterrorism is also performed under the highest level of biosafety.

Though not a sterilization method, low temperature can be employed as a physical agent to control microbial populations. At low temperatures, such as in a refrigerator or freezer, enzyme activity diminishes, and microbial reproduction slows considerably. The mobility of molecules through membranes is also reduced, and the rate of chemical reactions in the cytoplasm is lowered. Note, however, that low temperatures only slow microbial growth; they do not kill microbes, except if ice crystals manage to tear cells apart. It is well to remember that freezing or refrigerating foods preserves the foods but does not completely eliminate microbial populations.

■ Chemical Methods of Control

The practices of disinfection, antisepsis, and chemical control of microbes are not new. The Bible often refers to cleanliness and prescribes certain dietary laws to prevent consumption of possibly contaminated foods. The Egyptians used resins and aromatic chemicals for embalming even before they had a written language, and other ancient peoples burned sulfur for deodorizing and sanitary purposes. Over the centuries, necessity demanded that chemicals be used for food preservation, and spices were useful as preservatives and as masks for foul odors in foods. Indeed, Marco Polo's trips to the Orient for new spices were made more out of necessity than for adventure.

Medicinal chemicals came into widespread use in the 1800s. As early as 1830, for example, the *U.S. Pharmacopoeia* listed tincture of iodine as a valuable antiseptic, and soldiers used it in plentiful amounts during the Civil War. Copper was valuable for preventing fungal disease in plants, and mercury was sometimes used for treating syphilis, as Arabian physicians had suggested centuries before. Moviegoers have probably noticed that American cowboys practiced the art of disinfection by pouring whiskey into wounds—between drinks, that is.

Disinfection and antisepsis received a considerable boost in the 1860s with the work of Joseph Lister, a physician from the University of Glasgow, Scotland. Lister was aware of Louis Pasteur's work with airborne microbes, and he resolved to use chemical compounds to prevent infections that invariably followed surgery. After experimenting with several chemicals, Lister finally decided on phenol (carbolic acid), having read in a newspaper that it was used in a nearby town for treating sewage. He soaked his instruments and ligatures in phenol and sprayed it in the air near the patient (FIGURE 11.6). Although his initial attempts were unsuccessful, Lister remained optimistic, and he continued to improve his procedures until he achieved outstanding success (he became one of the most famous surgeons of his time because of the high survival rate of his patients). Before his death in 1912, he was knighted for his work.

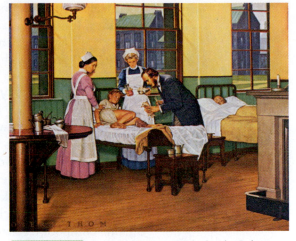

FIGURE 11.6   **Joseph Lister.** A painting by Robert Thom depicting Joseph Lister using antiseptic methods in the surgical treatment of a leg wound.

General Principles

To interrupt the spread of microbes, chemical controls are applied in such diverse locations as the hospital environment, the food-processing plant, and the typical household. While physical methods are used to achieve sterilization, chemical methods are generally used to lower microbial populations and kill most, but not necessarily all, pathogens. The chemicals used are called disinfectants or antiseptics. Disinfectants are chemical compounds formulated for use on inanimate (lifeless) objects, while antiseptics are meant for use on the surface of the body. Although certain chemical agents are strong enough to sterilize, exposure to the air following treatment reintroduces contamination. Therefore, for both antiseptics and disinfectants, sterilization is not a realistic objective.

No single chemical agent is ideal for controlling all microbes under all conditions. However, if an ideal chemical agent were to exist, it would possess an elaborate array of characteristics: It would have the capacity to kill all microbes; it would be soluble in water; it would be stable on standing and not lose its antimicrobial action over time; and it would be nontoxic to humans and animals. Furthermore, it would be uniform in composition so that all ingredients would be present in each application; it would not combine with organic matter other than in microbes; its toxicity to microbes would be highest at room or body temperature; it would efficiently penetrate surfaces; it would not corrode or rust metals or damage or stain fabrics; and it would be readily available in useful quantities and at reasonable prices. With such stringent requirements, it is not surprising that an ideal antiseptic or disinfectant does not exist.

Alcohols and Aldehydes

Among the important chemical agents for microbial control are the alcohols. The most widely used alcohol is ethyl alcohol, usually in a 70% solution. Ethyl alcohol denatures proteins and dissolves lipids like those in the cell membranes of microbes. It may be used as an antiseptic on the skin or as a disinfectant by immersing instruments in it for a minimum of 10 minutes. The chemical is effective against multiplying bacteria, but it has no effect on spores. Isopropyl (rubbing) alcohol is equally useful.

Aldehydes are organic compounds in which an end-of-chain carbon atom is linked to an oxygen atom and a hydrogen atom. Two aldehydes, formaldehyde and glutaraldehyde, are useful as microbial control agents, but in closed environments (because the chemicals are toxic). Both aldehydes react with amino groups (in proteins) and nucleotides, linking them together and changing the structure of the chemical compound. Following use, materials must be rinsed thoroughly with sterile water, dried in a special cabinet with sterile air, and stored in a sterile container.

Halogens and Heavy Metals

Halogens are extremely reactive elements as a result of the configuration of the electrons in their atoms. Two halogens, iodine and chlorine, are particularly useful as chemical control agents. Iodine can be used as a tincture of iodine (2% iodine in ethyl alcohol) or as iodine-detergent compounds known as iodophors (Betadine or Wescodyne, for example). In iodophors, the detergent loosens microbes from the skin surface and the iodine kills them.

Chlorine is used as chlorine gas to reduce the microbial content of water or as an organic compound called a chloramine. Like iodine, chlorine reacts with proteins, is effective against all types of microbes, and may be used in antiseptics or disinfectants depending on the formulation. Chlorine is also used in the form of sodium hypochlorite in a 5% concentration in household bleach. To disinfect clear water, the CDC recommends a half-teaspoon of bleach in 2 gallons of water, with 30 minutes of contact time before consumption. Other uses for chlorine are shown in FIGURE 11.7 .

Three heavy metals are useful as chemical antiseptics and disinfectants: Silver in the form of silver nitrite is employed as a general antiseptic. Mercury is still used in some antiseptics for treating skin wounds, such as Mercurochrome and Merthiolate,

FIGURE 11.7   Some Practical Applications of Disinfection with Chlorine Compounds.

but it is toxic. Indeed, a mercury compound named thimersol was widely used in vaccine preparations, but in 1999, the CDC recommended its removal as a safety measure.

Copper in the form of copper sulfate is used to control cyanobacteria in swimming pools and to restore the clarity of the water. Copper sulfate is also mixed with lime to form the famous Bordeaux mixture for controlling fungal growth. The benefits of the mixture was discovered in 1882 by Alexis Millardet of the University of Bordeaux, France. While strolling through a vineyard, Millardet noticed the telltale blue color of the mixture on the healthier grapevines along the road. The local vintner told him that it was used to make the grapevines look poisonous and ward off any potential grape thieves. But Millardet was more interested in the good health of the grapevines and he surmised that the copper sulfate–lime mixture was preserving the vines from fungal disease. Millardet conducted a series of experiments to test his theory, and the rest, as they say, is history.

Detergents and Phenols

Detergents are strong wetting agents and surface tension reducers; that is, they work their way between microbes and a surface and "lift" the microbes so that they can be removed with the wash water. They also dissolve the microbial cell membrane by reacting with its lipids, thereby causing leakage through the membrane and cell death. When used on cutting boards, detergents can reduce the possibility of cross contamination of foods and utensils. Wood and plastic cutting boards are compared in A Closer Look (below).

Phenol compounds (also known as phenolics) were among the first chemical agents used for microbial destruction, having been employed by Lister in his landmark experiments in the 1860s. Phenol derivatives include Lysol, hexylresorcinol

A CLOSER LOOK

Cutting Board Wars

He: "I'll get out a cutting board to cut up the salad."
She: "You might want to use the plastic one instead of the wooden one."
He: "Why's that?"
She: "Because bacteria get caught in the grooves of the wood."
He: "But I read somewhere that the wood draws moisture and bacteria so deep into the grooves they can't reach the food. The article said that plastic lets bacteria stay close to the surface so they get on the food."
She: "Well I heard that plastic cleans better than wood, so it's safer."
He: "Yeah, but wood has antibacterial powers that plastic doesn't have."
She: "Maybe so, but plastic doesn't have all those grooves and scars where bacteria can hide."
He: "How about this: I read about a guy in California who found that *Salmonella* infections are more likely if a household uses a plastic board."
She: "Well, I don't know about anybody in California, but I do know that plastic is easier to dry thoroughly. Wood stays moist, and that lets bacteria stay alive."
He: "How about we forget the salad and go out to McDonald's?"
She: "How about Burger King?"

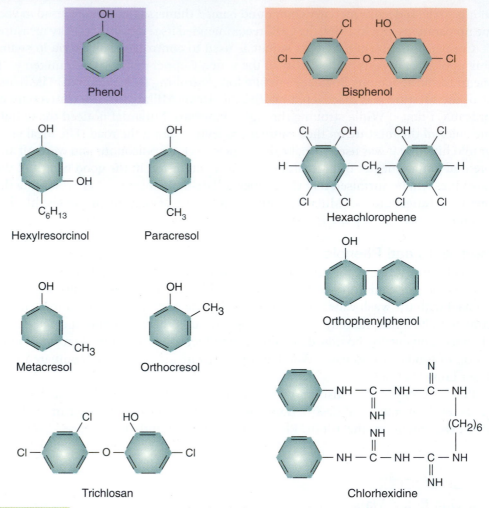

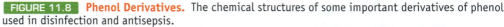

FIGURE 11.8 **Phenol Derivatives.** The chemical structures of some important derivatives of phenol used in disinfection and antisepsis.

(used in mouthwashes), and hexachlorophene (the active ingredient in pHisoHex handwash). Another important phenol derivative is chlorhexidine, used commercially in Hibiclens, a common handwash found in hospital and clinical facilities. Chlorhexidine also kills the bacteria in caries-related plaque between the teeth and at the gumline. (A toothpaste manufacturer has experimented with chlorhexidine in toothcare products.) FIGURE 11.8 displays the chemical structures of phenol and related compounds.

A phenol derivative in widespread use is trichlosan, which destroys bacterial cell membranes by blocking lipid synthesis. Trichlosan, known commercially as Irgasan and Ster-Zac, is fairly mild and nontoxic, and it is moderately effective against pathogenic bacteria (but less so against viruses and fungi). The chemical is incorporated in "antibacterial "soaps, lotions, mouthwashes, and toothpastes and in plastic and synthetic fibers used to make toys, food trays, underwear, kitchen sponges, utensils, and cutting boards. The negative side to overuse is that bacteria may develop resistance to the chemical (just as they have developed resistance to antibiotics). Indeed, a ninth-grade student presenting her experiment at a science fair in Massachusetts in 2000 was among the first to draw attention to this possibility.

Ethylene Oxide

The development of plastics for use in microbiology labs required a suitable method for sterilizing these heat-sensitive materials. In the 1950s, research scientists discovered the antimicrobial properties of ethylene oxide (EtO), making the plastic Petri dish and plastic syringe possible. Ethylene oxide is a small molecule with excellent penetration capacity and sporicidal ability. However, it is toxic and explosive, so it must be used in a tightly sealed chamber. The chamber must then be flushed with inert gas for 8 to 12 hours to ensure that all traces of ethylene oxide are removed; otherwise, the chemical will cause "cold burns" on contact with the skin.

Manufacturers use ethylene oxide to sterilize paper, leather, wood, metal, and rubber products, as well as plastics. In hospitals, it is used to sterilize catheters, artificial heart valves, heart-lung machine components, and optical equipment. The National Aeronautics and space Administration (NASA) uses the gas for sterilizing interplanetary space capsules. For sterilization purposes, ethylene oxide chambers have become the chemical counterparts of autoclaves.

Antibiotics

For many centuries, physicians believed that heroic measures were necessary to save patients from the ravages of infectious disease. They prescribed frightening courses of purges and bloodlettings, enormous doses of strange chemical concoctions, ice water baths, starvation, and other drastic remedies. Unfortunately, these treatments probably worsened an already difficult situation by reducing the body's defenses to the point of exhaustion.

But a revolution in medicine took place during the 1940s, when the antibiotics burst on the scene. Doctors were astonished to learn that they could kill bacteria in the body without doing substantial harm to the body itself. The practice of medicine experienced a period of powerful, decisive therapy for infectious disease, and physicians found that they could successfully alter the course of disease. The antibiotics effected a radical change in medicine, charting a new course that has been followed to the present day. Indeed, many historians consider the development of antibiotics to be the greatest medical event of the twentieth century.

The First Antibacterials

When the germ theory of disease emerged in the late 1800s, the newly discovered information about microbes added considerably to the understanding of infectious disease. Furthermore, it increased the storehouse of knowledge available to physicians. However, it did not change the fact that little, if anything, could be done for the infected patient. Tuberculosis continued to kill one out of every seven people; streptococcal pneumonia was a fatal experience; and meningococcal meningitis exacted a heavy toll in human life.

Against this backdrop, scientists dreamed of chemical substances that might be used to kill microbes in the body without damaging the body. One such investigator

FIGURE 11.9 Ehrlich and Hata. A painting by Robert Thom depicting Paul Ehrlich and Sahachiro Hata, the two investigators who developed arsphenamine for the treatment of syphilis. Ehrlich is shown writing a work order with the stubby colored pencil he habitually used. He and Hata conducted their experiments at the Institute of Experimental Therapy in Frankfurt, Germany.

was a chemist named Paul Ehrlich, pictured in FIGURE 11.9. In the early 1900s, Ehrlich envisioned antimicrobial chemical substances as "magic bullets" that could be used to seek out and destroy microbes in the body without damaging the tissues. He attempted to develop a chemical that would kill the bacterial agent of syphilis (then, as now, a serious disease). With his assistant Sahachiro Hata, Ehrlich synthesized hundreds of arsenic-based compounds and eventually, located the magic bullet, a compound named arsphenamine (traditionally referred to as "606" because it was the 606th compound Ehrlich synthesized). Although the compound showed promise, some physicians used it indiscriminately, resulting in overdoses for some patients and adverse reactions in the liver and kidneys. Moreover, World War I was beginning, and support for drug development was very limited. Soon, Ehrlich's magic bullet was forgotten.

Significant advances in drug therapy would not be made for another 20 years. Then, in 1932, a new chemical called prontosil showed promise against Gram-positive bacteria such as staphylococci and streptococci. Prontosil's discoverer was a German investigator named Gerhard Domagk. In February 1935, Domagk's daughter Hildegarde became gravely ill with a blood infection after pricking her finger with a needle. Domagk decided to gamble with his new drug, and he gave her an injection of prontosil. When her health improved dramatically, the efficacy of the drug was established. Four years later, Domagk was awarded the 1939 Nobel Prize in Physiology or Medicine (*in absentia*, because Chancellor Adolph Hitler forbade him to accept the award).

In the years thereafter, French investigators isolated sulfanilamide, the active substance in prontosil. At another period in history, scientists might have spent years researching the therapeutic value of sulfanilamide, but in the late 1930s, World War II was in progress in Europe, and sulfanilamide was a godsend to soldiers who might otherwise have died from wound-related infections. (FIGURE 11.10 shows how the drug works to kill bacteria.) The success of sulfanilamide did not go unnoticed by medical researchers, who began an intense search for new antimicrobial substances, hoping to find something as good as or better than sulfanilamide.

The Development of Penicillin

In 1939, Rene Dubos, a Rockefeller Institute researcher, reported that soil bacteria could produce myriad antibacterial substances. At the same time, a group of English scientists at Oxford University were investigating an antimicrobial substance discovered many years before by a fellow Englishman. The Englishman was Alexander Fleming, a physician at St. Mary's hospital in London.

In 1928, Fleming was performing research on staphylococci (FIGURE 11.11). Before going on vacation, he spread staphylococci on plates of nutrient agar and set them aside to incubate. On his return, he noted that one plate was contaminated by a green mold, which he identified as *Penicillium*. Fleming's attention was drawn to the clear area around the mold, an area where the staphylococci were unable to grow. Unsure

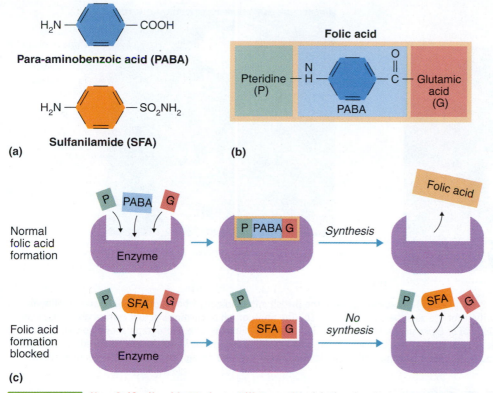

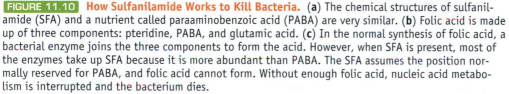

FIGURE 11.10 **How Sulfanilamide Works to Kill Bacteria.** (**a**) The chemical structures of sulfanilamide (SFA) and a nutrient called paraaminobenzoic acid (PABA) are very similar. (**b**) Folic acid is made up of three components: pteridine, PABA, and glutamic acid. (**c**) In the normal synthesis of folic acid, a bacterial enzyme joins the three components to form the acid. However, when SFA is present, most of the enzymes take up SFA because it is more abundant than PABA. The SFA assumes the position normally reserved for PABA, and folic acid cannot form. Without enough folic acid, nucleic acid metabolism is interrupted and the bacterium dies.

what was happening, he cultivated the *Penicillium* in broth and added a small drop of the broth to a culture of staphylococci. As he looked on with astonishment, the staphylococci disintegrated before his eyes. Other Gram-positive bacteria were equally susceptible to the mold broth, which he appropriately called "penicillin." Unfortunately, Fleming could not isolate the active substance in the broth.

We now fast-forward to England in 1939 and the research group at Oxford University. Spurred by the success of sulfanilamide, the group led by Howard Florey and Ernst Boris Chain searched the literature for antimicrobial substances and came upon the report by Fleming. Using the newest biochemical methods for separation and purification, Florey, Chain, and their associates isolated penicillin, the active principle in the mold broth, and they began human trials. But England was at war; German bombs were falling on London, and the researchers feared for their lives. They therefore turned to American companies to produce penicillin in industrial quantities (*see* A Closer Look on page 248). Soon production facilities in the midwestern United States were churning out penicillin by the ton, and the Age of Antibiotics was underway. The international scientific community recognized Fleming, Florey, and Chain with the 1945 Nobel Prize in Physiology or Medicine.

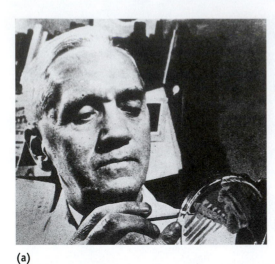

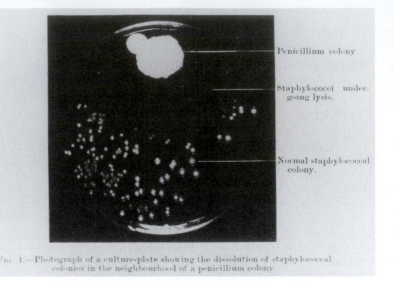

(a)   (b)

FIGURE 11.11 **The Discovery of Penicillin.** (**a**) Alexander Fleming, the British microbiologist who reported the existence of penicillin in 1928 but was unable to purify it for use as a therapeutic agent. (**b**) The photograph of Fleming's actual culture plate that was originally published in the *British Journal of Experimental Pathology* in 1929. Taken by Fleming, the photograph shows how staphylococci in the region of the *Penicillium* colony have been killed (they are undergoing lysis) by some unknown substance produced by the mold. Fleming called the substance penicillin. Ten years later, penicillin would be rediscovered and developed as the first modern antibiotic.

A CLOSER LOOK

Transporting a Treasure

Their timing could not have been worse. Howard Florey, Ernst B. Chain, Norman Heatley, and others of their team had rediscovered penicillin, refined it, and proven it useful in infected patients. But it was 1939, and German bombs were falling on London. This was no time for research into new drugs.

There was hope, however. Researchers in the United States were willing to attempt the industrial production of penicillin, so the British scientists made plans to move their lab across the ocean. There were many problems, to be sure, but one was particularly interesting—how to transport the vital *Penicillium* cultures. If the molds were to fall into enemy hands or if the enemy were to learn the secret of penicillin, all their work would be wasted. Then Heatley made a suggestion: They would rub the mold on the inside linings of their coats, deposit the mold spores there, and transport the *Penicillium* cultures across the ocean that way.

And so they did. On arrival in the United States, they set to work to reisolate the mold from their coat linings, then they began the laborious task of manufacturing penicillin. One of the great ironies of medicine is that virtually all the world's original penicillin-producing mold has been derived from those few spores in the linings of the British coats. Few coats in history have yielded so great a bounty.

Penicillins

Since the 1940s, penicillin has remained the most widely used antibiotic, primarily because of its low cost and numerous derivatives (for example, penicillin G, ampicillin, amoxicillin, methicillin, and others). All of these relatives, referred to as the penicillins, share the same fundamental structure, with a chemical complex called a beta-lactam nucleus at the core. For this reason, they are sometimes called beta-lactam antibiotics.

Penicillins are active primarily against a variety of Gram-positive bacteria, but some formulations also affect Gram-negative bacteria, especially the diplococci that cause gonorrhea and meningitis. Penicillin and its derivatives function during the synthesis of the bacterial cell wall—they block the formation of the peptidoglycan portion of the wall (Chapter 5), thus resulting in a weak wall that gives way to internal pressures, causing the microbe to swell and burst. Penicillin is therefore most useful when bacteria are multiplying rapidly, as they do during infection. It would not be effective against bacterial spores used in a bioterrorist attack because the spores would not be multiplying. However, once the spores reverted to multiplying bacteria in the body, the antibiotic would be useful against them.

Over the decades of exposure, many bacterial species have developed resistance to penicillin, as we shall discuss presently. These organisms are able to produce an enzyme called penicillinase, which converts penicillin into a harmless substance called penicilloic acid, shown in FIGURE 11.12A . This ability has probably always existed in some bacteria, but exposure to penicillin creates an environment in which penicillinase-producing forms are encouraged to emerge. As penicillin-susceptible bacteria die off, a form of natural selection takes place, and the rapid multiplication of penicillinase-producing bacteria results in populations that are resistant to the antibiotic.

Various penicillin derivatives are produced by altering the side groups attached to the beta-lactam nucleus (FIGURE 11.12B). One derivative, ampicillin, is absorbed more readily than penicillin is. Amoxicillin, another derivative, is more stable in stomach acid and does not bind to food, as many antibiotics are inclined to do. Other drugs in this family include carbenicillin and ticarcillin, both of which are used against organisms that have developed resistance to penicillin itself. All the penicillins may induce allergic reactions in a patient (Chapter 17), and some have been implicated in disturbance of the intestinal tract.

Cephalosporins and Aminoglycosides

A valuable alternative to the penicillin group of antibiotics is the cephalosporin group. The first cephalosporins were used against Gram-positive cocci. They included cephalexin (Keflex) and cephalothin (Keflin). The newer members of this family are also used against Gram-negative bacteria. They include such drugs as cefotaxime (Claforan), ceftriaxone (Rocephin), and ceftazidime (Fortaz). All the cephalosporins resemble the penicillins except that their beta-lactam nucleus is slightly different. For this reason, they resist enzymes that destroy penicillins. Like penicillins, they are produced by a mold (*Cephalosporium* species), and they interfere with cell wall synthesis. FIGURE 11.13 displays the effect of cefalexin on cholera bacilli.

Aminoglycoside antibiotics have traditionally been used against Gram-negative bacteria. One of the first aminoglycoside antibiotics to be discovered was streptomycin. In 1943, this antibiotic caused a sensation because it was found that it helped cure tuberculosis. Other drugs have replaced it since then, but the appearance of drug resistances in *Mycobacterium tuberculosis* have led physicians to reconsider the use of streptomycin for tuberculosis therapy. Unfortunately, long-term use of this antibiotic may result in hearing loss.

Aminoglycoside antibiotics are produced by members of a genus of moldlike soil bacteria called *Streptomyces*. These antibiotics, which function by inhibiting bacterial

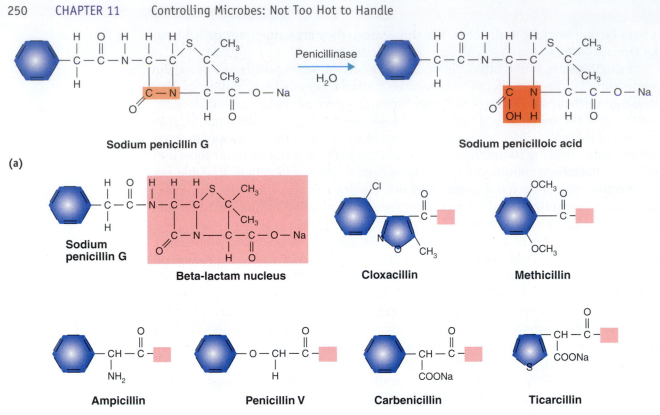

(a)

(b)

FIGURE 11.12 **Penicillin Resistance and Penicillin Derivatives.** (**a**) The action of penicillinase on sodium penicillin G is shown. The enzyme converts penicillin to harmless penicilloic acid by opening the beta-lactam nucleus and adding a hydroxyl group (OH) to the carbon and a hydrogen (H) to the nitrogen. (**b**) Some members of the penicillin group of antibiotics. The beta-lactam nucleus is common to all the penicillins, which differ in the side group attached to that nucleus.

protein synthesis, include gentamicin, used against urinary tract infections; neomycin, for intestinal tract and eye infections; and many other "mycins" that are useful against Gram-negative bacteria.

Broad-Spectrum Antibiotics

■ *Streptomyces*
strep-tŏ-mī'sĕs

In 1947, scientists discovered the first broad-spectrum antibiotic, that is, an antibiotic that kills numerous types of microbes. Isolated from a species of *Streptomyces*, this antibiotic is called chloramphenicol. Investigators discovered that it was inhibitory to Gram-positive as well as Gram-negative bacteria and to many species of rickettsiae, chlamydiae, and fungi. Although chloramphenicol is still used to treat certain diseases such as typhoid fever, it is very toxic and is considered a last-resort antibiotic. The drug causes bone marrow disturbances, resulting in red blood cells that lack hemoglobin.

Also in the 1940s, investigators found a broad-spectrum antibiotic named tetracycline. Tetracycline and its derivatives (minocycline, doxycycline, and others) remain the drugs of choice for many diseases caused by Gram-negative bacteria. The antibiotics are also effective against rickettsiae and chlamydiae. Because tetracycline has few side effects, many physicians prescribe this valuable antibiotic in trivial situations (such as for acne), and the overexposure encourages antibiotic-resistant bacteria to emerge, as Chapter 10 explains. Furthermore, the antibiotic is often used in excessive quantities, which destroy the normal bacteria in the large intestine, thus allowing yeasts to

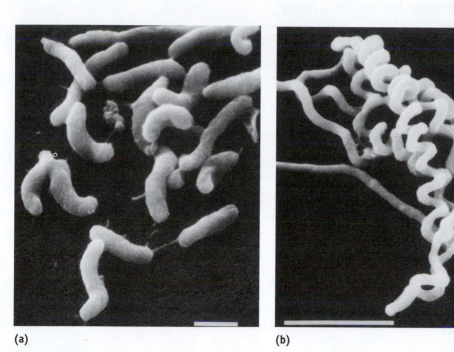

(a)   (b)

FIGURE 11.13 **The Effects of Cephalexin on *Vibrio cholerae*.** (**a**) A scanning electron micrograph of control cells of the bacteria grown in a medium free of antibiotic. The cells exhibit the typical vibrio shape, with a short curve and an incomplete spiral. (Bar = 1 μm.) (**b**) Experimental cells treated with 3.13 μg of cephalexin per mL. The cells have elongated and formed right-handed spirals that are complete. (Bar = 5 μm.)

flourish (a similar phenomenon happens in the vaginal tract when too much tetracycline is used). Overuse of tetracycline may also result in a yellow-gray-brown discoloration of the teeth or stunted bones in children. Like many other antibiotics, it interferes with protein synthesis in bacteria by binding to bacterial ribosomes.

Other Antibiotics

Another antibiotic with clinical importance is erythromycin. This product of a *Streptomyces* species inhibits protein synthesis in many Gram-positive bacteria. It is particularly useful as an alternative to penicillin when a patient is allergic to that antibiotic. Two other antibiotics in the same family as erythromycin are clarithromycin (Biaxin) and azithromycin (Zithromax), which are used against Gram-negative bacteria.

Vancomycin has contemporary significance as a last-resort antibiotic for use against Gram-positive bacteria having multidrug resistance. It is administered during serious cases of streptococcal, staphylococcal, and other infections due to Gram-positive bacteria. Because of its damaging side effects in the ears and kidneys, vancomycin is not routinely used for trivial situations. Unfortunately, resistance to this antibiotic has also been observed (Chapter 10), and substitute drugs have been sought. One possibility is for the combination of quinupristin and dalfopristin sold under the trade name Synercid. It is the first of the streptogramin class of antibiotics to be approved.

Rifampin is a synthetic antibiotic prescribed for tuberculosis patients. It is also administered to individuals who harbor *Neisseria* and *Haemophilus* species that cause meningitis and for protective purposes to individuals who have been exposed to these bacteria. Rifampin acts by interfering with RNA synthesis in bacteria. It may give an orange-red color to urine, feces, tears, and other body secretions and may cause liver damage.

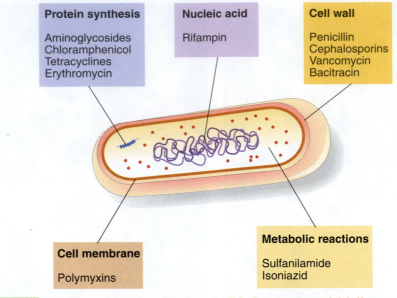

Protein synthesis

Aminoglycosides
Chloramphenicol
Tetracyclines
Erythromycin

Nucleic acid

Rifampin

Cell wall

Penicillin
Cephalosporins
Vancomycin
Bacitracin

Cell membrane

Polymyxins

Metabolic reactions

Sulfanilamide
Isoniazid

FIGURE 11.14 The Sites of Activity of Various Antibiotics on a Bacterial Cell.

Both bacitracin and polymyxin B are polypeptide antibiotics produced by *Bacillus* species. These antibiotics are generally restricted to use on the skin because, if taken internally, they are poorly absorbed from the intestine and may cause kidney damage. Bacitracin is available in pharmaceutical skin ointments and is effective against Gram-positive bacteria such as staphylococci. Polymyxin B is useful against Gram-negative bacilli, particularly those that cause superficial infections in wounds, abrasions, and burns. Bacitracin inhibits cell wall synthesis, while polymyxin B injures bacterial membranes. **FIGURE 11.14** summarizes the sites at which various antibiotics affect a bacterial cell.

Antiviral and Antifungal Antibiotics

Over the decades, scientists have developed a set of antibiotics that are useful against the viruses and the fungi. To control viruses, the viral replication process can be interrupted at any of several places. For example, the drug amantadine prevents attachment of influenza viruses to the host cell surface. Another licensed drug, acyclovir, interferes with the process of DNA production in viruses such as those of genital herpes and chickenpox. Chapter 6 explores other antiviral medications.

When dealing with fungal disease, the medical mycologist has relatively few drugs available. One example is nystatin, which is effective against yeast infections due to *Candida albicans*. It acts by reacting with sterol compounds and thereby changing the structure of the cell membrane of the yeast. Other useful antifungal antibiotics include griseofulvin for ringworm and athlete's foot and amphotericin B for serious fungal diseases of the internal organs. Another class of antifungal drugs are the imidazoles, which also interfere with sterol activity in the fungal cell membrane. Many of these compounds are used against *Candida albicans*. They include clotrimazole (Lotrimin) and miconazole (Monistat).

Antibiotic Resistance

During the last quarter of the twentieth century, an alarming number of bacterial species developed resistance to antibiotics. These resistant organisms are increasingly responsible for human diseases of the blood, intestinal tract, lungs, and urinary tract. They are of special danger to those in intensive care units and burn wards, as well as to children, the elderly, and those with compromised immune systems (because the body's natural defenses are especially weak in these individuals). Among the most difficult to treat are some common diseases that once succumbed to a single dose of antibiotics, such as bacterial pneumonia, streptococcal blood disease, and gonorrhea.

Microbes have acquired resistance to antibiotics in a number of ways. With some antibiotics, such as penicillin, resistance arises from the microbe's ability to destroy the antibiotic before the antibiotic destroys the microbe. Resistance can also arise from changes in the passageways of the microbial cell wall and membrane that restrict movement of the antibiotic into the cytoplasm. Alternatively, a microbe may alter the metabolic pathway on which an antibiotic normally acts and thereby bypass the drug inhibition. Resistance may also develop when a microbe changes the target of the antibiotic, such as by changing the structure of its ribosome (Chapter 10 explores these methods in more detail).

Resistance may occur in bacteria during the normal course of events, as happened with penicillinase-producing bacteria. Antibiotic abuse, however, also stimulates the selection of resistant organisms. For example, drug companies promote antibiotics heavily, and patients pressure doctors for quick cures, often encouraging a physician to write a prescription without ordering costly and time-consuming tests to pinpoint the cause of the illness. Another forcing ground for the emergence of antibiotic resistance is the hospital. Physicians sometimes use unnecessarily large doses of antibiotics to prevent infection before, during, and following surgery. This overuse increases the possibility that resistant bacteria will overgrow susceptible strains and spread to other patients.

Antibiotics may also be abused in developing countries. In many Central American countries, for instance, the most potent antibiotics are available without prescription. The problem of antibiotic abuse is particularly widespread in livestock feeds. An astonishing 30% of all antibiotics produced in the United States find their way into animal feeds, where they check disease and promote growth. These antibiotics also encourage the more resistant bacteria to predominate.

The problem of antibiotic resistance has been made even worse because it is now apparent that microbes can transfer their genes from one cell to another and even from one species to another. Researchers have demonstrated that plasmids that carry antibiotic-resistance genes can move among bacterial species. Indeed, bacteria may accumulate a number of antibiotic-resistance genes and become resistant to numerous antibiotics at one time. This has happened in species of *Shigella* and *Staphylococcus*, a problem explored in Chapter 10.

Dealing with emerging antibiotic resistance is a major problem confronting contemporary researchers. One alternative to the use of antibiotics may lie in methods to boost the immune system. Another approach may involve the use of bacteriophages to inhibit certain species of bacteria. Still another may involve innovative drugs that overcome the resistance. For example, biochemists have demonstrated that it is possible to prevent a bacterium from pumping an antibiotic out of its cytoplasm. Curbing the overuse of antibiotics through public and physician education and government regulation would be helpful as well (although such action will be particularly difficult, because the antibiotics market is currently worth about $23 billion). Indeed, some researchers suggest that antibiotics should be controlled as strictly as narcotics. The antibiotic roulette that is currently taking place should be a matter of discussion and concern for all individuals, whether scientists or students or average citizens.

■ A FINAL THOUGHT

In the last 100 years, there were two periods in which the incidence of disease declined sharply. The first was in the early 1900s, when a new understanding of the disease process led to numerous social measures, such as water purification, care in food production, control of insects, milk pasteurization, and patient isolation. Sanitary practices such as these made it possible to prevent virulent microbes from reaching their human targets.

The second period began in the 1940s with the development of antibiotics and blossomed in the years thereafter, when physicians found they could treat established cases of disease. Major health gains were made as serious illnesses came under control. An outgrowth of these successes has been the belief by many people that science can cure any infectious disease. A shot of this, a tablet of that, and then perfect health. Right? Unfortunately not.

Scientists may show us how to avoid infectious microbes and doctors may be able to control certain diseases with antibiotics, but the ultimate body defense relies on the immune system and other natural measures of resistance. Used correctly, the antibiotics provide that extra something needed by natural defenses to overcome pathogenic microbes. The antibiotics supplement natural defenses; they do not replace them.

The great advances in antibiotic research should be viewed with caution. Antibiotics have undoubtedly relieved much misery and suffering, but they are not the cure-all some people perceive them to be. In the end, it is well to remember that good health comes from within, not from without.

■ QUESTIONS TO CONSIDER

1. Instead of saying that food has been irradiated, processors indicate that it has been "cold pasteurized" Why do you believe they use this deception? Do you think it is ethical? What will it take for the food-processing industry to avoid the deception and use the correct term? Can you think of any other euphemism like this used about foods?
2. Of all the sterilization methods reviewed in this chapter, why do you think none has been widely adapted to the sterilization of milk? Which, in your opinion, holds the most promise?
3. While on a camping trip, you find that a luxury hotel has been built near the stream where you once swam and from which you drank freely. Fearing contamination of the stream, you decide that it would be wise to use some form of disinfection before drinking the water. The nearest town has only a grocery store, pharmacy, and post office. What might you purchase? Why?
4. With over 11 million children currently attending day-care centers in the United States, the possibilities for disease transmission among children have mounted

considerably. Under what circumstances may antiseptics and disinfectants be used to preclude the spread of microorganisms?

5. A brochure called *Operation Clean Hands* lists several times when individuals should wash their hands thoroughly. One list entitled "Before you" includes "prepare food" and "insert contact lenses." A second list entitled "After you" includes "change a diaper" and "play with an animal." How many items can you add to each list?

6. One of the novel approaches to treating gum disease is to impregnate tiny vinyl bands with antibiotic, stretch them across the teeth, and push them beneath the gumline. Presumably, the antibiotic would kill bacteria that form pockets of infection in the gums. What might be the advantages and disadvantages of this therapeutic device?

7. The antibiotic issue can be argued from two perspectives. Some people contend that because of side effects and microbial resistance, medical use of antibiotics will eventually be abandoned. Others see the future development of a superantibiotic, a type of "miracle drug." What arguments can you offer for either view? Which do you see as more likely to be true?

KEY TERMS

Informative facts are necessary for the expression of every concept, and the information for a concept is founded in a set of key terms. The following terms form the basis for the concepts of this chapter. On completing the chapter, you should be able to explain and/or define each one:

aminoglycoside	filtration
antibiotic	formaldehyde
antiseptic	glutaraldehyde
autoclave	iodine
Bordeaux mixture	ionizing radiation
broad-spectrum antibiotic	lyophilization
cephalosporin	pasteurization
chlorhexidine	penicillin
chlorine	penicillinase
desiccation	phenolic
detergent	sterilization
disinfectant	sulfanilamide
ethyl alcohol	trichlosan
ethylene oxide	ultraviolet radiation

http://microbiology.jbpub.com/book/microbes

The site features **eLearning**, an online review area that provides quizzes and other tools to help you study for your class. You can also follow useful links for in-depth information, read more stories of microbiology, or just find out the latest microbiology news.

II

Microbes and Human Affairs

In 1944, a doctor traveling in Europe noted that the kitchens of many homes had a loaf of moldy bread hanging in one corner. He inquired about the bread and was told that when a family member sustained a wound or abrasion, a sliver of the bread was mixed with water to form a paste, and the paste was applied to the skin. A wound so treated was less likely to become infected than one left untreated. Modern scientists speculate that the moldy bread probably contained an unidentified chemical that contemporary medicine would consider an antibiotic.

Was this the first time microbes had unwittingly helped out humans with an antibiotic? Apparently not. In 1980, a graduate student from Detroit was examining slides of bone tissue when she noticed a peculiar yellow-green glow coming from the tissue. Her colleagues identified the source of the emission as the antibiotic tetracycline. What made the finding remarkable was that the bone tissue was from a 2000-year-old mummy excavated from the floodplain of the Nile River. Scientists had believed that the people living there were unusually free of disease, and now they had a possible answer—perhaps a microbe had lived in the grain they used for making bread and had deposited its antibiotic substance.

It seems that microbes have been helping us out even when we did not realize it and long before any of us suspected. How they have woven their magic in our lives will be the major theme of Part II of *Microbes and Society*. We shall see evidence of their beneficence in Chapter 12 as we visit a restaurant and experience a menu from a microbial point of view. Next, Chapter 13 reveals how microbes contaminate foods and how scientists limit that contamination. In Chapter 14, we explore how microbes make possible a stunning variety of products ranging from oral contraceptives to drain openers to vitamins. Chapter 15 describes the chores microbes perform on the farm, showing how essential they are to producing our meat and dairy products and how modern technologists use them to hold off the frost, make insecticides, and create veggie vaccines. Our look at the bright side concludes in Chapter 16, which outlines how microbes continue to protect our environment while making possible life itself.

As we all know, there is a darker side to the microbes, and we spend time discussing that side in the final three chapters of this book: Chapter 17 describes the disease process and the means by which our bodies develop resistance; Chapters 18 and 19 present brief looks at familiar and not-so-familiar viral and bacterial diseases. Fortunately, the innovative scientists in our midst have found numerous medicines we can use to fight off these diseases (even when they have had to rely on loaves of moldy bread), and we can generally expect to go on enjoying the helpful benefits of the microbes for many years to come.

Microbes and Food: A Menu of Microbial Delights

12

Many of the dairy products we enjoy are created by adding certain species of bacteria. The inset shows *L. Acidophilus*, the bacterium used in the production of yogurt.

Looking Ahead

Although we often think of microbes as food contaminants, many species play vital roles in producing the foods we enjoy, as we shall discover in this chapter.

On completing this chapter, you should be able to . . .

- name many of the broad variety of foods that owe their existence to microbes.
- understand the fermentation process and how it is used in the production of numerous foods.
- list some of the microbes whose beneficial effects are employed by the food industry.
- summarize the industrial and microbial processes used to produce many foods we enjoy.
- describe how "spoiled" food can be advantageous to our health and act as a safeguard against microbial contamination.
- recognize that microbes contribute substantially to our way of life, and appreciate their many positive contributions.
- increase your vocabulary related to microbes and foods.

Over the decades, many food-related microbes have received bad press: They have been linked to numerous disease outbreaks associated with foods, and sometimes they even received the blame although no trace of a microbe could be found. One possible reason for this negative reputation is that people are usually afraid of things they cannot see, and microbes certainly meet that criterion. To be sure, a small minority of microbial species are associated with disease in foods, but these few species hardly provide sufficient rationale for condemning the thousands of harmless species.

This chapter tries to create some good press for the food-related microbes. Through its pages, we shall see that microbes are absolutely indispensable for producing and processing many of the things we eat and drink. Indeed, we shall encounter numerous examples of the substantial contributions microbes make to the quality of our gastronomic lives. (We consider the contributions of microbes to the production of meat and many dairy products in Chapter 15 because there is not enough space to do them justice here.)

We shall use a fine restaurant as our venue to explore the relationships between microbes and foods, for it is here that microbes work hard to please our palates and heighten our senses. Many of the foods we shall order are culturally unique. For example, in the United States, Germany, and several other countries, sauerkraut is consumed liberally with frankfurters, sausages, and other meats, or it is eaten separately as a vegetable. By contrast, in other parts of the world, people see sauerkraut as spoiled cabbage and are quick to discard it. Ironically, what is "spoiled" food in one culture is relished in another.

A recurring theme in our exploration of microbes and foods is fermentation. Fermentation refers not only to the process that produces alcoholic beverages, but to any partial breakdown of carbohydrates taking place in the absence of oxygen. Microbes are the tools of fermentation because they produce the enzymes necessary for the chemistry to take place. A variety of organic substances such as alcohols, acids, aldehydes, and ketones result from fermentation. These products add distinctive aromas and tastes to many foods, including those pictured in **FIGURE 12.1**. Moreover, they act as preservatives, making the foods safe to eat by holding in check any dangerous microbes that may be present.

Fermentation also brings nutritional benefits because it changes complex substances to more simple ones that can be easily used by the body. For example, we shall encounter many acids produced from complex carbohydrates. By converting the carbohydrate to an acid, the microbe preserves much of the energy originally present in the molecule. This energy can be released when the body metabolizes the acid. As another example, ethyl alcohol (in alcoholic beverages) is usually regarded as toxic, but it contains 90% of the caloric value present in glucose, the starting point of the fermentation that yields the alcohol.

Beginning Our Meal

Having explored a few bits of science, we are now ready to consider how microbes help us enjoy a meal. To begin, we enter the restaurant and are seated at our table by our charming hostess, who extends the manager's greetings. Then, a handsome waiter approaches and inquires if we would like to start with a glass of wine.

FIGURE 12.1 **Foods of Microbial Origins.** An array of foods and beverages produced by microbes.

A Glass of Wine

In ancient times, humans must have been awestruck by wine, for it was powerful enough to turn the mind, even though it began as mere grape juice. No doubt, early peoples marveled at the mystical abilities of this beverage and wondered how it came about. We may wonder, too, but for the moment, we must set aside the technicalities of fermentation (Chapter 9). We must first decide if we prefer white wine or red wine. For making red wine, black grapes are used, together with their skins and sometimes their stems. White wine can be fermented from black grapes or white grapes, but in both cases, the stems and skins are removed. Once we have made our selection, the waiter leaves to fill our order. Now we can consider how alcoholic fermentation works.

As long as oxygen is present in their environment, many microbes—including yeasts—will live contentedly. If the oxygen is removed, however, yeast cells shift their metabolism to a form of oxygen-free chemistry known as fermentation. Rather than shuttling pyruvic acid into the Krebs cycle and aerobic metabolism, yeasts use the acid as an acceptor of electrons and convert it to ethyl alcohol. This somewhat complex chemistry, which is the key to fermentation, is explored in depth in Chapter 9.

Before we go much further, we should note that few organisms other than yeasts can convert pyruvic acid to ethyl alcohol. Furthermore, this conversion is not particularly desirable for the yeast cells because they can gain more energy for their life processes by putting the pyruvic acid into the Krebs cycle than by using it for fermentation. But the yeast cells have no choice when there is no oxygen in their environment; without using fermentation at those times, they would quickly die. For yeasts, fermentation therefore brings an evolutionary advantage in the struggle to survive—and for vintners and brewers, fermentation is economically advantageous as the basis for the alcohol industry.

Yeasts are plentiful wherever there are orchards or vineyards, which is not unexpected because fruit is the source of carbohydrates for fermentation. The haze on an unprocessed apple is a layer of yeasts (FIGURE 12.2), and if those yeasts penetrate the skin and enter the soft flesh of the apple, they will ferment its carbohydrates and produce alcoholic apple juice (i.e., applejack). The same process occurs with other fruits or grains: Yeasts growing in peach juice will produce peach-flavored ethyl alcohol; yeasts in potato will produce unflavored ethyl alcohol; yeasts in orange juice will produce orange-flavored ethyl alcohol; and yeasts in barley grain will produce barley-flavored ethyl alcohol, the alcohol in beer that we shall discuss later.

But the most common fermentation is performed by yeasts in grapes. Species of *Saccharomyces* (especially *S. ellipsoideus*) are crushed with grapes during natural fermentation or are added to grape juice during controlled fermentation. These yeasts

■ *Saccharomyces*
sak-ä-rō-mī′sēs

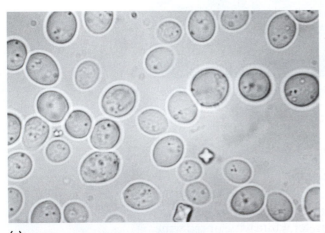

(a)

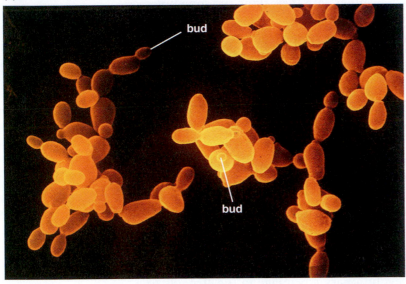

(b)

(c)

FIGURE 12.2 **The Yeast and Its Product.** (a) A photomicrograph of the yeast *Saccharomyces* used in wine fermentation (×1000). (b) A colored scanning electron micrograph of the organism. Note that several cells are budding. (c) A red wine, fruit of the vine and product of fermentation.

then multiply and perform their chemistry. First, the grape juice—known as must—bubbles intensely and froths from the carbon dioxide produced during the Krebs cycle, an aerobic process that requires oxygen (Chapter 9). Indeed, the word "fermentation" is derived from Latin stems that mean "agitate" or "stir up." The carbon dioxide fills all the air spaces in the juice, and, so long as the juice remains still, the environment becomes oxygen-free, or anaerobic. When the oxygen is depleted, the yeast cells shift their metabolism from aerobic metabolism to fermentation and start producing ethyl alcohol. But there is a limit to this process—when the percentage of alcohol in the wine reaches about 13 percent, the population of yeast cells declines as the microbes die off. That is why no natural wine has an alcoholic content exceeding 13 percent. Some healthful benefits of wine are explored in A Closer Look below.

It is difficult to locate the roots of winemaking, but archeological evidence indicates that the techniques were well developed at least 6000 years ago. Apparently the Egyptians taught the art of winemaking to the Greeks, who established the first notable vineyards in regions bordering the Mediterranean Sea. The Romans organized winemaking into a highly efficient operation to supply the needs of their troops, and their later conquests spread the art of winemaking to Northern Europe. Wine eventually became a preferred alternative to water, which was usually contaminated with microbes and often caused horrible intestinal diseases. The Romans thought so highly

A CLOSER LOOK
Answering the Paradox

The medicinal properties of wine received a boost in 1996 with reports that wine, especially red wine, is the answer to a perplexing question that emerged some years before. The story began in 1992 when researchers from the Bordeaux region of France performed a scientific survey and noted that French and other Mediterranean peoples eat large amounts of fatty foods, yet suffer a relatively low incidence of coronary artery disease. Soon thereafter, a *60 Minutes* report pointed out that fatty meats, creams, butters, and sauces have little apparent effect on French hearts. The "French paradox" was born.

In 1996, researchers answered the paradox with studies on the phenol-based compounds in red wine. Scientists at numerous research centers found that phenolics, as they are called, inhibit the oxidation of low-density lipoproteins (the LDLs in the blood) and, by doing so, prevent the buildup of cholesterol and blood platelets in the arteries of the heart. Chemists point out that red wine contains more phenolics than white wine, and far more than beer. Indeed, the most abundant phenolic in red wine, catechin (CAT-eh-kin) is a well-known antioxidant. The phenolics are also present in many foods (e.g., raisins and onions), but not in the quantity found in the skins of grapes used for red wine. (And they concentrate even more after fermentation has taken place.)

But red wine is not for everyone. Indeed, alcohol should be avoided by pregnant women, people taking medication, those under the legal drinking age, and anyone with a family history of alcoholism. For these individuals and anyone else wishing to stay away from alcohol, the good news is dealcoholized wines. These alcohol-free beverages offer the opportunity to take advantage of a natural health ingredient while enjoying a glass of nature's bounty. "Ah, a glass of wine and a healthy heart."

of wine that they adopted the Greek god of wine (Dionysus) but gave him their own name (Bacchus).

Among grapes, the species *Vitis vinifera* is recognized as producing the highest-quality fruit for winemaking. Characteristics of the grapes are determined by soil and climate conditions, such as temperature, humidity, and amount of water. Thus, wine varies, and there are "vintage years" and "poor years." Although the production of alcohol by fermentation requires only a few days, the aging process may go on for weeks or months. Wooden casks are commonly used for aging, and the wine develops its unique flavor, aroma, and bouquet with help from organic molecules in the wood.

The basic fermentation process is varied to obtain a broad variety of wine types. To produce a dry wine, for instance, most or all of the sugar in the grape juice is allowed to break down. Should a sweet wine be desired, some sugar is left unfermented. For a sparkling wine, such as champagne, a second fermentation takes place inside the bottle: Sugar cubes are added to the wine after the first fermentation, and the yeast is encouraged to continue fermenting the sugar within the bottle. Carbon dioxide builds up and adds the sparkling bubbles to the wine. Incidentally, a thick bottle is needed for champagne because the gas pressure would cause ordinary glass to break, and a wire cage is used to prevent the cork from popping out.

Most table wines average about 10 to 12 percent alcohol. Exceptions are the fortified wines, such as port, sherry, and madeira. These wines have an alcohol content that approaches 22 percent. They are produced by adding brandy or other spirits to the wine following fermentation. Because fortified wines are generally considered dessert wines, we shall pass on them for now.

The Appetizers

As we anticipate our meal, our palates will first be stimulated by fermented olives and cheese brought to our table by our waiter. Olives (FIGURE 12.3) have traditionally represented the abundance of life, and in many cultures, the olive branch is a symbol of peace. Unfortunately, the natural taste of olives is quite bitter (a characteristic now known to be caused by a substance called oleuropein).

Tradition and microbial fermentation resolved the bitterness problem well before the chemistry was understood. In regions of Western Europe, unripened olives were soaked in lye (sodium hydroxide) to neutralize their bitter taste, then washed and covered with brine. Next, the olives were sealed in casks, and bacteria normally present on their skins were encouraged to ferment their carbohydrates. Some weeks later, when the fermentation was complete, the tasty Spanish, or "green," olive resulted.

In Greece, olives were eaten without the benefit of fermentation, and so they had to be preserved for later use. This was accomplished by allowing the olives to ripen on the tree, then picking them and exposing them to the air for weeks. During this interval, chemical conversions of the tannin compounds in the olive skins created black deposits,

FIGURE 12.3 **Olives.** A variety of fermented olives to please the palate.

yielding Greek, or "black," olives. The Italians modified the process by placing the black olives in salt and encouraging the naturally occurring microbes in the olive skins to carry out fermentation.

Our Spanish, Greek, and Italian olives are accompanied by crackers and assorted cheeses. Cheese results when microbes interact with casein, the major protein in milk. In the dairy plant, the microbes produce enzymes that join with added enzymes to curdle the casein. The curds are then separated as "unripened cheese" or the familiar cottage cheese (the remaining fluid is called whey, as we note in Chapter 15).

To prepare different kinds of ripened cheese, milk curds are washed, and salt is added to flavor the curds and prevent spoilage. Then, cultures of microbes are added to the curds. For example, if a fine Swiss cheese is to be made, two different bacteria are mixed in. During the aging process, these microbes bring about the unique chemical changes in the available proteins and carbohydrates: Species of *Lactobacillus* produce acids to lend sourness, and species of *Propionibacterium* produce a variety of organic compounds, as well as carbon dioxide. The organic compounds give Swiss cheese its distinctively nutty flavor, and the carbon dioxide seeks out weak spots in the curd and accumulates as the holes, or eyes, in the cheese. `FIGURE 12.4` displays the complete process used to produce such cheese.

■ *Propionibacterium*
prō-pē-on′ē-bak-ti-rē-um

However, we may instead select a mold-ripened cheese for our tasting pleasure. Among our choices in this category are Camembert and Roquefort cheeses. Camembert, a soft cheese, is made by dipping milk curds into the fungus *Penicillium camemberti*. As the fungus grows on the outside of the curds, it digests the proteins and softens the cheese. To make Roquefort, a blue-green veined cheese, the curds are rolled with spores of the blue-green fungus *Penicillium roqueforti*. The fungus penetrates cracks in the curds and grows within them, thereby creating the distinctive blue-green veins in the cheese.

■ *Penicillium*
pen-i-sil′lē-um

These are just a sampling of the numerous cheeses that owe their existence to microbes. We consider other cheeses as well as many other dairy products in Chapter 15.

Of Salad and Bread

For our salad course, we select an assortment of mixed greens and add some other healthful and nutritious vegetables, such as tomatoes, red onions, and cucumbers. On our salad, we order an oil-and-vinegar dressing (a vinaigrette). Although microbes cannot claim to have produced the greens or the oil, they are essential for producing the vinegar, which is a fermented food. Vinegar has traditionally been made by the souring of wine (the word "vinegar" is derived from the French word *vinaigre*, which means "sour wine"). Apple cider vinegar, for instance, was once apple cider wine fermented from apple cider. And clear white vinegar began as potato starch, which was fermented to potato wine, then converted to clear vinegar. This vinegar has no taste other than the natural sour taste of acetic acid. By contrast, balsamic vinegar acquires its sweet flavor from the wood barrels in which it is aged; the barrels are made of balsam fir. And the flavor of wine vinegar is determined by compounds present in the original wine as well as products of bacterial growth.

The Germans were among the first industrial producers of vinegar. As early as the 1800s, people in the German countryside practiced the art of converting fruit juice to vinegar. Then, as now, they fermented the juice and sprayed the wine into a tank

(a)

(b)

(c)

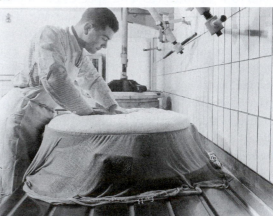

(d)

(e)

(f)

FIGURE 12.4   The Production of Swiss Cheese. (**a**) The milk is mixed with the enzyme rennet and heated in a large kettle. (**b**) After the curds form, the cheesemaker cuts the curds into small pieces using a series of copper wires called a cheese harp. A square piece of cheesecloth composed of coarse-weave hemp is passed under the curds. (**c**) The four corners of the cheesecloth are tied together, and the curds are removed from the kettle using a block-and-tackle. (**d**) The curds are deposited in a wooden hoop, and their surface is kneaded lightly with the palms of the hands. The hoop forms the cheese into the familiar wheels. (**e**) Press boards are placed on top of the cheese mass until no more whey emerges. (**f**) The cheese is transported to a warm cellar, where it is salted and set aside to ripen. Bacteria ferment the cheese for about 2 weeks at a temperature of 50°C and a relative humidity of 90 percent. The cheese is then inspected and graded.

called a vinegar generator. Traditionally and in the present, the tanks are filled with wood shavings and gravel containing naturally occurring cultures of the bacterium *Acetobacter aceti*. The bacterium grows and multiplies on the wood shavings and gravel, and as alcohol percolates through, the bacterial enzymes convert the alcohol into acetic acid. The fresh vinegar recirculates several times through the tank before collecting at the bottom; it has an acetic acid content of about 3 to 5 percent.

Of course, we have not forgotten the bread basket. For this part of our meal, we once again turn to the yeasts, especially *Saccharomyces cerevisiae*. This microbe is added to flour and water, the other two basic ingredients for making all the types of breads pictured in   FIGURE 12.5 . The yeast's ability to produce naturally leavened bread depends largely on gluten, the major protein in wheat flour. Gluten is a tough, gray, elastic protein that, when wet, resembles chewing gum. It helps form the elastic network that traps the gas produced by the yeast. The result is a firm, spongy texture in wheat-based bread or a more dense texture in bread made with low-gluten flour such as rye flour.

FIGURE 12.5   **Breads.** An array of breads resulting from the action of yeasts.

When yeast is added to flour and water (plus sugar, salt, and other ingredients at the whim of the breadmaker), it subsists on carbohydrates in the dough and produces substantial amounts of carbon dioxide through its aerobic metabolism and the chemistry of the Krebs cycle (Chapter 9). The carbon dioxide expands the dough, and it rises seemingly by magic. (Imagine how awestruck ancient peoples must have been by this phenomenon!) Enzymes also produced by the yeast break down some of the gluten in the flour and help give bread its spongy texture (the bread is kneaded to redistribute the gluten and rid the dough of any large gas pockets). In addition, yeast cells manufacture some ethyl alcohol in the dough during fermentation, but the high temperature of baking causes most of the alcohol to vaporize, and it is driven off. Yeast for breadmaking can be purchased at grocery stores in the form of "active dry yeast" or a "yeast cake" or carried over from the previous batch of "mother dough." In addition, all large commercial bakeries have laboratories where strains of *S. cerevisiae* are maintained and further developed.

■ *Acetobacter aceti*
a-sē′tō-bak-tèr a-set′ē

Yeasts are not necessarily the only microbes growing in the dough. In some cases, for example, cultures of bacteria are added to produce a sour taste. These bacteria, usually of the genus *Lactobacillus*, degrade a portion of the starch in the flour and produce lactic, acetic, carbonic, and other acids. (Good starter cultures are closely guarded trade secrets.) One result is San Francisco–style sourdough bread.

Yeast and flour can come together in an almost infinite range of variations, and the baker can lend some ingenuity to the process. Vienna bread, for example, is baked in a high-humidity oven and develops a flaky crust. Semolina flour is used to make semolina bread; potato flour is used for potato breads and rolls. Bagels are boiled in water before baking, as A Closer Look (on page 268) explains; pizza dough is modified to give it high elasticity; and pumpernickel bread is produced from rye flour and yeast-fermented molasses.

A CLOSER LOOK
Petrified Donuts

They've been called everything from Brooklyn jawbreakers to Jewish English muffins, from bugles to beagles, from rolls with holes to petrified donuts. They are the ubiquitous bagels, which have become a popular choice for a late-night snack or a Sunday-morning breakfast. Part of their success is due to their low calorie count (about 225 for a 3-ounce bagel), their low fat and sugar content, and their crusty, chewy texture (a result of boiling them before baking). It doesn't hurt that bagels can be adapted for making pizza, burgers, or French toast, or that they come in almost any imaginable flavor, including blueberry, jalapeño, veggie, or (my personal favorite) "everything."

Legend has it that bagels were "invented" in 1683 by a loyal baker who wanted to pay homage to the king of Austria. The king, it seems, was an avid horseman, so the baker fashioned strips of dough into the shape of a *bügel*, the German word for stirrup. He boiled his circular gifts to keep them moist, then baked and delivered them to the palace. The king was delighted, and the rest, as they say, is history.

As bagelmania has continued to spread, people find new flavors, new uses, and new adaptations for these bakery treats. And they find new jokes. For example, "If seagulls fly over the sea, which birds fly over the bay?"

The origins of bread are shrouded in the fog of history. Undoubtedly, our distant ancestors once ate a mixture of flour and water cooked over a fire to a crackerlike consistency. You can imagine their delight when they discovered the immense improvement brought about by yeast. Today, the unmistakable aroma of fresh bread emanates from our kitchens, appealing to the palate and enticing the appetite.

Continuing Our Meal

Assuming we have not overindulged during the preliminary courses, we are now ready for our main course. Note that microbes have contributed mightily to our dining experience thus far. That trend will continue as our culinary adventure unfolds.

The Main Course

Our microbial menu contains two choices for the main course: an Asian dish and a European dish. The Asian dish consists of slices of aged beef marinated in soy sauce, then stir-fried in a mixture of soy sauce and spices. Scientifically speaking, the beef is a product of microbial growth (although an indirect product), as we discuss in Chapter 15. Cattle would have little ability to manufacture protein without the intervention of microbes. Indeed, the process of converting carbohydrates in plants to proteins in beef cattle is intimately dependent on microbes. Microbes live in enormous numbers in every cow's stomach, where they break down plants and synthesize simple nutrients that can be used by the animal. Without these microbes, beef cattle would be unable to make more of themselves.

For marinating and cooking the beef, our chef is using soy sauce, which is a product of fermentation. To make soy sauce, manufacturers begin with a starter material called koji. Koji begins as a soupy mixture of soybeans, rice, and wheat bran that is inoculated with the fungus *Aspergillus oryzae* and incubated at slightly over 30°C (85°F). The fungus grows in the mixture (FIGURE 12.6) and breaks down complex proteins and carbohydrates into smaller molecules. Various species of bacteria continue the fermentation that produces the koji.

The koji is then added to more soybeans mixed with roasted wheat and steamed wheat bran. The fungus continues to grow in the mixture, and after three days, the initial fermentation is complete. Over the course of another year, the mixture continues to age as bacteria produce some lactic acid and yeasts add a small amount of alcohol. The liquid pressed from the mixture is soy sauce. Note that the fungus has contributed the predominant flavor and aroma to the soy sauce. Thus, our Asian entree—with its microbe-derived beef and microbe-derived soy sauce—has totally microbial roots.

The European choice for the main course consists of a variety of sausages accompanied by appropriate vegetables. Sausages generally consist of dry or semi-dry fermented meats. They include pepperoni from Italy, thuringer from Germany, and polsa from Sweden. To produce sausages, curing and seasoning agents are added to ground meat. The meat mixture is then stuffed into casings and incubated at warm temperatures. Microbes multiply and produce a mixture of acids from the meat's carbohydrates, thereby giving the sausage its unique taste.

To accompany our German sausage, we order "sour cabbage," or sauerkraut, as it is better known. Sauerkraut is a well-preserved and tasty form of cabbage that is an excellent source of vitamin C. Indeed, it is said that the British took sauerkraut on their ocean voyages to help prevent scurvy (as an alternative to the more expensive citrus fruits). Modern researchers have further verified the healthful benefits of sauerkraut: In 2000, scientists from the University of Illinois noted that Polish women who had immigrated to the United States had less likelihood of developing breast cancer than non-Polish immigrant women. Scientists found that typically Polish foods such as sauerkraut and fermented products of other cabbage family members (e.g., broccoli, cauliflower, and Brussels sprouts) contain compounds that block the activity of estrogen, a possible stimulator of breast cancer. While the research is not complete, it is extremely provocative.

Species of *Leuconostoc*, a coccus, and *Lactobacillus*, a bacillus, are essential to the production of sauerkraut. These are both Gram-positive bacteria naturally found in the leaves and tissues of a head of cabbage. Sauerkraut preparation begins by shredding the cabbage and adding about 3 percent salt. The salt ruptures the walls of the cabbage plant cells and releases their juices, while adding flavor. Then the shredded and salted cabbage is packed tightly into a closed container to eliminate oxygen and stimulate anaerobic conditions. About a day later, the *Leuconostoc* species begin multiplying rapidly. The cocci ferment the carbohydrates and produce lactic and acetic acids. After several days, the pH content of the cabbage is an acidic 3.5. Now, species of acid-tolerant *Lactobacillus* take over. They ferment the carbohydrates further and

FIGURE 12.6 **The Production of Soy Sauce.** *Aspergillus oryzae* is added to the koji used to produce soy sauce. The fungus begins the fermentation, which is later completed by bacteria during the aging process.

■ *Aspergillus oryzae*
a-spér-jil′lus ôri-zī

■ *Leuconostoc*
lü-kū-nos′ tok

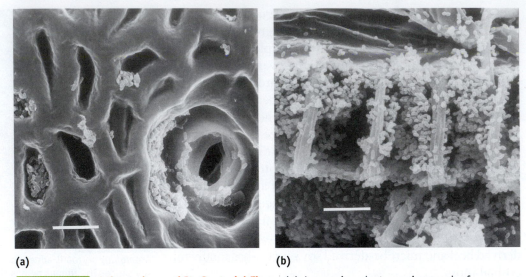

(a)   (b)

FIGURE 12.7   **A Cucumber and Its Bacterial Flora.** (**a**) A scanning electron micrograph of *Lactobacillus plantarum* on the surface tissues of a cucumber. The openings are stomata, through which gases are exchanged. Note the accumulation of lactobacilli at these stomata. (Bar = 10μm.) (**b**) A longitudinal section through the vascular tissue of a brinedcucumber showing *Leuconostoc* species along the tubular walls. (Bar = 10 μm.)

produce additional lactic acid to reduce the pH to about 2.0 and give the sauerkraut its tangy sourness.

Having a taste for sour foods, we have also ordered a variety of pickled cucumbers to accompany the sausages. For all types of dill, sour, and sweet pickles, the fermentation is essentially similar: Manufacturers begin by placing cucumbers of any type or size in a high-salt solution, where the cucumbers change color from bright green to a duller olive-green. Then the fermentation begins in an aging tank. The first bacterium to emerge is an *Enterobacter* species. This Gram-negative rod produces large amounts of carbon dioxide gas, which takes up the air space in the tank and establishes anaerobic conditions. The next bacteria to proliferate are *Leuconostoc* and *Lactobacillus* species, shown in FIGURE 12.7 . They produce large amounts of acid to soften and sour the cucumbers. Yeasts also grow in the aging tank and establish many of the flavors associated with pickled cucumbers. Various herbs and spices are added to finish the process. Virtually any vegetable can be substituted for cucumbers for an equally tasty result.

A Refreshing Beer

The distinctive flavors of the sausage-and-sauerkraut entree are wonderfully complemented by the sparkle and tang of a glass of beer. Not surprisingly, we find that microbes once more play a part in our meal.

Much of the chemistry of wine fermentation applies equally well to the production of beer. The origins of wine trace back to antiquity (FIGURE 12.8), and beer has been known to exist since thousands of centuries before the birth of Christ. Historians tell us that as early as 3400 B.C., Egyptians were placing a tax on beer produced in the ancient city of Memphis on the Nile. In later generations, the Greeks brought the art of brewing beer to Western Europe, and the Romans refined the process further. Writers of that period point out that the main drink of Caesar's legions was beer. As we noted previously, the water was so polluted that beer and wine were beverages of choice, not only because of their taste but also because soldiers could avoid intestinal illness by drinking them.

Nor did the history of brewing end there. During the Middle Ages, the monasteries were the centers of beermaking. Historians report that by the year 1200, taverns and breweries were commonplace throughout the towns of Great Britain. Centuries would pass, however, before beer would make its appearance in cans. That auspicious event took place in the United States in 1935 in the city of Newton, New Jersey. The six-pack was the logical successor.

The word "beer" is derived from the Anglo-Saxon *baere*, which means "barley." This terminology developed because beer is traditionally a product of barley fermentation. The process begins by predigesting barley grains, a process called malting. During malting, the barley grains are steeped in water, and naturally occurring enzymes in the barley digest the starch into smaller carbohydrates, among them maltose (also known as malt sugar).

FIGURE 12.8   Ancient Wine Making. An Egyptian wall painting from ca. 1419–1380 BC shows harvesting and pressing of grapes.

Next, the malt is ground with water in the process known as mashing, and the liquid portion, or wort, is removed. At this point, the brewmaster adds dried flowers (hops) of the hop vine *Humulus lupulus*. Hops give the wort its characteristic beer flavor, while adding color and stability. Then the fluid is filtered, and a species of *Saccharomyces*, the yeast, is added. Initially, there is intense frothing as carbon dioxide is produced during aerobic respiration and the Krebs cycle. Then the frothing subsides, and the yeast shifts its metabolism to fermentation. Alcohol production begins. The final alcoholic content of beer is approximately 4 to 5 percent.

■ *Humulus lupulus*
hŭ′mū-lus lū′pū-lus

Various species of yeasts are used to produce different types of beer. For example, *S. cerevisiae* gives a dark cloudiness to beer; it is called a top yeast because it is carried to the top of the vat by the extensive carbon dioxide foam. The beer it produces is an English-type ale or stout. A different species, *S. carlsbergensis*, causes a slower fermentation and produces a lighter, clearer beer with less alcohol. This microbe is called a bottom yeast because there is less frothing, and the yeast settles to the bottom. Fermentation with this yeast is carried out at a cooler temperature (approximately 15°C) than is used for ale production (approximately 20°C). The beer produced by bottom yeast is Pilsener, also called lager. Almost three-quarters of the beer produced in the world is lager beer. To produce "ice beer," a still cooler temperature is used for the fermentation.

After about seven days of fermentation, the yeast has produced a "young beer." This young beer is transferred to wooden vats for secondary aging, also called lagering. This process may take an additional six months, during which time the beer develops its characteristic flavor and taste. FIGURE 12.9 summarizes the process.

If the beer is to be canned, it is usually pasteurized at 60°C (140°F) for a period of 55 minutes to kill the yeasts. Alternatively, the beer can be filtered before canning

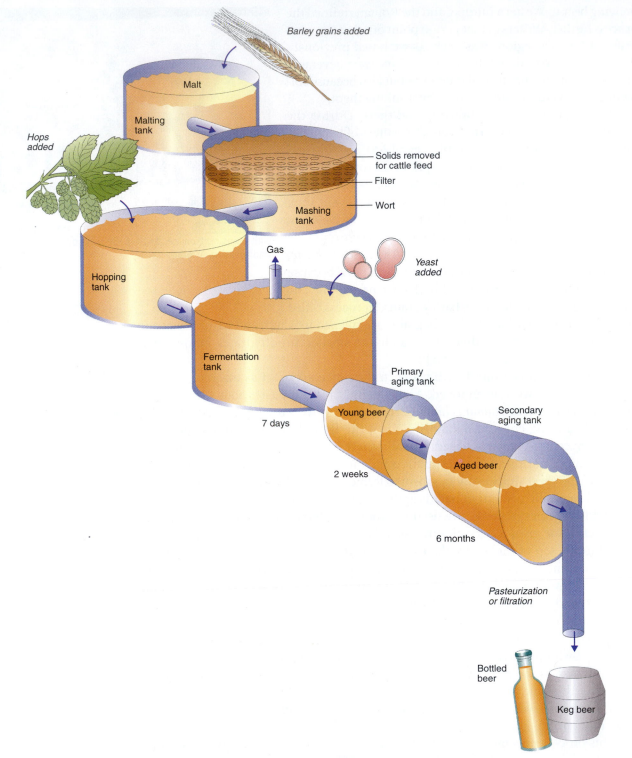

FIGURE 12.9 **The Process for Producing Beer.** Barley grains are held in malting tanks while the seeds germinate to yield fermentable sugars. The digested grain, the malt, is then mashed in a mashing tank, and the fluid portion, the wort, is removed. Hops are added to the wort in the next step, followed by yeast growth and alcohol production during fermentation. The young beer is aged in primary and secondary aging tanks. When it is ready for consumption, it is transferred to kegs, bottles, or cans.

to remove the yeasts; in this case, the beer is called "draft beer" (because true draft beer is not subjected to heat). If the beer is to be delivered directly to an alehouse, it is placed in casks and immediately chilled, making pasteurization unnecessary (this is true draft beer).

With the Asian dish, we might be inclined to try sake. Though many people consider sake a wine, it is more correctly a type of beer because it is fermented from rice (a grain rather than a fruit). To produce sake, steamed rice is mixed with the fungus *Aspergillus oryzae* and set aside. During the incubation period, enzymes from the fungus break down the starches in the rice to simpler sugars. Fermentation by yeasts follows. The final product has the alcoholic concentration of a wine, about 13 percent, and is commonly drunk at room temperature or warmer.

But we cannot linger too long, for it is time for dessert.

Dessert

Try as we might, we cannot really credit microbes for the luscious assortments of cakes, pastries, and cookies on the restaurant's dessert cart. We might note that some of the fruit fillings and toppings owe their flavors to slight degrees of fermentation, but this would be stretching the point a bit.

However, we can draw an association between microbes and the flavors of cocoa and coffee. To prepare cocoa, the cocoa beans must first be separated from the pulp that covers them in the pod. Enter the microbes. Manufacturers add a controlled culture of bacteria and yeasts to the beans, and the microbes ferment the pulp and soften it so that it is easily shed. The fermentation also lends a bit of taste to the resulting cocoa. In deed, part of the color, flavor, and aroma of chocolate is due to the action of microbes on cocoa. Chocolate lovers should recall the role that microbes play when they sit down to a luscious chocolate dessert.

Coffee, as we all know, is made from coffee beans. Like the cocoa bean, the coffee bean is surrounded by a fleshy pulp. Microbes are used to ferment this pulp and assist in its removal. In addition, microbes are used to help remove the outer skin of the coffee bean by digesting a protein called pectin that comprises a large portion of that skin. This protein is broken down by pectin-digesting enzymes produced by various fungi and by bacteria, including *Erwinia dissolvens*. Bacteria that produce lactic acid bacteria assist the pectin removal but do not appear to add to the final flavor of the coffee.

To express appreciation for our visit, the restaurant manager has sent over a tray of chocolate-covered cherries to enjoy with our coffee. Unknowingly, he has exposed us to yet another microbial product. The soft cherries owe their production to invertase, an enzyme obtained from yeasts and species of *Bacillus*. Sweet cherries are pitted, then mixed with the enzyme and dipped into a vat of chocolate for their coating. Over the next few days, the enzyme softens the cherries and produces the tasty liquor that surrounds them inside the chocolate coating. Delightful!

And in Conclusion

To conclude our meal, we have requested a snifter of brandy. Brandy is one of a broad variety of distilled spirits that contain considerably more alcohol than beer or wine. Each type of spirit has a proof number, which is twice the percentage of alcohol in the product. A 90 proof brandy, for example, is 45 percent alcohol.

To produce brandy or other distilled spirits, wine fermentation must first take place. For example, a raw product such as cherries is fermented by yeasts to produce cherry wine. The wine is aged and matured in casks. Once it is ready, manufacturers heat the wine at approximately 80°C (the word "brandy" was originally derived from the Dutch word *brantewijn*, which means "burnt wine"). The heat drives off the alcohol in fumes, which are trapped in a condenser and converted back to a more concentrated liquid, which is cherry brandy. Wooden casks are then used to mature the brandy and introduce unique flavors from organic compounds in the wood. Brandy is the ultimate expression of fermentation and among the most expensive fermented products available.

A sip of brandy pleases the palate yet again, but it also tells us that we must be leaving soon. We have arrived at the end of our gastronomic adventure, which has been marked by good company, fine food, and intriguing insights into the microbial world. And there is one final bit of good news: We don't have to pay the tab because there is none.

■ A FINAL THOUGHT

When we were children, most of us were taught that Marco Polo traveled to China during the 1200s to obtain spices and explore new trade routes. "Why spices?" we may have thought. The fact is that spices were not a luxury at that time. They were essential for improving the smell and taste of leftover food, which ranged from awful to sickening. Refrigeration was unknown, and canning had yet to be invented. What couldn't be eaten quickly was often lost to spoilage. Spices had some antimicrobial qualities, to be sure, but they also made food palatable.

An alternative to spices was fermentation. By fermenting food, people allowed it to spoil naturally. The food might taste bad, but they gradually came to accept the bad taste and found fermentation to be a worthwhile way of preventing harmful spoilage and preserving food for later use. Indeed, fermented food was regularly on board when ships took off for distant parts of the globe. Even Marco Polo probably brought some along on his journeys.

Thus, we see that microbes can spoil food, but if we use fermentation to control that spoilage, we can make a bad thing good. Indeed, as this chapter shows, some of those "spoiled" foods are really quite good!

■ QUESTIONS TO CONSIDER

1. One day, the students in a microbiology class presented the instructor with a basket of "microbial cheer" in recognition of her efforts on their behalf. From your knowledge of this and other chapters, can you guess some of the things that the basket contained?
2. Yeasts are sometimes known as the "schizophrenic microbes." What do you think that means?

3. Sometimes even the most careful food preparation can lead to tragedy. For many generations, humans have made sausages, but on occasion the sausages have become contaminated with *Clostridium botulinum*, and individuals have developed a deadly disease known as botulism (indeed, the word is derived from the Latin *botulus* for "sausage"). How do you think sausage becomes contaminated, and what characteristics of *C. botulinum* make it attracted to sausage?

4. This chapter has surveyed many foods of microbial origin, but the survey is incomplete. From your reading of other chapters and your general knowledge, what foods of microbial origin can you add to the list discussed in this chapter?

5. Wine is part of many cultures world wide, but one of the great mysteries of the human experience is how wine was discovered. Write a scenario depicting the circumstances under which wine was first experienced by a human.

6. Suppose that you decide to enter the pickling business. You intend to pickle tomatoes, peppers, and a host of other foods. How would you proceed with the science end of your new business?

7. During the feast of Passover, Jewish people have with their meals a type of unleavened bread. What is that bread called, and how does it differ from usual bread? What traditions dictate use of this bread at Passover time?

KEY TERMS

The key terms in this chapter are the various foods that owe their existence to microbes. In each case, you should name the food and explain how one or more microbes contribute to its production.

beer	pickled cucumbers
brandy	ripened cheese
bread	sauerkraut
chocolate	sausage
cocoa	soy sauce
coffee	vinegar
fermented olives	wine

http://microbiology.jbpub.com/book/microbes

The site features **eLearning**, an online review area that provides quizzes and other tools to help you study for your class. You can also follow useful links for in-depth information, read more stories of microbiology, or just find out the latest microbiology news.

13

Food Preservation and Safety: The Competition

◼ Looking Ahead

Although microbes are valuable allies to humans, they can also be formidable competitors because they multiply in many of the foods consumed by humans. Unfortunately, this competition can lead to illness, as we shall see in this chapter.

On completing this chapter, you should be able to . . .

- understand some general concepts associated with food spoilage, such as shelf life and microbial load, and list some sources of food contamination.
- summarize how the characteristics of foods contribute to spoilage and name various microbes and organic compounds that result in undesirable tastes and smells in foods.
- identify some of the problems encountered in preventing spoilage in meats, fish, and other foods and understand how the foods may become contaminated.
- discuss the unique problems associated with spoilage in milk and dairy products and list some microbes responsible for these problems.
- explain how heat and low temperatures can be used to preserve the quality of foods and eliminate contaminating microbes.

- compare the advantages and disadvantages of such food preservation methods as drying, osmotic pressure, radiation, and chemical preservatives.
- list some suggestions for preventing illness arising from consumption of contaminated food.

It was the annual company picnic, and the softball game was finally over. Now the serious eating could begin. The picnic table was overflowing with goodies—salads, stuffed eggs, barbecued chickens, and lots of desserts. Unfortunately, there was also an unwelcome guest at the picnic—a species of *Salmonella*.

During the next three days, it became apparent that the picnic would be remembered, but for all the wrong reasons: Over half the picnic attendees suffered abdominal cramps, diarrhea, headaches, and fever. When public health inspectors questioned the company employees, they found a common thread: All the sick attendees had eaten the barbecued chicken.

The inspectors then questioned the woman in charge of the chickens. She led them to a local supermarket, where barbecued chickens were sold "ready to eat." Inspectors learned that store employees had cooked the chickens, then put them back in their original trays. It didn't take long for microbiologists to find evidence of *Salmonella* in the trays. Further questioning revealed that the woman had bought the chickens in the morning and stored them for the next 7 hours in the trunk of her car, believing they would remain cool. They did not. In fact, they quickly reached ideal growth temperatures for bacteria, and by dinnertime, at 7:00 P.M., they were teeming with *Salmonella*.

The lessons of this incident are clear: Keep foods cold for storage; store them in clean, fresh trays after cooking; and eat them quickly after preparing them (even if it means playing softball *after* dinner, not before).

Incidents like this one highlight how most foods, even cooked foods, provide favorable conditions for microbial growth. The organic matter in food is plentiful, the water content is usually sufficient, and the pH is either neutral or only slightly acidic. To a food manufacturer or restaurant owner, contaminating microbes may spell economic loss or a bad reputation. To the consumer, they often mean illness or, in extreme cases, death.

The primary focus of this chapter is to examine the types of microbes that contaminate foods and to point out the consequences of contamination. We shall also examine food spoilage and the methods used to prevent it.

Food Spoilage

They say it occurred in 1878 at Moon's Lake House Restaurant in fashionable Saratoga Springs, New York. Cornelius Vanderbilt, the wealthy entrepreneur, was visiting the resort, and one evening he took issue with the fried potatoes. They were too thick, it seemed, and the chef, a certain George Crum, should have known better. But Vanderbilt was magnanimous—Crum would have another chance to make a more acceptable dish with thinner potatoes.

Crum did not take the criticism well, and he plotted revenge. If Vanderbilt wanted thin potatoes, then thin potatoes he would get. Crum sliced the potatoes superthin, dropped them into boiling oil, and fried them to a crisp. Then he sent the potatoes out to his picky patron, fully expecting to be working elsewhere the next day. You can imagine his surprise when he heard sounds of delight from the dining room—Vanderbilt loved the crunchy potatoes. Within weeks, people were coming from miles around to sample these "Saratoga chips," as they were first called. The chips later became famous simply as potato chips.

Microbiologically speaking, the best thing about potato chips is that they are as dry as a bone. Dry foods resist spoilage, and today's potato chips are as resistant to microbial contamination as George Crum's original ones were. Drying is one of the many methods of food preservation that we survey in this chapter as we examine the continuing competition between microbes and humans for the foods we eat.

General Principles

Food scientists have estimated that approximately one third of all food manufactured in the world is lost to spoilage. Since the greatest percentage of this spoilage is due to microbial contamination, the microbiologist occupies an important place in the food manufacturing process. The microbiologist's work helps to ensure that food is safe to eat and will not transmit disease. Moreover, by reducing microbial contamination, he or she increases the wholesomeness of food and extends its shelf life (the time it can remain available for sale). Preservation methods are high on the list of tasks addressed by the microbiologist, and in cases where food preservation has been unsuccessful, the microbiologist is responsible for detecting the offending microbes and working toward their elimination.

The microbial content of foods usually has qualitative as well as quantitative aspects. The qualitative aspects refer to *which* microorganisms are present, while the quantitative aspects refer to *how many* of them are present. In some cases, the quantitative requirements for a food product must be zero—a can of vegetables, for example, is expected to have no microbes in it. By contrast, hamburger meat is usually considered acceptable if it contains up to 1000 staphylococci per gram of meat because cooking kills staphylococci and because staphylococci do not normally multiply in the human gastrointestinal tract, where they could cause illness. However, it should be noted that staphylococci do multiply in contaminated meat and produce diarrhea-inducing toxins (as **FIGURE 13.1** indicates). Thus, meat processors must work to keep the amount of staphylococci as low as possible.

The shelf life of a particular food depends largely on the quantity of microbes present in it and the conditions at which it is held. For instance, because of their dry condition, non-perishable foods (e.g., potato chips and pasta products) normally have low numbers of microbes and long shelf lives. Semiperishable foods, such as breads and citrus fruits contain more moisture, and higher numbers of microbes will develop in them with time. Perishable foods, such as fish and eggs, present the most favorable conditions for microbial growth and eventually contain the highest numbers of microbes. These foods have the shortest shelf lives.

In order to provide information about a product, a manufacturer adds a code that summarizes an identifying set of characteristics. For instance, the code on a product may read B32D4: The B signifies that the product was produced in plant B; the 32 indicates that it was produced on the thirty-second day of the year; the D applies to the D shift; the 4 refers to the fourth line at the processing plant.

In addition, the product may be marked "Good until . . ." or "Must be sold before . . ." or "Sell by. . . ." These three designations refer to the shelf life and signify

1. On September 25, a chef purchased a precooked packaged ham, baked it at 400°F for 1.5 hours, and sliced the ham while it was hot on a commercial slicer. The chef reported having no cuts, sores, or infected wounds on her hands; however, she admitted that she did not dismantle and clean the meat slicer according to recommended procedures and that she did not use an approved sanitizer.

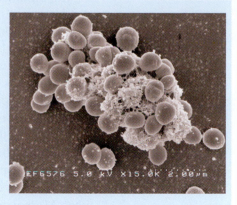

2. All 16 pounds of sliced ham was placed in a large, single-use plastic container, covered with foil and refrigerated overnight. The ham was to be served at a retirement party the next day. Rice pilaf was prepared the day of the party by a different person.

3. On September 26, 125 people attended the party where the cold ham and rice pilaf were served. On September 27, 18 of the guests became ill and reported nausea, vomiting, diarrhea, weakness, sweating, chills, fatigue, myalgia, headache, and fever.

4. One sample of leftover cooked ham and one sample of leftover rice pilaf were analyzed to identify staphylococcal contamination; both samples were positive for *Staphylococcus aureus*.

5. To reduce *Staphylococcus aureus* food poisoning, the amount of manual handling should be minimized. Slicers should be dismantled when cleaned and sanitized. Ham should be sliced when it is cold. Rapid cooling should be permitted by storing food in small, loosely covered containers.

FIGURE 13.1   A Case of *Staphylococcus* Food Poisoning Associated with Precooked Ham, Florida, 1997. Ham is the most common source of staphylococcal food poisoning. The salt content of precooked and packaged ham is approximately 3.5%, which is ideal for the growth of *Staphylococcus*.

the amount of time that will pass until the product is unacceptable to most consumers in taste, smell, or general quality. It should be noted, however, that the date on a product is not a safety date; it merely refers to a date by which consumers should eat the food for its best quality. Manufacturers point out that eating a food product that is past its "sell by" date will not necessarily cause illness. With enough time, spoilage microbes will multiply even if a food product is stored correctly, but they are not the ones that make us sick.

The microbial content of a particular food is known as its microbial load. Consumers generally expect that the microbial load is low in all foods, but in some cases, it is surprisingly high. For example, a teaspoon of yogurt contains billions of harmless bacteria, the microbes that have converted condensed milk into yogurt. In the body, these bacteria join those already in the large intestine and help to keep pathogens under control. Also, the microbial load in foods such as pickles and sauerkraut tends to be high because these foods contain the bacteria responsible for the fermentation used to produce them (Chapter 12).

Thus, the microbial load of a food can be high, as long as the food does not contain pathogens. Milk, for instance, is pasteurized to remove all pathogenic microorganisms, but its microbial load can still be as high as 1000 bacteria per milliliter (20 drops) and it will be safe to drink. In fact, one type of milk known as acidophilus milk contains cultures of *Lactobacillus acidophilus* added at the dairy plant. The bacteria enhance good health by helping to break down otherwise indigestible foods in our small and large intestines and by producing certain vitamins used in our metabolism.

Microbial contamination of foods is usually impossible to avoid, and the general trend is to control or minimize the contamination by good management processes, acceptable sanitary practices, rapid movement of the foods through processing plants, and well-tested preservation procedures. Contaminating microbes enter foods from various sources: Airborne organisms fall onto fruits and vegetables, then penetrate to the soft tissues beneath when the skin or rind is broken. Crops bring bacteria from the soil to the processing plant. Shellfish, such as clams and oysters, concentrate microbes in their tissues by catching them in their filtering apparatus as they strain their food from contaminated water. And rodents and arthropods transport microbes on their feet and body parts as they move about among foods.

Human handling is also a source of food contamination. For instance, when meat is handled carelessly by a butcher, bacteria from the animal's intestines mix with the meat. Moreover, if the water used to clean meat, poultry, and fish products has not been treated to eliminate microbes, they will enter the food. Of even more concern are raw vegetables purchased at supermarkets because they have been exposed to the contaminated hands of numerous shoppers. (Indeed, one need not be a rocket scientist to identify dozens of ways in which food can become contaminated with microbes.)

The practice of rework also increases the possibility for contamination in a food product. Rework consists of taking an unacceptable final product and returning it to the beginning of the manufacturing process. For example, icing may fail to cover a cake and when the cake is sent back, any bacteria picked up toward the end of the cake-making process are introduced to the beginning of the process—where they can probably multiply more easily. Discarding the unacceptable product or selling it at reduced prices at an outlet store are preferable alternatives to rework.

Food contamination can also be direct and premeditated, an act of bioterrorism. In 1984, for example, followers of a philosophical and religious leader named Bhagwan Shree Rajneesh contaminated the salad bars at restaurants in The Dalles, Oregon. The Rajneeshees sprayed broth containing *Salmonella typhimurium* onto the dressings and salad fixings at ten establishments in the community. A total of 751 people became seriously ill with cramps, nausea, weakness, and diarrhea. Fortunately, none died.

The investigation to identify the source of the outbreak went on for a year until, in 1985, cult leaders revealed the conspiracy. The group had been trying to win a local election to further its expansion plans, and it used bioterrorism to infect the electorate, hoping that its supporters would carry the vote. So far as is known, this was the first instance of bioterrorism in contemporary times.

The Conditions for Spoilage

Because food is basically a culture medium for microorganisms, its chemical and physical properties have a significant effect on what types of microbes it will harbor. Water, for instance, is a prerequisite for life, and to support microbial growth, food must be moist, with minimum water content of 18 to 20%. Microbes do not grow in foods such as potato chips, dried beans, rice, and flour because of their low water content (FIGURE 13.2).

Another important factor affecting microbial growth is pH. Most foods fall into the slightly acidic range on the pH scale, and numerous species of bacteria find these conditions acceptable. In foods with a pH of 5.0 or below, acid-loving fungi are the predominant microbes. Citrus fruits, for example, generally escape bacterial spoilage but yield to fungal growth and spoilage.

Moist foods	Neutral foods	Unrefrigerated (25°C)	Ground or sliced meat

Foods that spoil quickly

Dry foods	Acidic foods	Refrigerated (5°C)	Whole meat

Foods that resist spoilage

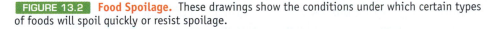

FIGURE 13.2   Food Spoilage. These drawings show the conditions under which certain types of foods will spoil quickly or resist spoilage.

Still another property that contributes to a food's tendency to spoil is its physical structure. A steak is not likely to spoil quickly because microbes cannot penetrate the surface of the meat easily. However, an uncooked hamburger deteriorates rapidly because microbes grow within the loosely-packed structure of the moist ground meat.

Oxygen and temperature are still other considerations. Vacuum-sealed cans of food do not support the growth of aerobic bacteria, nor do fresh vegetables or bakery products support anaerobes (however, the reverse holds true). Similarly, a refrigerator is usually too cold for the growth of human pathogens, but the warm hold of a ship and a hot, humid warehouse or storeroom are more conducive to the proliferation of spoilage microbes. Moreover, it is common knowledge that contamination is more likely in cooked food held at warm temperatures than in cooked food that is refrigerated. The memory of a certain company picnic engraves this principle on the mind.

Spoilage in foods is often due to the chemistry brought about by the contaminating microbes. Yeasts, for instance, live in apple juice and convert the carbohydrate into ethyl alcohol, a product that gives unpreserved juice an alcoholic taste. Certain species of bacteria convert food proteins into amino acids, then break down the amino acids into foul-smelling end-products. Digestion of the amino acid cysteine, for example, yields hydrogen sulfide, which imparts a smell of rotten eggs to food. Other possible products of microbial metabolism are acid, which causes food to sour, and gas, which causes sealed cans to swell.

Furthermore, pigment production by microbes can impart color to contaminated food. Indeed, in numerous historical incidents, the bright red pigment produced by *Serratia marcescens* growing in bread has been interpreted as a sign of blood. For example, in the dark, damp churches of medieval times, sacramental wafers used in Holy Communion often became contaminated with *S. marcescens*. In more than one instance, the "blood" on the Host was thought to be a miracle. One such event inspired the artist Raphael to paint his awe-inspiring masterpiece *The Mass of Bolsena*.

■ *Serratia marcescens*
ser-rä'tē-ä mär-ses'sens

Meats

Spoilage in meat and meat products is almost expected because of the nature of the food. Meat cannot be pasteurized like milk, and the extraordinarily high nutrient supply in meat supports a huge variety of microbes. In addition, the many steps in meat processing permit numerous opportunities for microbial contamination. It is not surprising, therefore, that meat and meat products are often related to episodes of disease. One example was the 1993 outbreak of hemolytic diarrhea traced to hamburger meat contaminated by *Escherichia coli* O157:H7. Over 500 patrons of Jack-in-the-Box restaurants were involved. A 1998 outbreak of the same disease necessitated a recall of 25 million pounds of hamburger meat in the infamous Hudson Foods outbreak. The responsible organism is pictured in **FIGURE 13.3** .

In the processing of fresh meat, contamination may be traced to such things as cutting boards used in preparing the meat, conveyor belts that transport the meat, improper temperature control while holding the meat, and failure to distribute meat and meat products quickly. Fecal bacteria can enter from the animal's intestinal contents via knives and cutting blocks. An early indication of meat spoilage is the loss of red color and the appearance of a brown color with surface slime. Such bacteria as species of *Lactobacillus* and *Leuconostoc* are often responsible for these changes. Although they are not dangerous to health, these microbes can give an abnormal taste and smell to the meat even after it has been cooked.

■ *Leuconostoc*
lü-kū-nos'tok

There is also the problem of the so-called choke points, or places where animals come together. In the United States, for instance, beef cattle produce calves on about 9000 farms; the young animals are then sent to about 50,000 feedlots for developing, and then to about 80 plants for slaughter. Microbes can spread at any of these points. Moreover, animal feeds often contain the entrails of sheep, poultry, and other animals added as protein sources; this presents another opportunity for microbes to spread. The so-called mad cow disease (Chapter 18) is believed to have spread from sheep to cattle in this way.

Processed meats, such as luncheon meats, sausages, and frankfurters, pose special hazards because they are handled often and contain a variety of meat products. Also, natural sausage and salami casings made from animal intestines may contain residual bacteria, especially spores of the bacterium that causes botulism. As early as the 1820s, people recognized the symptoms of "sausage poisoning" and coined the name botulism from the Latin *botulus* for "sausage." It should be pointed out, however, that fermented meats, such as sausages, are rarely contaminated because they contain a variety of organic chemicals that are toxic to spoilage microbes.

Cured meats such as ham, bacon, and corned beef are often treated with large doses of salt to draw water out of microbes and thereby kill them. However, lactobacilli tolerate the salt and ferment carbohydrates in the cured meat to lactic acid, which sours the meat and causes an "off" taste. In addition, the lactobacilli produce gas that causes the package to puff and slime. Moreover, these microbes form hydrogen peroxide, which changes the red meat pigments to green ones and gives the meat a spoiled look.

Fish

It is interesting to note that the microbes in fish are naturally adapted to the cold environment in which fish live. Therefore, cooling does not affect them as much as it does microbes in meat; freezing is preferred. Shellfish, such as clams, oysters, and mussels, are of particular concern because they obtain their food by filtering particles from the water and, therefore, concentrate in their tissues the pathogens that cause hepatitis A, typhoid fever, cholera, and other intestinal illnesses.

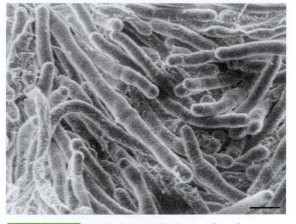

FIGURE 13.3 *Escherichia coli.* A scanning electron micrograph of *Escherichia coli.* A strain of this organism known as *E. coli* 0157:H7 causes serious intestinal disease and was involved in the Hudson Foods outbreak in 1998. (Bar = 10 μm.)

The tissues of fish tend to be looser than those of meats and tend to spoil faster. Microbiologists can often trace spoilage in fish to the water from which it was taken or in which it was held. Fish taken from polluted waters usually have high microbial counts, and it is not surprising that fish caught close to land are more contaminated than fish from the deep sea. Very often, spoilage organisms concentrate in the gills of fish, because these structures trap microbes as water is strained. Another source of contamination is the boxes used for storing and transporting fish. Cracks, pits, and splinters in the wood trap microbes and spread them among the fish.

Salting, heating, and drying are used as preservation methods for fish, but the spoilage rate of fresh fish is usually rapid. The characteristic odor associated with spoiled fish is generally due to the breakdown of organic compounds within fish muscle tissue by marine bacteria. For example, when a fish is taken out of its natural habitat, it dies quickly, and the natural process of decomposition begins. Marine bacteria grow and use a compound in fish muscle tissue called trimethylamine-N-oxide (TNO). They convert the TNO to trimethylamine, which gives rotting fish its dreadful smell.

Poultry and Eggs

Contamination in poultry and eggs may be due to human contact, but it often stems from bacteria that have infected the bird (FIGURE 13.4). Members of the genus *Salmonella*, which includes over 2400 pathogenic species, commonly cause diseases in chickens and turkeys and then pass on to consumers via poultry as well as egg products. For example, an outbreak of salmonellosis in New York in the 1990s was traced to contaminated eggs used to make cheese lasagna and stuffed shells. Processed foods such as chicken pot pie, whole egg custard, mayonnaise, egg nog, and egg salad may also be sources of salmonellosis (Chapter 19 discusses this disease).

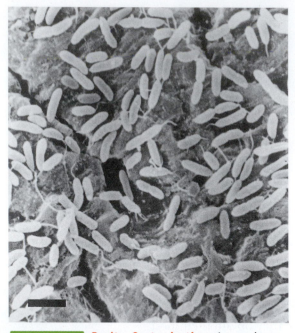

FIGURE 13.4 **Poultry Contamination.** A scanning electron micrograph of unidentified flagellated bacteria growing on the skin of a chicken carcass. Bacteria such as these contaminate frozen food and grow to large numbers when the food is thawed and held for long periods before cooking. (Bar = 2 μm.)

Eggs are normally sterile when laid, but the outer waxy membrane, as well as the shell and inner shell membrane, can be penetrated by bacteria. Bacteria of the genus *Proteus* cause black rot in eggs when they break down the amino acid cysteine and produce hydrogen sulfide gas, which leads to black deposits and gives rotten eggs their horrid smell. Spoilage in eggs also occurs as the yolk develops a blood-red appearance from growth of pigment-producing *Serratia marcescens*. Scientists note that the primary location of egg contamination is in the yolk rather than the white because the yolk is more nutritious and the white has an inhospitable pH of approximately 9.0.

Part of the problem of infected chicken and poultry products traces to the use by high-tech chicken farms of machines to remove the eggs from the hen as they are produced. In so doing, manufacturers prevent normal bacteria from entering the young chicks and keeping pathogenic microbes in check. One solution to this problem is explored in Chapter 1. Another solution may be to use pasteurized eggs, as A Closer Look on page 285 discusses.

Milk and Dairy Products

Milk is an extremely nutritious food. It is a solution of proteins, fats, and carbohydrates, with numerous vitamins and minerals. Milk has a pH of about 7.0 and is an excellent growth medium for humans and animals, as well as microbes. About 87% of milk is water. Another 2.5% is a protein called casein, a mixture of three different long chains of amino acids. A second protein in milk, lactalbumin, is a whey protein; that is, it remains in the clear fluid (the whey) after the casein is removed during cheese production.

Carbohydrates make up about 5% of milk. The major carbohydrate is lactose, sometimes referred to as milk sugar (*lactus* is Latin for "milk"). Rarely found elsewhere, lactose is a disaccharide that is digested by relatively few species of bacteria, and these are usually harmless. The last major component of milk is butterfat, a mixture of fats that can be churned into butter. When bacterial enzymes digest these fats into fatty acids, the milk or butter becomes rancid and develops a sour taste.

Milk spoilage often develops in the kitchen refrigerator or the supermarket dairy case. *Lactobacillus* or *Streptococcus* species that have survived pasteurization multiply and ferment the milk's lactose. Soon, large quantities of lactic and acetic acids accumulate and change the structure of the casein, causing it to solidify as a sour curd. Sweet curdling (so called because little acid is present) results when bacterial enzymes attack milk's casein. As its weak chemical bonds break, casein loses its three-dimensional structure and curdles. Sweet curdling is an essential first step in the production of cheese, as we explore in Chapter 15. The remaining liquid, the whey, contains lactose, minerals, vitamins, lactalbumin, and other components. It is used to make ricotta, as well as processed cheeses and "cheese foods."

Ropiness in milk develops from glycocalyx-producing organisms, such as species of *Alcaligenes* and *Enterobacter*. These Gram-negative rods multiply in milk, even at low temperatures, and deposit gummy material that appears as stringy threads and slime.

"Bring on the Caesar Salad!"

Some people have it Hawaiian style with pineapple and ham; some have it Italian style with pasta and tomatoes; and some enjoy it Southwestern style with roasted chili peppers. Some toss it with grilled chicken or flank steak or grilled calamari or Cajun scallops. Regardless of the addition, however, the salad remains the same—it's Caesar salad.

For history buffs, Caesar salad was "invented" on July 4, 1924, in the mind of Caesar Cardini, the proprietor of a restaurant (Caesar's Place) in Tijuana, Mexico. Cardini was desperate for a fill-in during a particularly busy day, so he threw together some Romaine lettuce, Parmesan cheese, lemon, garlic, oil, and raw eggs. His customers were enchanted.

Over the years, the reputation of the salad grew. There was only one problem, however—the eggs. For true Caesar salad, raw eggs are used to add creaminess to the dressing. But that became a problem when an increasing incidence of *Salmonella* infections were traced to raw eggs. Eggless Caesar dressings (with mayonnaise or heavy cream) were tried, but they just weren't the same.

Caesar salad fans, take heart, the pasteurized egg is on the way. Purdue University microbiologists have found that *Salmonella* can be eliminated from eggs by heating them in hot water or a microwave oven, then maintaining them at 56°C (134°F) in a hot-air oven for 1 hour. And a New Hampshire company (Pasteurized Eggs L.P.) is touting the benefits of its new machine for destroying eggborne *Salmonella*. The machine heats eggs slowly, then directs them to successive baths in water ranging from 62°C to 72°C. Indeed, the process works so well that the U.S. Department of Agriculture has issued a new stamp certifying that eggs treated by the company meet standards for egg pasteurization established by the Food and Drug Administration.

Agricultural officials estimate that almost 50 billion eggs are produced for American consumption each year, and that over 2 million are infected with *Salmonella*. In the years ahead, that second number should dwindle as egg pasteurization becomes standard practice (as milk pasteurization already has). Then it will be safe to sample the cookie dough, or to have eggs over easy, or to enjoy Caesar salad the way it was meant to be enjoyed.

Milk spoilage also results from the red pigment deposited in dairy products by *Serratia marcescens* and from the gray rot caused by certain *Clostridium* species. Spoilage due to wild yeasts is usually characterized by a pink, orange, or yellow coloration in the milk.

Milk is normally sterile in the udder of the cow, but contamination occurs as it enters the ducts leading from the udder. Species of soilborne *Lactobacillus* and *Streptococcus* are acquired there, together with various pathogenic microbes from dust, manure, and polluted water. For example, bacteria of the genus *Campylobacter* (FIGURE 13.5) can cause mild to serious diarrhea if they contaminate milk. Milk may acquire other contaminants from dairy plant equipment and unsanitary handling. Pasteurization lowers the microbial content considerably, but it does not sterilize the milk, as we shall see presently.

Breads and Bakery Products

In the production of bakery products, ingredients such as flour, eggs, and sugar are generally the sources of spoilage microbes. Although most contaminants are killed

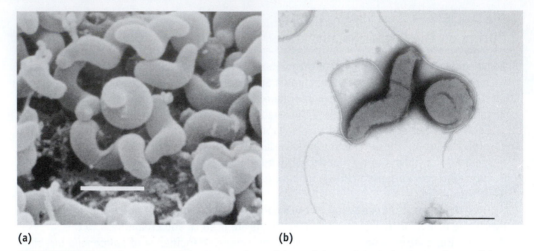

(a)   (b)

FIGURE 13.5   *Campylobacter jejuni.* Two views of *Campylobacter jejuni,* the cause of campylobacteriosis. (**a**) A scanning electron micrograph of *C. jejuni* taken from a colony of cells. (Bar = 1 μm.)
Note the curved shape of most of the organisms and the coccus shape of several. (**b**) A transmission
electron micrograph of negatively stained cells, showing the flagellar arrangement of both types of
cells. (Bar = 0.5 μm.)

during baking, some bacterial and fungal spores can survive the high oven temperatures. Members of the genus *Bacillus* are notorious in this respect; they proliferate
and their glycocalyx material accumulates, giving the bread a soft, cheesy texture with
long, stringy threads. The bread is said to be ropy.

Cream fillings and toppings in bakery products provide excellent chemical and
physical conditions for microbial growth. For instance, custards made with whole
eggs may be contaminated with *Salmonella* species, and whipped cream may contain
microbes such as species of *Lactobacillus* and *Streptococcus*. The acid produced by
these bacteria results in a sour taste. High-sugar environments in chocolate toppings
and sweet icings support the growth of fungi.

Grains

Two types of grain spoilage are important in microbiology. The first type is caused by
the ascomycete *Aspergillus flavus*. This fungus produces aflatoxins, a series of toxins
that accumulate in stored grains (such as wheat) as well as peanuts, soybeans, and
corn. Scientists have implicated aflatoxins in liver and colon cancers in humans. The
toxins are consumed with grain products and meat from animals that feed on contaminated grain. They can also be consumed with foods or water that have been
sprayed by bioterrorists with laboratory-produced aflatoxins. Indeed, for this reason,
government officials consider aflatoxins to be a candidate for use as a bioweapon,
much as botulism toxin could be used.

The second important type of grain spoilage is caused by *Claviceps purpurea*, the
cause of ergot poisoning (ergotism), which manifests itself in neurological symptoms.
Rye plants are particularly susceptible to this type of spoilage, but wheat and barley grains
may also be affected. The toxins deposited by *C. purpurea* may trigger convulsions and
hallucinations when consumed (the drug LSD is derived from the toxin). Chapter 8 discusses one researcher's view of how this toxin may have set off the French Revolution.

■ *Aspergillus flavus*
a-spėr-jil'lus flă'vus

■ *Claviceps purpurea*
kla'vi-seps pür-pü-rē'ä

Food Preservation

Centuries ago, humans battled the elements to keep a steady supply of food at hand. Sometimes, there was a short growing season; at other times, locusts descended on their crops; at still other times, they underestimated their needs and had to cope with scarcity. However, experience taught humans that they could prevail through difficult times by preserving foods. Among the earliest methods was drying vegetables and strips of meat and fish in the sun. Foods could also be preserved by salting, smoking, and fermenting. One benefit of the new technologies was that individuals could trek far from their native lands, and soon they took to the sea and moved overland to explore new places.

The next great advance did not come until the mid-1700s. In 1767, Lazaro Spallanzani attempted to disprove spontaneous generation by showing that beef broth would remain unspoiled after being subjected to heat. Nicholas Appert, a French winemaker, took note and applied this principle to a variety of foods (see A Closer Look below). But neither Appert nor his contemporaries was quite sure why the food was being preserved. The significance of microbes as agents of spoilage awaited Pasteur's classic experiments with wine several generations later.

Through the centuries, preservation methods have had a common objective: to reduce the microbial population and maintain it at a low level until the food can be

A CLOSER LOOK
To Feed an Army

Part of Napoleon's genius was understanding the finer points of warfare, including how to feed an army. Recognizing that thousands of men-at-arms were a glut on the land, he broke up his army into smaller units that foraged on their own as they moved. When the time for battle neared, he reassembled his forces and engaged the enemy.

The shortcomings of this system became painfully clear when Napoleon crossed into northern Italy in 1800 and engaged the Austrians at the Battle of Marengo. Aware that the French army was scattered about the countryside, the Austrians charged before Napoleon could bring his forces together. Disaster was averted only when units arrived on the flanks to repel the Austrians. Napoleon had learned an important lesson: The next time he went to war, food would go with him.

Included in Napoleon's plan to resurrect France was a ministry that encouraged industry by offering prizes for imaginative inventions. A winemaker named Nicholas Appert attracted attention with his process of preserving food. Appert placed fruits, vegetables, soups, and stews in thick bottles, then boiled the bottles for several hours. He used wax and cork to seal the bottles and wire wine cages to prevent inadvertent opening of the bottle. By 1805, Appert had set up a bottling industry outside Paris and had a thriving business.

The Ministry of Industry encouraged Appert to publish his methods and submit samples of bottled foods for government testing. The French navy took bottles of Appert's food on long voyages and reported excellent food preservation. In 1810, Appert was awarded 12,000 francs for his invention. Two years later, Napoleon assembled hundreds of cannons, thousands of men, and millions of bottles of food, and marched off to war with Russia.

eaten. Modern preservation methods still have that objective. Though today's methods are sophisticated and technologically dynamic, advances have been counterbalanced by the increased complexity of food products and the great volumes of food that must be preserved. Thus, the food preservation problems that early humans faced do not differ fundamentally from those confronting modern food technologists. As we shall see in this section, many preservation methods are old standbys, while others are thoroughly modern.

Heat

Heat kills microorganisms by changing the physical and chemical properties of their proteins. In a moist heat environment, proteins are denatured and lose their specific three-dimensional structure, taking on a different three-dimensional form or reverting to a two-dimensional form. As their structural proteins and enzymes undergo this change, the microbes die. Chapter 11 explores various heat methods for controlling microbes. You might find a brief review helpful in relating heat to food preservation.

The most useful application of heat in food technology is in canning (FIGURE 13.6). Shortly after Appert used bottles in establishing the value of heat in preservation, an English engineer named Bryan Donkin substituted iron cans coated with tin. Soon he was supplying canned meat to the British navy. In the United States, the tin can was virtually ignored until the Civil War period, but in the years thereafter, mass production of canned food began. Soon the tin can was the symbol of prepackaged convenience.

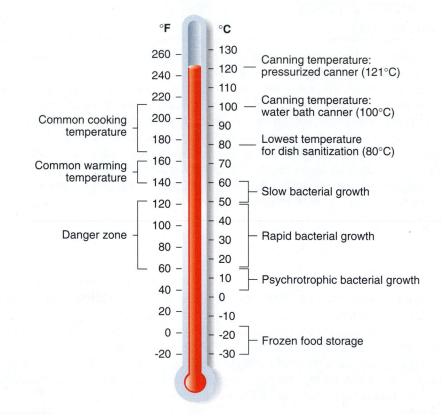

FIGURE 13.6   **Important Temperature Ranges in Food Microbiology.**

Modern canning processes are complex. Machines wash, sort, and grade the food, then subject it to steam heat for 3 to 5 minutes. This last process, called blanching, destroys many enzymes in the food and prevents any further cellular metabolism. Next, the food is processed (peeled and cored, for example) and then put into the can. The air is evacuated from the can, which is then placed in a pressurized steam sterilizer similar to an autoclave at a temperature of 121°C or lower, depending on the food's pH, density, and heat penetration rate.

The sterilizing process is designed to eliminate the most resistant bacterial spores, especially those of the genera *Bacillus* and *Clostridium*. However, the process is considered commercial sterilization, which is not as rigorous as the true sterilization used for medical instruments. Moreover, if a machine error leads to improper heating temperatures, or a small hole allows airborne bacteria to enter, or a proper seal does not form, contamination may result. The contamination is usually obvious, since most spoiled canned food has a putrid odor. Most contamination of canned food is due to facultative or anaerobic bacteria (such as *Clostridium*) that produce gas and cause the ends of the can to bulge. Common contaminants also include coliform bacteria, a group of Gram-negative nonsporeforming rods that ferment lactose to acid and gas. (It should be noted that can swelling can also be due to a reaction between food acids and the can's metal, a reaction that leads to accumulation of hydrogen gas.)

Growth of acid-producing microbes presents a different problem because spoilage cannot be discerned from the can's shape. Food has a flat-sour taste from the acid and has probably been contaminated by a *Bacillus* species, a species of coliform, or another acid-producing bacterium that survived the heating. This may happen when the population of bacteria was extremely high at the outset and the heat was unable to destroy all the microbes.

Pasteurization

The process of pasteurization was developed by Louis Pasteur in the 1850s to eliminate bacteria in wines. His method was first applied to milk in Denmark about 1870 and was widely employed by 1895. Although the primary object of pasteurization is to eliminate pathogenic bacteria from milk, the process also lowers the total number of bacteria and thereby reduces the chance of spoilage.

The more traditional method of pasteurization, the holding method, involves heating the milk in a large bulk tank at 62.8°C (145°F) for 30 minutes. Concentrated products, such as cream, are heated at the higher temperature of 68.3°C (155°F) to ensure successful pasteurization. The more modern method of pasteurization is called the flash method. In this process, machines pass the milk through a hot cylinder at 71.7°C (161°F) for a period of 15 to 17 seconds. Next, the milk is cooled rapidly, in part by transferring its heat to the incoming milk. A newer method called ultrapasteurization is used in some dairy plants. In this process, milk and milk products are subjected to heat at 82.2°C (180°F) for 3 seconds. Pasteurization is also important for fruit juices, as FIGURE 13.7 shows.

The bacteria that survive pasteurization may spoil the milk, but they generally cause no other harm. *Streptococcus lactis*, for instance, grows slowly in refrigerated milk, and when its numbers reach about 20 million per milliliter, enough lactic acid has been produced to make the milk sour. The shelf life of the milk indicated on the

1 On September 14, 2006, the Food and Drug Administration (FDA) issued an alert to consumers that an outbreak of *E. coli* O157:H7 in multiple states that may be associated with bagged fresh spinach. The FDA advised consumers not to eat fresh spinach or fresh spinach-containing products until further notice. States that had reported *E. coli* O157:H7 illness by September 14th, 2006, included: Connecticut, Idaho, Indiana, Michigan, New Mexico, Oregon, Utah, and Wisconsin.

2 By September 21st, 157 cases of *E. coli* O157:H7 associated illness had been reported, and the FDA, working closely with both the CDC and the State of California, determined that the spinach implicated in the outbreak was grown in the California counties of Monterey, San Benito, and Santa Clara.

3 By November 12, 2006, a total of 13 product samples were identified in the laboratory to be contaminated by *E. coli* O157:H7 and a total of six product recalls were initiated. The CDC reported a total of 204 *E. coli* O157:H7–associated illness including three deaths. The 26 affected U.S. states were: Arizona (8), California (2), Colorado (1), Connecticut (3), Idaho (8), Illinois (2), Indiana (10), Kentucky (8), Maine (3), Maryland (5), Michigan (4), Minnesota (2), Nebraska (11), Nevada (2), New Mexico (5), New York (11), Ohio (26), Oregon (6), Pennsylvania (10), Tennessee (1), Utah (19), Virginia (2), Washington (3), West Virginia (1), Wisconsin (50), and Wyoming (1). In addition, Canada had one confirmed case in the Province of Ontario.

4 The FDA was able to trace the source of contamination to cow feces by matching the *E. coli* O157:H7 strain found in these cows to one of four implicated spinach fields in Monterey and San Benito counties, CA.

5 There is long history of *E. coli* O157:H7 outbreaks from leafy greens grown in California. The FDA and the State of California now expect the food industry to develop a comprehensive plan designed to minimize the risk of another outbreak due to *E. coli* O157:H7.

FIGURE 13.7   Multistate Outbreak of *E. coli* O157:H7 Infections Associated with Consumption of Fresh Spinach. The strain of *E. coli* named O157:H7 has been associated with severe illness and even death when consumed by humans. This recent outbreak was associated with the consumption of uncooked spinach in the late summer and early fall of 2006.

carton is an estimate of when that is likely to occur. Microbes such as the streptococci that survive pasteurization are described as thermoduric (heat-enduring). Pasteurization is virtually useless against thermophilic (heat-loving) bacteria, since these microbes are able to grow at 60°C to 70°C and higher. Fortunately, they do not grow at refrigerator temperatures or in the human body as pathogens because conditions are too cool. Pasteurization has no effect on bacterial spores.

Although most milk in the United States is normally pasteurized, milk can be sterilized by exposure to pressurized steam at 140°C (252°F) for 3 seconds. This ultra-high temperature results in milk with an indefinite shelf life, so long as the container remains sealed. Small containers of coffee cream are often prepared this way.

Low Temperatures

In a refrigerator or freezer, the lower temperature reduces the rate of enzyme activity in microbes and thus slows their rate of growth and reproduction, while extending the shelf life of food. Although the microbes are not killed, their numbers are kept low, and spoilage is minimized (ironically, the food is preserved by "preserving" the microbes). Well before the invention of refrigerators, the Greeks and Romans had partially solved the problem of keeping things cold. They dug snow cellars in the basements of their homes, lined the cellars with logs, insulated them with heavy layers of straw, and packed them densely with snow delivered from far-off mountaintops ("The iceman cometh!"). The compressed snow turned to a block of ice, and foods left in this makeshift refrigerator would remain unspoiled for long periods.

Modern refrigerators at 5°C (41°F) accomplish the same goal by providing a suitable environment for preserving food without destroying its appearance, taste, or cellular integrity. However, psychrotrophic (cold-tolerant) microbes survive in the cold, and given enough time, they will cause meat surfaces to turn green, eggs to rot, fruits to become moldy, and milk to sour. Staphylococci are problematic: As they multiply in foods, they deposit their toxins, and if the toxins are not destroyed by heating, they will cause staphylococcal food poisoning; diarrhea, nausea, and cramps can be anticipated.

When food is placed in a freezer at –5°C (23°F), ice crystals form and tear and shred microbes, thereby killing substantial numbers. However, many survive, and since the ice crystals are equally destructive to food cells, the microbes multiply quickly when the food thaws. Rapid thawing and cooking are therefore recommended for frozen foods. Moreover, food should not be refrozen; during the time it takes to thaw and refreeze, microbes can deposit sufficient toxins to cause food poisoning the next time the food is thawed. Microwave cooking, which requires minimal thawing, may eliminate some of these problems.

Deep freezing at –60°C results in smaller ice crystals, and although the physical damage to microorganisms is less severe, their biochemical activity is reduced considerably. Some food producers blanch their product (apply moist heat briefly) before deep freezing, a process that reduces the number of microbes. A major drawback of freezing is freezer burn, which develops as food dries out from moisture evaporation. Another disadvantage is the considerable energy cost. Nevertheless, freezing has been a mainstay of preservation since Clarence Birdseye first offered frozen foods for sale in the 1920s. Approximately a third of all preserved food in the United States is frozen.

Drying

The rationale for drying foods is best expressed thusly: "Where there is no water, there is no natural life." In past centuries, people used the sun for drying, but modern technologists have developed sophisticated machinery for this purpose. For example, the spray dryer expels a fine mist of liquid such as coffee into a barrel cylinder containing hot air. The water evaporates quickly, and the coffee powder falls to the bottom of the cylinder.

Another machine for drying is the heated drum. Machines pour onto the drum's surface liquids such as soup, and the water evaporates rapidly, leaving dried soup to be scraped off. A third machine utilizes a belt heater that exposes liquids such as milk to a stream of hot air. The air evaporates any water and leaves dried milk solids. Unfortunately, sporeforming and capsule-producing bacteria resist drying.

During the past few decades, freeze-drying, or lyophilization, has emerged as a valuable preservation method. In this process, food is deep frozen, and then a vacuum pump draws off the water. (Water passes from its solid phase [ice] to its gaseous phase [water vapor] without passing through its liquid phase [water].) The machinery used

FIGURE 13.8 **An Industrial Lyophilizer.** This lyophilizer removes 500 pounds of product moisture in 24 hours of freeze-drying a product. Vacuum and heat are used to draw off water from ice without passing through the liquid phase.

is pictured in **FIGURE 13.8**. The dry product is sealed in foil and easily reconstituted with water. Hikers and campers find considerable advantages in freeze-dried foods because of their light weight and durability. However, there is a disquieting note: Lyophilization is also a useful method for storing, transporting, and preserving bacterial cultures (Chapter 11).

Lyophilization can be used by bioterrorists to dry the broth in which bacteria have grown. The broth is placed in flasks, then subjected to the freezing and drying processes to produce a fine, light powder. Although it can be used in this form by bioterrorists, more sophisticated methods involve still another step in which the lyophilized powder is converted to a biodust. This step keeps the spores from clumping together and ensures the fine particles necessary to penetrate to the lung tissues or pass through cracks in the skin. The production of the biodust is referred to as "weaponizing" the biological agent of bioterrorism.

Osmotic Pressure

When living cells are immersed in large quantities of a compound such as salt or sugar, water diffuses out of cells through cell membranes and into the surrounding environment. The flow of water is called osmosis, and the force that drives the water is termed osmotic pressure.

Osmotic pressure can be used to preserve foods because water flows out of microbes as well as food cells. For example, in highly salted or sugared foods, microbes dehydrate, shrink, and die. Jams, jellies, fruits, maple syrups, honey, and similar products are preserved this way. Foods preserved by salting include ham, cod, bacon, and beef, as well as certain vegetables such as sauerkraut, which has the added benefit of large quantities of acid. It should be noted, however, that staphylococci tolerate salt and may survive the exposure.

Chemical Preservatives

For a chemical preservative to be useful in foods, it must be inhibitory to microbes while easily broken down and eliminated by the body without side effects. These requirements are enforced by the U.S. Food and Drug Administration (FDA). The criteria have limited the number of chemicals used as food preservatives to a select few.

A key group of chemical preservatives are organic acids, including sorbic acid, benzoic acid, and propionic acid. These compounds damage microbial membranes and interfere with the uptake of certain essential organic substances such as amino acids. Sorbic acid, which came into use in 1955, is added to syrups, salad dressings, jellies, and certain cakes. Benzoic acid, the first chemical to be approved by the FDA for use in foods, protects beverages, catsup, margarine, and apple cider. Propionic acid is incorporated in wrappings for butter and cheese, and it is added to breads and bakery products, where it prevents the growth of fungi and inhibits the ropiness commonly due to *Bacillus* species. Other natural acids in foods add flavor while serving as preservatives. Examples are lactic acid in sauerkraut and yogurt and acetic acid in vinegar.

The process of smoking with hickory or other woods accomplishes the dual purposes of drying food and depositing chemical preservatives. By-products of smoke, such as aldehydes, acids, and certain phenol compounds, effectively inhibit microbial growth for long periods of time. Smoked fish and meats have been staples of the human diet for many centuries.

Sulfur dioxide has gained popularity as a preservative for dried fruits, molasses, and juice concentrates. Used in either gas or liquid form, the chemical retards color changes on the fruit surface, while reducing microbial spoilage. Another sulfur compound, sulfurous acid, is used to prevent growth of the lactobacilli that sour wine. The two sulfur compounds are commonly known as sulfites. Unfortunately, the FDA estimates that over 1% of the U.S. population is sensitive to sulfites, including over a million individuals who suffer from asthma.

Radiation

Though much of the public is apprehensive about exposing foods to radiation, various forms of radiation have received FDA approval for preserving foods (FIGURE 13.9). For instance, gamma rays are used to extend the shelf life of fruits, vegetables, fish, and poultry from several days to several weeks. This form of radiation also increases the distance fresh food can be transported and significantly extends the storage time for food in the home. Gamma rays are a high-frequency type of electromagnetic energy emitted by the radioactive isotopes cobalt-60 and cesium-137. Health officials are quick to note that such radiation does not cause food to become radioactive. The radiation kills microbes by reacting with and

FIGURE 13.9   **Food Irradiation.** The FDA has approved irradiation as a preservation method for numerous foods, including many fruits and vegetables as well as poultry and red meats.

destroying microbial DNA and other key organic compounds (Chapter 11). Opponents to the use of radiation point out that it also breaks chemical bonds in foods and causes new ones to form, thereby raising the possibility of new and toxic chemical compounds.

Interest in radiation for food preservation grew during the 1950s under President Eisenhower's Atoms for Peace program. For the next quarter-century, the FDA conducted extensive tests to determine whether the process was safe. In March 1981, it approved radiated foods such as spices, condiments, fruits, and vegetables for sale to American consumers. Gamma radiation of pork to prevent trichinosis (a disease caused by a worm parasite) won approval in 1985, and irradiated strawberries appeared in the marketplace in 1992. Irradiation of red meat was approved in 1997.

Food processors wishing to employ irradiation must constantly confront a leery public, some of whom still have visions of Hiroshima and Nagasaki. Food technology plants using isotopes store the pellets encapsulated twice in stainless steel pencil-like tubes arranged in racks under water. When food is to be irradiated, the racks are withdrawn from the water, and the food is passed through the radiation field. At this time, the gamma rays penetrate the food and cause microbial death. Nutritional losses are similar to those occurring with cooking and/or freezing. And scientists hasten to point out that the food does not glow.

■ Preventing Foodborne Disease

As the story opening this chapter illustrates, food can be a method for transferring microbes as well as a culture medium for their growth. Unsuspecting consumers are affected by either the microbes or the toxins they have produced in the food. When the microbes themselves are transferred to consumers, a food infection is established; when their toxins are consumed, a food poisoning (or food intoxication) occurs.

Food infections are typified by typhoid fever, salmonellosis, cholera, and shigellosis, all of which are of bacterial origin. Amoebiasis and giardiasis represent foodborne diseases caused by protozoa. Viral infections are exemplified by hepatitis A. Food poisonings include staphylococcal food poisoning, botulism, and clostridial food poisoning. Full discussions of these diseases are presented elsewhere (Chapters 18 and 19), and we shall not examine them in depth here. Food infections and poisonings are nasty experiences, and this section presents some suggestions on how to avoid them.

Helpful Suggestions

In the United States, public health microbiologists estimate that between 2 and 10 million people are affected by foodborne disease annually. Many episodes require medical attention, but the vast majority of patients recover rapidly without serious complications. In many cases, the incident might have been avoided if some basic precautions had been taken. For example, unrefrigerated foods are a prime source of staphylococci and *Salmonella* species, so perishable groceries such as meats and dairy products should not be allowed to warm up while other errands are performed. Also, a thermometer should be used to ensure that the refrigerator temperature remains below 5°C (41°F) at all times.

Another way to avoid foodborne disease is to be aware that skin infections are often caused by the same staphylococci that cause food poisoning; it is therefore prudent to cover any skin infections while working with foods. Moreover, the hands should always be washed thoroughly before and after handling raw vegetables (or salad fixings) to avoid cross-contamination of other foods. It is wise to cook meat from a frozen or partly frozen state; if this is not possible, the meat should be thawed in the refrigerator. Cutting boards should be cleaned with hot, soapy water after use, and old cutting boards with cracks and pits should be discarded.

Studies indicate that leftovers are implicated in most outbreaks of foodborne disease. It is therefore important to refrigerate leftovers promptly and keep them no more than a few days. Thorough reheating of leftovers, preferably to boiling, also reduces the possibility of illness.

Many instances of foodborne disease occur during the summer months, when foods are taken on picnics, where they cannot be refrigerated. As a general rule, foods containing eggs, such as custards, cream pies, pastries, and deli salads, should be left off the picnic menu. For outdoor barbecues, one dish should be used for carrying chicken, hamburgers, or steaks to the grill and another dish for serving them. Many of these principles apply equally well to fall and winter tailgate parties.

Over 90% of botulism outbreaks reported to the CDC are traced to home-canned food. To prevent this sometimes fatal foodborne disease, health officials urge that homemakers use the pressure method to can foods. Reliable canning instructions should be obtained and followed stringently. Canned foods suspected of contamination should not be tasted to confirm the suspicion, but should be discarded immediately. If there is any doubt, health officials advise boiling the food for a minimum of 10 minutes and thoroughly washing the utensil used to stir the food. Bulging or leaking cans must be discarded in a way that will not endanger other people or animals. Despite these precautions, botulism can occur under unexpected conditions, as **FIGURE 13.10** indicates.

HACCP Systems

Fueled by consumer awareness, the entire food industry has been placed under a food-safety spotlight. Among the most important food-safety systems is Hazard Analysis Critical Control Points (HACCP), a set of scientifically based and federally enforced safety regulations for the seafood, meat, and poultry industries. Each establishment tailors the HACCP system to its individual product, processing, or distribution conditions, identifying points at which the safety of a product could be affected. These points are called critical control points, or CCPs (in the jargon of food technology). The CCPs are supervised to ensure that any hazards associated with the operation are contained or, preferably, eliminated (the key principle is prevention). When all possible hazards are controlled at the CCPs, the safety of the product can be assumed without further testing or inspection.

HACCP systems are overseen by the U.S. Food and Drug Administration (FDA) and the U.S. Department of Agriculture (USDA). The standard regulations require food processors to monitor and control eight key sanitation areas: (1) the safety of water that contacts food or is used to make ice; (2) the condition and cleanliness of utensils, gloves, outer garments, and other food contact surfaces; (3) the prevention of cross-contamination of food from raw products and unsanitary objects; (4) the maintenance of hand-washing and toilet facilities; (5) the protection of foods and food surfaces from adulteration with lubricants, fuel, pesticides, sanitizing agents,

1 On April 7, a chef at a Greek restaurant wrapped a large number of potatoes in aluminum foil and baked them in the oven for about an hour, at roughly 400°F. Leaving them in their aluminum wrappings, he set them aside for use the next day. The baked potatoes were left at room temperature. A total of 18 hours would pass before they would be used.

2 The next day, April 8, the chef removed the potatoes from their wrappings and mashed them into a Greek dip called skordalia, which he then stored in a refrigerator.

3 That afternoon and evening, many restaurant patrons enjoyed the dip as an appetizer to their meals.

4 On April 10, a father and son, who had eaten at the restaurant, reported to the local hospital suffering from labored vision, difficult breathing, numbness, and general weakness. An alert physician diagnosed botulism.

5 Investigators found evidence of botulism toxin in the leftover dip. They concluded that *Clostridium botulinum* spores were on the potato skin and had germinated during the 18 hours of storage. Twenty-two cases were eventually found among 235 patrons. All recovered.

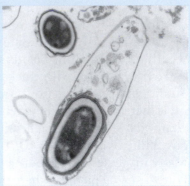

FIGURE 13.10 **An Outbreak of Botulism.** This incident occurred during April 1994 in El Paso, Texas. It was the largest U.S. outbreak of botulism in 11 years.

and other contaminants; (6) the proper labeling, storage, and use of toxic compounds; (7) the control of employee health conditions that could result in food contamination; and (8) the exclusion of pests from the food plant.

The HACCP system is a risk-reduction process originally developed in the 1960s for foods used in space travel ("space foods"), but not applied to the food industry until the 1990s. It places the responsibility for food safety on the shoulders of the food industry, but it also focuses consumer attention on food handling and safety issues. Several well-publicized outbreaks of foodborne disease and product recalls raised questions about the safety and quality of foods, and HACCP systems represent an attempt to restore consumer confidence. Improved epidemiological investigations and increased surveillance are expected to add to that confidence in the ensuing years. Indeed, January 26, 1998, was an important date for consumers. On that day, HACCP systems began at 312 of the largest meat and poultry processing plants in the United States; by the time you read this, the remaining 6100 plants will have been phased in. Hopefully, the result will be added safety in our foods. If so, then the competition between humans and microbes will tip in our favor.

A FINAL THOUGHT

Since 1925, only five deaths from botulism have been attributed to commercially canned food in the United States. During this period, almost 100 billion cans of food were produced for sale to consumers.

I believe that these figures are a testament to the high standards achieved by the canning industry. They also represent an achievement of which we consumers can be justifiably proud. I say "we consumers" because we are the ones who understand that foods can be a vehicle for disease, and we refuse to tolerate a manufacturer's ignorance or negligence. Working through our representatives in government agencies, we exact heavy penalties from companies whose products are tainted. For example, 25 million pounds of hamburger meat were recalled in 1998 when the meat was found to be contaminated.

The next time you shop at the supermarket, stop and take note of the broad variety of foods we consume, and consider that we buy and eat these foods with full confidence that none will make us ill. It is a confidence that is not shared by peoples in other parts of the world.

QUESTIONS TO CONSIDER

1. Chicken and salad are two items on the dinner menu at home, and you are put in charge of preparing both. You have a cutting board and knife for slicing up the salad items and cutting the chicken into pieces. Which task should you perform first? Why? What other precautions might you take to ensure that dinner is not remembered for the wrong reason?

2. A writer in a food technology magazine once suggested that refrigerators be fitted with ultraviolet lights to reduce the level of microbial contamination in foods. Would you support this idea?

3. It is a hot Saturday morning in July. You get into your car at 9:00 A.M. with the following list of chores: Pick up the custard eclairs for tonight's dinner party, drop off clothes at the cleaners, buy the ground beef for tomorrow's barbecue, deliver the kids to the Little League baseball game, pick up a broiler at the poultry farm. Microbiologically speaking, what sequence should you follow?

4. It is 5:30 P.M. and you arrive on campus for your evening college class. You stop off at the cafeteria for a bite to eat. Which foods might you be inclined to avoid purchasing?

5. To avoid *Salmonella* infection when preparing eggs for breakfast, the operative phrase is "scramble or gamble." How many foods can you name that use uncooked or undercooked eggs and that can represent a health hazard?

6. Which principles of preservation ensure that each of the following remains uncontaminated on the pantry shelf: vinegar, olive oil, brown sugar, tea bags, spaghetti, hot cocoa mix, pancake syrup, soy sauce, rice?

7. Foods from tropical nations such as Mexico tend to be very spicy, with lots of hot peppers, spices, garlic, and lemon juice. By contrast, foods from cooler countries such as Norway and Sweden tend to be much less spicy. Why do you think this pattern has evolved over the ages?

◼ KEY TERMS

Informative fact are necessary for the expression of every concept, and the information for a concept is founded in a set of key terms. The following terms form the basis for the concepts of this chapter. On completing the chapter, you should be able to explain and/or define each one:

blanching   microbial load
canning   osmosis
choke points   pasteurization
commercial sterilization   rework
flash method   ropiness
HACCP   shelf life
holding method   sour curd
lyophilization

◼ http://microbiology.jbpub.com/book/microbes

The site features **eLearning**, an online review area that provides quizzes and other tools to help you study for your class. You can also follow useful links for in-depth information, read more stories of microbiology, or just find out the latest microbiology news.

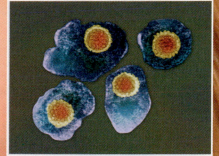

Biotechnology and Industry: Microbes at Work

14

Looking Ahead

The use of microbes and the awe-inspiring breakthroughs in microbiology and DNA research have resulted in a wealth of innovative biotechnological applications in industry. But this is not surprising, for microbes have been used in industry for many decades to synthesize a broad variety of valuable products. We shall combine the new with the old in this chapter as we survey the goods and services that microbes provide to enhance the quality of our lives.

On completing this chapter, you should be able to . . .

- understand how the processes of biotechnology have yielded numerous valuable pharmaceutical products to help lessen human disease and suffering.
- appreciate the molecular basis of such biotechnological innovations as antisense molecules and DNA vaccines.
- explain how DNA amplification methods and gene probes underlie new microbial detection methods for use in diagnostic procedures and research.
- understand some of the complex methods of biotechnology scientists use to work with genes and appreciate some of the problems that must be resolved to achieve success.

The advances in biotechnology have lead to vaccines that help prevent childhood diseases like the chickenpox.

- list a number of microbial products that have practical value in our everyday lives.
- describe the broad variety of industrially important microbes and survey the processes by which they synthesize commercial materials.
- compare new and traditional directions in industrial microbiology and gain an appreciation of future trends.

During the 1950s and 1960s, biochemists made quantum leaps in microbiology and molecular biology as they clarified the role of DNA in gene expression and discovered how DNA influences heredity (Chapter 4). But the best was yet to come. In the 1970s, scientists set aside their tendency to observe how DNA acts and, instead, began to manipulate the process. In rapid succession, they devised methods to cut and splice DNA fragments to yield recombined molecules; they placed the recombined molecules into living cells; and they induced the cells to do their bidding. In so doing, they gave birth to the era of genetic engineering and biotechnology. (Many of the events leading to this era area described in Chapter 10.)

The advent of biotechnology brought a new frontier to explore, a frontier where the wildest imaginations could be accommodated. Researchers discovered, for example, that they could transplant animal genes into bacterial cells and coax the animal genes to encode their proteins in the bacterial cytoplasm. Their experiments kindled hopes that plants, noted for their ability to produce carbohydrates, could be transformed into protein producers that would help solve the worldwide problem of hunger. And, as we shall see, their research laid the foundations for entirely new genetic treatments for infectious diseases. On the horizon lay inexpensive sources of bioenergy, mass production of pharmaceuticals, and novel kinds of vaccines. There appeared to be no limit to what could be accomplished.

Today, much of what scientists predicted has come to pass. We have already witnessed how genetically engineered bacteria can produce such pharmaceutical products as insulin, human growth hormone, interferon, and blood clotting factors. Close to 50% of the soybean crop in the United States is planted with genetically engineered seeds. Many hundreds of individuals are now receiving gene therapy. Genetic and infectious diseases can be detected by DNA probes, and the project to map the entire human genome has been successful. The gains in genetic engineering and biotechnology have matched the prognostications of the most optimistic futurists. You and I stand at the threshold of this revolution, with the power to use it for the good of all humanity.

In this chapter, we shall examine the promises and tools of biotechnology in the context of the microbes that are its essential underpinnings. Some of the principles underlying the new technologies are complex, to be sure, but we must be able to comprehend them if we are to understand the advances being reported daily in the public media.

Microbes and Gene Technology

Every so often in scientific endeavor, a window opens—and, suddenly, the theoretical becomes possible, then inevitable. Discoveries emerge in rapid succession, and powerful new insights drive researchers to unimagined heights. With the advent of genetic engineering and biotechnology, such a window opened in microbiology, and scientists began to see impossible dreams become reality. For example, they have produced pigs that manufacture human hemoglobin; they have engineered rot-resistant tomatoes; and they have mapped each of the genes in the nucleus of a human cell.

The DNA molecule is central to the imaginative techniques of modern biotechnology. It is now possible to reproduce DNA in a test tube, fragment it, determine its composition, change its structure, exchange pieces of it, and map its genes. The principles learned from these extraordinary breakthroughs have been applied to the production of pharmaceutical products, the development of vaccines and therapeutic drugs, the diagnosis of infectious disease, the use of gene therapy, and the start of an agricultural revolution. At the foundation of all these advances is an understanding and control of DNA expression in the synthesis of protein, as Chapter 4 describes. This chapter extends Chapter 4 by illustrating the myriad products and uses of biotechnology. Extraordinary as these products and uses are, it must be remembered that they are only the first fruits of an industry whose potential is still too great to comprehend. Indeed, as early as the 1980s, writers described genetic engineering as "the most awesome skill acquired by man since the splitting of the atom."

Pharmaceutical Products

Since the 1980s, radical changes have been occurring in the treatment of such diseases as diabetes and hemophilia. Diabetes (FIGURE 14.1) usually results from the failure of the pancreas to produce enough insulin, while hemophilia is related to the liver's inability to synthesize an essential clotting factor. Both insulin and the clotting factor are proteins. While other sources can be used to replace what the body lacks, these can be expensive and can bring microbial contamination and allergic reactions.

Biotechnology has been used to develop safer modes of treatment for both diseases. To produce insulin, the genes that encode this protein are obtained from human cells and, in a test tube, are attached to the plasmids of *Escherichia coli*. Then a promoter site derived from the *lac* operon (Chapter 4) is placed next to the structural genes. The plasmids are further modified by adding a short segment of DNA between the promoter site and the structural genes. The new segment, called a signal sequence, directs *E. coli* cells to attach a signal peptide to one end of the insulin molecule. Consisting of 24 amino acids, the signal peptide encourages the bacterium to secrete its insulin to the surrounding environment, where the biotechnologist can collect it.

Now the plasmids are inserted in *E. coli* cells by the process of transformation (Chapter 10). As the bacteria grow and multiply, the human genes use the bacterial resources to produce human insulin. In July 1980, 17 volunteers at Guy's Hospital in London were the first to receive Humulin, the trade name for bacteria-derived human insulin. In fact, they were the first to receive any pharmaceutical product of biotechnology. A historic first step had been taken.

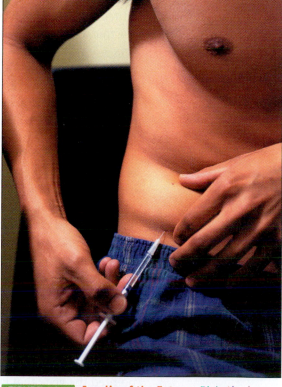

FIGURE 14.1   **Insulin of the Future.** Diabetics have traditionally used insulin derived from animal pancreatic tissue to prevent the development of disease symptoms. In generations ahead, the major source of insulin will be genetically altered bacteria.

Clotting Factor VIII is required in all humans to stimulate normal blood clotting. This factor is not produced in sufferers of hemophilia A, the most common inherited blood disorder in the United States (about 1 in 10,000 males is affected). To help these individuals, Factor VIII was historically obtained from units of whole human blood, but microbial contamination was a problem, and contamination with human immunodeficiency virus (HIV) was a particularly gruesome tragedy during the 1980s. Then biotechnologists successfully located on the X chromosome the human gene that encodes Factor VIII. The gene consists of 186,000 base pairs organized in 26 expressing units (exons). The scientists isolated the messenger RNA (mRNA) encoded by the gene and worked backward using the enzyme reverse transcriptase to synthesize complementary DNA molecule. This molecule is identical to the gene, but without the nonexpressing units (introns). Then they inserted the DNA in hamster kidney cells (because bacterial cells would be "confused" by the complexity of the gene), and to their delight, the gene was soon encoding Factor VIII protein in the hamster cells. By 1993, the Food and Drug Administration (FDA) had licensed the protein for use in patients with hemophilia.

In many individuals, the pituitary gland fails to secrete a sufficient amount of human growth hormone (HGH), a protein that stimulates body growth by encouraging protein synthesis. Those so affected suffer dwarfism, a disorder that can be treated by injections of HGH. Unfortunately, obtaining HGH has required isolating it from the pituitary glands of cadavers, a procedure that carries the risk of contamination. Biotechnologists can now produce HGH by splicing the HGH gene into a bacterial plasmid and forcing it into *E. coli* cells. There is a problem, however, because the HGH gene contains a signal peptide to which *E. coli* responds erratically. **FIGURE 14.2** illustrates how biotechnologists worked to circumvent this problem. Treatments with the synthetic hormone (known commercially as Protropin) encourage growth spurts that permit children to reach a normal height for their age. As you can imagine, the synthetic hormone can be abused by athletes dreaming of becoming NBA centers.

Among other new therapeutic products of biotechnology are tissue plasminogen activator (tPA). Used to dissolve blood clots in the coronary vessels of the heart, tPA is a protein-digesting enzyme that stimulates other body enzymes to break down a clot. In the early 1980s, biotechnologists synthesized the gene that encodes tPA using the gene's mRNA as a template (a model). Then they attached the synthetic gene to a bacterial plasmid containing a signal sequence, promoter site, and terminator site. The synthetic plasmids were added to animal cell cultures, and high levels of tPA were soon isolated from the cellular environment. In 1987, the FDA licensed tPA under the trade name Activase for use in persons displaying signs of heart attack or stroke. More than a few individuals owe their continued good health to this microbe-derived protein.

Interferon, another therapeutic product of biotechnology, is a group of over 20 proteins produced by various body cells after stimulation by viruses (Chapter 6). These proteins trigger a reaction that protects adjacent cells against infection by viruses, presumably by preventing their entry into the cells. In 1980, Swiss and Japanese biotechnologists pinpointed the gene that encodes one of the interferons (alpha-interferon) and spliced the gene into *E. coli* plasmids. The successful isolation of interferon from the bacterial cultures resulted in government approval for the use of various forms of interferon to treat leukemia, genital warts, and an AIDS-associated skin cancer called Kaposi's sarcoma.

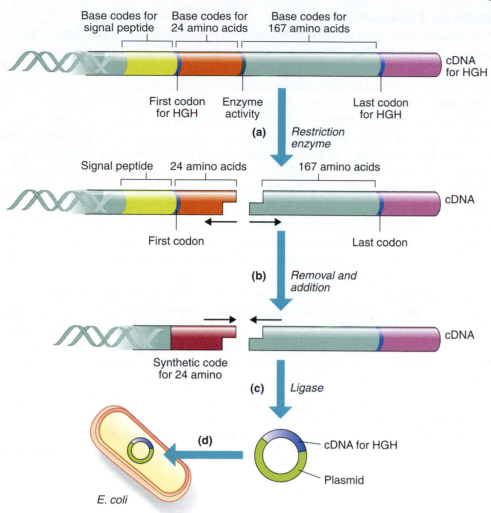

FIGURE 14.2   HGH Production. The production of human growth hormone (HGH) is a complex process that begins with a complementary DNA (cDNA) molecule. The bases encoding the signal peptide must be removed from the cDNA with a restriction enzyme; to do so, (**a**) the code for an additional 24 amino acids must also be deleted. (**b**) Then, a synthetic segment of DNA containing the code is added back, and (**c**) the gene is inserted into a plasmid for (**d**) insertion into a bacterium. Soon the bacterium will be producing HGH.

Among other genetically engineered pharmaceuticals is a synthetic version of the hormone erythropoietin, known commercially as Epogen (or "Epo"). This hormone is produced by genetically modified *E. coli* cells. Normally produced by the kidneys, the hormone stimulates stem cells in the bone marrow to quickly mature into red blood cells, thus increasing the blood's red cell count. (Unfortunately, some athletes use the synthetic hormone to increase their red blood cell counts and enhance the oxygen-carrying capacity of their bodies, a form of "blood-doping.") The hormone is yet another product of innovative research in biotechnology. In the next section, we examine a series of drug molecules that could not have been imagined before the advent of biotechnology.

Antisense Molecules

As success piled on success, the public soon became accustomed to exhilarating break-throughs in biotechnology. They were hardly prepared, however, for the antisense mol-ecules. An antisense molecule is a type of genetic projectile that enters an infected cell and combines with and destroys its complementary messenger RNA molecule, thereby interfering with protein synthesis. This interference can be used to inhibit cancer, in-terrupt genetic disease, or stop viral replication such as that associated with AIDS.

Molecular biologists begin the synthesis of antisense molecules by obtaining the mRNA molecule used in a protein's production. (The poly-A tail is often used as an identifier of such a molecule, as Chapter 4 notes.) Then they put the enzyme RNA polymerase to work, combining it with nucleotides and other substances to synthe-size a complementary mRNA molecule; that is, an mRNA molecule with bases com-plementary to those of the original mRNA. The original mRNA has "sense" because it encodes a protein; the new, complementary mRNA molecule has "antisense" be-cause its base sequence complements (is the opposite of) that of the sense molecule.

Soon scientists realized that they could use antisense molecules to treat infectious disease. Consider this: When a person has HIV infection, DNA-containing proviruses on a chromosome encode mRNA molecules for use in replicating enzymes and virus parts (as Chapter 6 describes). Antisense molecules enter the infected cell and unite with the mRNA molecules, thereby rendering them useless, as FIGURE 14.3 depicts. Without the mRNA molecules, the proviruses cannot encode the proteins for making more of themselves. The replication cycle will soon come to an end. Researchers remain optimistic that the molecules will usher in a new generation of antiviral treatments.

Antisense molecules could also be used to interrupt the activity of oncogenes, the stretches of DNA that cause a normal cell to become a cancer cell (Chapter 6). In the 1990s, for example, researchers used antisense molecules to block the expression of

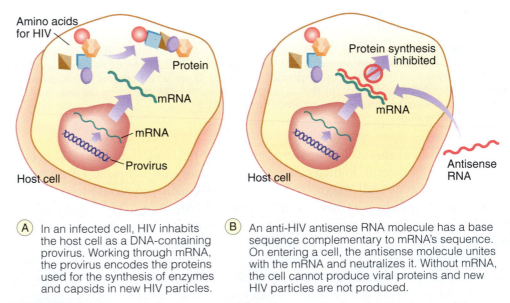

(A) In an infected cell, HIV inhabits the host cell as a DNA-containing provirus. Working through mRNA, the provirus encodes the proteins used for the synthesis of enzymes and capsids in new HIV particles.

(B) An anti-HIV antisense RNA molecule has a base sequence complementary to mRNA's sequence. On entering a cell, the antisense molecule unites with the mRNA and neutralizes it. Without mRNA, the cell cannot produce viral proteins and new HIV particles are not produced.

FIGURE 14.3   Antisense Technology and AIDS. Antisense molecules are being developed that can interrupt the replication of the human immunodeficiency virus (HIV) in AIDS patients.

an oncogene found in cells from the human larynx. Scientists spliced the DNA that encodes the antisense molecule into those cells, and soon they produced antisense molecules. When the antisense molecules reacted with the sense molecules, the oncogene's activity was impeded, and the cancer cells failed to multiply further.

Another way of producing an antisense molecule is to activate the strand of DNA that normally remains inactive during protein synthesis, a type of DNA role-reversal. Biotechnologists are now able to splice a promoter site at the beginning of the DNA strand they wish to activate. The normally inactive DNA strand encodes a natural antisense molecule that is secreted and taken into neighboring cells, where it blocks the function of a normal mRNA molecule. Such a process is being tested to treat patients with Crohns disease, an inflammation of the intestinal tissues due to an unusual protein produced in the affected cells. If the protein production can be turned off via an antisense molecule, then the inflammation can be reduced.

Research on antisense molecules entered the practical arena in 1998, when fomivirsen, an antisense medication, was approved by the FDA for treating eye infections due to cytomegalovirus (Chapter 18). Eye infections by this virus are a prelude to blindness in AIDS patients, and the new medication was greeted as a welcome application of molecular genetics (actually, "reverse genetics"). Finally, one observer wrote, "We are making sense of antisense."

Pharmaceutical products such as antisense molecules portend a new wave of therapies in the effort to control disease. But billions of dollars could be saved if the diseases were prevented in the first place. Traditional vaccines have gone a long way toward accomplishing this saving, and the new biotechnology-based vaccines may bring on a renaissance of vaccine technology, as we examine in the next section.

Biotech Vaccines

Vaccines have traditionally consisted of whole microbes, inactivated or attenuated (weakened) viruses, or microbial fragments. These vaccine components stimulate the immune system to produce antibodies that protect the body against future attack by the microbes. Although many vaccines have been very successful (the small pox and polio vaccines are notable in this regard), some carry an element of risk because they are accompanied by side effects. On rare occasions, some even cause the disease they are meant to protect against. Biotechnologists have therefore continued to search for new ways of producing vaccines.

One successful product of their search is the vaccine for hepatitis B. The key element of this vaccine is a protein in the surface capsid (coat) of the virus. Biotechnologists have identified the gene that encodes the protein and have cloned it and inserted it into plasmids of the harmless yeast *Saccharomyces cerevisiae*. Scientists have also spliced in promoter genes and terminator sites, as well as a gene that permits the yeast cells to grow only in the absence of the amino acid leucine. Those that grow in the leucine-deficient medium are the ones that can produce the viral capsid proteins—enough of the proteins for vaccine use.

Saccharomyces cerevisiae sak-ä-rō-mī'sēs se-ri-vis'ē-ī

In 1987, the remarkable new vaccine for hepatitis B was licensed by the FDA. Currently marketed as Recombivax by one company and Engerix-B by another, this vaccine is used to immunize millions of health care workers worldwide. A far cry from traditional vaccines, the new vaccine contains no microbes of any sort, and it is useful for anyone coming in contact with blood or body secretions (for example, police officers, firefighters, emergency services personnel, and medical laboratory workers). Many pediatricians recommend it for newborns.

Another biotech vaccine of the future will be one to protect against AIDS. Researchers have identified and cloned the genes that encode two of the major glycoproteins found in the envelope of HIV: glycoprotein 120 (gp120) and glycoprotein 41 (gp41). The genes encoding these glycoproteins have been successfully inserted in *Escherichia coli* cells, and the bacteria have responded by producing usable amounts of gp120 and gp41. Purified and injected into volunteers, the vaccine of gp120 and gp41 proteins elicits antibodies that bind to the gp120 and gp41 molecules on the surface of the HIV envelope and thereby prevent the virus from binding to its host T cells. FIGURE 14.4 demonstrates this action. The vaccine works in laboratory animals, but numerous other problems must be resolved before it becomes part of the anti-AIDS regimen.

Another possible carrier for a biotech vaccine is a bacterium. Biotechnologists have cultivated a harmless strain of the tubercle bacillus that carries a gene from the parasite *Leishmania tropica* (a protozoal parasite that causes a blood disease of the in-

■ *Leishmania tropica*
lish'mä-nē-ä trop'i-kä

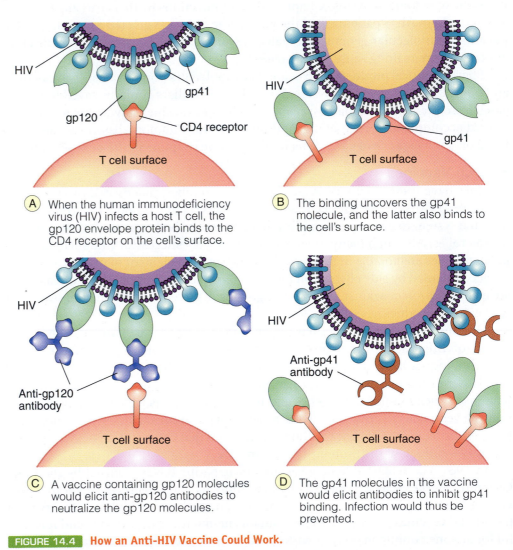

A When the human immunodeficiency virus (HIV) infects a host T cell, the gp120 envelope protein binds to the CD4 receptor on the cell's surface.

B The binding uncovers the gp41 molecule, and the latter also binds to the cell's surface.

C A vaccine containing gp120 molecules would elicit anti-gp120 antibodies to neutralize the gp120 molecules.

D The gp41 molecules in the vaccine would elicit antibodies to inhibit gp41 binding. Infection would thus be prevented.

FIGURE 14.4   **How an Anti-HIV Vaccine Could Work.**

ternal organs). The bacterium displays the parasite gene on its surface, and when the bacterium is injected into a patient, the gene induces an immune response. Because *Leishmania* parasites affect millions of people worldwide, public health workers are encouraged by the early success of this research.

A vaccine of a totally different type can be prepared by altering the genome of the pathogenic organism itself. For instance, *Vibrio cholerae*, the bacterium that causes cholera, can be altered by removing from its chromosome certain genes that encode its toxic protein. This toxin causes the body to release large amounts of salts, and the unrelenting diarrhea of cholera follows. When reengineered, *V. cholerae* cells are unable to produce the protein, but they can be used as a vaccine to stimulate an immune response. Thus, the pathogen is turned against itself through the innovative use of biotechnology. The ingenuity of gene engineers is further displayed by the use of vegetables and fruits as vaccine carriers (these "veggie vaccines" are described in detail in Chapter 15).

Even though millions of people worldwide suffer debilitating and even fatal infections related to fungi, public health agencies do not have a single fungal vaccine available. (Part of the reason is that fungi are difficult to manipulate genetically because their DNA is contained within a nucleus.) That may change in the future. In 2000, researchers at the University of Wisconsin announced the development of a genetically crippled strain of *Blastomyces dermatitidis*, the fungus that causes serious respiratory disease in immunocompromised individuals. The scientists disabled the gene that helps the fungus attach to host tissues. When the crippled fungus was injected into laboratory animals, it elicited a strong antibody response. This research effort laid the groundwork for the development of vaccines against more common fungi, such as those responsible for athlete's foot and candidiasis ("yeast infection").

■ *Blastomyces dermatitidis* blas-tō-mī′sēz dėr-mä-tit′i-dis

DNA Vaccines

Part of the renaissance in vaccine technology are the DNA vaccines. These vaccines consist of plasmids engineered to contain a protein-encoding gene, as A Closer Look on page 308 discusses. Unlike replicating viruses or live bacteria, plasmids offer a measure of safety because they are not infectious or replicative, nor do they encode any proteins other than those specified by their genes. However, a DNA vaccine is difficult to make because the plasmid must have numerous features (in addition, of course, to the vaccine gene) to be effective in its new cell. Among these are a promoter site, a convenient site for inserting the vaccine gene, a polyadenine (poly-A) tail as a terminator sequence, an origin of replication (Chapter 10), and a marker gene so that biotechnologists can follow the uptake of the vaccine gene.

One advantage of DNA vaccines is that no special formulation is necessary; that is, animals can be immunized simply by injecting plasmids suspended in saline (salt) solution. Furthermore, delivery of the plasmids to target cells can be accomplished by various means: by nasal spray; by injection into the muscle, vein, or skin; or by a so-called gene gun (a propulsion device that shoots DNA-coated gold beads into the skin). Deploying plasmids as DNA vaccines has the added advantage of stimulating both antibody-mediated immunity (AMI) and cell-mediated immunity (CMI), because DNA vaccines encode proteins that are released from the cell (for AMI) and proteins that fix themselves to the target cell surface (for CMI). These forms of immunity are discussed at length in Chapter 17. There is also hope that DNA vaccines

A CLOSER LOOK
A Happening

It was another of those remarkable moments in science, an unexpected observation that opened the door to a whole new type of vaccine. It happened in 1989 at Vical Inc., a California biotechnology company. Biochemist Philip Felgner and his research group were experimenting with plasmids, the ultramicroscopic ringlets of cytoplasmic DNA that carry many non-essential genes in bacteria. Felgner wanted to learn whether the plasmids could carry genes into a mouse if they were wrapped in lipid-containing bodies called liposomes. He could hardly imagine what he was about to discover.

Felgner's protocol was simple. Some plasmids were packaged in the tiny, spiral liposomes, then injected into the muscle of a mouse. As a control, some plasmids were left unpackaged and injected into another mouse. The latter plasmids, by all expectations, should have remained inert or been destroyed.

But they were not destroyed; nor did they remain inert. Instead, the cells receiving the naked plasmids began synthesizing the protein encoded by those plasmids. Somehow, the plasmids had remained intact, found the necessary biochemical machinery, and induced the cell to begin making protein. And, adding to the wild results, the proteins were stimulating the mouse's immune system to produce antibodies. The realization slowly dawned on Felgner and his group: They had discovered a new way to immunize an animal; they had produced a DNA vaccine.

Scientists are constantly taught never to anticipate their results when performing an experiment. The trick is to formulate a hypothesis, devise a reasonable set of experimental conditions, turn on the juice, and observe nature's truths. Usually the process is slow and plodding. But every now and then, an astonishing outcome makes all the dreary days worthwhile. It has happened innumerable times in science, and it happened once again in 1989 in San Diego.

Oh, by the way, the plasmids packaged in liposomes also encoded protein.

can be used to immunize infants still in the womb. In 2000, for example, researchers at the University of Saskatchewan treated fetal lambs with a DNA vaccine containing herpesvirus genes and found that the vaccine induced antibody buildup against herpesviruses present in the ewe.

DNA vaccines also appear to elicit a strong immune response, possibly because of other genes in the plasmid, and they are more stable than conventional vaccines at low and high temperatures (which makes shipping easier). DNA vaccines have been used experimentally to protect against such human diseases as influenza, salmonellosis, HIV infection, herpes simplex, and hepatitis B. Before the age of biotechnology, such vaccines were not possible. Neither were the wealth of new diagnostic procedures and detection methods we examine next.

DNA Analyses and Microbial Detection

One of the more remarkable applications of biotechnology occurs in the diagnostic laboratory when DNA samples are analyzed to determine whether target genes and/or DNA segments are present. These applications make it possible for scientists to identify microbes directly (instead of spending long hours cultivating them or looking for their telltale antibodies). Furthermore, these methods allow the verification of numerous inherited diseases whose presence could previously be predicted only by guesswork.

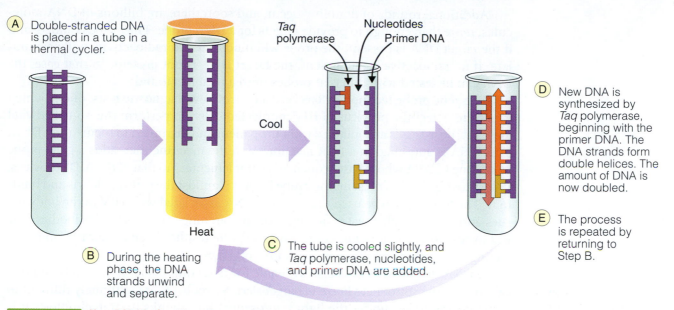

A Double-stranded DNA is placed in a tube in a thermal cycler.

Heat

B During the heating phase, the DNA strands unwind and separate.

Cool

Taq polymerase   Nucleotides   Primer DNA

C The tube is cooled slightly, and *Taq* polymerase, nucleotides, and primer DNA are added.

D New DNA is synthesized by *Taq* polymerase, beginning with the primer DNA. The DNA strands form double helices. The amount of DNA is now doubled.

E The process is repeated by returning to Step B.

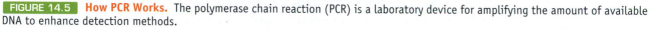

FIGURE 14.5 **How PCR Works.** The polymerase chain reaction (PCR) is a laboratory device for amplifying the amount of available DNA to enhance detection methods.

Two essential elements in DNA analyses are the DNA probe and the polymerase chain reaction. The DNA probe (also called the gene probe) is a synthetic fragment of single-stranded DNA that hunts down a complementary DNA fragment in a mass of cellular material and signals when it has located that fragment. Developed in the 1980s, the DNA probe reacts specifically with a target DNA fragment having complementary base pairs, much like a left hand matches up with its complementary right hand. As millions of the two kinds of DNA molecules bind, a radioactive substance tagged to the probe accumulates and signals that binding has taken place. To construct a DNA probe, the molecular biologist isolates the mRNA encoded by the target DNA; this mRNA is then used with the enzyme reverse transcriptase (to be discussed presently) to synthesize a complementary fragment of DNA. This fragment is the DNA probe.

In order to use DNA probes effectively, billions of DNA molecules must be available. Making these DNA molecules is the job of the polymerase chain reaction (PCR). In a highly sophisticated apparatus (often described as a molecular equivalent of a photocopier), the PCR takes a strand of DNA and multiplies it billions of times. To begin the process, a sample of nuclear material (DNA) is collected. In the PCR machine, the double helix is unwound at a high temperature to yield single strands of DNA. Then a special, heat-resistant DNA polymerase enzyme called *Taq* polymerase is added. (The enzyme is obtained from the thermophilic bacterium *Thermus aquaticus.*) Now the machine mixes in nucleotide molecules containing the four bases (adenine, thymine, guanine, and cytosine) together with a strand of primer DNA—the primer recognizes a complementary section of the sample DNA, binds to it, and serves as a starting point for the elongation. The *Taq* polymerase brings together the nucleotides to form a strand of DNA that complements the single-stranded DNA, adding the nucleotides onto the primer DNA one at a time. Next, the mixture is cooled, whereupon the new and old strands of DNA twist together to form double-stranded DNA molecules. At this point, the original number of DNA molecules has doubled, as **FIGURE 14.5** illustrates.

Additional rounds of doubling occur, and soon there are billions of DNA molecules, more than enough to provide targets for a DNA probe. The probe is added, and if the target DNA is present, the probe will bind to it and radioactivity will accumulate. If no radioactive signal is sent, the target DNA is not present. In that case, the DNA can be tested with different probes until a match is found.

The gene probe technology has bred a host of new diagnostic tests, one of which is used to detect the provirus of HIV in infected cells. To perform this so-called viral load test, host T cells are obtained from the patient and disrupted to secure their DNA. This DNA is amplified in a PCR machine, and the machine then adds a gene probe consisting of DNA whose base sequence is complementary to that of the AIDS provirus. If the proviral DNA is present, the probe locates it amidst the cellular debris and binds to it. The test can also be used to detect the RNA associated with HIV in the body tissues and blood. Such a test is far superior to searching for anti-HIV antibodies (the ELISA test) because it is a direct test for HIV. The test has been a major boon in detecting HIV early and following the course of AIDS.

A DNA diagnostic test is also available for detecting Lyme disease, a blood disease due to the spirochete *Borrelia burgdorferi*. Spirochetes are extremely difficult to cultivate and to see under the light microscope. But, thanks to biotechnology, it is possible to secure blood from a patient, amplify the spirochetal DNA (if present), and then use a DNA probe to detect it. A DNA diagnostic test for human papilloma virus (HPV) is also in use. HPV causes genital warts in humans, and possibly cervical cancers, so early diagnosis and treatment are desirable. The new technologies make this possible and reduce the misery caused by these diseases.

A slightly different DNA diagnostic procedure has been used to identify the cause of a viral epidemic even before the virus was isolated or cultivated from patients. In 1994, an epidemic of viral disease occurred in the southwestern United States, where four states come together (Arizona, Colorado, New Mexico, and Utah). Named the four-corners disease, the disease spread among Native Americans, who suffered hemorrhaging, lung infection, and kidney problems (FIGURE 14.6). To detect the responsible virus, scientists painstakingly mixed patient antibodies with numerous types of laboratory viruses, one at a time, until a rare virus known as the hantavirus was pinpointed. Then, to develop a diagnostic test, they obtained cultures of hantaviruses and produced large quantities of its DNA in a PCR machine. Next, they synthesized a DNA probe using the hantaviral DNA as a template. Finally, to identify the disease in a patient, they took diseased tissue and used the probe to search for matching DNA. When a match was identified, they began preventive treatment, while continuing to chart the course of the epidemic. Soon afterward, a DNA probe was used to solve another microbial mystery, as A Closer Look on page 311 explains.

DNA analysis can also be used to help public health officials "connect the dots" when an epidemic surfaces. In 1998, for example, a cluster of sick people in Los Angeles became infected with *Shigella sonnei*, a Gram-negative rod that causes serious intestinal disturbances. Simultaneously,

■ *Borrelia burgdorferi*
bôr-rel′ē-ä burg-dôr′fē-ē

FIGURE 14.6   **Four-Corners Disease.** In 1994, an outbreak of hantavirus disease occurred in the region of the southwestern United States where four states intersect. Using the latest biotechnology methods, researchers were able to detect the responsible virus in patients even before the agent was isolated.

another cluster of *Shigella*-infected patients was reported in Minnesota, thousands of miles away. Researchers performed analyses with gene probes on the bacteria and found that their DNAs were virtually identical. Health officials then began to look beyond the local areas for a common source, and they ultimately traced the origin of the outbreaks to infected parsley imported from Mexico. The parsley had become contaminated with water used for irrigation purposes.

DNA analyses with gene probes can also be used to track the source of bacteria used in bioterrorism. For example, in 1985, a DNA analysis was used to link *Salmonella* found in the laboratories of the Rajneeshees to the *Salmonella* that infected patients at salad bars in Oregon during an incident of bioterrorism (Chapter 13). And in 2001, the bioterrorist attacks with anthrax were studied by doing DNA analyses of the anthrax bacilli found in letters sent to public officials and media personalities. By comparing the DNA profiles, public health microbiologists were able to ascertain that the source of the anthrax was probably the same for all the attacks and that the bacilli most likely came from a laboratory in the United States.

In the genetics laboratory, DNA analyses have great value in detecting the potential for inherited diseases. For instance, a DNA probe test is used to determine whether a person carries the gene associated with cystic fibrosis. Cystic fibrosis is an abnormality in which thick, sticky mucus clogs the respiratory passageways and tissues. Normally, a protein carries various ions out of the respiratory tract cells. But in patients with cystic fibrosis, a defective gene encodes a malformed protein, and ions remain in the cells. As ions concentrate, water accumulates as well, leaving little water in the passageway, and a sticky mucus results. A DNA probe can now detect the defective gene and help identify individuals who might wish to begin lifestyle changes to lessen the severity of the disease. DNA analyses are also used to detect such genetic diseases as Duchenne's muscular dystrophy, Huntington's disease, and familial Alzheimer's disease.

A CLOSER LOOK
"Not Guilty"

Poor Christopher Columbus! Historians have accused you of bringing smallpox to the New World—and measles and whooping cough and tuberculosis and almost every other infectious disease. One can almost imagine that the *Santa Maria* was a hospital ship!

Well, rest easy, Chris, for scientists have cleared you of bringing at least one disease—tuberculosis. Your defense is based on 1995 research by Arthur Aufderheide (pronounced OFF-der-hide) from the University of Minnesota. Some years before, Aufderheide was studying the remains of a mummified woman from Peru when he noticed in her lung tissues several lumps reminiscent of tuberculosis. He enlisted the help of a molecular biologist, who extracted DNA from the lumps and amplified it so that there was enough DNA to identify. The DNA turned out to be identical to that of *Mycobacterium tuberculosis*, the tubercle bacillus.

Why was that important? Well, Chris, the mummy was a thousand years old—that's right, one thousand years. Apparently both the woman and the tubercle bacilli were already here hundreds of years before you arrived. It's even possible you might have taken some back with you to Europe. . . . Oops! Sorry!

DNA Fingerprinting

A form of DNA analysis called DNA fingerprinting has become a valuable investigative technique in forensic medicine. This technology derives from observations reported in 1985 by the English geneticist Alec J. Jeffreys. Jeffreys noted that between the genes of a chromosome there exist stretches of DNA whose base sequences are repeated many times over. For example, in some individuals, a particular base sequence is repeated 2 times, while in other individuals the sequence is repeated 10 times, and in other individuals 200 times. The nature of the bases in the sequence is not as important as the number of times the sequence is repeated because the number of repeats determines the length of the sequence, which is of primary importance in the analysis. These fragments are known as variable-number tandem repeats (VNTRs, pronounced "vinters"). Scientists have found that the pattern of VNTRs for an individual is unique. This pattern is the basis for the DNA fingerprint.

To develop a DNA fingerprint, the technologist obtains a tissue sample (such as a few blood cells or a sample of semen). The DNA is extracted from the cells by standard laboratory techniques, and restriction enzymes break the DNA into segments by reacting at points flanking the VNTRs. The VNTRs are separated by a process called electrophoresis, and DNA probes carrying radioactive markers are used to determine the pattern of VNTRs. The pattern is somewhat like a bar code used to price goods in the supermarket.

DNA fingerprinting is often performed with DNA samples obtained from a crime scene and from a suspect. By comparing the patterns of VNTRs, a DNA technologist can place the suspect at the crime scene or eliminate him or her. Indeed, the technologist can announce in court that the chances of a crime scene's DNA matching a suspect's DNA are 999,999 in 1,000,000. An analysis of this sort can also be used to match any two people's DNA. For example, DNA obtained from relatives was used to help identify victims of the terrorist attack on the World Trade Center in September 2001.

New methods of diagnostic medicine and DNA analyses have helped improve the lives of tens of thousands of individuals, while providing medical detectives with new devices for seeking the truth. But many long hours have been spent bringing the technologies to practical use, as we explore in the pages to follow.

The Methods of Biotechnology

On the surface, the methods of genetic engineering and biotechnology appear so simple that one might wonder why it took so long for scientists to develop them. It appears, for example, that to manufacture a human protein all one has to do is identify the gene, insert it into a bacterium, and stand back while the bacterium does its thing. In our naiveté, we might suggest numerous other applications of genetic engineering, and we might be inclined to ask "Why don't they just . . .?"

But scientists working in the world of genetic engineering and biotechnology must apply a highly sophisticated blend of biology, chemistry, physics, and mathematics. The technology is available to those who have the educational background to use it and the monetary support to make it work. Genetic engineering and biotechnology are among the more elegant endeavors of science, and the processes are far more complex than the explanations we have considered might make them appear.

One of the first problems is obtaining the correct gene for use in genetic engineering. For example, if the objective is to reengineer a bacterium to produce human in-

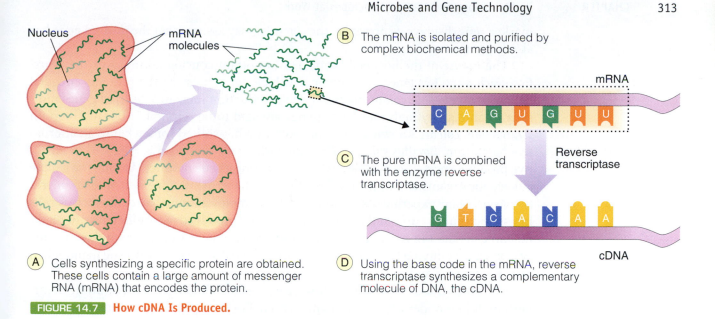

A Cells synthesizing a specific protein are obtained. These cells contain a large amount of messenger RNA (mRNA) that encodes the protein.

B The mRNA is isolated and purified by complex biochemical methods.

C The pure mRNA is combined with the enzyme reverse transcriptase.

D Using the base code in the mRNA, reverse transcriptase synthesizes a complementary molecule of DNA, the cDNA.

FIGURE 14.7   How cDNA Is Produced.

sulin, the gene that encodes insulin must first be isolated. However, the insulin gene would be extraordinarily difficult to locate among the tens of thousands of human genes. This problem may be resolved by hunting for insulin-related mRNA, which is expected to be in abundance in insulin-producing cells such as pancreas cells. To isolate this mRNA, the pancreatic cells are disrupted and subjected to exhaustive chemical and physical treatments to exclude as much extraneous material as possible. Then, the mRNA molecules can be obtained by taking advantage of its poly-A tail (the chain of 150 to 200 nucleotides that all contain the base adenine). To do so, fragments of DNA molecules containing only thymine are bound to cellulose particles and mixed with material from pancreas cells. Because adenine and thymine are complementary bases, the poly-A tail of the mRNA molecules binds to the poly-T molecules on the particles, and the particles can be separated from the mixture. Then, the mRNA molecules can be collected from the cellulose particles and concentrated. Those mRNA molecules encoding insulin are separated from the rest.

Once the correct mRNA molecules have been purified, a molecule of DNA is synthesized by using the enzyme reverse transcriptase. Reverse transcriptase uses mRNA as a template and synthesizes a molecule of complementary DNA (cDNA) as FIGURE 14.7 illustrates. Then, to produce a double-stranded DNA molecule, the enzyme DNA polymerase is employed. DNA polymerase uses the cDNA molecule as a template to synthesize a complementary DNA molecule. The two strands bind and twist into a double helix, thus yielding the gene that encodes insulin production.

Once the gene is obtained, it is inserted into a vector, such as the plasmid used by Cohen (Chapter 10). Among the other useful vectors are retroviruses. These are RNA-containing viruses that enter host cells and encode DNA, which attaches to a chromosome and remains there permanently. A useful vector should have an identifying marker gene, so that it can be located in host cells following insertion. A gene conferring antibiotic resistance is a suitable marker because cells having the marker will live in the presence of an antibiotic, while cells lacking the marker will die. If the marker gene displays its presence, the scientist may assume that the desired gene has also been inserted.

The nature of the host cell is another important consideration. Over the years, *Escherichia coli* has emerged as a major vehicle for expressing the proteins encoded by inserted genes. *E. coli* displays a high reproduction rate, and inserted genes multiply at an equally high rate. (The genes are said to be "cloned.") Because certain strains of *E. coli* are known to be toxic, attention has shifted somewhat to an alternative bacterium, *Bacillus subtilis*. This nonpathogenic organism is regarded as an active protein exporter, a valuable asset for collecting a gene product. Among the simple eukaryotic organisms, the host cell of choice is the fermentation yeast *Saccharomyces cerevisiae*. Many experiments in biotechnology employ mammalian cells, such as cells in tissue cultures. Large and complex proteins are produced more efficiently in mammalian cells than in bacterial cells because bacterial cells lack the complex enzyme systems for modifying the protein into its final form. However, cultivating mammalian cells is a tedious and expensive task.

Once the gene is inside suitable host cells, the biotechnologist must be aware of certain conditions under which it can express itself. For example, the gene must have available a promoter site, a termination site, and a ribosomal binding site. Another consideration is the presence of introns and exons, as we discuss in Chapter 4. Bacteria do not excise the non-functioning introns; therefore, if a human gene were to be expressed by a bacterium, there would be no mechanism for removing the introns. This problem can be circumvented by using complementary DNA (cDNA), a gene that has no introns.

Collecting the gene product also presents challenges because the protein may be secreted by the host cell (which is preferable) or retained in the cytoplasm. To enhance secretion, biochemists attach stretches of bases called signal sequences to the vector. The protein encoded by the signal sequence will attach to the gene product and usher it out of the cell. Using signal sequences is another bit of biochemical magic developed by genetic engineers during the past decades. The problem of modifying the final protein must also be addressed because protein production rarely comes to an end with amino acid connection. A mix of intuition, insight, and luck is needed to circumvent these and numerous other problems that confront biotechnologists almost daily.

You may conclude that our discussion of microbes, DNA, and biotechnology has come to an end. Not by a long shot! Numerous other examples of microbial accomplishments are sprinkled throughout this book, especially in Chapters 15 and 16, where we discuss the extraordinary microbial-related advances occurring in agriculture and environmental science. For the time being, we shall shift our attention to the more traditional uses of microbes in industrial processes.

Microbes and Industry

The uses of microbes in biotechnology are exciting, novel, and imaginative. But we hasten to point out that these are not the first uses of microbes for human welfare. Far to the contrary, microbes were used to benefit society well before people realized that they were taking advantage of these tiny organisms. Consider these examples:

Modern linen manufacturers use pectinase, an enzyme from a *Clostridium* species, to macerate flax plants and separate their fibers, a process called retting. Pectinase decomposes the pectin, a cementlike protein holding the cellulose fibers together in flax plants. After separation, the cellulose fibers are spun into linen. In past generations, the process was much different: The retting process began by bundling flax plants and drying them in stacks. Then the stacks were placed in a long trench several feet deep, covered with water, and weighted down with stones to exclude as much air as possible. After a few days, the water turned black, and an unmistakable stench signaled that retting was taking place. Two weeks later, the flax was so soft and pliable that the fibers could be easily removed by pounding with wooden blocks. Unknown to the ancient industrialists, *Clostridium* species from the soil were producing pectinase in the trenches.

The old method for cleaning (or "bating") hides was equally messy: Animal skins were mixed with dog or fowl manure and set aside to cure in a warm, damp place. Fragments of tissue and hair gradually dissolved in the muck, and soon the hide had lost its stiffness. Modern microbiologists point out that protease enzymes from fecal bacteria were responsible for bating the hides. It was another smelly process, to be sure, but like the method for retting, it was reliable.

Now we fast-forward to the modern era, where microbes are still performing valuable services for society. A major difference is the format of the industrial plant, a highly mechanized and finely tuned factory in which microbes are the key workers. Microbes synthesize the broad variety of enzymes to catalyze an array of chemical conversions. In addition, they have a relatively high metabolic rate that allows rapid conversions; they have large surface areas that permit quick absorption of nutrients and rapid release of end products; and they multiply at a high rate, keeping the "workforce" young and active.

But, numerous problems must be confronted before the chemistry of the test tube can be scaled up to an industrial fermenter containing 100,000 gallons of fluid (FIGURE 14.8 shows a small example). For example, heat, humidity, and nutrient levels must be controlled and waste products must be removed regularly. Stock cultures, generally the trade secrets of industrial corporations, must be carefully maintained throughout, and microbial media must be strictly regulated. In some cases, the desired product is retained within the cells, which means that the microbiologists must find ways to rupture the cells and recover the product. (Indeed, these same problems need to be resolved by bioterrorists wishing to produce industrial quantities of microbes or their toxins for use as bioweapons.)

Nevertheless, with time, experience, and intellect, industrial microbiologists have brought forth a range of industrial products that we use on a daily basis. Many products discussed in other chapters (such as fermented foods, pharmaceutical products, laboratory test reagents, and dairy products) result from industrial processes. To them, we add some other important products, discussed in the upcoming section.

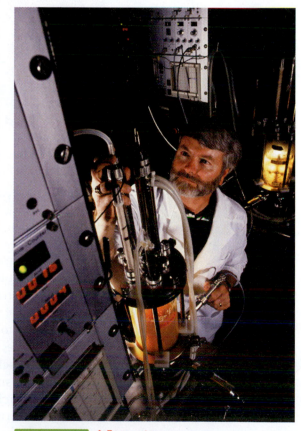

FIGURE 14.8   **A Fermenter.**

(a)

(b)

(c)

(d)

FIGURE 14.9 **An Array of Products Containing Enzymes Produced Industrially by Microbes.** (**a**)Household cleaning products such as spot remover and laundry pre-soaker. (**b**) Stone-washed jeans. (**c**) Fruit juice. (**d**) Chocolate covered cherries.

Industrial Enzymes

Enzymes are well known as the biological catalysts that bring about metabolic reactions, while themselves remaining unchanged. Over the decades, microbiologists have found that enzymes can be extracted from microbes and put to practical uses (FIGURE 14.9). For example, an enzyme from the fungus *Trichoderma* breaks down the plant cellulose in jeans, giving them a stone-washed appearance. In 2001, the market value of all industrially produced enzymes was estimated to be $1.4 to $1.6 billion.

Among the most useful industrial enzymes is amylase. This enzyme breaks down starch molecules into smaller sugars, including the disaccharide maltose. Several species of microbes, including *Bacillus subtilis* and *Aspergillus niger*, produce large amounts of amylase during their normal cycles of growth. Bakers add the amylase to dough to promote the breakdown of starch to sugar, after which the sugar is used by

■ *Trichoderma*
trik'-ō-dėr-ma

yeasts. Moreover, amylase assists the digestion of starch at the beginning of the process of beer production. The enzyme can also be used to produce gasohol, with corn starch as the starting material; and many spot removers contain amylase to breakdown the starches in plant material that soils clothing.

The protein-digesting enzymes of industrial significance are epitomized by the proteases. Proteases digest proteins into smaller peptides and amino acids. In the baking industry, proteases encourage the breakdown of gluten proteins in flour to simpler proteins, thereby increasing the nutritional value of the bread. Proteases are also used in the tenderizers sprinkled on meat before cooking. Here they break down the protein fibers and release the juices. In laundry products, proteases are used as spot removers for anything that contains protein, such as egg, blood, and milk. And, as we noted earlier, proteases help remove the organic matter from leather during the bating process.

Rennin is an enzyme used to curdle the proteins in milk, an early step in the manufacture of cheese. Historically obtained from the stomach lining of a calf, rennin is now produced by microbes grown in industrial plants. And for breaking down compounds in aromatic ("strong-smelling") vegetables, the fungus *Aspergillus niger* produces galactosidase. This enzyme is the major ingredient in Beano®, a commercial product discussed in Chapter 7.

■ *Aspergillus niger*
a-spėr-jil′lus nī′jėr

Other useful microbial enzymes include pectinase, which breaks down the pectins in plant tissues (as in retting flax); it is also used to clarify fruit juice by digesting the pectins that cloud it. Streptokinase, obtained from species of streptococci, finds valuable use as a clot-dissolving enzyme for patients with coronary blockages; and invertase, obtained from yeast cultures, acts on sucrose and breaks it down to glucose and fructose, as in chocolate-covered cherries (Chapter 12).

Organic Acids and Vitamins

A quick check of various soft drink labels will attest, citric acid is one of the most widely encountered acids found in consumable items. Hundreds of thousands of tons of citric acid are produced annually in the United States (the 1999 market value of citric acid production was $1.5 billion), and this organic compound is used in such diverse products as soft drinks, candies, frozen fruits, and wines. Citric acid is also used to tan leather, to electroplate metals, and to activate slow-flowing oil wells (these wells are often clogged with iron deposits, which can be broken down by pumping citric acid into the hole). Most citric acid is produced by the normal metabolism of the fungus *Aspergillus niger*. The citric acid is excreted because the fungus lacks the enzyme to break it down further through the Krebs cycle (Chapter 9). Thus, the fungus unwittingly becomes a partner in an industrial process.

Gluconic acid is an equally important acid. Also produced by *Aspergillus niger*, it is employed in factories to cleanse bottles before filling them with beer, milk, or soft drinks. Gluconic acid also finds its way into dishwashing detergents and soaps, where it prevents mineral deposits that lead to "bathtub ring." Metal cleaners contain gluconic acid, as do many foods such as baking powder, cheeses, and sausages. Moreover, gluconic acid is combined with calcium to form calcium gluconate, which is used to transport calcium into the human body. (Calcium gluconate is also added to the feed of laying hens to provide calcium to strengthen the eggshells.) Production of gluconic acid remains a multimillion-dollar industry.

Still another microbial product is lactic acid. Several *Lactobacillus* species produce this acid from the whey portion of milk derived from cheese production. Lactic acid is used as a flavoring and preservative agent in many foods, as many labels attest. It is also employed to finish fabrics, prepare hides for leather, and dissolve lacquers.

Among the widely used amino acids produced by microbes is lysine. Lysine, a product of *Corynebacterium* species, is used as a food additive, particularly in bread. Because this amino acid is not synthesized by the body, it must be obtained in foods (it is an essential amino acid). This nutritional requirement creates a strong consumer demand for lysine. Indeed, lysine is a key U.S. export item.

■ *Corynebacterium*
kôr´ē-nē-bak-ti-rē-um

Glutamic acid, also produced by *Corynebacterium* species, is another essential amino acid. Hundreds of millions of pounds of glutamic acid are produced annually by microbes in U.S. factories, and sales of this amino acid exceed billions of dollars. Glutamic acid is also used as a flavoring agent in Asian cuisines in the form of monosodium glutamate.

Two important vitamins, riboflavin (vitamin B_2) and cyanocobalamin (vitamin B_{12}), are products of microbial growth. Riboflavin is a product of *Ashbya gossypii*, a mold that produces 20,000 times the amount of riboflavin it needs for its metabolism. Cyanocobalamin is produced by growing selected species of *Pseudomonas*, *Propionibacterium*, and *Streptomyces* in a cobalt-supplemented medium. The vitamin prevents pernicious anemia in humans and is used in bread, flour, cereal products, and animal feeds.

■ *Ashbya gossypii*
ash´bē-ä gos-sip´ē-ē

■ *Pseudomonas*
sū-dō-mō´näs

■ *Propionibacterium*
prō-pē-on´bak-ti-rē-um

Other Microbial Products

Some microbial products lie outside the spheres we have discussed, but merit consideration because they are quite important in everyday life.

Gibberellins are a series of plant hormones that promote growth by stimulating cell elongation in the stem. Botanists use the hormones to hasten seed germination and flowering, and agriculturists find them valuable for setting blooms in the plant. This increases the yield of fruit and, in the case of grapes, enhances the size. Gibberellins are produced during the metabolism of the fungus *Gibberella fujikuroi*, and they may be extracted from these organisms for commercial use.

Steroids are lipid (fatty) compounds having a structure that includes four rings of carbon atoms, with various other chemical groups attached. Included in the group are cholesterol, sex hormones, and cortisone, a drug used by doctors to relieve the symptoms of arthritis and treat other types of joint inflammations. Steroids can be produced by chemical synthesis, but the process is arduous and expensive. For example, the industrial manufacturing of cortisone from its chemical components requires 37 chemical conversions. Enter the microbes. Scientists have found that they can reduce the number of steps to 11 and leapfrog over several difficult steps by using the fungus *Rhizopus nigricans*. This fungus takes progesterone, an intermediary in the process, and converts it to another intermediary closer to cortisone. Before the use of *Rhizopus*, a gram of cortisone cost $200 to produce; by employing the fungus, the price was reduced to $6 per gram. Further modifications, including the use of *Aspergillus* species, have reduced the cost to $0.68 per gram.

The chemical skills of microbes have also been harnessed to produce the hormones used in oral contraceptives, which were formerly produced from Mexican yams (now in short supply). Microbiologists use microbes to convert fish oils and wool grease to contraceptive steroids such as estrogens.

Microbes are even used to produce perfume: Musk oil is manufactured by chemical conversions from ustilagic acid, a product of the fungus *Ustilago zeae*; ironically, this fungus causes smut diseases in crop plants (Chapter 8). Moreover, numerous pharmaceutical products are derived from the poisons synthesized by the fungus *Claviceps purpurea*. These derivatives are prescribed to induce labor, treat menstrual disorders, and control migraine headaches. Most individuals taking these drugs would prefer to remain blissfully ignorant of their source.

■ *Ustilago zeae*
ū-sti-lä′gō zē′ī

■ *Claviceps purpurea*
kla′vi-seps pür-pü-rē′ä

Antibiotics

Antibiotics are chemical products (or their synthetic derivatives) of microbes used to kill other microbes. To this writing, over 5000 antibiotic substances have been described and approximately 100 antibiotics are available to medical practitioners. Although a significant number are products of *Penicillium* or *Bacillus* species, most antibiotics are produced by species of *Streptomyces*, shown in FIGURE 14.10 . Antibiotic production is carried on in huge, aerated, stainless steel tanks similar to those used in brewing. A typical tank may hold 30,000 gallons of medium. New technologies employ small fragments of hyphae or cells, submerged in the medium and agitated with a constant stream of oxygen. After several weeks of growth, the microbes are removed, and the antibiotic is extracted from the medium for further conversion to the desired product.

The traditional screening method for discovering new antibiotics was rather straightforward: Microbes were isolated (usually from the soil) and cultivated in the laboratory at various temperatures and acidity levels as well as in a variety of selective and differential media. Then biochemists would attempt to harvest, purify, and identify the organic products of the microbes' metabolism and see if those had any

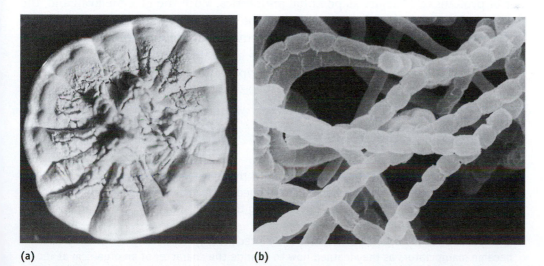

(a)   (b)

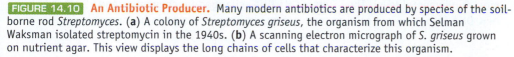

FIGURE 14.10 **An Antibiotic Producer.** Many modern antibiotics are produced by species of the soil-borne rod *Streptomyces*. (**a**) A colony of *Streptomyces griseus*, the organism from which Selman Waksman isolated streptomycin in the 1940s. (**b**) A scanning electron micrograph of *S. griseus* grown on nutrient agar. This view displays the long chains of cells that characterize this organism.

antibiotic activity. By these methods, novel compounds were discovered, but only rarely. Indeed, to increase the odds of finding new antibiotic producers, company employees traveling to distant countries were given cellophane bags with instructions for bringing home soil samples. Some companies even funded collection trips to exotic locales.

But with the advances of biotechnology, the search for antibiotics has taken on a new dimension. It has become clear that the great majority of microbial species in the soil cannot yet be cultivated in the lab. This restriction has not hindered scientists, however, because they can now analyze the DNA molecules in a handful of soil and develop a biochemical image of the soil's microbial population, while synthesizing new antibiotics.

The process goes like this: The DNA in a soil sample is amplified by the polymerase chain reaction, and large fragments of DNA are isolated. Then the fragments are integrated into plasmids, and the plasmids are inserted in *E. coli* cells. When the bacteria grow in the lab, they synthesize their own proteins plus the proteins encoded by the foreign DNA fragments. To determine whether any of these proteins are antibiotics, the *E. coli* cells are cultivated with various pathogens to see if the pathogens die. If a "hit" occurs, the biochemists set to work to identify the new protein produced in the *E. coli* cells. The work is eased considerably by the knowledge that the DNA encoding the protein is already isolated on a plasmid. The encoding gene can be further pinpointed by established biochemical methods and then sequenced and analyzed.

Knowing the complete genome of an organism further assists the drug discovery process. As more and more microbial genomes are elucidated (Chapter 4), molecular biologists become better equipped to study microbial biology. The patterns generated from genome analyses give clues to the function of a previously unknown gene and open the possibility that it encodes an antibiotic. Before the genome was uncovered, scientists had a limited number of genes to work with and therefore a small number of proteins to consider as potential antibiotics. With the genome available, the barrier has been lifted, and new proteins can be hunted down. Moreover, a wealth of new gene targets have been identified as objectives of antibiotic action. In the future, we can anticipate a variety of new antibiotics generated from increasingly efficient drug-hunting programs. Indeed, we can anticipate quite a substantial harvest of the fruits of biotechnology.

■ A FINAL THOUGHT

For many decades, humans stood by and watched the "game of life." They marveled at the wonders of nature; they searched out and cataloged the plants, animals, and microbes of the world; and they spent exhaustive hours trying to understand how these organisms fit into the scheme of things.

Then dawned the age of DNA science and biotechnology. Now many of the observers became manipulators as they learned how to change the character of an organism at its most fundamental level. They isolated its DNA, changed its chemistry, and inserted the new DNA into recipient cells to see what would happen. Bacteria began producing human hormones and antiviral proteins, and consumers tried to make sense of "antisense molecules." In the offing were new diagnostic tests and vaccines undreamed of a generation before. Microbes were at the center of biotechnology, and their widespread use added to the positive press they enjoyed for their industrial contributions.

We stand at the brink of an adventure that will carry us through the twenty-first century and beyond. The implications of biotechnology are so colossal that the human mind has yet to imagine all of them. Biotechnology will continue to impact human lives for many centuries to come, and we can proudly tell our grandchildren that we were there at the beginning. It's a wonderful time to be studying microbiology.

QUESTIONS TO CONSIDER

1. Certain bacteria produce many thousands of times more of specific vitamins than they require. Some biologists suggest that this makes little sense because the excess is wasted. Can you suggest a reason for this apparent overproduction in nature?

2. During the past decade, publications such as *Discover* magazine and *Scientific American* have carried articles about a renaissance in vaccines and the increased reliance future generations will have on vaccines to preserve health. What evidence of such a renaissance do you see in this chapter, and why do you suppose we will rely on vaccines even more in the future?

3. A biotechnologist suggests that one day it may be possible to engineer certain harmless bacteria to produce antibiotics and then to feed the bacteria to people who are ill with infectious disease. The bacteria would then serve as antibiotic producers within the body. Would you favor research of this type?

4. An article in a local newspaper recently carried this headline: "Making Sense of Antisense." Suppose you were writing the article. What would you say?

5. Textbook writers tend to oversimplify complex issues in attempting to make them understandable to students. This chapter has a section on the methods of biotechnology to show the complexity of this discipline, but space limitations have required that a host of issues be omitted. (For example, how do scientists "store" genes?) From your knowledge of microbiology and of science in general, what other issues must biotechnologists confront, and how might these problems be circumvented?

6. How many times in the last 24 hours have you had the opportunity to use or consume the industrial product of a microbe?

7. Although the products of biotechnology have been of great benefit in numerous arenas, they have also been abused. One example is the use of erythropoietin by athletes to increase their red blood cell counts and give them an unfair advantage at competitive events. What other abuses of products of biotechnology can you think of?

■ KEY TERMS

Informative facts are necessary for the expression of every concept, and the information for a concept is founded in a set of key terms. The following terms form the basis for the concepts of this chapter. On completing the chapter, you should be able to explain and/or define each one:

amylase	hemophilia A
antisense molecule	interferon
Crohn's disease	lysine
diabetes	oncogenes
DNA fingerprinting	pectinase
DNA polymerase	polymerase chain reaction (PCR)
DNA probe	protease
DNA vaccine	rennin
dwarfism	reverse transcriptase
erythropoietin	signal sequence
galactosidase	tissue plasminogen activator
gibberellins	vector
glutamic acid	viral load test
hantavirus	VNTRs

■ http://microbiology.jbpub.com/book/microbes

The site features **eLearning**, an online review area that provides quizzes and other tools to help you study for your class. You can also follow useful links for in-depth information, read more stories of microbiology, or just find out the latest microbiology news.

Microbes and Agriculture: No Microbes, No Hamburgers

15

Looking Ahead

Microbes play a key role in agriculture—they are essential to the production of meat, dairy products, and numerous other foods. Moreover, they help farmers fight crop pests, and they are instrumental in the cycling of elements in the soil. These are but a few of the countless places where microbes exert a powerful influence on agriculture, as the pages ahead will show.

On completing this chapter, you should be able to . . .

- explain the chemistry performed by microbes in nitrogen metabolism as they bring this important element into the cycle of life.
- describe how microbes bridge the gap between carbohydrates and proteins and help ruminant animals produce meat and meat products.
- name many of the dairy products that owe their existence to the microbes and explain the chemistry by which these products are manufactured.
- recognize the roles played by microbes in the new biotechnology methods used in agriculture.
- explain how microbes can be employed by biotechnologists to develop new agricultural insecticides.

Bacteria, yeasts and molds can spoil crops in a silo. It sounds odd but to prevent spoilage, farmers add more bacteria. They use a silage inoculant containing *Lactobacillus plantarum*. *L. plantarum* produces lactic acid, which lowers the pH levels on the crops. The lower pH inhibits the growth of other microorganisms.

- appreciate the role of biotechnology in the production of new types of foods for our consumption.
- describe the innovative vaccines of the future that will use vegetables or fruits instead of needles.

....

The man from Delft stood in front of the assembled farmers and made an outrageous proposal: " Don't plant your crops in the same field as last year," he said. "Leave the field alone for the next two years; let it lie fallow." The year was 1887; the man was Martinus Willem Beijerinck; the country was The Netherlands. Because agricultural land was scarce in that small country, the proposal was outrageous.

Beijerinck was a local bacteriologist. While his colleagues were investigating the germ theory of disease and its implications, Beijerinck was out in the fields. He had observed that land is very productive when it has been freshly cleared of brush and trees and newly planted, but less productive after several years of use. Moreover, he had noted that fields yield bountiful crops when the farmer has been away for a couple of years. And now he thought he had the answer to these mysteries: great populations of bacteria.

Beijerinck was an agricultural expert, to be sure, but he also had a solid background in chemistry, something other botanists lacked. He was of the opinion that atmospheric nitrogen is essential for plant growth, but he could not figure out how the nitrogen gas became part of the plants' cells—what bridged the gap between atmosphere and protein? Then it dawned on him that bacteria supply the link, and that the little lumps and bumps on plant roots were the answer. Time and again he had observed great hordes of bacteria in the lumps and bumps (they were called nodules). He did not often see the nodules on tended crops, but they always seemed to be in abundance on wild plants growing in untended fields. FIGURE 15.1 displays the nodules and their inhabitants.

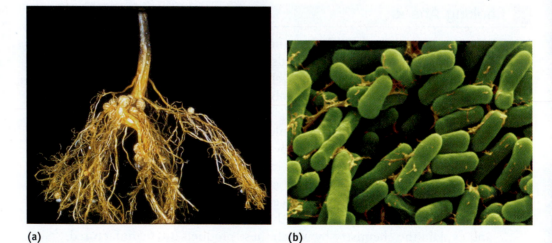

(a)   (b)

FIGURE 15.1   **Nitrogen-Fixing Bacteria.** Martinus Beijerinck, the Dutch scientist who in the 1880s postulated that bacteria enhance the nitrogen content of the soil. (**a**) Nodules containing nitrogen-fixing bacteria on the roots of a cowpea, a legume plant. Species of *Rhizobium* within the nodules incorporate atmospheric nitrogen into nitrogen-containing compounds. When the bacteria die, the compounds are utilized by the legume to synthesize amino acids. (**b**) A false-color scanning electron micrograph of a *Rhizobium* species comprising a root nodule.

Beijerinck performed a number of laboratory experiments that strengthened his views: He isolated bacteria from the nodules and injected them into seedlings of different kinds of plants. Invariably, the nitrogen content of the soil rose dramatically. So, his advice that day in 1887 was direct and straightforward: Leave the field alone for a spell; plant elsewhere; let wild plants and clover grow in the field so that their bacteria enrich the nitrogen content of the soil; and later, when the field is finally planted, the crop yield will be worth the wait.

Indeed, he was right. Modern farmers know that every so often, it is important to let a field lie fallow and "refresh" itself. To the macrobiologist, the nodules on the roots of wild plants are the important parts. To the microbiologist, it's what's inside that counts; for inside the nodules, bacteria perform the essential chemistry of the nitrogen cycle and bring this key element into plant proteins. As we shall see in this chapter, this chemical transition would not occur without the bacteria, and our supply of protein would quickly diminish. Indeed, life as we know it would probably come to an end.

Microbes on the Farm

Next time you are at your local library, stop by the CD collection and borrow a copy of Beethoven's *Sixth Symphony*, known as the *Pastorale*. Drop it into your CD player, and sit back and enjoy one of the most beautiful pieces of music ever composed. Envision a tree-shaded meadow, wild flowers and clover everywhere, and a herd of cows gathered in the midday sun under a sprawling oak. The idyllic scene seems too good to be true. Certainly it can't have much to do with microbiology!

But it does, for this is where music and microbiology mix: In the nodules of the clover, microbes are busy at work bringing nitrogen into the nitrogen cycle; and in the stomachs of the cows, microbes are taking plant materials the cows eat and changing them to protein to make meat and dairy products. In both cases, microbes are continuing their unceasing efforts on our behalf. In the end, this music of life gives us a major share of the food we enjoy. (By the way, Beethoven's *Sixth Symphony* is featured in Walt Disney's 1940 animated feature, *Fantasia*.)

In this section, we shall look into the nodules of the plants and the stomach of the cow to see how all this happens. We shall discuss how microbes bridge the gap between atmospheric nitrogen and the nitrogen in protein. Then we shall shift our attention to the cows and see how microbes help these animals make more of themselves. In both cases, we see microbes at their best, doing things that are natural for them and indispensable for us. These processes are worthy of a fine piece of musical harmony.

Connecting the Nitrogen Dots

In soils and waters, elements such as carbon, oxygen, and hydrogen are cycled and recycled by a broad variety of plants, animals, and microbes (as we discuss in Chapter 16); but the cycle of nitrogen is largely the province of microbes alone. Scientists recognize that the productivity of many ecosystems is limited by the supply of nitrogen available to organisms in that system, and one part of the nitrogen cycle (i.e., the conversion of nitrogen gas to ammonia and organic nitrogen compounds) is performed almost exclusively by a limited group of microbial species. Without these species, the entire process would grind to a halt.

But why is nitrogen metabolism so important? Nitrogen is an essential element of many organic compounds, including all the amino acids (and, consequently, all proteins) as well as all the nitrogenous bases of nucleic acids. Moreover, between 9 and 15% of the dry weight of a typical cell is nitrogen. To fill cellular needs, scientists estimate that microbes make available approximately 200 million tons of nitrogen compounds each year. Ironically, even though nitrogen is the most plentiful gas of the atmosphere (about 80% of the air is nitrogen gas), neither plants nor animals can use this nitrogen to synthesize the compounds they must have to survive. Instead, animals and plants depend on microbes to bring nitrogen into the cycle of life.

The trapping of nitrogen gas from the atmosphere and its incorporation in organic compounds is known as nitrogen fixation. Two types of microbes accomplish nitrogen fixation: free-living microbes (ones that live independent lives) and symbiotic microbes (ones that live in relationships with other organisms). Both types of microbes produce nitrogenase, an enzyme that uses nitrogen to synthesize ammonia. Through a complex series of reactions, nitrogenase takes a nitrogen atom and adds three hydrogen atoms to yield ammonia (NH_3). Nitrogenase is very sensitive to oxygen and is inactivated by this gas. Therefore, nitrogen fixation takes place where conditions are anaerobic.

The free-living nitrogen-fixers include species of *Azotobacter*, *Beijerinckia* (named for Beijerinck), and *Clostridium*, as well as several genera of cyanobacteria such as *Nostoc* and *Anabaena*. In soils and waters, these microbes manufacture ammonia as well as a variety of amino acids, as FIGURE 15.2 illustrates. In the Arctic region, cyanobac-

■ *Azotobacter*
ä-zo'to-bak-tėr

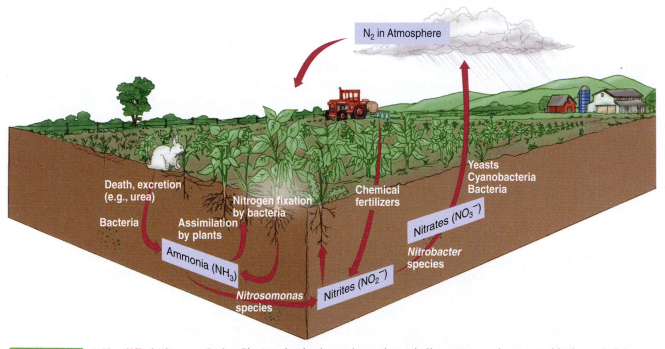

FIGURE 15.2 **A Simplified Nitrogen Cycle.** Plant and animal proteins and metabolic wastes are decomposed by bacteria into ammonia. The ammonia may be utilized by plants, or it may be converted by *Nitrosomonas* and *Nitrobacter* species to nitrates, which are also used by plants. Some nitrate is broken down to atmospheric nitrogen. This nitrogen is returned to the leguminous plants by nitrogen-fixing microorganisms as nitrates, which are then converted to ammonia. Animals consume the plants to obtain proteins that contain the nitrogen.

teria are the most important nitrogen-fixers in the ecosystem. These microbes synthesize ammonia, then convert it to amino acids and proteins. Later, when the cyanobacteria die, their cells are decomposed by other bacteria, and the nitrogen compounds are converted back to ammonia, which is made available to plants.

The symbiotic nitrogen-fixers are typified by species of *Rhizobium* and *Bradyrhizobium*. These microbes live in nodules on the roots of legume plants such as soybeans, alfalfa, peas, clover, and beans (legumes are plants that bear their seeds in pods). The symbiotic relationship between the microbe and plant is quite complex: It involves a chemical attraction between the bacterium and the cells of the plant root, a binding to the plant root, the curling and branching of rootlets, and the entry of bacteria into root hairs. In quick succession, an infection thread develops; the plant cell transforms into a tumorlike growth, which is the nodule; and the bacteria assume distorted forms known as bacteroids.

Bacteroids cannot live independently once they have entered the symbiotic relationship with the plant. Indeed, they derive life-sustaining compounds from the legume plant while using their nitrogenase to fix atmospheric nitrogen for the plant's benefit. Modern farmers know that soil fertility is enhanced dramatically when legumes are cultivated in a field. So much organic nitrogen is captured, in fact, that the net amount of nitrogen (and the net worth of the soil) increases considerably after a crop of legumes is harvested. Moreover, no artificial nitrogen-containing fertilizer is necessary when legumes have been cultivated in the field or when a crop such as clover or alfalfa is plowed under. In addition, nutritionists know full well the value of legumes in the diet, and humans are indebted to microbes for the high protein content of various kinds of peas and beans. Because cattle and other animals feed on clover and other legume plants, humans also depend on microbes for many indirect protein products of nitrogen fixation, such as milk and hamburgers.

But nitrogen fixation is only one aspect of microbial involvement in the nitrogen cycle. After much of the ammonia has been deposited in the soil, it is converted to valuable nitrate ions through the process of nitrification. In the first step of nitrification, bacteria such as *Nitrosomonas* species convert the ammonia molecules to nitrite ions (NO_2^-). Then the nitrite ions are converted by species of *Nitrobacter* to nitrate ions (NO_3^-), which are relished by plants for making their organic materials. Nitrification occurs under aerobic conditions and is used by microbes as an energy-yielding process. In fact, so much energy is obtained that the bacteria can grow without other energy sources.

From the plant and animal point of view, one of the less beneficial aspects of the nitrogen cycle is denitrification. In this process, some species of microbes break down nitrate ions and produce nitrogen gas, which they give off into the atmosphere. This process removes nitrogen from the cycle of life and can thus be seen as detrimental to most living things on Earth. Operating through nitrogen fixation, however, microbes bring nitrogen back into the cycle and make it available to the organisms that depend on it for life.

The importance of nitrogen fixation by symbiotic bacteria has not escaped the attention of biotechnologists. Scientists foresee the day when they can use gene alterations to improve the nitrogen-fixing chemistry in *Rhizobium* and other bacteria and increase the efficiency of the bacteria-plant interaction. Agricultural scientists are currently seeking ways to transfer the genes for nitrogen fixation from *Rhizobium* species to a bacterium such as *Agrobacterium tumefaciens*. This bacterium inhabits the tissues

■ *Agrobacterium tumefaciens*
ag'rō-bak-ti're-urn
tü'me-fāsh-enz

of a variety of plants such as tobacco, petunias, and tomatoes, and the knowledge base about it and its plant hosts is well developed, as we shall discuss later in this chapter. It is conceivable that nitrogen fixation could one day be extended to these plants. Another possibility is to engineer a *Rhizobium* species that would be able to assume a symbiotic relationship with a nonleguminous plant, such as wheat, rice, or corn. Such an advance would go a long way to solving world hunger problems.

In the most optimistic of molecular biology circles, some look to the day when the genes for nitrogen fixation can be removed from microbes and transferred directly to animals. Then the bacterial link in the nitrogen connection could be severed, and animals could extract their nitrogen directly from the atmosphere. The prospect of an animal synthesizing its own amino acids from atmospheric nitrogen fires the imagination of even the most innovative futurists.

Those Remarkable Ruminants

Ruminants are grazing animals such as cattle, sheep, deer, goats, camels, buffaloes, and numerous other "cud-chewing" species. These animals have adapted to a diet of plants, grasses, and other carbohydrate-rich plants; plants whose cells contain much energy-rich starch and whose cell walls consist primarily of cellulose, another energy-rich polysaccharide composed of chains of glucose molecules. Despite an almost purely carbohydrate diet, ruminants manage to make protein and, consequently, more of themselves. We, in turn, rely on many of these animals for meat and dairy products. (We shall use a cow as the prototypical ruminant in the paragraphs ahead.)

The cow's secret for converting carbohydrate to protein is intimately bound up with microbes. Although cows can use starch without trouble, they cannot digest cellulose into its valuable glucose molecules because they cannot produce cellulase, a cellulose-digesting enzyme. (Indeed, no vertebrate animal, including humans, can synthesize cellulase.) Cellulase, it turns out, is provided by microbes, including those surveyed in A Closer Look on page 329.

What happens is this: The cow's large stomach is divided into several compartments, the first and biggest of which is called the rumen. The rumen has a constant temperature, as well as a slightly acidic and oxygen-free environment. It teems with numerous species of anaerobic bacteria and other microbes that produce cellulase. The microbial cellulase digests the cellulose in plants, producing the disaccharide cellobiose and, most importantly, the monosaccharide glucose. Then the glucose molecules are fermented by microbes to simple organic acids such as propionic acid and acetic acid along with gases such as methane and carbon dioxide. Some of the organic acids pass through the rumen wall and enter the cow's bloodstream, by which they are transported to its body cells. Here they are key energy sources fueling the cow's metabolism. Gases from the rumen are eliminated when the cow belches, a process that occurs regularly and vigorously as it reclines under an oak tree in the midday sun.

But energy alone does not make cow muscle or dairy products. Quite to the contrary, the energy is used to make more microbes. It seems that while performing their biochemical magic on cellulose, the microbes continue to grow and multiply at fabulously high rates. A single milliliter (20 drops) of rumen fluid may contain up to 10^{12} microbes—that is, *a million billion* microbes. The rumen is literally a fermentation tank and home to incomprehensible numbers of bacteria, protozoa, fungi, and

A CLOSER LOOK

A Laxative for Termites

Americans spend millions of dollars annually to protect their homes and properties against wood-eating termites. The notorious reputation acquired by these insects and their voracious appetite for wood are linked by a quirk of nature to protozoa.

The termite's intestine is home to a species of protozoa belonging to the genus *Trichonympha* (tri-co-NIM-fah). These multiflagellated organisms are among the few species of living things that can produce the enzyme cellulase. The cellulase is released into the intestinal cavity of the termite, where it breaks down cellulose, the principal component of wood. The protozoan thus lives in the stable environment of the termite's intestine and returns the favor by digesting the termite's next meal. To the ecologist, this symbiotic relationship is called mutualism. To the homeowner, it spells disaster.

But there may be hope in the future. In the tropics, a method of termite extermination has proven so successful that American companies are examining its feasibility. The method involves placing a paste of ground-up plant material, known to be a termite laxative, into termite tunnels. As termites eat the paste, they develop diarrhea and excrete their protozoal inhabitants. Thereafter, the termites are unable to digest cellulose, and they starve to death. With insecticides coming under fire for their hazard to the environment, the research into termite laxatives holds substantial promise for the future.

other microbes. In addition to the organic acids we noted above, the microbes synthesize protein for the cow, using normally toxic urea as a nitrogen source. Even the microbes themselves eventually become a protein source for the cow as they are broken down and their cellular components used to form its muscle.

But there is more. As microbes continue to multiply furiously in the rumen, they soon overgrow the space and pass into the second part of the stomach, the reticulum. Here the microbes are squeezed together with undigested plant material to form balls of cud material. At its leisure, the cow regurgitates the cud, mixes it with copious amounts of saliva, and chews it to crush the fibers. Then the cud is reswallowed and returned to the rumen, where it is redigested by more cellulose-digesting microbes. Next, it passes back into the reticulum for regurgitation and rechewing. Finally, the plant-microbe mass passes into the omasum and the abomasum (together, the "true" stomach of the cow), where protein-digesting enzymes break down the protein to release the amino acids. The entire system is shown in FIGURE 15.3 .

From this point on, the physiology of the cow is similar to that of other vertebrates (including humans). Further digestion of protein in the intestines yields amino acids and a bevy of vitamins and other growth factors that pass into the cow's bloodstream and are transported to its muscles and organs and fashioned into its cells. At the fundamental level, the ruminant animal has satisfied its requirement for protein by eating microbes. Humans are next in line for this "microbial steak."

Other ruminants obtain their protein in the same way as the cow. However, nonruminants such as horses, rabbits, and guinea pigs are slightly different. They are also grass-eaters, but they have a sidepocket in their gut called the cecum. In this organ,

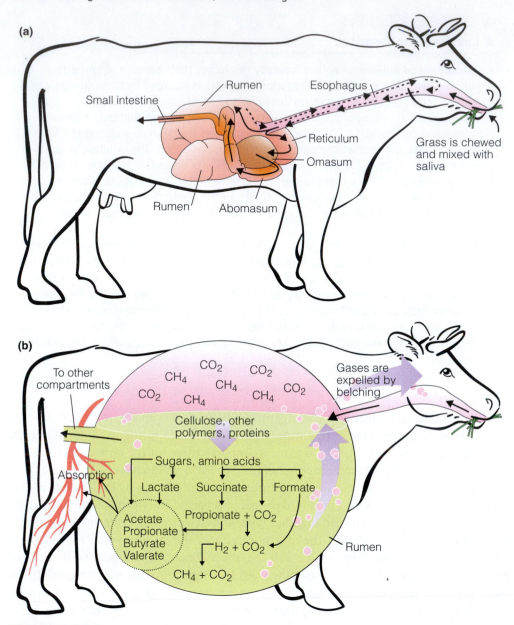

(a)

Rumen

Esophagus

Small intestine

Reticulum

Omasum

Grass is chewed
and mixed with
saliva

Rumen

Abomasum

(b)

CO_2 CO_2

To other
compartments

CH_4

CH_4

Gases are
expelled by
belching

CO_2

CO_2

CO_2

CH_4

CH_4

Cellulose, other
polymers, proteins

Absorption

Sugars, amino acids

Lactate   Succinate   Formate

Acetate
Propionate
Butyrate
Valerate

Propionate + CO_2

Rumen

$H_2 + CO_2$

$CH_4 + CO_2$

FIGURE 15.3   **Microbes and the Rumen.** (**a**) A schematic diagram of the rumen and other components of the digestive tract of the cow. Arrows show the pathway of food; dashed lines represent the
pathway of regurgitated food when the cow chews its cud. Note the large size of the rumen relative to
the other compartments of the stomach. (**b**) Some of the chemical reactions taking place in the rumen.

a rich microbial population performs the same task as the microbes in the rumen.
Moreover, instead of regurgitating the cud, certain "cecum animals" such as rabbits
eat their fecal pellets, thereby giving the food a second pass through the intestine.
This activity, called coprophagy, also gives the animal an opportunity to obtain the
vitamins produced by the bacteria in the cecum. At one time in its evolution, the human appendix may have been the equivalent of the animal's cecum.

Before we leave our ruminant friends, we note another place where microbes impact on their lives. Cows, goats, and other barnyard animals are fond of silage, a product of microbial fermentation. Silage is fermented in the tall cylindrical structure often located adjacent to a barn (FIGURE 15.4). It is an enormous fermentation tank into which the farmer loads grasses, grains, legumes, and any other plants available. Under tightly packed, anaerobic conditions, numerous species of bacteria consume the plant material and carry on fermentation to reduce the carbohydrates to a more digestible and tastier form. A host of acids and other organic compounds preserve the silage from microbial decay and produce an animal food that is both economical and pleasing to the palate (the animal palate, that is).

At the Dairy Plant

The milk from cows provides an excellent growth medium for the growth of microbes. It is rich in various proteins, in-

FIGURE 15.4 Silos. Silos act as fermentation tanks in which microbes convert plant material to digestible animal feed.

cluding casein, the complex protein that gives it its white color. It also contains carbohydrates, primarily the disaccharide lactose ("milk sugar"). Moreover, it has a variety of fats, vitamins, and minerals.

Allowed to stand at room temperature, unpasteurized milk rapidly undergoes a natural souring brought about by bacteria fermenting the lactose and producing lactic acid. (If excess lactic acid is present, the protein will curdle.) This acidification provides a natural barrier to the growth of other microbes because the acidic environment is generally inhospitable. Although the milk is sour to taste and somewhat offensive to smell, dairy farmers have put this biochemical process to work to produce a variety of fermented milk products. Over the centuries, people have come to accept these dairy products for their resistance to deterioration as well as their interesting and unique flavors.

One example of a fermented milk product is buttermilk. Thought to originate in the East European country of Bulgaria, buttermilk is produced by two species of bacteria: *Lactobacillus bulgaricus* (named for Bulgaria) and *Leuconostoc citrovorum*. At the dairy plant, these bacteria are added to skim milk, whereupon the *Lactobacillus* produces lactic acid from the lactose and sours the milk, while the *Leuconostoc* synthesizes polysaccharides to make the milk slightly thick. Sour cream is produced in the same way as buttermilk, except that cream is used instead of skim milk. Biochemists tell us that the acid in sour cream retards microbial decay; oblivious to that bit of science, we just know that sour cream tastes good on a baked potato.

Another popular fermented milk product is yogurt. Also thought to originate with Bulgarian tribes, yogurt is a popular food today throughout North America and Europe. Among the principal bacteria used in yogurt production are *Streptococcus thermophilus* and *Lactobacillus bifidus* (other species such as *S. lactis* and *L. acidophilus* can also be used). Milk is first concentrated by adding dried milk protein, then the bacteria (the "active cultures") are added by mixing in a sample of previously prepared "mother" yogurt. At a high temperature (about 60°C or 166°F), the streptococci are the first to

■ *Lactobacillus bulgaricus*
lak-tō-bä-sil′lus
bul-gā′ri-kus

■ *Leuconostoc citrovorum*
lü-kū-nos′tok
sit-rō-vôr′um

■ *thermophilus*
thĕr-mo′fil-us

A CLOSER LOOK
Making Your Own Yogurt

The popularity of yogurt as a nutritious low-calorie food has prompted many to try their hand at making it at home. Here is a recipe that works well.

Heat one quart of milk to about 170°F (77°C), stirring often and using a thermometer to check the temperature. This heating will evaporate some of the liquid and reduce the bacterial population. Let the milk cool to about 130°F (55°C), then add one cup of powdered milk and one-third cup of unflavored commercial yogurt. Mix thoroughly and pour into small containers with lids. Styrofoam coffee cups may be used.

For the incubation step, you will need a small cooler of the type used for picnics. Fill the cooler with several inches of water at 130°F (55°C). Now place the containers in the cooler, close the lid tightly, and let the containers stand for about 6 to 8 hours. (A large pan of hot water in the oven also works well, or the cups can be wrapped with hot towels.) During this time, the bacteria will multiply and the yogurt will thicken. Refrigerate, add fresh or frozen fruit, and enjoy.

ferment the lactose. Then the lactobacilli take over the fermentation and produce more acid and the characteristic yogurt texture and consistency. Yogurt can be made easily at home (see A Closer Look above), and although it is really a "spoiled milk" product, we consider it a healthful addition to our diet.

■ *acidophilus*
a-sid-o'fil-us

Acidophilus milk, another fermented milk product, takes its name from the organism used to make it—the bacterium *Lactobacillus acidophilus*. Milk is first pasteurized (in some cases, sterilized); then a pure culture of the lactobacilli is added to the milk ("sweet acidophilus milk"). Alternately, the bacteria can be incubated in the milk for a short time to ferment it slightly ("sour acidophilus milk"). In the intestine, the lactobacilli establish themselves and become part of the normal microflora (the microbial population typically found there). They hold disease organisms in check and assist the digestion of foods, while contributing significant amounts of vitamins to our metabolism. Acidophilus milk is often recommended to persons who suffer bouts of diarrhea as a way of "calming" the intestine and reestablishing the normal microflora. This use is particularly important following antibiotic therapy because the normal microflora are often killed by the drug treatment.

A variety of other fermented dairy products shown in FIGURE 15.5 are produced in different parts of the world. They vary according to the source of the milk, temperature of incubation, and species of microbes participating in the fermentation. For example, kefir (first made in the Caucasus Mountains of the Ukraine) is produced using lactobacilli, streptococci, and the yeast *Saccharomyces kefir*. Both acid and alcohol are produced during the fermentation, and a unique effervescent quality is present in the sour milk because it contains fermentation gases (much like champagne). A fermentation like this, which produces both acid and alcohol, is referred to as a mixed fermentation.

■ *Saccharomyces*
sak-ä-rō-mī'sēs

Although we do not often think of butter as a microbial product, microbes contribute to its production. One popular method for preparing butter begins with starter cultures of streptococci and *Leuconostoc* species added to pasteurized sweet cream. The streptococci sour the cream slightly by producing lactic acid, and the *Leuconostoc*

(a)

(b)

FIGURE 15.5 **Fermented Milk Products.** The products shown, (**a**) yogurt and (**b**) sour cream on a baked potato, are produced by the action of microbes on various components of the milk.

species synthesize a substance called diacetyl. Diacetyl gives butter its characteristic aroma and taste. Once the reactions have been completed, the slightly sour milk is churned to aggregate the fat globules into butter.

Of Curds and Whey

Legend has it that one day a Bedouin traveler (a young boy it is said) filled his goatskin pouch with milk and set off on a journey. At midday, he stopped to eat, only to find that the contents of his pouch had turned to large white lumps. Frustrated but hungry, he decided to taste the lumps and found them to be quite good. He washed down the lumps with the leftover fluid and continued on his way . . . having discovered cheese.

Cheese is one of the most important products of microbial action on milk. Most cheese is produced from the protein casein in milk, a fact reflected in the translation of the Latin *caseus* to "cheese." Cheese has historically been a way of maintaining the nutrient quality of milk by preserving the protein for a long period (over the winter, for example, when milking the cows was difficult).

Cheese is manufactured from milk in three steps: curd formation, curd treatment, and curd ripening. In the initial step, casein separates from the milk fluids when the acid produced during microbial fermentation causes the casein to lose its three-dimensional structure and "curdle" out. This results in a sour curd. Alternately, the curd can be produced by altering the casein with the industrial enzyme rennin (which was traditionally obtained from the lining of a calf's stomach but is now produced by microbes). The result is a sweet curd. In both cases, the casein forms an insoluble clot (the curd), plus a butter-yellow liquid called whey. Once the curd has settled as a solid mass, it is separated from the whey by draining. The whey can then be used to make "cheese food." Italians have traditionally used whey to make ricotta cheese.

Unripened cheese such as cottage cheese or cream cheese is made directly from milk curds. Such cheese varies in fat and calories depending on the starting material. For instance, cottage cheese is produced from skim milk (low fat and calories), while cream cheese is derived from whole milk enriched with cream (high fat and calories). Both types of cheese spoil quickly because they lack salt, which normally inhibits microbial growth, and their pH is close to neutral. Another reason they spoil rapidly is that they have not been preserved by ripening, as we discuss next.

In some societies, the ripening of milk curds is considered a form of spoilage; in others, it is the zenith of microbial accomplishments for the palate. The French, for example, relish Camembert cheese, but many Americans consider its smell repulsive. To begin the ripening process, weights are used to put pressure on the curds and make them dense and compact. The curds are then salted, shaped, and inoculated with bacteria and fungi. The temperature and humidity conditions for ripening are controlled carefully, and as the microbes grow, the desired cheese gradually emerges, changing its texture with time and developing its aroma and flavor. Ripened cheeses fall into different categories based on such things as how the curd was formed (sour curd or sweet curd), its texture (hard, semihard, or soft), the ripening microbe (mold or bacteria), the milk source (goat, cow, mare, or ewe), and the country of origin.

Hundreds of different kinds of hard ripened cheeses are available, with equally varied names. For example, cheddar cheese was originally made in Cheddar, England. To produce this cheese, the curds are heated and pressed often to remove as much whey as possible (a process called cheddaring). The curds are wrapped with paraffin or plastic film to prevent surface contamination, and bacteria that produce lactic acid proliferate in the curd, their enzymes bringing about the chemical changes that give the cheese its cheddarness. Pecorino Romano cheese is Italian cheese made from sheep's milk (*pecora* is Italian for "sheep"; "Romano" refers to Rome). Several other hard cheeses are explored in Chapter 12.

Soft ripened cheeses are typified by Limburger and Camembert. In these cheeses, microbial enzymes break down the protein curds. Camembert cheese, for instance, is produced by inoculating the curd at its surface with the blue-green mold *Penicillium camemberti*. As the mold grows, its protein-digesting enzymes break down the casein into smaller peptides and amino acids that nourish the mold. In the process, the interior of the cheese gradually changes from a chalky paste to an oozing cream that delights the palate.

Although many soft ripened cheeses are made around the world, one type remains the province of the French. In southwestern France in a village of the same name, Roquefort cheese is still made as it was centuries ago: Ewe's milk is cured in limestone caves where the temperature and humidity are constant. Spores of the mold *Penicillium roqueforti* fall onto the curds, and as the fungus grows, it seeks cracks in the curd and spreads its hyphae throughout the cheese, giving it the characteristic blue-green veins. The local residents turn out 30 million pounds of Roquefort cheese per year and insist that the only cheese to be called Roquefort is the one coming from their town. Gorgonzola, Danablue, and blue cheese are mold-ripened cheeses, but none is Roquefort. **FIGURE 15.6** pictures the impressive variety of cheeses that microbes produce.

Cheese is one of history's longest established foods: Ancient writings indicate that it was being made as long ago as 2000 B.C. The Greeks considered it to be of divine origin, and it was often part of their sacrifice to the gods. Roman soldiers carried it with them when they went off to battle, and by the Middle Ages, it was a staple of the European diet. When the English settlers came to America, they brought a number of cows and proceeded to make large wheels of cheese. Immigrants to the Midwest brought cheese-making skills to Wisconsin and adjoining states, and it was only a matter of time before the "cheesehead" became a familiar sight at Green Bay Packers football games.

FIGURE 15.6   **Various Cheeses Produced by Microbes.**

This discussion of cheese closes our explorations of the traditional uses of microbes on the farm. For more centuries than we can imagine, microbes have been connecting the nitrogen dots, performing their vital tasks within the ruminant stomach, and stimulating our palates with nutritious and delicious dairy products. Now we move to the current era and explore other roles of microbes on the contemporary farm. Microbes have become the source of novel insecticides; they have made herbicides more effective; they have been used to enhance frost resistance among crop plants; they been used to create new foods; and they have provided the means to make farm animals more efficient producers of dairy products. We shall see how in the next section.

Biotechnology on the Farm

The techniques of genetic engineering and biotechnology have opened numerous possibilities for improving agriculture and the quality of life, possibilities that were unimagined before the breakthroughs in molecular biology that occurred in the last half of the twentieth century. Biotechnology has found ways to dramatically increase crop yields, substantially decrease plant disease, and use plants for novel purposes. For example, one biotechnology company is producing plantibodies, that is, human antibodies synthesized by genetically engineered corn plants. Scientists at Agracetus Inc. have found a way to force the genes that encode human antibodies into the chromosomes of corn plants. After harvesting the corn, the human antibodies are purified from the kernels. They can then be chemically bound to anticancer drugs to deliver the drugs to tumor tissues.

The possibilities of biotechnology in agriculture are enormous. For instance, productivity may be improved by increasing the size of edible plants, and the nutritional value of foods may be enhanced by increasing their amino acid content. Some biotech botanists even suggest that plants, notable for carbohydrate production, may be converted to protein producers. One company, for example, is working with potato plants and genes obtained from the cells of chickens. Its researchers are attempting to splice the genes into the chromosomes of the potato cells to augment their ability to synthesize protein. On a more mundane level, other biotechnologists are trying to develop a potato that resists discoloration when peeled. They have removed from potato plants the enzyme responsible for the color change by removing the gene that encodes the enzyme. Such a potato can be peeled without "bruising," that is, without changing color. We shall see other examples of such plant innovations in the paragraphs ahead.

DNA Into Plant Cells

One of the breakthroughs in agricultural technology was the ability to cultivate plants from a single plant cell. Such a cell can be genetically modified and then cultivated in a medium of carefully balanced plant nutrients and hormones. It will develop into a random mass of cells called a callus. Some days or weeks later, the forerunners of roots, stems, and leaves appear on the callus. Shortly thereafter, the plantlet is transferred to a container until ready for planting outdoors. The result is a transgenic plant.

Inserting fragments of DNA into plant cells requires the proper instrumentation, a high degree of dexterity, and a bit of luck. One widely used procedure is microinjection, in which a microscopic syringe penetrates the cell wall and cell membrane of a plant cell and delivers DNA fragments into the nucleus. An alternative method uses a type of genetic shotgun called the biolistic. This is a cylinder with a cartridge and a nylon projectile carrying millions of tiny spheres coated with DNA. The spheres pass through the plant's cell wall and membrane into the cytoplasm.

A valuable method for DNA delivery is the Ti plasmid (short for tumor-inducing plasmid) obtained from the bacterium *Agrobacterium tumefaciens*. This microbe causes plant tumors and a disease called crown gall. Crown gall develops when the bacterium releases its Ti plasmid in the plant cell and the plasmid inserts itself into one of the cell's chromosomes. The plasmid then encodes proteins that stimulate bacterial reproduction and lead to the visible crown gall pictured in **FIGURE 15.7**. For biotechnology experiments, Ti plasmids are removed from bacterial cells and modified to remove the tumor-inducing genes and substitute foreign genes. Then the plasmids are inserted in *A. tumefaciens* cells, and the bacteria are inoculated into the plant cell. The plasmids then transport the foreign genes into the plant cell chromosome. Marker genes are used to determine whether insertion has been successful.

Bacterial Insecticides

To be useful as an insecticide, a microbe should act only on a particular targeted pest and should act rapidly. It should be stable in the environment and easily dispensed, as well as inexpensive to produce. Finding a microbe that fits these criteria has been an ongoing challenge for biotechnologists.

In the early 1900s, a scientist named G. S. Berliner found that cells of a newly identified *Bacillus* species were able to kill moth larvae (a larva is a wormlike preadult stage of an insect). Berliner named the organism *Bacillus thuringiensis* after Thuringia, the German province where he lived. The bacterium remained relatively obscure until the 1980s, when scientists learned that it produces toxic proteins while forming its endospore. When the toxic proteins are deposited on leaves, they are in-

■ *thuringiensis*
thur-in-jē-en'sis

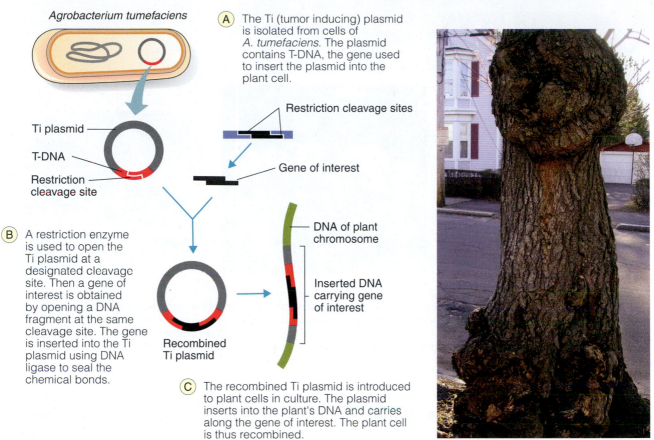

Agrobacterium tumefaciens

(A) The Ti (tumor inducing) plasmid is isolated from cells of *A. tumefaciens*. The plasmid contains T-DNA, the gene used to insert the plasmid into the plant cell.

Ti plasmid

T-DNA

Restriction cleavage site

Restriction cleavage sites

Gene of interest

(B) A restriction enzyme is used to open the Ti plasmid at a designated cleavage site. Then a gene of interest is obtained by opening a DNA fragment at the same cleavage site. The gene is inserted into the Ti plasmid using DNA ligase to seal the chemical bonds.

Recombined Ti plasmid

DNA of plant chromosome

Inserted DNA carrying gene of interest

(C) The recombined Ti plasmid is introduced to plant cells in culture. The plasmid inserts into the plant's DNA and carries along the gene of interest. The plant cell is thus recombined.

FIGURE 15.7 *Agrobacterium tumefaciens* **as a Vector in Genetic Engineering.** *A. tumefaciens* induces tumors in plants and causes a disease called crown gall. A lump of tumor tissue forms at the infection site, as the photograph shows. (b) The catalyst for infection is a Ti plasmid carrying DNA. This plasmid is used to carry a foreign gene into plant cells.

gested by caterpillars (the larval forms of butterflies, moths, and related insects). In the caterpillar gut, the protein breaks down the cells of the gut wall, presumably by causing pores to form in the intestinal membranes. As gut liquid diffuses between the cells, the larvae undergo paralysis; deadly infection by other bacteria soon follows.

Bacillus thuringiensis (commonly shortened to Bt) is apparently harmless to plants and the great majority of animals. It is produced by harvesting the bacteria at the onset of spore formation and drying them to make a commercially available dusting powder such as Dipel. The product is used against tomato hornworms, corn borers, and gypsy moth caterpillars. Its success has encouraged further research and led to the discovery of a new and more powerful strain called *B. thuringiensis israelensis* (commonly known as Bti), first isolated in Israel. **FIGURE 15.8** shows the organism.

But bacteria sprayed onto plants can soon wash off, so the protective effect of Bt as an insecticide can be limited. Therefore, biotechnologists have sought to provide long-term protection by inserting the toxin-encoding Bt gene directly into plant cells. To accomplish this, they have identified and cloned the gene and spliced it into the Ti plasmid of *Agrobacterium tumefaciens*. Then they have used the Ti plasmid to ferry the Bt gene into plant cells.

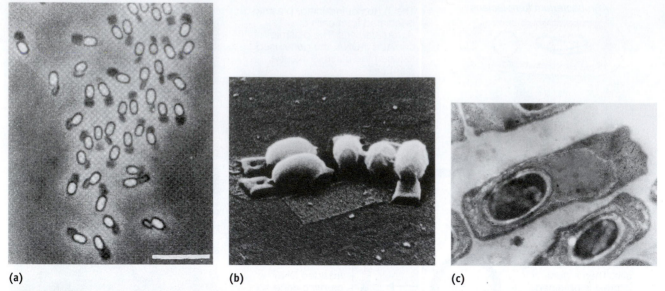

(a)   (b)   (c)

FIGURE 15.8   The Crystals and Spores of *Bacillus thuringiensis*. Photographs of a strain of the Gram-positive rod *Bacillus thuringiensis* showing its spores and the crystals associated with the insecticidal ability of this organism. (**a**) A phase-contrast photomicrograph of *B. thuringiensis* spores attached to the insecticidal crystals. (Bar = 5 µm.) (**b**) A scanning electron micrograph of the spore-crystal complex after isolation from the cells. (**c**) A transmission electron micrograph of a sporulating cell that contains a crystal-spore complex surrounded by a membrane called the exosporium.

The first seeds with inserted Bt genes were put in U.S. soil in 1996, and by 2001, over 35% of all corn planted in the United States consisted of genetically modified corn, popularly known as Bt corn. Losses due to corn borers have dropped dramatically. Genetically modified cotton (Bt cotton) has also been planted on millions of acres in the United States. Although both crops display resistance to pests, scientists have noted a resistance to Bt toxin emerging in some caterpillars. Later in this chapter, we shall explore some other considerations relating to genetically modified crops.

Still another approach has been tried to incorporate gene-related resistance in corn plants. Researchers have spliced the Bt gene into a bacterium that normally lives harmlessly with corn plants. The scientists then forced some of the gene-altered bacteria into corn seeds. When the seeds were sown, the toxin-producing bacteria flourished along with the plant, and when an insect ate the plant, it also consumed the toxin. This approach is advantageous because only the insect attacking the plant is subjected to the toxin. Research on this approach is ongoing.

Another useful *Bacillus* species is *Bacillus sphaericus*, a microbe whose toxin kills at least two species of mosquitoes. To increase the efficiency of the toxin, researchers inserted the gene that encodes it into the bacterium *Asticcacaulis excentris*. Using *A. excentris* as a gene carrier has several advantages: It is easier to grow in large quantity; it tolerates sunlight better than *B. sphaericus*; and it floats in water, where mosquitoes feed (the heavier, spore-laden *Bacillus* species sinks). Scientists were delighted to observe insecticidal activity by *A. excentris* against mosquitoes that transmit such microbial diseases as malaria and viral encephalitis.

Many bacteriologists are also investigating the insecticidal potential of the bacterium *Photorhabdus luminescens*. The microbe's toxin, known as Pht, attacks the gut lining of insect larvae (as Bt does), and it is contained in large cytoplasmic crystals

■ *sphaericus*
sfe'ri-kus

■ *Asticcacaulis excentris*
as-tik-ä-cäu'lis
ek-sen'tris

■ *Photorhabdus
luminescens*
fō-tō-rab'dus
lü-mi-nes'senz

(also as Bt is). However, the spectrum of activity of Pht is wider than that of Bt, and it includes numerous species of caterpillars as well as cockroaches. Normally, *P. luminescens* lives in the intestines of soilborne worms called nematodes. The worms invade tissues of insects in the soil, and the toxins produced by *P. luminescens* kill the insects. The toxin-encoding Pht genes have been isolated, and efforts are underway to introduce them to plant cells.

Viral and Fungal Insecticides

Viruses also show promise as agricultural pest-control agents, in part because they are more selective in their activity than bacteria. Pure cultures of viruses can be used, or biotechnologists can harvest infected insects, grind them up, and use them to disseminate the virus. Among the insects successfully controlled with viruses are the cotton bollworm, the cabbage looper, and the alfalfa caterpillar.

Researchers have also developed an insecticide using a toxin found in the venom of a scorpion. The toxin, which paralyzes the larvae of moths, is encoded by a gene cloned from the cells of the scorpion. To make the insecticide, this gene is attached to a baculovirus, a type of virus having a high affinity for insect tissues. To conduct a field trial, the virus is sprayed on lettuce and cotton plants infested with moth larvae. At the conclusion of the trial, the plot is sprayed with a 1% solution of household bleach to destroy any remaining viruses.

Viruses and their genes have also been enlisted to protect grapevines, although in a seemingly odd way. The grape fan-leaf virus (GFLV) is transmitted by nematodes and is commonly found in the soil of many regions of France, where it infects various kinds of grapevines. The virus causes the leaves of the grapevine to shrivel ("fanleaf"). Soon, the leaves lose their chlorophyll and become yellow; eventually, the grapevine dies.

Beginning in 1993, French biotechnologists used GFLV to protect grapevines that produce champagne grapes. They identified and cloned the genes that encode the proteins of the viral capsid and inserted them in *A. tumefaciens* cells. Then they infected the champagne grapevines with this bacterium. Soon the plant cells were producing the capsid proteins of GFLV, and as they did so, the plants developed resistance to the virus. Although researchers are uncertain why this happens, they postulate that the capsid proteins are taking up the spots on the plant cell surfaces where the pathogenic viruses would normally attach during the replication process. Tying up these sites apparently brings resistance to the virus.

Even a fungus is being employed in the pest wars. California researchers have used the mold *Lagenidium giganteum* to protect soybean and rice plants against crop-damaging mosquitoes, and they have sprayed the mold in mosquito-infested settings such as wetlands. The fungus grows on the surface of the mosquito larvae and forces its spores into their internal organs. Within a day or two, the larvae die. Marketed as Laginex, the fungal preparation has been approved for certain specified uses in the United States by the Environmental Protection Agency (EPA).

■ *Lagenidium giganteum*
la-je-ni'dē-um
ji-gan'tē-üm

Biotech Replacements

A slightly different strategy to help farmers is being studied by biotechnologists at the University of California, who are trying to endow insects with so-called suicide genes. The scientists are working with cotton bollworms, the pest caterpillars that thrive in cotton bolls and systematically destroy the cotton plants. They are attempting to introduce into laboratory-reared bollworms a lethal gene that would activate in the insects' offspring. Once activated, the gene would encode enzymes that stimulate chemical reactions leading to insect death. Scientists hope to raise millions of genetically modified bollworms and release them to the environment. Theoretically, the transgenic bollworms will mature to moths that mate with normal wild moths and produce offspring that die from the effect of the gene.

The goal of genetically disarming agricultural pests received a boost during the 1990s when Greek scientists produced a transgenic medfly. Medflies (i.e., *Medi*terranean fruit *flies*) are notorious destroyers of fruit and coffee crops throughout the world. Researchers relied on techniques learned from long years of experience with laboratory fruit flies. They located a DNA sequence that transports a foreign gene from the surface of medfly sperm and egg cells into the interior. Then, they attached a marker gene for eye color and found the marker expressed in the offspring. The next step is to insert genes useful for controlling the pest, as is done for the bollworm.

Molecular biologists have also found that microbes can be used in the fields to enhance frost resistance in plants. In many cases, frost develops on a plant in association with a group of so-called ice-nucleating bacteria. At freezing temperature (0°C or 32°F), these microbes encourage frost development by forming ice crystals on a variety of vegetable crops, as well as fruit and nut trees. Biotechnologists have attempted to replace the ice-nucleating bacteria with bacteria that form ice crystals at lower temperatures. *Pseudomonas syringae* is among the more prevalent ice-nucleating species of bacteria. Researchers have taken a laboratory strain of this microbe and successfully removed the gene that encodes the proteins involved with ice formation. In doing so, they produced the first "ice-minus bacteria," that is, bacteria lacking the ice-nucleating gene. When sprayed on plants, these bacteria gradually crowd out the ice-nucleating strains of *P. syringae*. Ice forms on the plants at a lower temperature, meaning that the growing season can be extended considerably. Experiments are ongoing with potato and strawberry plants to determine how practical the ice-minus bacteria will be. Researchers also foresee using such bacteria in ice cream and cake production to prevent the formation of ice crystals during the freezing process.

Resistance to Herbicides

Herbicides are weed-killing chemicals used to clear the land of all plant growth before a field is sown. However, some weed seeds usually remain among the crop seeds. Then weeds and crops grow together, and the weeds often rob the crop plants of vital nutrients, while crowding them out. It is generally agreed that it would be advantageous to increase the resistance of crop plants to herbicides, so that herbicide could be sprayed on the field during the growing season.

One commonly used herbicide is glyphosate, the active ingredient in commercial products such as Roundup and Tumbleweed. Glyphosate inhibits the activity of enzymes that synthesize essential amino acids in plant chloroplasts. By a coincidence of nature, *Escherichia coli* cells possess a gene that encodes an enzyme that has the same synthetic function but is more resistant to glyphosate. Biotechnologists have isolated and cloned this gene and attached it to the Ti plasmid of *Agrobacterium tumefaciens* (discussed above). Then they used the plasmid to deliver the gene to the cells

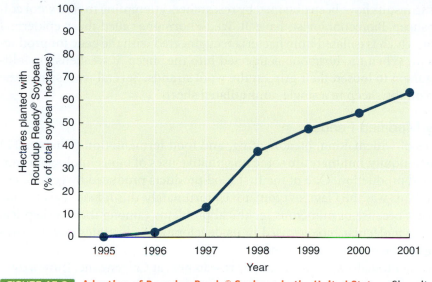

FIGURE 15.9 **Adoption of Roundup Ready® Soybean in the United States.** Since its introduction in 1996, U.S. farmers have rapidly adopted genetically modified soybean to tolerate glyphosate. *Source:* Data are from USDA and National Center for Food and Agricultural Policy.

of tobacco and soybean plants. Here the genes encode the more resistant *E. coli* enzyme and make the plants more resistant to glyphosate than the weeds. Thus, when glyphosate is sprayed on the field, the weed plants die but the crop plants resist the glyphosate and live. By 2001, over 60% of the U.S. soybean crop was planted with genetically modified (i.e., "Roundup-resistant") seeds (**FIGURE 15.9**).

Scientists have performed similar experiments with cotton and other plants and the herbicide bromoxynil. Bromoxynil-tolerant plants are being cultivated in experimental plots throughout the United States. Critics of the process suggest that the herbicide-tolerance genes could pass from the crop plants to the weed plants and make the herbicide less useful. Some research evidence indicates that this transfer does indeed occur, and efforts are currently underway to assess the effects of the transfer.

Pharm Animals

Many of the products of gene-altered microbes have found their way into animals as scientists continue to experiment to improve the quantity and quality of our foods. Most experiments have involved inoculations with hormones produced by gene-altered bacteria. Animals treated with these hormones and other pharmaceutical products have acquired the catchy name "pharm animals."

Among the first animals treated with a synthetic bacteria-derived hormone was the dairy cow. As early as 1983, the gene for bovine growth hormone (BGH) was identified and cloned in *Escherichia coli*. Soon the cells were reproducing the hormone *en masse*. Injected into beef and dairy cattle, BGH (also known as somatotropin) promotes growth of bone and muscle and increases milk production. Increases in milk production of up to 25% have been observed, and the FDA has declared that milk produced by hormone-treated cows is safe to drink. Pigs have also been treated with the synthetic hormone to reduce the amount of untrimmable fat in their muscle.

Interesting results have been obtained by researchers attempting to ease wool collection from sheep. Biotechnologists have utilized a hormone called sheep epidermal growth factor, which is isolated from bacteria reengineered with the gene for producing the protein. When the hormone is injected into the sheep, it weakens the follicles, causing them to loosen their grip on the wool strands. Several weeks later, the fleece comes off the sheep in a single, unmutilated sheet.

Genetically Modified Foods

The processes of winemaking, breadmaking, and food fermentation (Chapter 12) trace back to antiquity, but there are some imaginative uses of biotechnology in contemporary food production. One of the first food products produced by genetic engineering methods was the FlavrSavr tomato. Unfortunately, this tomato never made it to the market because of numerous technical problems (among them a skeptical public), but the molecular biology used in its development illustrates how innovative thinking can be put to practical use.

The FlavrSavr tomato was "invented" by researchers at Calgene Inc. (now a division of Monsanto Inc.). The tomato contains a gene that delays rotting, thereby allowing farmers to leave it on the vine to ripen several days longer and ship it without refrigeration. The secret behind the tomato is biochemical control of an enzyme called polygalacturonase (PG). This enzyme digests pectin, a protein in tomato cell walls. Pectin digestion causes a tomato to decay ("to rot"), a process the producers would prefer to avoid during storage and distribution.

In the late 1980s, researchers at Calgene identified the gene that encodes PG and successfully removed the gene from tomato plant cells. Then they inserted the gene in the bacterium *Escherichia coli* and used the new methods of biotechnology to produce a mirror image of the gene, a DNA molecule with complementary bases. Such a gene encodes an mRNA molecule that is complementary to the normal mRNA molecule. This complementary mRNA molecule acts as an antisense molecule (Chapter 14), and it binds to and destroys the normal mRNA molecule (the sense molecule) that carries the genetic message for PG. Working backwards from the mRNA, the scientists synthesized an antisense DNA molecule—essentially, an antisense gene.

The researchers cloned this antisense gene in *E. coli* cells and inserted it into tomato plant cells. Here the gene encoded the antisense mRNA, which sought out and destroyed the normal (sense) mRNA, thereby halting the synthesis of PG. Without PG, pectin digestion failed to take place, and the tomato took considerably longer to decay. **FIGURE 15.10** illustrates the process. The FlavrSavr tomato was anticipated to be the first of a new wave of genetically engineered foods, but a very expensive approval process, the lack of a correct picking apparatus, the failure to produce a mixture of varieties, and a skeptical public contributed to its demise.

Failure of the biotech tomato did not deter scientists. They soon produced a new type of canola oil. Canola oil, extracted from canola seeds, is used in such things as ice cream, detergents, and facial creams. It is also valuable as a cooking oil because it contains relatively few saturated fats, a characteristic that appeals to nutritionists. Biotechnologists improved canola oil by adding new genes: In the California bay laurel tree, they located a gene that encodes lauric acid, and they introduced the gene to canola plant cells. Lauric acid (laurical) is a fatty acid not normally found in canola

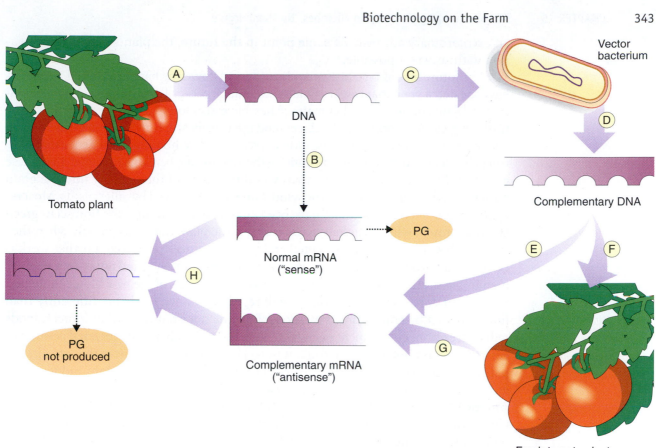

Tomato plant

DNA

Vector bacterium

Complementary DNA

Normal mRNA ("sense")

PG

Complementary mRNA ("antisense")

PG not produced

Fresh tomato plant

FIGURE 15.10 **The FlavrSavr Tomato.** (**a**) DNA technologists begin by isolating the gene that encodes the "rotting enzyme" poly-galacturonase (PG). (**b**) In tomato plants, this gene encodes a normal mRNA molecule (the "sense" molecule), which is translated to PG. (**c**) DNA technologists transfer the PG gene to a vector bacterium and (**d**) induce it to produce a DNA molecule that is complementary to the PG gene. (**e**) This gene encodes an mRNA molecule that complements the normal mRNA molecule. The complementary mRNA molecule is "antisense."(**f**) The complementary DNA is inserted into another tomato plant. (**g**) Here the DNA encodes the antisense mRNA molecule. (**h**)The antisense mRNA unites with the sense mRNA molecule, and thereby neutralizes it. The mRNA message is not translated, and PG is not produced. Without PG, rotting does not occur in the tomato.

oil. It gives added nutritional value to the oil and has the feature of not breaking down easily during cooking, a factor that appeals to homemakers and chefs.

As we noted in this section's introduction, biotechnologists are continuing to work on a host of other genetically modified foods, including a bruise-resistant potato that will retain eye-appeal when peeled. Moreover, a type of yellow squash has been genetically altered to resist two plant viruses (appropriately, the squash is called "Freedom 2"). And in 2001, scientists at the University of California introduced the salt-tolerant tomato. This tomato resulted from the addition of a gene from the mustard plant *Arabidopsis*. The gene encodes a protein that shuttles sodium ions into cytoplasmic sacs so that the ions cannot be used to form sodium chloride, or salt (salt-tolerant plants use this gene to survive high-salt environments, but crop plants apparently lack or turn off the gene). By some estimates, salt-laden soil limits the use of over a quarter of the Earth's irrigated land, and thus, salt-tolerant plants would be an extraordinary advance. At some point in the future, the plants may make irrigation with seawater possible.

The future may also yield plants whose time of flowering has been altered. Scientists have located a gene that can be mutated to induce earlier flowering, a helpful adaptation for fruits grown in cold climates. They have also found a way to delay flowering through gene alteration and thereby extend the time before leafy crops such as spinach and lettuce send up their flowers. Other researchers are homing in on genes that control plant height in grain plants by limiting the activity of giberellins (Chapter 14). These genes could lead to dwarf plants that would put more of their energy into the grain rather than the stalk, while limiting wind damage experienced by taller plants. Most recently, plants such as the potato have been engineered to produce the florescent green jellyfish protein when the plant needs water. If the plants are watered only when they glow, over watering can be prevented and water resources preserved. Further work is being conducted to produce plants that report on their nitrogen and potassium requirements. These are but a few of the imaginative applications of the new biotechnologies.

The FDA has announced that it will regulate biotech foods no differently than foods created by conventional methods. But this does not mean that biotech foods will be free from government oversight. Such food must have a nutritional content of equal value to nonengineered food, it must contain only substances that are already part of the human diet, and it must contain no substances that might induce allergies. In 2000, to help allay consumer fears regarding genetically modified foods, the American Society for Microbiology (ASM), representing over 42,000 microbiologists worldwide, took the unusual step of issuing a statement supporting the safety of biotech foods. Said the society's press release: "ASM is sufficiently convinced to assure the public that plant varieties and products created with biotechnology have the potential of improved nutrition, better taste, and longer shelf-life . . . and ASM is not aware of any acceptable evidence that food produced with biotechnology and subject to FDA oversight constitutes high risk or is unsafe."

Veggie Vaccines

Researchers continue to report encouraging results on the use of transgenic vegetables and fruits as carriers of microbial vaccines. Whimsically called veggie vaccines, fruit and vegetable carriers are desirable because they are inexpensive to store and are readily accepted in developing countries, where vaccines are needed most. The vaccines could be produced locally (thus avoiding transportation and refrigeration problems), and vaccine production would go on indefinitely because the plants could be propagated continuously. Moreover, use of the vaccines would not necessitate exposing people to potentially contaminated syringes. And the vaccines would be safer than vaccines containing whole organisms or their parts.

Researchers at New York's Boyce-Thompson Institute were among the first to splice microbial genes into plant cells. In 1997, a group of biotechnologists reported success in incorporating in potato plant cells a benign segment of toxin genes of pathogenic *Escherichia coli*. When chunks of the potato were eaten by volunteers during trials, an antibody response against the toxin was provoked. Unfortunately, the potatoes had to be eaten uncooked because cooking reduced the amount of immune-stimulating proteins. To resolve this dilemma, researchers tried to amplify the amount of toxin protein produced by the potato and administer "booster " feedings. Both modifications met with limited success.

Future research efforts will employ more palatable plants such as tomatoes and bananas. Tomatoes are appealing vaccine carriers because of their short growing season. Bananas have attracted interest because children like them (children are the main recipients of vaccines); they grow in many tropical countries (where vaccines are essential to public health systems); and they don't have to be cooked before consumption. Indeed, scientists have already spliced genes from hepatitis B viruses into banana plant cells and are attempting to induce the plant to produce immunizing proteins that will stimulate an antibody response. Ultimately, they hope to engineer a "vaccine banana" whose genes can encode these proteins in a human recipient.

As expected, using vegetables and fruits as vaccine carriers carries a unique set of problems that must be addressed: How can researchers ensure the passage of immune-stimulating substances through the acid of the stomach and the enzymes of the small intestine? What methods can be used to encourage passage of the substances from the gastrointestinal tract to the bloodstream, where the immune reaction occurs? Will the vaccine components be strong enough to elicit a protective antibody response? Will vaccine components combine with body proteins and elicit allergic reactions? Despite these obstacles, biotechnologists have great hopes for transgenic vaccines. Unfortunately, many years of laboratory tinkering and many clinical trials will pass before the banana replaces the needle as a vehicle for delivering vaccines.

A FINAL THOUGHT

Although still in its infancy, biotechnology has breathtaking possibilities for improving human health and nutrition. Genetically modified foods stand at the forefront of this revolution. But no revolution is without controversy, and the critics of genetically modified foods are numerous and loud. They point to the potential for allergic reactions in consumers, the possibility of gene transfers in the environment to create "superweeds," and the death of unintended animal victims of insecticides in nature. Another concern of some is the lost profits in the pesticide and herbicide businesses.

Supporters of genetically modified foods respond that these risks are present but are slight when compared to the benefits. They point to a burgeoning world population that must be fed—a population of 6 billion people in 2000, and an estimated 9 billion in 2050—and note that biotechnology can dramatically improve productivity where food shortages arise from pest damage and plant disease. They point to the steady decline in the world's arable land, a decline that will accelerate precipitously in the years ahead, and they note that biotechnology can raise overall crop yields in developing countries by over 25%. And they mention the already successful biotech foods such as vitamin-enhanced "golden rice" that make a teaspoon of food more nutritious than ever.

Supporters also cite the safeguards already in place. They note the rigorous oversight of biotechnology exerted in the United States by the FDA, the USDA, and the EPA. But they also concede that making life easier for farmers will not outweigh real or imagined risks and that the public needs to hear about the benefits of biotech foods from credible experts. A recent survey found that two out of three Americans approve of genetically modified foods and have confidence in the FDA. To build and sustain that confidence during this period of transition, the biotechnology industry will have to offer the public something of value. A good starting point is the fact that American farmers who use Bt cotton have cut back their use of insecticides by 2 million pounds per year.

■ QUESTIONS TO CONSIDER

1. This chapter relates much about the innovative and imaginative work being done in agricultural biotechnology. Suppose you were the head of a lab and had the choice of pursuing any project you wished. What would that project be? Why would you pursue it? What would be the chances of your success?

2. A carton of yogurt usually carries the message "contains active cultures." However, knowledgeable individuals know that the wording should really be "contains live bacteria." Why do you suppose yogurt manufacturers perpetuate the minor deception, and what can be done to get them to tell the truth?

3. It is intriguing to stop and wonder how what got to be got to be. How, for example, did microbes find their way into the rumen of cattle and evolve to become as important as they now are? And what if they had found their way inside some other type of animal? Suppose they had found their way into the human intestine? What answers might you give to these questions?

4. Italians are fond of a dish called *pásta e fagiòle*. The dish consists of pasta with beans, chick peas, or other legume. Italians know that if they eat enough *pásta e fagiòle*, they don't have to eat quite so much meat, which is often in limited supply. Microbiologically speaking, why does this idea make sense?

5. Scientists have noted that one of the side effects of the use of Bt toxin has been a reduction in the population of Monarch butterflies. Why do you think this may have happened, and what might you as a concerned citizen do about it?

6. *War of the Worlds* is a classic novel by H. G. Wells, made into several movies. The story details the invasion of Earth by aliens from the planet Mars. All the might and resources of earthlings are exhausted as the aliens make their way through American cities destroying everything in their path. When all appears lost, the aliens suddenly die and their space vehicles crash to Earth. After a pregnant pause, the narrator solemnly explains, "In the end, the invaders were exposed to 'germs' in the Earth's atmosphere and died." Although *War of the Worlds* is fiction, nevertheless, a case can be made for microbes contributing mightily to the well-being of society. In what ways has this chapter supported their case?

7. Futurists tell us that a great variety of genetically modified foods will soon be appearing in the supermarkets of the world. If you had your choice, what sort of improvements would you like to see in foods? What foods would you change to improve characteristics such as flavor, shelf life, and nutritional quality? What genetic changes would be necessary to effect these improvements?

◼ KEY TERMS

Informative facts are necessary for the expression of every concept, and the information for a concept is founded in a set of key terms. The following terms form the basis for the concepts of this chapter. On completing the chapter, you should be able to explain and/or define each one:

acidophilus milk
bacteroids
baculovirus
Beijerinck
biolistic
bromoxynil
Bt corn
callus
denitrification

glyphosate
nitrification
nitrogen fixation
plantibodies
ripened cheese
ruminant
sour curd
yogurt

◼ http://microbiology.jbpub.com/book/microbes

The site features **eLearning**, an online review area that provides quizzes and other tools to help you study for your class. You can also follow useful links for in-depth information, read more stories of microbiology, or just find out the latest microbiology news.

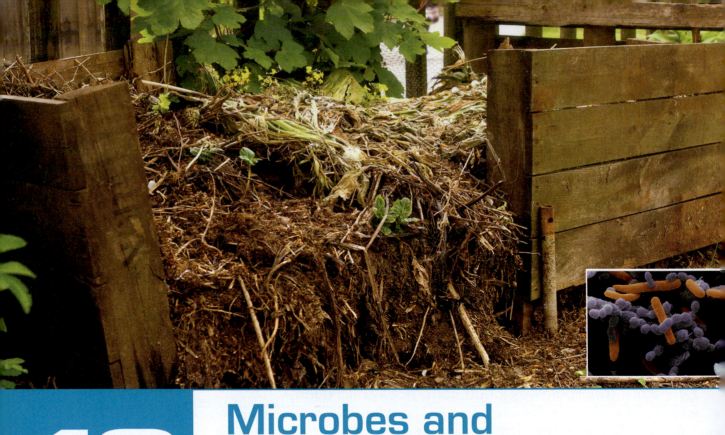

16

Microbes and the Environment: No Microbes, No Life

Compost is the remnant of aerobic bacterial metabolism of organic materials and makes an excellent fertilizer for plants. Composting can be done on the household or industrial scale and is an excellent way to reduce landfill garbage and produce fertilizer at the same time. The inset shows a picture of some of the many different types of bacteria involved in forming compost.

▮ Looking Ahead

It may appear brash to suggest that microbes are among the most important participants in the web of life on Earth, but the pages of this chapter will provide ample opportunities for you to see the wisdom of that assertion.

On completing this chapter, you should be able to . . .

- appreciate the fundamental importance of microbes in the recycling of carbon on Earth.
- understand how the metabolism of microbes is essential to the sulfur, nitrogen, and phosphorus cycles in the environment.
- explain the steps taken by municipalities to dispose of organic waste and note where microbes play essential roles in that disposal.
- summarize how microbes help eliminate environmental pollutants and industrial waste through the process of bioremediation.
- compare the microbial floras of various water environments and explain how microbes contribute to water pollution.

- identify several traditional and modern methods for detecting microbes in water supplies.
- describe the methods by which municipalities lower the microbial contents of water and ensure its safety for drinking purposes.

During the 1960s, the renowned scientist and writer Buckminster Fuller described our planet as "Spaceship Earth." Fuller envisioned Earth as a self-sustaining entity whirling through space, completely isolated from everything but sunlight. He hoped to convey the notion that some of Earth's resources—such as oil and gas—are nonrenewable, but most others are renewable and must be recycled.

An essential element of Fuller's vision is that Spaceship Earth depends for life-support on life itself: The living things aboard the great craft produce its oxygen, cleanse its air, adjust its gases, transfer its energy, and recycle its waste products—all with great efficiency.

High on the list of key passengers are the microbes. As we shall see in this chapter, many important resources of Spaceship Earth could not be recycled without microbial intervention. In fact, without microbes, the plants and animals of Earth would have depleted its resources countless eons ago, and the great experiment of life would have failed miserably. That it did not fail is testament to the power and adaptation of microbes: They have managed to fill every conceivable niche on Earth; they have evolved so they can sustain themselves on anything Earth has to offer; and they have adapted to participate in the intricate web of metabolic activities that permits Spaceship Earth to continue on its long journey through the universe.

The Cycles of Nature

Take a pinch of rich soil and hold it in your hand. You are face to face with an estimated *billion* microbes representing over 10,000 different species. If you were to try cultivating them in the laboratory, perhaps only a few species would grow. The remaining thousands of species would remain unknown until scientists developed the right combination of nutrients and environmental conditions to cultivate them. But our knowledge about them is immaterial to how important they are to us: They recycle our nutrients, refresh our environment, clean our waste, and stand as vital links between life as we know it and a world we cannot begin to imagine. Indeed, because microbes are invisible, their importance in the environment may go unsuspected. Yet, without microbes, life could not continue, as we shall see in the pages ahead.

The Earth Ecosystem

Because of their ubiquitous distribution and their diverse chemical activities, microbes provide the basic underpinnings to the cycles of elements on Earth. But they do not operate alone. Rather, they act as part of an ecosystem, a community of all the organisms in a defined space. Certain ecosystems (such as lakes) are fairly small, but other ecosystems (such as the entire North American continent) are enormous.

A CLOSER LOOK
"All for One and One for All"

Here's an intriguing thought: Suppose a bacterium were not a single-celled creature, but rather, a subunit of a global organism consisting of the entire bacterial world.

Such a view may not be quite so preposterous, according to Sorin Sonea of the University of Montreal. Sonea points out the vast majority of bacteria live in communities of mixed strains whose metabolisms complement one another. Such an association may be viewed as a sort of multicellular eukaryote. The communities of bacteria appear to improvise in nature and adapt to changes in the environment; for example, certain strains emerge and others disappear. The difficulty in cultivating various strains in the laboratory is a reflection of their metabolic dependence on one another.

Sonea continues his hypothesis by pointing out that a bacterium can only survive in nature as part of a temporary metabolic association. Such an association makes the bacterium different from the relatively independent eukaryotic cell. Essentially, then, each bacterium is part of a widely dispersed planetary system of subunits, all metabolically linked and all dependent on one another for the common good.

Do you agree with Sonea's hypothesis, or can you find fault with it? What evidence can you offer to support or reject the hypothesis? You might also enjoy investigating the Gaia hypothesis, first stated by James Lovelock. (Gaia is the mythical goddess of Earth.) Lovelock envisions Earth as one "superorganism" and living things as one metabolizing creature that is continually reshaping the landscape to fit its needs. The existence of one "superbacterium" might conceivably fit into this pattern. Certainly, the idea is novel enough to merit more than a passing moment of attention.

Whatever its size, an ecosystem is a type of superorganism having the ability to respond to and modify its environment. Although the lake may be easier to see as an ecosystem, the continent is no less an ecosystem. It has even been suggested that all the microbes on Earth may constitute a single ecosystem (see A Closer Look above).

Within the ecosystem that is the Earth, the sum total of all living organisms is known as the biota. Influenced by physical characteristics such as rain and tides, wastes and sewage, and oxygen and pH, the biota includes unknown numbers of species of microbes as well as hundreds of thousands of species of plants and animals. How these organisms interact with one another and with their physical and chemical environments is the subject matter of environmental microbiology.

The physical space or location where a species of microbe lives is called its habitat. Habitats may be freshwater lakes, marine environments, or soils. Soil is a complex mixture of mineral particles, along with organic matter from decaying organisms and plants, animals, and microbes. Within this habitat, microbes occupy an important niche as the preeminent recyclers of nutrients.

Microbes live together with other organisms in a community. The two major elements of a community are energy and chemical substances. Energy enters the ecosystem through sunlight and is used by photosynthetic organisms (green plants, for example) for the synthesis of organic matter. These organisms, known as producers, are then used as nutrients by other organisms called consumers. Much of the energy is lost as heat during respiratory processes going on in these organisms, and as primary consumers are in turn eaten by secondary consumers, the energy continues to

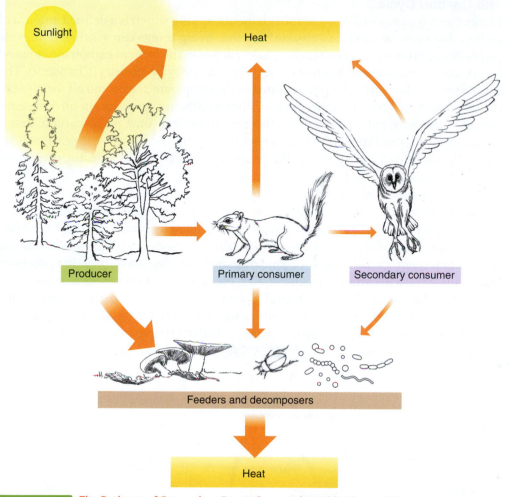

FIGURE 16.1 **The Pathway of Energy in a Forest Community.** This diagram illustrates the entry of energy into the community and its gradual loss during transfers between levels in the ecosystem. The width of the arrow represents the relative quantity of energy lost during each transfer.

be lost. Microbes and other feeders, collectively known as decomposers, attack the consumers after they die and use the available energy for their metabolism. Eventually, the energy that entered the ecosystem as sunlight is dissipated, primarily as heat energy (FIGURE 16.1).

But the dissipated energy is constantly being replaced by new energy from the sun, and life will continue as long as the chemical substances are recycled. In the case of carbon, for example, the chemical processes of respiration release carbon from carbohydrates in the form of carbon dioxide. The latter can then be used in photosynthesis and recycled back into the ecosystem. Nitrogen, sulfur, and other elements are also released during metabolic processes, and thus are made available for reentry into the ecosystem. This is the basic concept that underlies the biogeochemical cycles that occur in the environment. In the next several pages, we shall emphasize the roles that microbes fill and reemphasize the insight first enunciated by Louis Pasteur over 100 years ago: "The influence of the very small is very great, indeed."

The Carbon Cycle

Planet Earth is composed of numerous elements, among which is a defined amount of carbon that must be recycled constantly so that living things can synthesize organic compounds. Photosynthetic organisms take carbon in the form of carbon dioxide and form carbohydrates using the sun's energy and chlorophyll pigments (Chapter 9). The vast jungles of the world, the grassy plains of the temperate zones, and all the algae in the seas display the results of this process. Photosynthetic organisms, in turn, are consumed by animals, fish, and humans; these creatures use some of the carbohydrates as energy sources and convert the remainder to cell parts. Although some carbon is released by respiration and returns to the atmosphere as carbon dioxide, the major portion of the carbon is returned to the soil when an organism dies.

In the soil, microbes exert their influence as the primary decomposers of organic matter. Working in their countless billions, bacteria, fungi, and other microbes consume organic matter and release carbon dioxide for reuse by plants. This activity is carried on as a concerted action by a huge variety of microbes, each with its own nutritional pattern of protein, carbohydrate, or lipid digestion. Aerobic decomposition processes lead to the release of carbon dioxide through the reactions of glycolysis and the Krebs cycle, as we discuss in Chapter 9. A similar process occurs in the oceans, known to marine biologists as "carbon sinks." Here the cyanobacteria (Chapter 5) and zooplankton (Chapter 7) are important participants in the cycle that brings carbon dioxide back into the system through photosynthesis. FIGURE 16.2 displays a simplified carbon cycle.

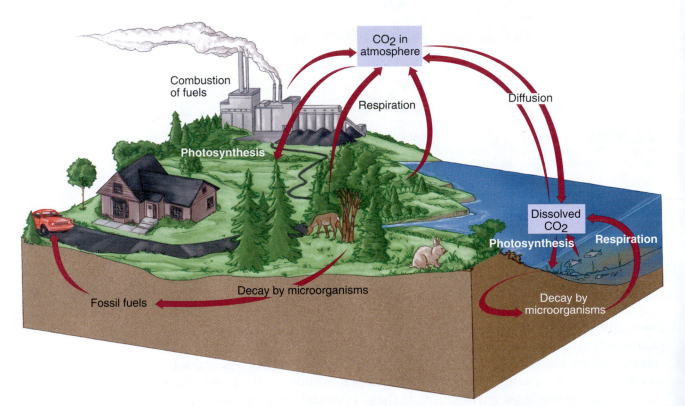

FIGURE 16.2   A Simplified Carbon Cycle. Photosynthesis represents the major process by which carbon dioxide is incorporated into organic matter, and respiration accounts for its return to the atmosphere. Microorganisms are crucial to all decay in soil and ocean sediments.

But not all the processes in the carbon cycle are aerobic. To the contrary, chemical decompositions take place in anaerobic environments such as the muddy bottoms of lakes and waterlogged soils, and fermentation reactions result in organic acids and other small molecules. These are later broken down to even smaller molecules, including methane, hydrogen, and carbon dioxide. Various species of archaea use carbon dioxide and hydrogen to form more methane (CH_4). Although much of the methane is lost to the environment, some is used by other archaea and by methanotrophic bacteria as a source of carbon for synthesizing their organic compounds and, consequently, making more of themselves. Protozoa and other microbes (Chapter 7) then consume the carbon compounds and participate in the food webs that bring the carbon back to larger animals.

The archaea even display their carbon-recycling abilities below the sea. In 2001, scientists reported that in the sediments of the ocean floor, these microbes devour as much as 300 million tons of methane each year and make the carbon available to nearby bacteria living symbiotically with them. Without the action of archaea, the methane would probably enter the atmosphere and enhance the greenhouse effect to the point that the Earth might become too hot for life to continue.

It is clear that without microbes, the Earth would be a veritable garbage dump of accumulating animal waste, dead plants, and organic debris. Operating in soils and waters of the environment, microbes break down animal and plant waste and contribute the greatest share of organic matter to the environment. In the soil, much of the organic matter remains unused, in part, because the microbes are killed by the toxic products of the decay. This left-over organic matter then combines with mineral particles to form humus, a dark-colored material that retains air and water and is excellent for plant growth.

Microbes accomplish a similar feat under controlled conditions, where they decompose manure and other natural waste materials and convert them to compost for crop fertilization. Although both conventional and organic farms use manure and compost for soil fertilization, only certified organic farmers are required to have a plan detailing the methods for building soil fertility with these materials. Furthermore, certified organic farmers are prohibited from using raw manure for at least 60 days prior to harvesting crops for human consumption.

Scientists believe that coal originated millennia ago in marshes and lakes as a result of microbes living and reproducing under anaerobic conditions. Petroleum may have originated at the bottoms of oceans in the same way. During the burning of the coal or petroleum, carbon dioxide is released back into the atmosphere. Indeed, petroleum buried underground where there is no oxygen does not decay, but when it is brought to the surface, it is used as a food source by aerobic microbes that attack its components and break them down. This microbial growth is an important aspect of petroleum microbiology. For example, jet fuels can support a flourishing microbial population, and the microbes tend to clog fuel strainers and interrupt the flow of fuel to the jet engines, a potentially disastrous situation.

But there is more. Microbes also break down the carbon-based chemicals produced by industrial processes, including herbicides, pesticides, and plastics. (We shall study how this process is put to use in bioremediation.) Moreover, many species of microbes trap carbon dioxide from the atmosphere and form carbohydrates through photosynthesis. And in 2001, scientists demonstrated for the first time that bacteria release the carbon locked up in organic compounds in sedimentary rock called shale, thereby filling another gap in the global cycle of carbon. Through activities like these, the microbes represent a fundamental underpinning of organic creation.

The Sulfur Cycle

The element sulfur (S) is incorporated in most living things and used to form various amino acids and coenzymes. For example, the sulfhydryl group (—SH) is a key component of the amino acids cysteine and cystine. In protein, it serves as a bridge between various parts of the molecule by binding nearby amino groups together, much like the rungs of a ladder hold together the uprights.

A major share of sulfur exists in nature as sulfate ions (SO_4^{2-}). Certain bacteria use sulfate ions as electron acceptors and convert the sulfur to hydrogen sulfide (H_2S). Bacteria of the genus *Desulfovibrio* accomplish this conversion under anaerobic conditions in water-logged soils and muds. The hydrogen sulfide they produce reacts with iron ions to yield iron disulfide (FeS_2), which is black. Anaerobic muds tend to be black for this reason. In addition, hydrogen sulfide has the odor of rotten eggs, and swamps, marshes, and landfills often give off this odor. Hydrogen sulfide is also known to corrode iron and steel pipes buried in soils, creating tiny holes and pits in the pipes, as we shall note presently.

■ *Desulfovibrio*
dē′sul-fō-vib-rē-ō

Although hydrogen sulfide is toxic to many microbes, certain photosynthetic bacteria use it as an electron donor. For instance, the purple sulfur and green sulfur bacteria convert hydrogen sulfide to elemental sulfur during their metabolism. The sulfur is often deposited in granules in the bacterial cytoplasm, later to be oxidized to sulfate ions. These bacteria grow in shallow streams and ponds where light is available for photosynthesis and plenty of hydrogen sulfide is to be found in the black, sulfide-rich mud below.

Other bacteria, the colorless sulfur bacteria, convert hydrogen sulfide into sulfate ions under aerobic conditions and derive energy from this conversion. For example, members of the genera *Thiobacillus* and *Beggiatoa* generate ATP (adenosine triphosphate) in this way. Colorless sulfur bacteria grow in acid-mine waters, where they break down the iron disulfide in pyrite ("fool's gold") and produce sulfuric acid. Draining from a mine, this sulfuric acid poses a serious environmental problem as an insoluble yellow precipitate (FIGURE 16.3). The precipitate colors polluted streams and rivers, making them unsightly and toxic. (To be sure, this is one place where microbes do not endear themselves to us.)

■ *Thiobacillus*
thī-ō-bä-sil′lus
■ *Beggiatoa*
bej′jē-ä-tō-ä

Species of *Beggiatoa* have also been linked to a deep-sea phenomenon. In the late 1970s, oceanographers were searching for evidence of undersea volcanoes when they discovered a series of rifts (cracks) in the ocean floor. Along the rifts, hydrothermal vents were spewing black smoke. To their amazement, the area was rich in animal life, particularly giant clams and 2-meter-long tubeworms. It was unclear how these animals could survive in the hellish environment until sulfur-utilizing bacteria were located in their tissues. Apparently, the bacteria metabolize sulfur compounds in the smoke and use the energy they obtain to turn carbon dioxide into organic matter for use by the animals. In return, the tubeworms, clams, and other animals provide a hospitable environment in their specialized organs, where the bacteria can carry on their biochemical processes while receiving the sulfur and oxygen they need. And all this takes place in an extraordinary ecosystem where the pressure reaches 250 atmos-

Acid Mine Drainage.

pheres (versus 1 atmosphere at the Earth's surface) and the temperature is 300°C (remember, water boils at 100°C).

Sulfur bacteria again made the news in 1999, but this time for their size. In a find described as "dazzling," German investigators reported the existence of a spherical, sulfur-metabolizing bacterium measuring 0.75 millimeter in diameter—a huge size for a microbe. These microbes were the first visible bacteria ever recorded. Found on the seafloor off the coast of Namibia, Africa, they consume both sulfur and nitrogen compounds, thereby linking the ecological cycles of the two elements. The bacteria apparently obtain energy by stripping electrons from hydrogen sulfide molecules and adding the electrons to nitrate ions (NO_4^{2-}). Their immense size is due to a nitrate-filled sac that encompasses 98 percent of the cell's interior, with the cytoplasm occupying a thin layer around the sac and resting against the cell membrane. The investigators named the microbe *Thiomargarita namibiensis*, the "sulfur-pearl of Namibia."

The Nitrogen Cycle

We discuss the nitrogen cycle at length in Chapter 15, but we must mention it here because the impact of microbes is considerable. Indeed, the cyclic movement of nitrogen is of paramount importance to life on Earth because this element is an essential part of nucleic acids and amino acids. Somewhat paradoxically, nitrogen is the most common gas in the Earth's atmosphere, but neither animals nor any but a very few species of plants can use it in its gaseous form. For nitrogen to enter the cycle of life, microbes must intercede.

■ *Thiomargarita namibiensis*
thī′ō-mär-gä-rē-tä na′mi-bē-n-sis

FIGURE 16.4 **A Field of Soybean Plants.** Microrganisms in the nodules of soybean roots convert atmospheric nitrogen into ammonia and enrich the soil.

■ *Nitrosomonas*
nī-trō-sō-mō′näs

■ *Azotobacter*
ä-zo′to-bak-tër

After organic waste is deposited in the environment, the process of putrefaction (protein digestion) by soil microbes yields a mixture of amino acids. Many of these amino acids are converted to inorganic compounds and ammonia. Then, nitrogen recyclers enter the picture: Much of the ammonia is converted to nitrite ions by *Nitrosomonas* species, a group of aerobic bacteria. The nitrite ions are then converted to nitrate ions by species of *Nitrobacter*, another group of aerobic bacteria that obtain energy from the process.

Nitrate is a crossroads substance: Some is used by plants for their nutritional needs, but most is broken down by various species of microbes and liberated to the atmosphere as nitrogen gas. A reverse trip back into living things is an absolute necessity for life as we know it to continue (**FIGURE 16.4**). The process is called nitrogen fixation. Once again, microbes play a key role because they possess the enzyme systems that trap atmospheric nitrogen and convert it to compounds useful to plants. We discuss this process in detail in Chapter 15. Free-living nitrogen-fixing microbes include bacteria of the genera *Bacillus*, *Clostridium*, and *Azotobacter*, as well as types of cyanobacteria and certain yeasts. Symbiotic species such as *Rhizobium* are also involved.

The importance of microbes in the nitrogen cycle cannot be overstated: Without the vital connections that microbes supply, there would be little opportunity for nitrogen to enter the cycle of life; nor would there be any way for nitrogen to be blended into the nutrients that living things must have. There would be few amino acids, few proteins, few enzymes, few structural materials, few of anything built around nitrogen. Of course, it is possible that living things could have evolved differently and then today's organisms would have a meager dependence on proteins, but that is pure speculation. Is life possible without nitrogen? Yes, it is possible, but it would not be what we are used to. Which is why we should always say "life *as we know it*."

The Phosphorus Cycle

The element phosphorus (P) has a place in the chemistry of living things in such compounds as nucleic acids, coenzymes, and the all-important energy molecule ATP (adenosine triphosphate). One portion of the phosphorus cycle begins when exposed rocks are worn away by rushing water (rain, streams, waterfalls, and so forth). The phosphorus enters the sea in the form of dissolved phosphates, ready to enter the cycle of life. Microbes come into play right at the outset. Unicellular algae and other microbes take up the phosphorus as they multiply in the waters. These relatively simple producers manufacture the all-important nucleic acids and other phosphorus-rich compounds. The phosphorus compounds are concentrated as the algae are consumed by other microbes which are then eaten by shellfish and fin fish. These animals feed upon one another and use the phosphorus for organic compounds and such body parts as bones and shells. The death of these animals returns the phosphorus to the sea.

Another portion of the cycle begins with phosphorus mined from rocks and used in fertilizer for the soil. Field crops incorporate the phosphorus and are eaten by animals. In time, animal waste and decomposition as well as soil erosion bring the phosphates to the sea, where some solidifies to rock. In this form, it would be unavailable for eons of time. Fortunately, microbes live in the sea. As above, the unicellular algae and other microbes trap much of the phosphates for reuse in the cycle of life. In doing so, they keep the phosphorus cycling and the "wheel of life" turning. FIGURE 16.5 shows the two portions of the phosphorus cycle.

The Oxygen Cycle

To appreciate the oxygen cycle, consider that the next breath you take probably contains oxygen molecules produced by microbes. Billions of years ago, cyanobacteria started releasing oxygen molecules into the atmosphere, and they have preformed that task unceasingly to this moment. Although we often thank green plants for their contribution to our oxygen needs, it would be more appropriate to extend kudos to the cyanobacteria, for they are our prime benefactors in this regard. As we note in Chapters 5 and 9, cyanobacteria absorb water molecules during photosynthesis, break them down to retrieve the protons, and release the leftover oxygen atoms into the air as oxygen gas. To be sure, plants perform the same type of chemistry, but the amount of oxygen they give off is dwarfed by what the cyanobacteria release.

Animals, plants, and other microbes complete the oxygen cycle during the chemistry of respiration. At the end of the respiratory process (described in Chapter 9), these organisms use oxygen as a final electron acceptor; they combine it with protons and form molecules of water. Bound in water molecules, the oxygen becomes useless for respiration, and our world would change dramatically if it were not released as the free gas. Not to worry—the plants do their share, but the major players in the game of photosynthesis are the cyanobacteria. No cyanobacteria, no oxygen, no life (as we know it).

It should be clear that microbes occupy a prominent position in the elemental cycles of life. To be sure, they keep organic matter fresh and vital as they sustain Spaceship Earth. But the work of microbes does not end here. Microbes also have a key place in preserving Earth's environment. They are the waste digesters, water purifiers, and soil cleansers *par excellence*. In these respects, they help society solve its age-old problem of how to keep the environment a fit place to live. We rely on microbes even more than we realize, as the next section illustrates.

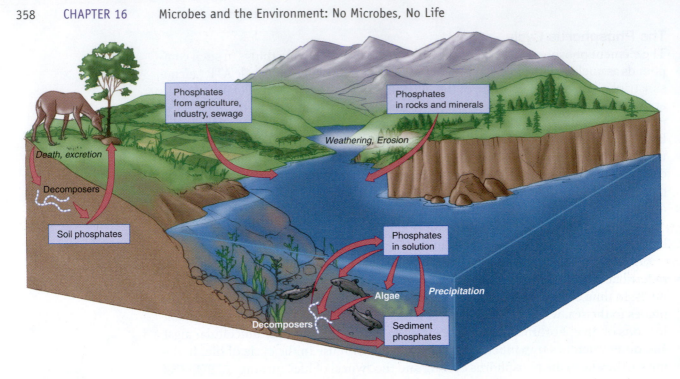

FIGURE 16.5   The Phosphorus Cycle in Nature. Phosphorus enters the cycle as phosphate from various points, and waterborne microbes incorporate it into nutrients at the base of the food chain.

Preserving the Environment

The traditional adage is "The solution to pollution is dilution." This somewhat simplified view recognizes the fact that industrial, agricultural, and human waste can be added to bodies of water and the water will distribute the waste and remain pure. For a small village situated along a quick-flowing river, the adage may be true; but for an industrialized society, the intervention of microbes is a necessary corollary, as we discuss next.

Sanitary and Waste Facilities

The word "sanitation" is derived from the Latin *sanitas*, meaning "health." Sanitation came into full flower in the mid-1800s, as a relatively modern phenomenon. Before that time, living conditions in some Western European and American cities were almost indescribably grim: Garbage and dead animals littered the streets; human feces and sewage stagnated in open sewers; rivers were used for washing, drinking, and excreting; and filth was rampant. Conditions like these fueled the Great Sanitary Movement of the mid-1800s, as we explore in A Closer Look on page 359.

Sewers are at least as old as the Cloaca Maxima of Roman times. Until the Great Sanitary Movement, however, they were nothing more than elongated pipes emptying into nearby rivers and streams. After the Movement, sewers and sewage treatment facilities became much more efficient, and during the 1900s, sophisticated facilities

A CLOSER LOOK
The Great Sanitary Movement

In the early 1800s, the steam engine and its product, the Industrial Revolution, brought crowds of rural inhabitants to European cities. To accommodate the rising tide, row houses and apartment blocks were hastily erected, and owners of already crowded houses took in more tenants. Not surprisingly, the bills of mortality from typhoid fever, cholera, tuberculosis, dysentery, and other diseases mounted in alarming proportions.

As the death rates rose, a few activists spoke up for reform. Among them was an English lawyer and journalist named Edwin Chadwick. Chadwick subscribed to the then-novel idea that humans could eliminate many diseases by doing away with filth. In 1842, well before the germ theory of disease was established, Chadwick published a landmark report indicating that poverty-stricken laborers suffered a far higher incidence of disease than those from middle or upper classes. He attributed the difference to the abominable living conditions of workers and declared that most of their diseases were preventable. His report established the basis for the Great Sanitary Movement.

Chadwick was not a medical man, but his ideas captured the imagination of both scientists and social reformers. He proposed that sewers be constructed using smooth ceramic pipes, and that enough water be flushed through the system to carry waste to a distant depository. In order to work, the system required the installation of new water and sewer pipes, the development of powerful pumps to bring water into homes, and the elimination of older sewage systems. The cost would be formidable.

Chadwick's vision eventually came to reality, but it might have taken decades longer were it not for an outbreak of cholera. In 1849, cholera broke out in London and terrified so many people that public opinion began to sway in favor of Chadwick's proposal. Another epidemic occurred in 1853, during which an engineer named John Snow proved that water was involved in transmission of the disease. In both outbreaks, the disease reached the affluent as well as the poor, and more than half of the sick perished. Construction of the sewer system began shortly thereafter.

As the last quarter of the 1800s unfolded, it became clear that Europe's sanitary movement needed a "smoking gun" to hammer home its point. That incident came in 1892 when a devastating epidemic of cholera erupted in Hamburg, Germany. For the most part, Hamburg drew its water directly from the polluted Elbe River. Just west of Hamburg lay Altona, a city where the German government had previously installed a water filtration plant. Altona remained free of cholera. The contrast was further sharpened by a street that divided Hamburg and Altona. On the Hamburg side of the street, multiple cases of cholera broke out; across the street, none occurred. The sanitarians could not have imagined a more clear-cut demonstration of the importance of water purification and sewage treatment.

were built in most large cities. All operate under the same basic principle: Water is separated from the waste, and the solid matter is broken down by microbes to simple compounds that can safely be returned to the soil and water.

The modern demographic patterns of dense populations, industrial-scale operations, and huge agricultural projects result in enormous amounts of waste that would otherwise overwhelm the self-purification capabilities of lakes, rivers, and streams.

Towns and cities must maintain satisfactory water quality by providing waste treatment plants to enhance and encourage the natural recycling abilities of nature. Microbes are central to the waste treatment process.

Primary waste treatment is normally carried on in settling basins and tanks. The solids are drawn off in the tank and, over a period of days, are subjected to decomposition by incomprehensible numbers of microbes. The objective is to reduce the biological oxygen demand (BOD) of the water. (BOD refers to the amount of oxygen used by microbes present in the water; it is a measure of the pollution of the water.) Aluminum and iron compounds may be added to drag the decomposed waste to the bottom. The liquid remaining contains most of the dissolved organic matter, which can now be subjected to secondary treatment.

In secondary treatment, helpful microbes supply enzymes that digest the organic matter remaining in the water after primary treatment. Oxidation lagoons, large lake-like expanses of water, are used to hold the water for months while bacteria and other microbes break down the organic matter into simple compounds and mineral products. In some countries, great populations of algae are grown in these lagoons and used as food sources. Under controlled conditions, the waste may be totally converted to simple salts such as carbonates, nitrates, phosphates, and sulfates. At the conclusion of the process, the microbes die naturally, the water clarifies, and the lagoon is emptied into a nearby river or stream.

An alternative to the oxidation lagoon is the trickling filter system. In this system, water from the primary treatment process is sprinkled over a bed of rocks coated with dense populations of bacteria such as *Zooglea ramigera*. Bacterial populations like these constitute a biofilm. A biofilm is an immobilized population of microbes caught in a tangled web of ultramicroscopic fibers adhering to a surface, as FIGURE 16.6 depicts. Within this microcosm, microbes interact and form communities that store nutrients, share metabolic by-products, and resist predators.

■ *Zooglea ramigera*
zō'ō-glē-ä ram-i-gėr'ä

As the water percolates through the biofilm, the microbes quickly break down the organic matter into simpler products, and the BOD is reduced. One drawback to this system is the need to wash out the filter bed periodically. Cold temperatures also reduce the effectiveness of the microbial metabolism.

Another variation for secondary treatment is to use biodiscs, which are closely spaced plastic disks coated with microbes and submerged in the water. Rotating the disks keeps the population aerated and in contact with the organic matter, and its decomposition continues without buildup of the microbial population. Still another method for secondary treatment is the activated sludge tank. The solid material from primary treatment is piped here, aerated, and supplemented with sludge from a previous batch. Microbes multiply vigorously, and great populations of bacteria, yeasts, molds, and protozoa thrive. Most of the organic matter is broken down, and a suspension of microbes and minerals settles to be drawn off as sludge. Microbial popu-

FIGURE 16.6   Biofilm Contamination. A scanning electron micrograph of a biofilm of *Pseudomonas aeruginosa* adhering to material used in urinary catheters in hospital settings. Experiments indicated that large numbers of these cells remained alive after exposure to high concentrations of tobramycin, an antibiotic to which the bacillus is normally susceptible. (Bar = 5 μm.)

lations kill most pathogens in the sludge, because pathogens are among the more fragile species present.

The sludge can be further treated in an anaerobic sludge digester, where bacteria are once again put to work. A variety of anaerobic bacteria are the major agents of digestion. Large amounts of methane and other putrid gases are produced during the anaerobic process, and they can be used as energy sources to drive the machinery of the treatment facility. Unfortunately, the digested sludge contains so much organic matter that it cannot be released into waters. Therefore, it must be placed in a sanitary landfill site; in some cases, it is dried and sold as a soil conditioner.

The tertiary treatment of waste removes any organic matter that has not been degraded during the primary and secondary process. It is particularly designed to remove phosphorus and nitrogen salts that might otherwise enrich aquatic systems and give rise to sudden blooms of algae. Filtration through activated carbon further purifies and clarifies the water, and chlorine gas can be added to disinfect the water and make it suitable for consumption. **FIGURE 16.7** shows an example of a municipal system for waste treatment.

Agricultural waste poses yet another problem for sanitary microbiologists. Most animal waste is currently handled through efficient systems in which manure is piped into clay-lined oxidation lagoons and allowed to remain while microbes rapidly decompose the waste. Problems arise, however, from the accumulation of ammonia from urine and the buildup of phosphorus in the soil. Moreover, the stench tends to be overpowering, except during the late summer when photosynthetic bacteria thrive and turn the water a deep purple. Microbiologists have isolated a species of *Rhodobacter* from the purple water, and they have shown that these bacteria can break down many odoriferous (odor-causing) compounds, including volatile fatty acids and phenols. In the future, seeding such bacteria in the lagoons may increase their efficiency and minimize their noxious odors. Still other innovative methods for preserving the environment are described in the next section.

■ *Rhodobacter*
rō-dō-bac'tėr

Bioremediation

Recruiting bacteria and other microbes to break down synthetic waste is an immensely appealing idea. It signals a willingness by technologists to work with nature and adapt to its sophisticated sanitation systems, rather than trying to reinvent them. Putting microorganisms to work in this manner is the crux of bioremediation.

Although the word "bioremediation" is recently coined (and somewhat imposing), the concept is not new. In the 1800s, for example, night-soil men would travel from house to house and for a small fee collect sewage and excrement. After making their rounds, they would scatter their collections on fields to be broken down by naturally occurring soil bacteria. As we have seen, modern waste disposal systems have replaced the night-soil men, but a new concern is the accumulation of industrial waste contaminating the land. Bioremediation seeks to exploit helpful microbes to degrade the pollutants in this waste.

The advantages of bioremediation were displayed following a major oil spill from the tanker *Exxon Valdez* in 1987 along the Alaska coastline (**FIGURE 16.8**). Previous studies showed that where oil is spilled, the bacteria that degrade oil (for example, *Pseudomonas* species) are already present, and all technologists have to do is encourage their growth.

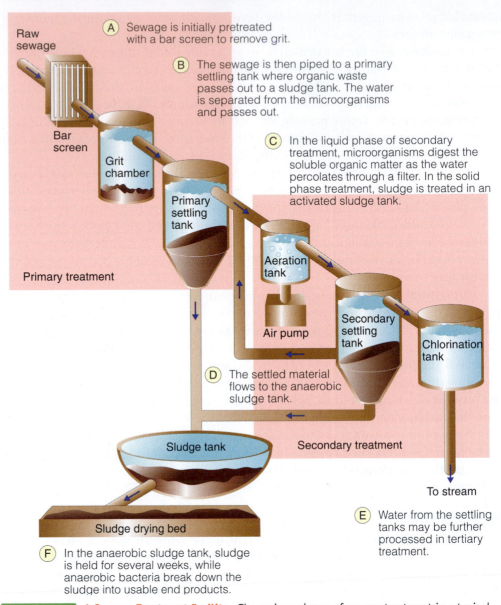

Raw sewage

A Sewage is initially pretreated with a bar screen to remove grit.

B The sewage is then piped to a primary settling tank where organic waste passes out to a sludge tank. The water is separated from the microorganisms and passes out.

Bar screen

Grit chamber

C In the liquid phase of secondary treatment, microorganisms digest the soluble organic matter as the water percolates through a filter. In the solid phase treatment, sludge is treated in an activated sludge tank.

Primary settling tank

Primary treatment

Aeration tank

Air pump

Secondary settling tank

Chlorination tank

D The settled material flows to the anaerobic sludge tank.

Sludge tank

Secondary treatment

To stream

Sludge drying bed

E Water from the settling tanks may be further processed in tertiary treatment.

F In the anaerobic sludge tank, sludge is held for several weeks, while anaerobic bacteria break down the sludge into usable end products.

FIGURE 16.7 **A Sewage-Treatment Facility.** The various phases of sewage treatment in a typical municipal plant are shown.

Thus, when the oil spill occurred, technologists "fertilized" the oil-soaked water with nitrogen sources (e.g., urea), phosphorus compounds (e.g., laureth phosphate), and other mineral nutrients to stimulate the growth of indigenous (naturally occurring) microorganisms. Areas treated in this way were cleared of oil significantly faster than nonremediated areas. Indeed, the oil degraded five times faster when microbes were enlisted in the cleanup.

Bioremediation can also be applied to help eliminate polychlorinated biphenyls (PCBs) from the environment. PCBs were used widely in industrial and electrical machinery before their threat to environmental quality was realized. The compounds contain numerous chlorine atoms and chlorine-containing groups, and researchers have discovered that certain anaerobic bacteria remove those atoms and groups and reduce the compounds to smaller molecules. Aerobic bacteria then take over and reduce the molecular size still further. Field demonstrations in New York's Hudson River showed the value of the combination of anaerobic and aerobic degradation.

Many years ago, trichloroethylene (TCE) was a much-used cleaning agent and solvent. At the time, scientists did not realize that TCE would diffuse through the soil and contaminate underground wells and water reservoirs (aquifers). To combat the problem and degrade the TCE, scientists are exploiting various strains of helpful bacteria that metabolize methane and produce a methane-digesting enzyme, which also breaks down TCE. Technologists pump methane and other nutrients into the TCE-contaminated water, and as the bacteria grow, they eliminate the TCE. The deliberate enhancement of microbial growth yields an environmental cleanup.

FIGURE 16.8 **The *Exxon Valdez* Oil Spill.** In 1987, the oil tanker *Exxon Valdez* ran aground on the shoreline of Alaska. During the ensuing cleanup efforts, numerous species of bacteria demonstrated their ability to break down the oil. This was one of the first large-scale attempts to use microbes in bioremediation.

Among the big news stories of 1999 was research on a "superbug" able to withstand 3000 times more radiation than humans can. The bacterium, a tetracoccus called *Deinococcus radiodurans*, was found some years earlier in a tin of irradiation-sterilized ground beef. Because *Deinococcus radiodurans* can survive exposure to radiation, these bacteria are now being used to clean toxins at nuclear waste sites. Genetic engineering methods have been used to produce a strain of *D. radiodurans* that can degrade the toxic mercury compounds common to these sites while withstanding high levels of radiation. The strain utilizes a gene cluster from *E. coli* to reduce mercury compounds to the less toxic elemental mercury found in thermometers. In 2000, scientists announced the addition of genes that encode enzymes to break down the toxic chemical toluene.

From the 1940s through the 1960s, a major component of weaponry was the explosive compound 2,4,6-trinitrotoluene (TNT). This compound has contaminated the soil around weapons plants. Scientists have found that they can reduce the level of contamination by adding molasses to the soil to encourage bacterial growth. In a pilot study, researchers mixed water with TNT-laced soil and added molasses at regular intervals. In a matter of weeks, the TNT concentration plummeted.

Moreover, plants are apparently able to conscript bacteria for the task of environmental cleanup: Following the Persian Gulf War of 1991, oily devastation remained in areas where Iraqi troops had blown up oil well heads. Within four years, however, plant life returned, aided in large measure by bacteria belonging to the genus *Arthrobacter*. When researchers dug into the oil-soaked desert, they found healthy plant roots surrounded by reservoirs of these oil-degrading bacteria.

■ *Arthrobacter*
är-thrō-bak′tėr

For many years, "haul and bury" was the prevailing method for disposing of synthetic waste. As the public becomes increasingly intolerant of that approach, the importance of bioremediation will become apparent. Technologists are testing microbes for their ability to degrade flame-retardants, phenols, chemical warfare agents, and numerous other waste products of industry. Indeed, one prominent researcher has called bioremediation "a field with its own mass and momentum."

The Pollution Problem

During the 1980s, health workers in Africa asked local villagers to identify their single greatest need. The villagers did not refer to food, shelter, or medical care; rather, their answer was an almost unanimous "water." During that same period, the World Health Organization surveyed various populations to determine the availability of water to them for daily use. The results of the survey were startling: Approximately three of every four humans alive today do not have enough water to drink, or if water is available, the supply is contaminated. Indeed, a 2001 study concluded that as many as 3.3 billion people are denied access to clean water because of contamination.

Clean water is essential for bathing, cooking, and cleaning, as well as for industry, irrigation, recreation, and the cultivation of fish and shellfish. Each day, in the United States, each individual uses about 100 gallons of water. Water pollution begins with the addition of sewage and industrial wastes discharged by companies near the water supply. Dairy and food-processing plants are notable for their waste discharges, as are meat-packing plants, petroleum refineries, and paper mills. Moreover, studies reported in 2000 indicate that an emerging problem concerns pharmaceutical contamination involving a broad mix of drugs including anticancer agents, psychiatric drugs, and anti-inflammatory compounds. Microbes multiply furiously in this nutrient-rich "witch's brew" of pollutants, resulting in polluted water, as **FIGURE 16.9** illustrates. Polluted water is not fit for consumption, nor should it be used for any other purposes.

Toxic substances that destroy marine and freshwater species are polluting because the decomposition of aquatic life sparks the bloom of bacteria, while yielding horrible odors. As bacteria use up the available oxygen, other protozoa, fish, small arthropods, and plants die and accumulate on the bottom. Anaerobic species of bacteria such as *Desulfovibrio* and *Clostridium* thrive in the sediments and produce gases that give the water a stench reminiscent of rotten eggs.

An important problem associated with water pollution is the corrosion of water pipelines. This corrosion is often due to hydrogen sulfide that bacteria produce from sulfate ions during the sulfur cycle. Hydrogen sulfide converts the iron in water pipelines into iron disulfide, leading to deterioration of the pipe, as we have noted previously. Anaerobic bacteria are usually involved, so such corrosion is most obvious on pipes buried in mud.

Water can also be a vehicle for the transport of pathogenic microbes that create human misery. Although the blame generally lies with humans who fail to deal properly with their sewage and waste, natural circumstances sometimes bear the brunt of the blame. Heavy rains, for example, wash pathogens from the soil and into rivers used for drinking water. Whether from human or natural sources, pollution must be detected, and to this end, public health departments apply a series of tests to identify health hazards and sound the alert. We shall describe some of these methods next.

Microbe Detection Methods

A major public health concern is the presence of pathogenic viruses, bacteria, protozoa, and other microbes in water supplies. Numerous infectious diseases such as typhoid fever, cholera, and hepatitis A can be transmitted in water. It is impossible to test water for all pathogens, so certain indicator organisms are used. Among the most

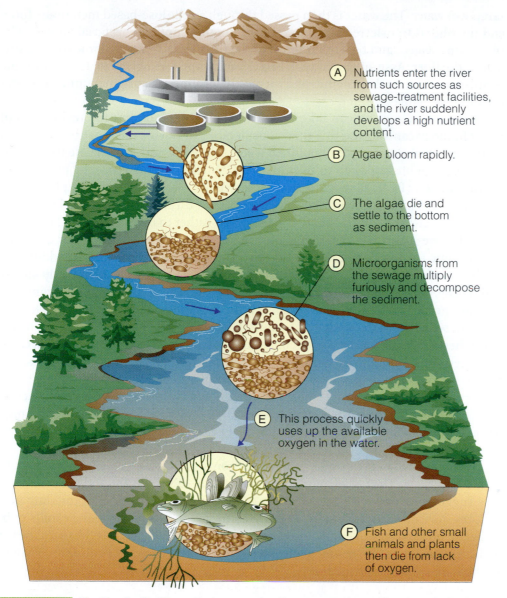

A Nutrients enter the river from such sources as sewage-treatment facilities, and the river suddenly develops a high nutrient content.

B Algae bloom rapidly.

C The algae die and settle to the bottom as sediment.

D Microorganisms from the sewage multiply furiously and decompose the sediment.

E This process quickly uses up the available oxygen in the water.

F Fish and other small animals and plants then die from lack of oxygen.

FIGURE 16.9 **The Death of a River.** Environmental pollutants spur the proliferation of microbes, and oxygen loss leads to tragic results.

frequently used indicators are coliform bacteria, a group of Gram negative rods normally found in human and animal large intestines. If coliform bacteria are present, then the likelihood is high that the water has been contaminated with intestinal microbes, including intestinal pathogens. Among the well-known coliform bacteria is *Escherichia coli*, found in the intestines of virtually all humans. The bacterium is easy to cultivate in the laboratory, and numerous tests have been devised for its detection.

The membrane filter technique is a popular laboratory test in water microbiology because it is straightforward and can be used in the field. A technologist holds a specially designed collecting bottle against the current and takes a 100-milliliter (mL) sample of water. The water is then passed through a cellulose-based membrane filter, and the filter is transferred to a plate of medium that supports bacterial growth. The plate is incubated, and bacteria trapped in the filter form visible colonies on the surface of the filter. Assuming that each bacterium trapped will give rise to a colony, the technologist counts the colonies and thereby determines the original number of bacteria in the sample of water.

A detection method can take several days to complete, but the techniques of biotechnology can shorten this time considerably. In one of the most sophisticated methods available, a sample of polluted water is filtered, and bacteria are trapped on the filter, then broken open to release their DNA. To detect *E. coli*, the polymerase chain reaction (PCR) is used to amplify any *E. coli* DNA that may be present (as Chapter 14 describes). Then, a gene probe specific for *E. coli*'s DNA is added. Essentially, the technologist attempts to identify *E. coli* genes, rather than *E. coli* cells. The process not only saves time, but is extremely sensitive. It has been estimated, for example, that a single *E. coli* cell can be detected in a 100-mL sample of water. In addition, a wealth of other microbes, including many pathogenic species, can be directly identified by DNA probe analysis, thereby eliminating the search for indicator organisms.

Another novel method uses cells from the horseshoe crab *Limulus polyphemus*. In 1968, scientists discovered that a solution of disrupted *Limulus* blood cells forms a gel when the debris is mixed with endotoxins (toxic substances retained *within* cells) produced by Gram-negative bacteria. The test, called the *Limulus* endotoxin assay, is performed by combining water with disrupted *Limulus* cells and incubating at 37°C for 60 minutes. Gel formation signals the presence of Gram-negative bacteria such as *E. coli*. Though the horseshoe crab has no known relatives and is apparently misnamed (it is a member of the spider family), it has seemingly found a useful niche in microbiology.

Ensuring the safety of water supplies for consumers is a high priority of the public health system. The use of amplification methods, DNA probes, and endotoxin assays suggests we are in an exciting new era of water-quality testing providing new measures of safety for the public. Rapid determination of a health risk permits public health officials to implement measures to benefit the most people. High on the list of these measures is the treatment of water to interrupt the spread of disease. We shall see how that is accomplished next.

Water Treatment

It is rare to locate a water source that does not need treatment before consumption. The general rule is that water must be treated to remove potentially harmful microbes and to improve its clarity, odor, and taste. In addition, such minerals as calcium and magnesium may be removed in order to "soften" the water, and fluoride may be added to enhance resistance to tooth decay.

To begin treatment, most municipal water supplies are subjected to coagulation. In this step, iron or aluminum compounds are added to the water to form insoluble precipitates. These precipitates trap organic matter and carry it out of the water. Next, the water is subjected to filtration. Filters can consist of fine grains of sand, coarse grains of sand, or sand subjected to pressure. Filtration rates are slower through fine grains of sand but the degree of filtration is greater than with coarse grains. Whichever

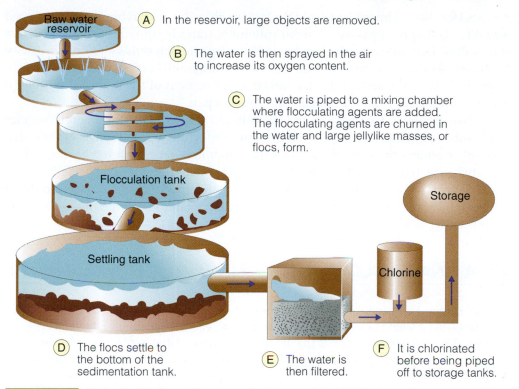

A In the reservoir, large objects are removed.

B The water is then sprayed in the air to increase its oxygen content.

C The water is piped to a mixing chamber where flocculating agents are added. The flocculating agents are churned in the water and large jellylike masses, or flocs, form.

D The flocs settle to the bottom of the sedimentation tank.

E The water is then filtered.

F It is chlorinated before being piped off to storage tanks.

FIGURE 16.10 **Water Purification.** The steps taken in a municipal facility to purify water for drinking purposes.

type of filter is used, roughly 99 percent of the organic matter and microbes have been removed from the water by the time this step is completed.

To ensure the safety of water while it is stored and transported, municipalities commonly add chlorine. Chlorine reacts with virtually all organic compounds and kills most microbes remaining after filtration. It is generally added as a gas from pressurized tanks or as calcium hypochlorite. A residual amount must remain in the water (usually 0.2 to 0.6 ppm, or parts per million) because chlorine reacts with any organic materials in the water and the residue ensures that enough is present to kill the microbes. **FIGURE 16.10** displays an overview of the process.

In some instances, it is necessary to treat water for drinking on an *ad hoc* basis. For example, raw sewage may find its way into water supplies and contaminate the water. Moreover, during drought conditions, sediment from the bottom of reservoirs may be stirred up, bringing bottom-dwelling microbes into the water and making it hazardous to health. Under conditions like these, consumers are generally advised to boil their water for a few minutes before drinking. This treatment kills all microbes with the possible exception of bacterial spores. However, with few exceptions, these spores do not represent a hazard to health. The water is safe to drink, but it is not sterile.

To disinfect clear waters, campers, backpackers, and hikers are advised to use commercially available chlorine or iodine tablets. If these cannot be obtained, the Centers for Disease Control and Prevention (CDC) recommends a half-teaspoon of household chlorine bleach in 2 gallons of water, with 30 minutes contact time before consumption.

And for those who wish to avoid the taste of chlorine, researchers have developed a novel pen-shaped apparatus. A small amount of water is poured into the top, where it dissolves a salt tablet to produce brine. Twisting the pen sends a jolt of electricity through the brine and creates highly reactive antimicrobial compounds. Poured into a quart of water, the activated brine kills over 99 percent of the microbes within 10 minutes. Researchers have dubbed it the "poison pen."

As we discuss in Chapters 18 and 19, polluted water can carry numerous species of microbes associated with infectious disease. Unfortunately, outbreaks of these diseases overshadow the myriad ways in which microbes benefit society through their activities in water, soil, and the environment. It is often difficult to set aside the pathogenic effects of microbes and concentrate on their "good works," but we should all try. Hopefully, this chapter has given you sufficient examples of microbes exhibiting their friendly sides.

A FINAL THOUGHT

As we continue our study of the microbes, we can see that it is easy to become paranoid about taking a breath of air, bite of food, or drink of water. After all, microbes lurk everywhere. I hope, however, that this chapter has shown you another side to the microbe story, a side that reflects the positive roles they play in the environment and in our lives.

Just consider, for example, the activity of microbes in the treatment of sewage. Through a complex network of processes, microbes transform the devilish cocktail of sewage into simple compounds that can be handled by the environment. Working with quiet and unceasing competence, they break down the vile mixture of human and animal feces, oily filth from roads, bloody effluent from slaughterhouses, an unimaginably ugly profusion of grot. This mammoth task is accomplished unfailingly and efficiently.

Nor does it end here. In the carbon, nitrogen, and sulfur cycles, microbes convert the basic elements of life into usable forms and replenish the soil to nourish all living things. By far, the great majority of microbes are engaged in socially constructive and wholesome activities. Microbes represent the most efficient way of breaking down organic matter, and, in the general scheme of things, this breakdown happens after organisms die, not before.

QUESTIONS TO CONSIDER

1. The poet John Donne once wrote: "No man is an island, entire of itself." This statement applies not only to humans, but to all living things in the natural world. What are some roles microbes play in the interrelationships among living things?

2. A student notes in her microbiology class that a particular species of bacteria actively dissolves fats, greases, and oils. Her mind stirs, and she wonders whether such an organism could be used to unclog the cesspool that collects waste from her house. What do you think she is considering? Will it work?

3. In the 1970s, a popular bumper sticker read: "Have you thanked a green plant today?" The reference was to photosynthesis taking place in plants. Suppose

you saw this bumper sticker: "Have you thanked a microbe today?" What might the owner of the car have in mind?

4. What information might you offer to dispute the following four adages common among campers and hikers? (1) Water in streams is safe to drink if there are no humans or large animals upstream. (2) Melted ice and snow is safer than running water. (3) Water gurgling directly out of the ground or running out from behind rocks is safe to drink. (4) Rapidly moving water is germ-free.

5. Some years ago, the syndicated columnist Erma Bombeck wrote a humorous book entitled *The Grass Is Always Greener Over the Septic Tank* (an adaptation of the expression "The grass is always greener on the other side of the fence"). Indeed, the grass is often greener over the septic tank. Why is this so? How |can you locate your home's cesspool or septic tank in the days following a winter snowfall?

6. Bioremediation holds the key to solving numerous types of environmental problems in the future. Yet the process has been used for generations without people realizing it. How many instances where bioremediation is currently in use can you point out?

7. The victims of disease, both animals and people, are buried in the soil, yet the soil is generally free of pathogenic organisms. Why?

KEY TERMS

Informative facts are necessary for the expression of every concept, and the information for a concept is founded in a set of key terms. The following terms form the basis for the concepts of this chapter. On completing the chapter, you should be able to explain and/or define each one:

biofilm
biological oxygen demand (BOD)
bioremediation
biota
carbon cycle
coliform bacteria
compost
consumers
cyanobacteria
decomposers

ecosystem
habitat
membrane filter technique
nitrogen cycle
nitrogen fixation
oxidation lagoon
primary waste treatment
producers
sulfur cycle

http://microbiology.jbpub.com/book/microbes

The site features **eLearning**, an online review area that provides quizzes and other tools to help you study for your class. You can also follow useful links for in-depth information, read more stories of microbiology, or just find out the latest microbiology news.

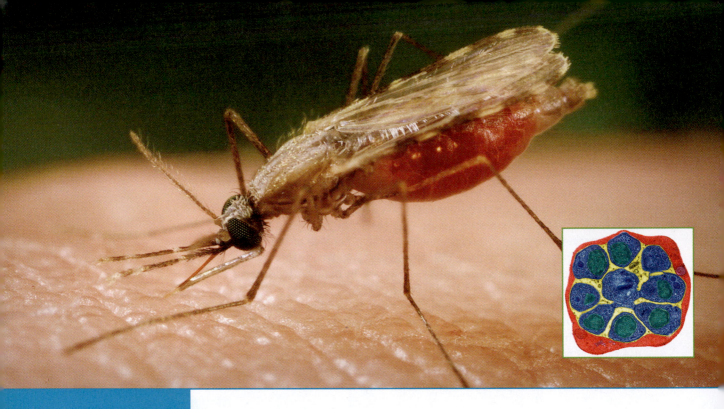

17

Disease and Resistance: The Wars Within

The *Anopheles gambia* mosquito is one of the leading carriers of *Plasmodium malariae*, the protozoa that causes malaria, in the world (the protozoa shown in the inset). Forty percent of the world's population, particularly in Africa, Asia, and South America are at risk for malaria. The disease was once common in the United States, but a national program eradicated the malaria, using DDT, by 1952.

Looking Ahead

For most of us, our earliest childhood experiences with microbes were related to their ability to cause infectious disease. Fortunately, we survived those early encounters by drawing on a number of resistance processes, as this chapter will illustrate.

On completing this chapter, you should be able to . . .

- appreciate some concepts relating to infectious disease, including the difference between infection and disease, the stages in the progress of disease, and the methods for transmitting disease.
- understand the relationships between microbes and their human hosts and list some important factors that contribute to the establishment of infectious disease.
- summarize some key aspects of nonspecific resistance, including chemical barriers to infection, phagocytosis, and fever.
- explain how antigens initiate the immune process.
- identify the parts of the immune system and recognize how the process of cell-mediated immunity works.

- describe the mechanism by which antibody-mediated immunity yields specific resistance, focusing on the activities of antibodies and the forms of immunity they engender.
- understand processes such as allergy and anaphylaxis through which the immune system can have deleterious effects on the human body.

In past centuries, the spread of disease appeared to be wildly erratic. Illnesses would attack some members of a population while leaving others untouched. A disease that for many generations had taken small, steady tolls would suddenly flare up in epidemic proportions. Strange, horrifying plagues descended unexpectedly on whole nations.

Scientists now know that humans live in a precarious equilibrium with the microbes that surround them. Most of the time, the relationship is harmonious, because the microbes have no disease potential or because humans have developed resistance to them. However, when human resistance is diminished by a pattern of life that gives microbes the edge, then disease sets in. For example, during the Industrial Revolution of the 1800s, many thousands of Europeans moved from rural areas to the cities. They sought jobs, prosperity, and a new life. They found endless labor, unventilated factories, and wretched living conditions—and they found infectious disease.

In this chapter, we explore the infectious disease process and the mechanisms by which the body responds to disease. We examine the host-parasite relationship and the factors that contribute to the establishment of disease, as well as the nonspecific and specific methods by which body resistance develops, with emphasis on the immune system.

The Host-Parasite Relationship

The idea of a tattoo seemed okay. All her friends had them, and a tattoo would send a unique and personal message. After all, she was already 22. It took some pushing from her friends, but she finally made it into the tattoo parlor that day in Fort Worth, Texas.

Two weeks later the pains started—first in her abdomen, then all over. Her fever was high, and a rash was breaking out; it looked like her skin was burned and peeling away. One visit to the doctor, then immediately to the emergency room of the local hospital. The gynecologist guessed it was an inflammation of the pelvic organs, so he gave her an antibiotic and sent her home.

But it got worse—the fever, the rash, the peeling, the pains. Back she went to the emergency room. This time the doctor admitted her to the hospital, gave her intravenous blood transfusions and antibiotics, kept her for 11 days, and discovered a severe blood infection due to *Staphylococcus aureus*. And there was an unusual diagnosis: toxic shock syndrome. "Don't women get that from using tampons improperly?" she asked. Most do, she was told, but a few get it from staphylococci entering a skin wound, a wound that can be made by a contaminated tattooing needle.

Indeed, *S. aureus* is commonly found on the human skin, where it "infects" the tissue. But that word does not necessarily imply disease, because infection refers to the relationship between two organisms, the host and the parasite, and the competition that takes place between them. A host whose resistance is strong remains healthy, and the parasite is either driven from the host or assumes a benign relationship with the host. By contrast, if the host loses the competition, disease develops. The term disease appears to have originated from Latin stems that mean "living apart," a reference to the separation of ill individuals from the general population. Disease is any change from the general state of good health.

The concept of infection is expressed in the body's normal microflora. The normal microflora is the population of microbes that infect the body without causing disease. Some organisms in the population establish a permanent relationship with the body; others are present for limited periods of time. In the large intestine of humans, for example, the bacterium *Escherichia coli* and the fungus *Candida albicans* are almost always found. FIGURE 17.1 illustrates five possible sources of microbes in the normal microflora.

A population of microbes may usually be found wherever the body is exposed to the external environment, including such places as the skin, respiratory tract, intestinal tract, and urinary tract. Unfortunately, not all microbes that the body is exposed to are benign, as we shall see in the next section.

Concepts of Infectious Disease

Pathogenicity refers to the ability of a parasite to gain entry into the host's tissues and bring about a physiological or anatomical change, resulting in poor health. This term

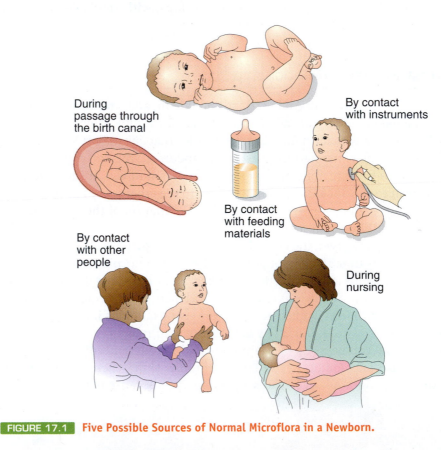

During passage through the birth canal

By contact with instruments

By contact with feeding materials

By contact with other people

During nursing

FIGURE 17.1 **Five Possible Sources of Normal Microflora in a Newborn.**

is derived from the Greek word *pathos*, meaning "suffering." The term pathogen has the same root and refers to any organism having pathogenicity.

A pathogen is also called a parasite, and the symbiotic relationship between host and parasite is referred to as parasitism. Parasites vary greatly in their pathogenicity. Certain parasites, such as the cholera, plague, and typhoid bacilli, are well known for their ability to cause serious human disease. Others, such as common cold viruses, are considered less pathogenic because they induce milder illnesses. Still other parasites are opportunistic. These microbes are benign until the normal body defenses are suppressed; they then seize the opportunity to invade the tissues and express their pathogenicity. An example of suppressed defenses is observed in individuals with acquired immune deficiency syndrome (AIDS). These patients are highly susceptible to disease caused by opportunistic organisms such as the protozoan *Pneumocystis carinii* and the fungus *Candida albicans*.

■ *Pneumocystis carinii*
nū-mō-sis′tis kär-i′nē-ī
(or kär-i′nē-ē)

Virulence is the concept used to express the degree of pathogenicity of a parasite. This term is derived from the Latin *virulentus*, meaning "full of poison." An organism that invariably causes disease, such as the typhoid bacillus, is said to be "highly virulent." By comparison, an organism that sometimes causes disease, such as *C. albicans*, is labeled "moderately virulent." Certain organisms, described as avirulent, are not regarded as disease agents. The lactobacilli and streptococci found in yogurt are examples of avirulent microbes.

In recent years, the term pathogenicity islands has been used to refer to the clusters of genes responsible for virulence. The genes encode many of the virulence factors (e.g., enzymes and toxins) we shall examine later in this chapter. These blocks of genetic information move into a benign organism and convert it to a pathogen by mechanisms we discuss in Chapter 10. In effect, they direct the host response in a way that favors survival of the pathogen over that of the host.

Infectious diseases may be described according to the level at which they occur in a population. An endemic disease, for example, is one that occurs at a low level in a certain geographic area (**FIGURE 17.2** presents an example). By comparison, an epidemic disease (or epidemic) breaks out in explosive proportions within a population, and a pandemic disease (pandemic) occurs worldwide. In the United States, measles is endemic, while influenza is epidemic; in the world, AIDS is pandemic.

The microbial agents of disease may be transmitted by a broad variety of methods conveniently divided into two general categories: direct methods and indirect methods. Direct methods of transmission imply physical contact with one who has the disease. Hand-shaking, kissing, and sexual intercourse are examples. Such diseases as gonorrhea and genital herpes are spread by direct contact. Direct contact may also mean exposure to droplets, the tiny particles of mucus and saliva expelled from the respiratory tract during a cough or sneeze. Diseases spread in this way include influenza, measles, pertussis (whooping cough), and streptococcal sore throat.

Indirect methods of disease transmission include the consumption of contaminated food or water and contact with fomites. Foods are contaminated during processing or handling or may be so because they were made from diseased animals. Poultry products, for example, are often a source of salmonellosis because *Salmonella* species frequently infect chickens. Fomites are inanimate objects that carry disease organisms. For instance, contaminated syringes and needles transport the viruses of hepatitis B and AIDS.

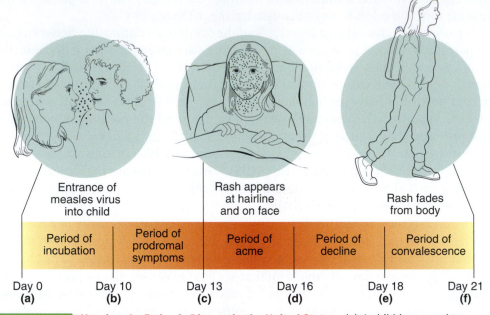

FIGURE 17.2   Measles: An Endemic Disease in the United States. (**a**) A child is exposed to measles viruses in respiratory droplets, and the period of incubation begins. (**b**) At the end of this period, the child experiences fever, respiratory distress, and general weakness as the period of prodromal symptoms ensues. (**c**) The period of acme begins with the appearance of specific measles symptoms, such as the body rash. (**d**) The period reaches a peak as the rash covers the body. (**e**) The rash fades first from the face and then the body trunk as the period of decline takes place. With the period of convalescence, the body returns to normal. (**f**) The child later returns to school.

Arthropods (e.g., arachnids and insects) are responsible for another indirect method of transmission (FIGURE 17.3). In some cases, the arthropod is a mechanical vector of disease because it transports microbes on its legs and other body parts. In other cases, the arthropod itself is diseased and serves as a biological vector. In malaria and yellow fever, for instance, microbes infect the arthropod (usually a mosquito) and accumulate in its salivary gland, from which they are injected during the next bite.

For a disease to perpetuate itself, a continuing natural source of disease organisms is necessary. These sources are called reservoirs. For smallpox, the sole reservoir of the viruses is humans—the World Health Organization was able to limit the spread of the virus and eradicate smallpox from the world in part by locating all human reservoirs and carrying out mass immunizations. A carrier is a special type of reservoir, an organism that has recovered from the disease but continues to shed the disease agents. For instance, people who have recovered from typhoid fever become carriers for many weeks after the symptoms of disease have left. Typhoid Mary, whose story is recounted in A Closer Look on page 376, is one of the most famous carriers in history. Animals may also be reservoirs of disease; house cats usually show no symptoms of toxoplasmosis but are able to transmit the disease among humans.

The Establishment of Disease

A microbial parasite must possess unusual abilities if it is to overcome host defenses and bring about the profound changes leading to disease. Before it can manifest these

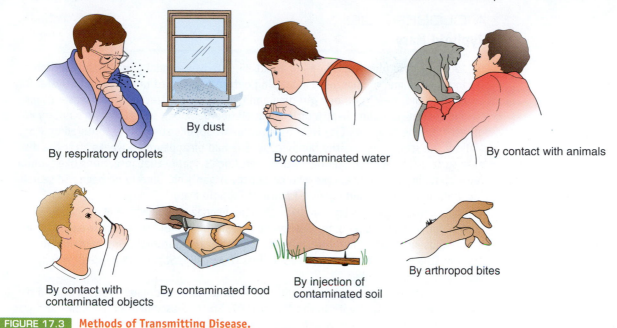

By respiratory droplets   By dust   By contaminated water   By contact with animals

By contact with contaminated objects   By contaminated food   By injection of contaminated soil   By arthropod bites

FIGURE 17.3   Methods of Transmitting Disease.

abilities, however, the parasite must first gain entry into the host in sufficient numbers to establish a population. Next, it must be able to penetrate the tissues and grow at that location. Disease is therefore a complex series of interactions between microbe and host.

The portal of entry refers to the site at which microbes enter the host. The site varies considerably for different microbes. For example, tetanus may develop if *Clostridium tetani* spores are introduced from the soil to the anaerobic tissue of a wound, but tetanus will not develop if the spores are consumed with food because they will not germinate in the human intestinal tract. This is why one can eat a freshly picked radish without fear of tetanus.

At the conclusion of its pathogenicity cycle, a pathogen must be able to leave the body through some suitable portal of exit. This is important because easy transmission permits the pathogen to continue its disease-spreading existence in the world. Contemporary microbiologists have suggested that if a microbe cannot find a suitable portal of exit or mode of transmission, it may be replaced in nature by less pathogenic species.

Dose refers to the number of parasites that must be taken into the body for disease to be established. Experiments indicate, for example, that the consumption of a few thousand typhoid bacilli will probably lead to the disease. By contrast, many millions of cholera bacteria must be ingested if cholera is to be established. One explanation is the high resistance of typhoid bacilli to the acidic conditions in the stomach, in contrast to the low resistance of cholera bacilli (most of which thus perish). Also, it may be safe to eat fish taken from water that contains hepatitis A viruses, but eating raw clams from that same water can be dangerous. Clams filter water to obtain nutrients, and in doing so, they concentrate any hepatitis A viruses in the water in their tissues. (Public health officials refer to raw clams as "hepatitis grenades.")

A CLOSER LOOK

Typhoid Mary

By 1906, typhoid fever was claiming about 25,000 lives annually in the United States. During the summer of that year, a puzzling outbreak occurred in the town of Oyster Bay on Long Island, New York; one girl died and five others contracted the disease. Eager to find the cause, public health officials hired George Soper, a well-known sanitary engineer from the New York City Health Department. Soper's suspicions centered on Mary Mallon, the seemingly healthy family cook. She had disappeared 3 weeks after the disease surfaced. Soper was familiar with Robert Koch's theory that infections like typhoid fever could be spread by people who harbor the organisms. Quietly he began to search for the woman who would become known as Typhoid Mary.

Soper's investigations led him back over the decade during which Mary Mallon cooked for several households. Twenty-eight cases of typhoid fever occurred in those households, and each time, the cook left soon after the outbreak. One epidemic in 1903 in Ithaca, New York, claimed 1300 lives. Ironically, Soper had gained his reputation during this episode.

Soper tracked Mary Mallon through a series of leads from domestic agencies and finally came face-to-face with her in March 1907. She had assumed a false name and was now working for a family in which typhoid had broken out. Soper explained his theory that she was a carrier, and pleaded that she be tested for typhoid bacilli. When she refused to cooperate, the police forcibly brought her to a city hospital on an island in the East River off the Bronx shore. Tests showed that her stools teemed with typhoid organisms, but fearing that her life was in danger, she adamantly refused the gall bladder operation that would eliminate them. As news of her imprisonment spread, Mary became a celebrity. Soon public sentiment led to a health department policy deploring the isolation of carriers. She was released in 1910.

But Mary's saga had not ended. In 1915, she turned up again at New York City's Sloane Hospital working as a cook under another new name. Eight people had recently died of typhoid fever, most of them doctors and nurses. Mary was taken back to the island, this time in handcuffs. Still she refused to have her gall bladder removed and vowed never to change her profession. Doctors placed her in isolation in a hospital room while trying to decide what to do. The weeks wore on.

Eventually Mary became less incorrigible and assumed a permanent residence in a cottage on the island. She gradually accepted her lot and began to help out with routine hospital work. However, she was forced to eat in solitude and was allowed few visitors. Mary Mallon died in 1938 at the age of 70 from the effects of a stroke. She was buried without fanfare in a local cemetery.

The ability of a parasite to penetrate tissues and cause structural damage is a virulence factor called invasiveness (FIGURE 17.4). Experiments reported by Stanley Falkow and his coworkers at Stanford University indicate that genes for invasiveness exist on the chromosome of certain bacteria. Falkow has reported that pathogens use their invasive capabilities to gain access to privileged niches denied to nonpathogens. For example, pathogenic *E. coli* 0157:H7 penetrate to sites in the small intestine and urinary system that are not available to nonpathogenic *E. coli* strains present in the small intestine.

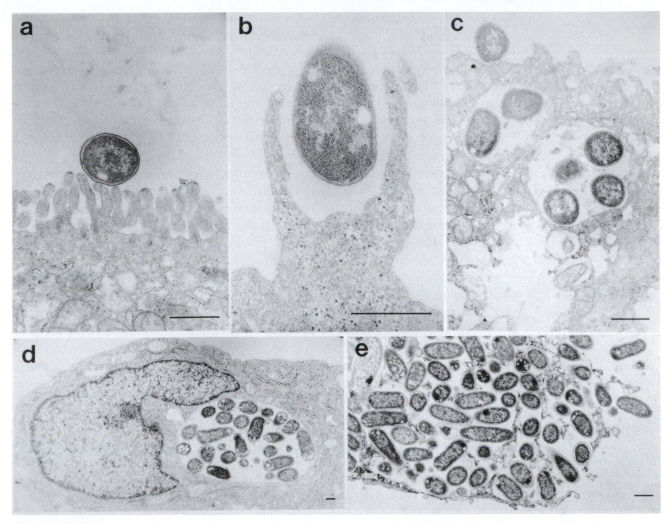

FIGURE 17.4 **Tissue Invasion.** A series of transmission electron micrographs display the interactions between invasive *Salmonella typhimurium* and epithelial cells of the human intestine. (**a**) At 30 minutes postinfection, bacteria are seen adhering to the tips of intestinal microvilli. (**b**) A bacterium is being engulfed by an epithelial cell. (**c**) At 1 hour postinfection, salmonellae can be observed within vacuoles in the cells. (**d**) At 12 hours postinfection, a number of vacuoles containing bacteria unite with one another, and bacteria multiply within this large vacuole. (**e**) At 24 hours postinfection, the epithelial cell is filled with salmonellae and is breaking down to release the bacteria. (Bar = 1 μm.)

The virulence of a microbe further depends on its ability to produce a series of enzymes that help it to counter body defenses. An example of a bacterial enzyme is the coagulase produced by virulent staphylococci. Coagulase catalyzes the formation of a blood clot from proteins in human blood. The clot sticks to the staphylococci, protecting them from phagocytosis. Many streptococci have the ability to produce the enzyme streptokinase. This substance dissolves fibrin clots used by the body to restrict and isolate an infected area. Streptokinase thus overcomes an important host defense and allows further tissue invasion by the parasites.

Toxins are microbial poisons that affect the establishment and course of disease by increasing a microbe's virulence. Two types of toxins are recognized: exotoxins and endotoxins. Exotoxins are protein molecules, manufactured primarily by Gram-positive bacteria. The exotoxin produced by the botulism organism, *Clostridium botulinum*, is among the most lethal toxins known (it is believed that 1 pint of the pure toxin would be sufficient to destroy the world's population). In humans, the toxin inhibits the release of a neurotransmitter called acetylcholine at the junction where nerve cells meet muscle cells. This inhibition leads to paralysis. It is not difficult to see why bioterrorists would be interested in using this toxin as a bioweapon.

The body responds to exotoxins by producing special antibodies called antitoxins. An antitoxin molecule combines with a toxin molecule and neutralizes it. In the laboratory, a chemical agent may be used to alter the toxin and destroy its toxicity without hindering its ability to elicit an immune response in the body. The altered toxin is called a toxoid. When toxoid molecules are injected into the body, the immune system responds with antitoxins that circulate and provide a measure of defense against the disease. Toxoids are used as vaccines to protect against diphtheria and tetanus. Indeed, a toxoid is being developed to protect individuals against the botulism toxin should it be used by bioterrorists.

Endotoxins are part of the cell wall of Gram-negative bacteria and thus are released only on disintegration of the cell. Endotoxins manifest their presence by certain signs and symptoms, including increased body temperature, substantial body weakness and aches, and general malaise. Damage to the circulatory system and shock may also occur, a condition called endotoxin shock. Together with exotoxins, endotoxins represent a competitive advantage for the pathogen, and these advantages must be addressed and counterbalanced if the host is to survive an episode of infectious disease. How the body accomplishes this is explored in the next section.

◼ Nonspecific Resistance to Disease

Think of the human body as a doughnut. Just as the hole passes through the center of the doughnut, so too the gastrointestinal tract passes through the center of the human body. The body secretes enzymes out into the hole, digests the food that is eaten, absorbs what it needs, and lets the remainder pass out of the hole. There is no natural opening to the internal tissues in the gastrointestinal tract; indeed, the fact that something is in the GI tract does not necessarily mean that it is in the body.

Close examination of other parts of the body (for example, the respiratory system) reveals similar dead ends, and it becomes clear that the body, like the doughnut, is a closed container. Only if the walls of this container are penetrated can most diseases be established. The skin and its extensions into the body cavities represent a major means of resistance to infection and disease. This resistance is nonspecific, because it exists in all humans and is present from birth. Also, it protects against all parasites. Other forms of resistance are specific and are centered in the immune system, as FIGURE 17.5 illustrates. These forms arise in response to particular parasites and are directed solely at a specific parasite, as we shall see later in this chapter.

Nonspecific resistance to disease involves a broad group of factors, many of which are still not defined. Such resistance depends on the general well-being of the individual and the proper functioning of the body's systems. Accordingly, it is affected by such determinants as nutrition, fatigue, age, sex, and climate. One form of nonspecific resistance, called species immunity, means that diseases affecting one species

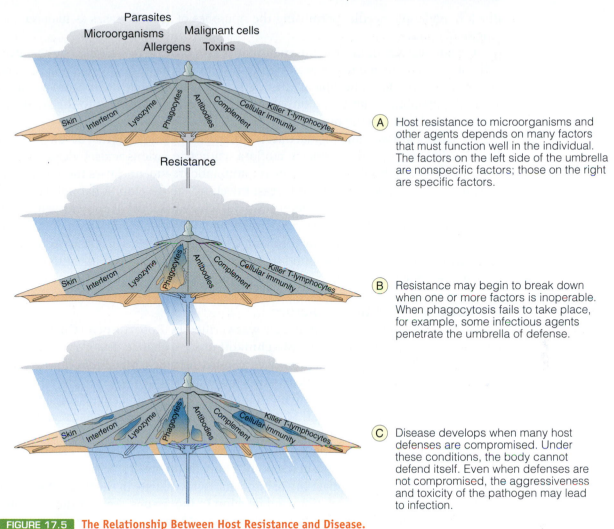

Parasites
Microorganisms   Malignant cells
Allergens   Toxins

Skin Interferon Lysozyme Phagocytes Antibodies Complement Cellular immunity Killer T-lymphocytes

Resistance

A Host resistance to microorganisms and other agents depends on many factors that must function well in the individual. The factors on the left side of the umbrella are nonspecific factors; those on the right are specific factors.

B Resistance may begin to break down when one or more factors is inoperable. When phagocytosis fails to take place, for example, some infectious agents penetrate the umbrella of defense.

C Disease develops when many host defenses are compromised. Under these conditions, the body cannot defend itself. Even when defenses are not compromised, the aggressiveness and toxicity of the pathogen may lead to infection.

FIGURE 17.5   **The Relationship Between Host Resistance and Disease.**

will not affect another. For example, humans do not contract hog cholera, and hogs do not contract AIDS. We shall consider other forms of nonspecific resistance in the next subsections.

Mechanical and Chemical Barriers

As we have noted, the intact skin and the mucous membranes that extend into the body cavities are among the most important means of nonspecific resistance. Not only does the skin provide mechanical protection, but its cells are constantly being sloughed off, and with them go any microbes present. In addition, the keratin (the protein in the dead skin cells) is a poor source of carbon for microbes; the sweat and the fatty acids in sebum (the oily substance in skin glands) contain antimicrobial agents; and, from the viewpoint of a microbe, the low water content of the skin makes it a veritable desert. Unless penetration of the skin and mucous membrane barriers occurs, infectious disease is rare. But penetration of the skin barrier is a fact of everyday life. A cut or abrasion allows staphylococci to enter the blood, and an arthropod bite acts like a hypodermic needle, permitting the microbes of such diseases as malaria and plague to enter.

Certain features of the mucous membranes also provide resistance. For instance, cells of the mucous membranes along the respiratory passageways secrete mucus, which traps microbes. The cilia of other cells nearby then move the mucoid particles along the membranes up to the throat, where they are swallowed into the stomach. Stomach acid destroys the microbes.

Resistance in a woman's vaginal tract is enhanced by the low pH that develops when *Lactobacillus* species in the normal microflora synthesize various acids. Indeed, the disappearance of lactobacilli during excessive antibiotic treatment causes the acid to disappear and encourages candidiasis ("yeast infection") to develop. A natural barrier to the intestinal tract is provided by stomach acid, which has a pH of approximately 2.0. (A cotton handkerchief placed in stomach acid would dissolve in a few moments.) Most microbes are destroyed in this environment. Other natural forms of resistance are provided by the enzyme lysozyme found in the saliva and the tears and the antiviral substance interferon produced in miniscule amounts by most body cells (Chapter 6).

Phagocytosis and Other Factors

Shortly after the germ theory of disease was verified by Robert Koch (Chapter 1), a native of the Ukraine named Elie Metchnikoff made a chance discovery indicating how living cells could protect themselves against microbes. Metchnikoff noted that certain cells in the larva of a starfish gather around a wooden splinter placed within the cell mass. He suggested that such cells actively seek out and engulf foreign particles in the animal's environment. Metchnikoff termed the process phagocytosis and published an account of it in 1884.

In contemporary microbiology, phagocytosis (literally, "cell-eating") is viewed as a form of nonspecific resistance in the body. The cells involved are called phagocytes, various large cells that originate in the bone marrow, circulate in the bloodstream for a while, then leave the circulation, and develop in the tissues. Among the most important phagocytes are the highly specialized macrophages found in the spleen, bone marrow, lymph nodes, and brain.

When a macrophage encounters a microbe, it encloses the microbe with a portion of its cell membrane, then infolds the membrane to form a phagocytic vesicle, or phagosome. The phagosome then pinches off and fuses with a lysosome, a cytoplasmic organelle that contributes enzymes for the microbe's digestion. Other lysosomal substances increase the permeability of capillaries, which brings more macrophages to the area. The process is completed as waste materials are expelled from the macrophage. **FIGURE 17.6** summarizes this process. When antibody molecules are available, the process is even more efficient.

Inflammation is a nonspecific defensive response that occurs when an irritant such as a microbe is present. The irritant sets into motion a process that limits the extent of the injury: Dilation of the blood vessels increases the flow of blood at the site of irritation. White blood cells and macrophages adhere to the vessels close to the injury and begin phagocytosis of the irritant. Soon the area will exhibit four characteristic signs of inflammation: rubor, a red color from blood accumulation; calor, warmth from the heat of the blood; tumor, swelling from the accumulation of fluid; and do-

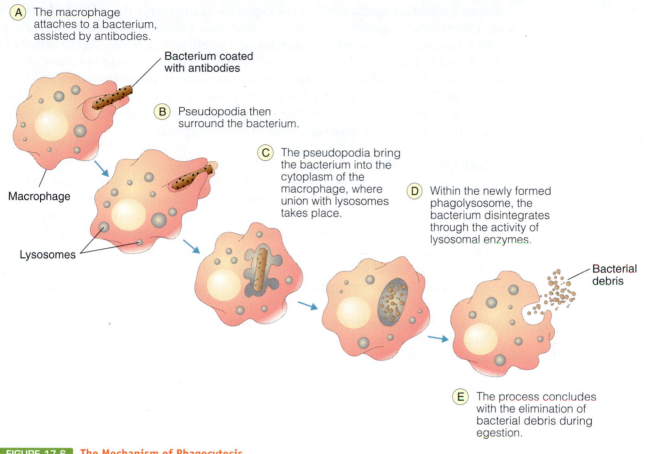

(A) The macrophage attaches to a bacterium, assisted by antibodies.

Bacterium coated with antibodies

(B) Pseudopodia then surround the bacterium.

(C) The pseudopodia bring the bacterium into the cytoplasm of the macrophage, where union with lysosomes takes place.

(D) Within the newly formed phagolysosome, the bacterium disintegrates through the activity of lysosomal enzymes.

Macrophage

Lysosomes

Bacterial debris

(E) The process concludes with the elimination of bacterial debris during egestion.

FIGURE 17.6   **The Mechanism of Phagocytosis.**

lor, pain from injury to the local nerves. Pus, a mixture of plasma, dead tissue cells, leukocytes, and dead bacteria, accumulates at the site.

Fever is an abnormally high body temperature that may act as a mechanism of nonspecific resistance to disease. Scientists believe that bacteria, viruses, and other microbes affect a region at the base of the brain called the hypothalamus and stimulate it to raise the body temperature several degrees. As the fever rises, cell metabolism speeds up and blood vessels constrict, thus denying blood to the skin and keeping its heat within the body. The increased level of cell metabolism also encourages rapid tissue repair and stimulates phagocytosis, while reducing the amount of bloodborne iron needed by parasites. However, if the body temperature rises above 40.6°C (105°F), convulsions and death may result.

Specific Resistance and the Immune System

Specific resistance refers to the reactions mounted by the immune system when a species of pathogenic microbe enters the body. Until the 1950s, specific resistance was virtually synonymous with vaccine use, and immunologists saw themselves as disease prevention specialists. But the explosion of interest in the biological sciences after World War II spilled over to immunology, and it soon became apparent that specific resistance is a phenomenon with broader implications, including the activity of T cells and B cells, phagocytosis, allergic reactions, and the effects of a variety of substances. In addition, the groundwork was laid for understanding the nature and function of antibodies. This work led to the maturing of immunology as one of the key scientific disciplines of our time. Our study of the immune system begins with a survey of the substances that stimulate the immune response.

Antigens

Antigens are chemical substances capable of stimulating the immune system and provoking an immune response. Most antigens are large, complex molecules (macromolecules), which are not normally found in the body and are consequently referred to as "nonself." The list of antigens is enormously diverse. It includes such things as milk proteins, bee venom proteins, hemoglobin molecules, bacterial toxins, and chemical substances found in microbial flagella, pili, and cytoplasm.

The most common antigens are proteins and polysaccharides. Proteins are potent antigens because their amino acids represent a great array of building blocks, allowing for a huge variety of combinations and, hence, diversity in three-dimensional structures. Polysaccharides are less potent as antigens than proteins because they have less chemical diversity and they break down rapidly in the body. In rare cases, lipids and nucleic acids can be antigens.

Antigens usually have a large molecular size and are easily phagocytized by macrophages, the necessary first step in the immune process. The antigen itself does not stimulate the immune system. Rather, that stimulation is accomplished by a small part of the antigen molecule called the antigenic determinant, or the epitope. An antigenic determinant (epitope) contains about six to eight amino acid molecules or monosaccharide units. An antigen may have numerous different antigenic determinants, and a structure such as a bacterial flagellum may have hundreds of different antigenic determinants.

Under normal circumstances, the body's own chemical substances do not stimulate an immune response. This failure to stimulate the immune system occurs because these substances are recognized as "self." Immunologists believe that prior to birth, the body's own proteins and polysaccharides inactivate immune system cells that otherwise might respond to them. In the fetal stage, these responsive cells are easily paralyzed. The body thereby develops a tolerance of "self" and remains able to respond only to "nonself." This theory is known as specific immunologic tolerance.

Origin of the Immune System

The immune system is a general term for the complex collection of cells, factors, and processes that provide an adaptive and specific response to antigens associated with microbes. The cornerstones of the immune system are a set of cells known as lymphocytes. These cells are distributed throughout the body and comprise the organs of the lymphoid system, including the lymph nodes, spleen, tonsils, and adenoids. Lymphocytes are small cells, about 10 to 20 μm in diameter, each with a large nucleus. Under the microscope, all lymphocytes look similar. However, two types of lymphocytes can be distinguished on the basis of developmental history, cellular function, and unique biochemical differences. The two types are B lymphocytes (or B cells) and T lymphocytes (or T cells). B cells are responsible for antibody-mediated immunity (AMI), while T cells are responsible for cell-mediated immunity (CMI), as we shall see presently.

The immune system arises in the fetus about 2 months after conception, as FIGURE 17.7 pictures. At this time, lymphocytes originate from stem cells, the primitive cells in the yolk sac and bone marrow. Some stem cells develop into lymphopoietic cells, which will become lymphocytes of the immune system. Lymphopoietic cells take either of two courses. Some of these cells move to an organ in the neck tissues called the thymus. This flat, bilobed organ lies below the thyroid gland near the top of the heart. Within the thymus, the lymphopoietic cells are either modified by the addition of surface receptor proteins or destroyed. Those that are destroyed are ones that would otherwise respond to "self." The modified lymphopoietic cells emerge from the thymus as T cells (T for thymus). Mature T cells are able to interact only with "nonself" antigens and are ready to engage in cell-mediated immunity; they are said to be immunocompetent. These cells colonize the lymph nodes, spleen, and other lymphoid organs.

The B cells take a different track. They mature in a site that has not been determined with certainty in humans. Much evidence suggests that the bone marrow is the maturation site, but some immunologists favor the liver, spleen, or tissue of the gut. In the embryonic chick, the maturation site has been identified as the bursa of Fabricius. For this reason, this type of cell has historically been known as the B cell, for bursa-derived. Like T cells, B cells mature with surface receptor proteins on their membranes and become immunocompetent. They then move through the circulatory system to colonize organs of the lymphoid system, where they join the T cells.

Before we proceed, we should take a brief look at the surface receptor proteins. These proteins enable B cells and T cells to recognize a specific antigenic determinant (epitope) and bind to it. In T cells, the surface receptor is composed of two chains linked to one another. In B cells, it is an antibody molecule, called IgD. About 100,000 surface receptors are found on each T cell or B cell.

The immune response originates with the entry of antigen-containing microbes into the body and their passage into the tissues. Here the microbes are phagocytized by macrophages and other phagocytic cells, and the microbes' antigens are broken down, releasing the antigenic determinants. The macrophages display the antigenic determinants on their surfaces and carry them to the lymphoid organs, where T cells and B cells are waiting.

Within the tissues of the lymphoid organs, the T cells and B cells are responsible for activity of the two arms of the immune system: B cells oversee antibody-mediated immunity, which is based in antibodies that react with microbes (e.g., bacteria and viruses), as well as with antigens dissolved in body fluids. T cells oversee cell-mediated immunity, which is based in special cells that interact with eukaryotic pathogens (e.g., fungi and protozoa), as well as with antigen-marked cells (such as virus-infected cells). In the next sections, we shall discuss both types of immunity, beginning with the cell-mediated type.

Cell-Mediated Immunity

The process of cell-mediated immunity (CMI) begins this way: When disease is occurring in the body, the pathogens display on their cell surfaces a series of antigenic determinants, like a set of "red flags." If the pathogens are infecting body cells, they alter

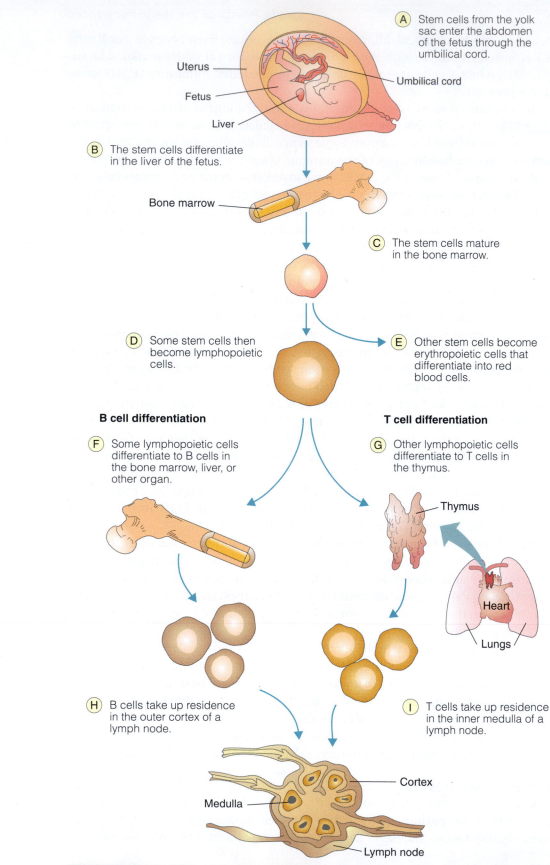

A Stem cells from the yolk sac enter the abdomen of the fetus through the umbilical cord.

Uterus

Fetus

Liver

Umbilical cord

B The stem cells differentiate in the liver of the fetus.

Bone marrow

C The stem cells mature in the bone marrow.

D Some stem cells then become lymphopoietic cells.

E Other stem cells become erythropoietic cells that differentiate into red blood cells.

B cell differentiation

F Some lymphopoietic cells differentiate to B cells in the bone marrow, liver, or other organ.

T cell differentiation

G Other lymphopoietic cells differentiate to T cells in the thymus.

Thymus

Heart

Lungs

H B cells take up residence in the outer cortex of a lymph node.

I T cells take up residence in the inner medulla of a lymph node.

Cortex

Medulla

Lymph node

FIGURE 17.7 **The Origin of the Immune System.**

these cells and cause them to display the "red flags" on their surfaces. A number of macrophages engulf the microbes or the infected cells and break down the antigens in their cytoplasm. The antigenic determinants are captured and returned to the macrophage's surface, nestled within MHC proteins, a set of organic molecules we shall discuss shortly. The macrophage now has the "red flags" on its surface; it is an antigen-presenting cell. (Remember that the infected body cell or microbe is displaying the same "red flags.") The antigen-presenting cell travels to an organ of the lymphoid system, and CMI is ready to continue.

When the antigen-presenting cell (the macrophage) enters the lymphoid organ, a hunt begins. With its exposed antigenic determinants, the macrophage mingles among the myriad groups of inactive T cells, searching for a group having the surface receptor proteins corresponding to its antigenic determinants. But a match with those proteins is not enough. The macrophage also has major histocompatibility proteins (MHC proteins) that must match MHC receptors on the surface of the T cell. Indeed, the T cell will recognize the antigenic determinants only if they are nestled within the correct MHC proteins. MHC proteins are found on virtually all cells of the human body. They define the uniqueness of the individual and are different for each person (the only exception is identical twins).

Once this matchup has been completed, the inactive T cell is ready to assume an activated form, as a cytotoxic T cell, as **FIGURE 17.8** illustrates. But the transition requires the assistance of a special type of T cell called a helper T cell. The correct helper T cells can be selected out of the crowd because the antigenic determinants and MHC proteins on the macrophage also match receptors on these T cells. Another surface receptor protein on the helper T cell is the CD4 receptor. This protein enhances the binding of the helper cell to the macrophage because it is attracted to an MHC protein of the macrophage. (Incidentally, this same CD4 receptor binds to HIV.)

Now things become lively: The chosen helper T cell multiplies and clones itself, and the resulting cells secrete a series of highly charged proteins known as lymphokines (or cytokines). These substances stimulate the cytotoxic T cells to enlarge and divide, yielding a host of cells capable of killing infected cells (some books call these cells "killer T cells").

Soon, the cytotoxic T cells leave the lymphoid tissue and enter the lymph and blood vessels. They circulate until they encounter their targets, the microbes or infected cells displaying the telltale antigenic determinants (the "red flags") on their surface. Now begins the cell-cell interaction characteristic of cell-mediated immunity. The receptor proteins of the cytotoxic T cell join with the antigenic determinants. Then the cytotoxic T cell releases a number of active substances, including a toxic protein called perforin. Perforin inserts into the membrane of the microbe or infected cell and dissolves it. Ions, fluids, and cell structures escape, thereby bringing about cell death and the so-called lethal hit.

Cytotoxic T cells also release lymphokines, small glycoprotein molecules used to enhance the defensive capabilities of the body (as we noted above). One lymphokine attracts macrophages to the infection site and activates their enzymes to encourage cellular digestion. Another lymphokine prevents macrophages from leaving the infection site. A third, called transfer factor, mobilizes other T cells in the area and encourages their conversion to cytotoxic cells. The latter continue the destruction of infected cells and augment the process of immunity.

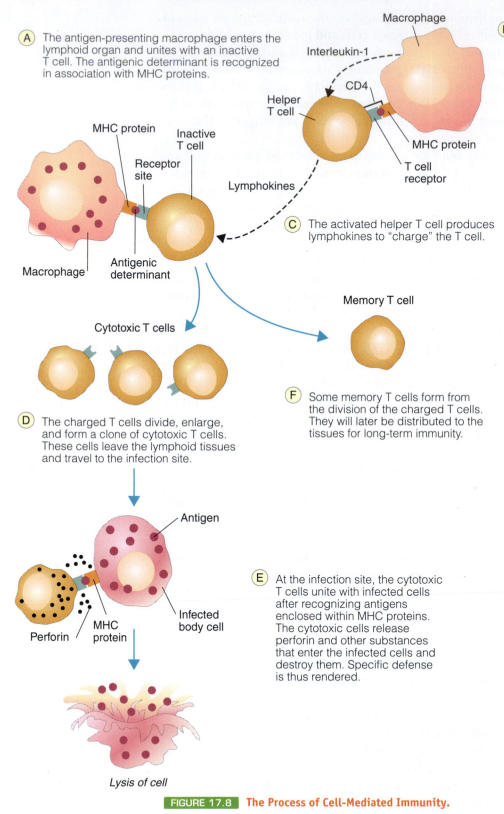

A The antigen-presenting macrophage enters the lymphoid organ and unites with an inactive T cell. The antigenic determinant is recognized in association with MHC proteins.

Macrophage

Interleukin-1

CD4

Helper T cell

MHC protein

T cell receptor

B Another macrophage locates a helper T cell and activates it. The cell's CD4 site assists the binding. Interleukin-1 from the macrophage augments the activation.

MHC protein

Inactive T cell

Receptor site

Lymphokines

Macrophage

Antigenic determinant

C The activated helper T cell produces lymphokines to "charge" the T cell.

Memory T cell

Cytotoxic T cells

F Some memory T cells form from the division of the charged T cells. They will later be distributed to the tissues for long-term immunity.

D The charged T cells divide, enlarge, and form a clone of cytotoxic T cells. These cells leave the lymphoid tissues and travel to the infection site.

Antigen

Infected body cell

Perforin

MHC protein

E At the infection site, the cytotoxic T cells unite with infected cells after recognizing antigens enclosed within MHC proteins. The cytotoxic cells release perforin and other substances that enter the infected cells and destroy them. Specific defense is thus rendered.

Lysis of cell

FIGURE 17.8   The Process of Cell-Mediated Immunity.

When the infected cells have been eliminated, another type of T cells form in the lymphoid tissues, cells that will provide resistance in the event the pathogen reenters the body in the future. These cells are called memory T cells. They distribute themselves to virtually all parts of the body and remain in the tissues to provide long-term immunity. Should the pathogen be detected once again in the tissues, the memory T cells will multiply rapidly, interact with the microbes or infected cells quickly, secrete lymphokines without delay, and immediately set into motion the process of providing CMI. This is one reason we enjoy long-term immunity to a given disease after having contracted and recovered from that disease.

Antibody-Mediated Immunity

While cell-mediated immunity is centered in attacking cells, antibody-mediated immunity (AMI) depends on the activity of antibodies, a series of protein molecules circulating in the body's fluids. (A body fluid is also known as a "humor," and antibody-mediated immunity was traditionally called humoral immunity.) In AMI, antibodies react with toxin molecules in the bloodstream, as well as with microbial antigens (on such structures as flagella, pili, and capsules) and with viruses in the body fluids. The interaction of antibody and antigen leads to elimination of the antigen by various means, as we shall discuss in this section.

Like cell-mediated immunity, antibody-mediated immunity begins with an encounter between phagocytes (bearing antigenic determinants) and cells of the immune system. In this case, however, the major type of participating immune system cell is the B cell, as first noted in the 1960s (see A Closer Look on page 388). Activation of the B cells begins when macrophages and other phagocytes find their way to the lymphoid organ and bring the antigenic determinants close to those B cells that can respond to them. As in CMI, a considerable search is involved as the antigenic determinants attempt to match with surface receptor proteins on the B cells, which are antibodies of the type IgD, as we noted above.

While this process has been going on, the macrophages have contacted helper T cells, which have recognized the macrophages as belonging to the body. This recognition occurs because the antigenic determinants are cradled within the correct MHC proteins on the macrophage surfaces, and the helper T cells have corresponding MHC receptors. Activation and cloning of the helper T cells follows, and they become keyed to the particular antigenic determinants, as FIGURE 17.9 displays. Lymphokines (cytokines) are produced by these helper cells, as the cells become more active. These activated helper T cells recognize the same complex of antigenic determinants and MHC proteins on the B cells, and they bind to the B cells and activate them with their lymphokines. This immunologic cooperation between the macrophage, the B cell, and the helper T cell continues the immune response.

Once activated, the B cells multiply rapidly, developing into plasma cells. Plasma cells are large, complex cells having no surface protein receptors. Their sole purpose is to produce antibodies. A plasma cell lives about 4 to 5 days, during which time it produces an incredible 2000 antibody molecules per second. In a matter of hours, the body is saturated with antibodies. Antibody molecules fill the blood, lymph, saliva, sweat, and all other body secretions. Because of their sheer force of numbers, some of them inevitably meet up with the antigens, bind to the antigens, and mark them for destruction, as we note in the next section.

A CLOSER LOOK

"You Gave Me Bad Chickens!"

For a young graduate student studying poultry science at Ohio State University, it was a rather embarrassing end to a routine laboratory experiment.

Timothy S. Chang had been assigned to show younger students how chickens develop immunity when injected with *Salmonella* cells. For the demonstration, he borrowed a dozen healthy chickens from Bruce Glick, a fellow graduate student. Glick had been using the chickens to study the functions of a mysterious gland in their intestines. Now Chang and his students carefully inoculated the chickens with *Salmonella*, waited a week, and then drew blood samples to test for evidence of *Salmonella* antibodies. To Chang's surprise and chagrin, 10 of the 12 chickens failed to show any sign of antibodies.

Chang went down the hall to Glick's lab and asked if he were playing some kind of joke. Puzzled at his friend's results, Glick checked his lab records and found he had removed the mysterious gland from the 10 "bad" chickens. The two animals that developed antibodies still had the gland intact. The year was 1954. The gland was the bursa of Fabricius, named for Hieronymus Fabricius, a seventeenth-century anatomist who discovered it. Glick had developed an interest in the obscure gland after observing it in the intestine of a goose and learning that its function was unknown. He had removed the gland in newborn chicks and found there was no effect on growth. His research was at a dead end until Chang's experiment gave him a clue to the true function of the gland. Glick and Chang repeated the experiment and published their results in 1956 in *Poultry Science*, a specialty journal.

The accidental discovery might have gone unnoticed except that zoologists at the University of Wisconsin spotted it and repeated the experiments, confirming the findings. As word spread to immunologists, Robert A. Good at the University of Minnesota assigned a team of colleagues to develop the theory of "bursa-derived immunity." In 1965, Good's team delivered its report before the American Academy of Pediatrics. Several immunologists cautioned against drawing conclusions about humans from work with chickens, but many saw the discovery as an important first step to understanding the origin of antibodies.

Antibodies are extremely diverse, and the cells of a single species of bacterium may elicit hundreds of different kinds of antibodies able to unite with hundreds of different kinds of bacterial antigens. The term "immunoglobulin (Ig)" is used interchangeably with "antibody" because antibodies exhibit the properties of globulin proteins and are used in the immune response. In this text, we use "antibody" more frequently.

We shall study the interaction of antibodies and antigens shortly, but at this point it is important to note that certain B cells do not become plasma cells. Instead, they develop into memory B cells. Memory B cells remain in the lymphoid tissues for many years, in some cases for a person's lifetime. Should the antigens reenter the body, these memory cells will revert to plasma cells and produce antibodies with out delay. The antibodies flood the blood stream rapidly and bring about an immediate reaction with the antigens. The symptoms of disease rarely occur this time.

Structure of Antibodies. The basic antibody molecule consists of four polypeptide chains: two identical heavy (H) chains and two identical light (L) chains. These

A   The antigenic determinate-receptor complex has been taken into the selected B cell, and the antigenic determinant is now displayed on the B cell's surface within the MHC protein.

B   Meanwhile, an antigen-presenting macrophage has activated a helper T cell. Interleukin-1 from the macrophage assists the activation.

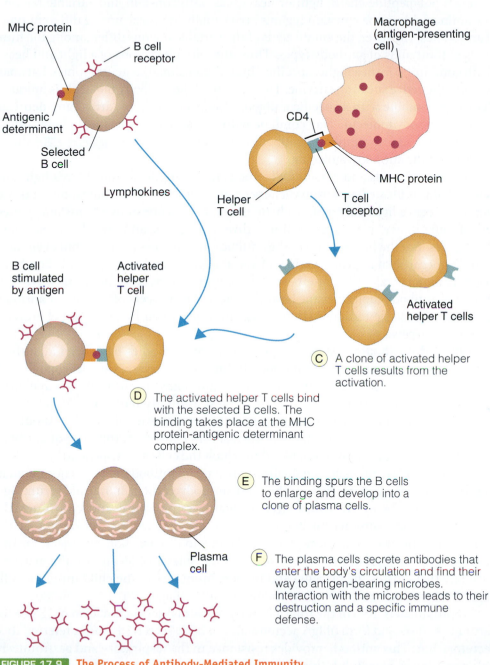

MHC protein

B cell receptor

Antigenic determinant

Selected B cell

Lymphokines

Macrophage (antigen-presenting cell)

CD4

MHC protein

Helper T cell

T cell receptor

Activated helper T cells

B cell stimulated by antigen

Activated helper T cell

C   A clone of activated helper T cells results from the activation.

D   The activated helper T cells bind with the selected B cells. The binding takes place at the MHC protein-antigenic determinant complex.

E   The binding spurs the B cells to enlarge and develop into a clone of plasma cells.

Plasma cell

F   The plasma cells secrete antibodies that enter the body's circulation and find their way to antigen-bearing microbes. Interaction with the microbes leads to their destruction and a specific immune defense.

FIGURE 17.9   **The Process of Antibody-Mediated Immunity.**

chains are joined together by chemical linkages to form a Y-shaped structure. Each heavy chain consists of about 400 amino acids, while each light chain has about 200 amino acids. An antibody molecule has two identical arms, each consisting of a heavy chain and a light chain.

Each polypeptide chain, light or heavy, has both constant and variable regions. The amino acids in the constant regions are virtually identical among different types of antibodies. However, the amino acids of the variable regions differ among the hundreds of thousands of antibody types. Thus, the variable regions of a light and heavy chain combine to form a highly specific, three-dimensional structure somewhat analogous to the active site of an enzyme. This portion of the antibody molecule is uniquely shaped to combine with a specific antigen. Moreover, since its "arms" are identical, a single antibody molecule may combine with two antigen molecules. These combinations may lead to a complex of antibody and antigen molecules. FIGURE 17.10 shows the parts of the antibody molecule.

In human B cells and plasma cells, the gene coding for an antibody's light and heavy chains is located on various gene segments on different chromosomes. For example, to form a light polypeptide chain, the body uses one of five "constant" genes, one of four "joiner" genes, one of three "diversity" genes, and one of over 80 "variable" genes. For the heavy chain of the antibody, the process is similar but even more complex because of a greater variety of variable genes and more possible combinations. Thus, an incalculable number of gene combinations account for the incredible variety of antibody molecules the body produces. The process of coding for antibodies is called somatic recombination; it was first described by 1987 Nobel laureate Susumu Tonegawa.

Types of Antibodies. At present, five types of antibodies have been identified. Using the abbreviation Ig (for immunoglobulin), the five classes are designated IgM, IgG, IgA, IgE, and IgD. IgM is the first type of antibody to appear in the circulatory system after B cell stimulation. It is the largest antibody molecule (M stands for macroglobulin), and because of its size, most of it remains in circulation. About 5 to 10% of the antibody components of normal serum consists of this type of antibody (serum is the fluid portion of the blood in which the cells are suspended).

IgG is the classical gamma globulin. This type of antibody is the major circulating one, comprising about 80% of the total antibody content in normal serum. IgG appears about 24 to 48 hours after antigenic stimulation and continues the antigen-antibody interaction begun by IgM. In addition, as a product of the memory B cells, it provides long-term resistance to disease. Booster injections of a vaccine raise the level of this type of antibody considerably. IgG is also the maternal antibody that crosses the placenta and renders immunity to the child until it is fully capable of producing antibodies at about 6 months of age.

Approximately 10% of the total antibody in normal serum is a form of IgA called serum IgA. A second form of IgA accumulates in body secretions and is referred to as secretory IgA. This antibody provides resistance in the respiratory and gastrointestinal tracts, possibly by inhibiting the attachment of parasites to the tissues. It is also located in tears and saliva and in the colostrum, the first milk secreted by a nursing mother. When consumed by a child during nursing, the secretory antibody provides resistance to gastrointestinal disorders.

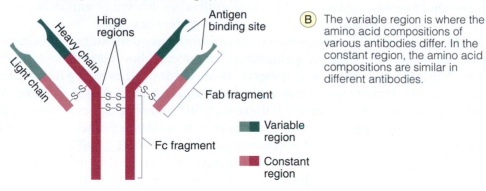

A The antibody molecule consists of four chains of protein: two light chains and two heavy chains connected by disulfide (S-S) linkages. The heavy chains bend at a hinge point.

Hinge regions

Heavy chain

Light chain

Antigen binding site

Fab fragment

Fc fragment

Variable region

Constant region

B The variable region is where the amino acid compositions of various antibodies differ. In the constant region, the amino acid compositions are similar in different antibodies.

C On treatment with papain enzyme, cleavage occurs at the hinge point, and three fragments result: two Fab fragments and one Fc fragment.

D This detailed view shows the disulfide linkages and the loops that form domains. All light chains have a single variable domain (V_L) and a single constant domain (C_L). Heavy chains contain a variable domain (V_H) and three or four constant domains (C_H).

Antigen binding site

V_H

C_{H_1}

V_L

C_L

C_{H_2}

C_{H_3}

Immobilized clump of antigens

E The reaction between antibody molecules and the antigenic determinants of an antigen shows the specificity that occurs. Note that the antibody molecules unite with the triangular antigenic determinants but not with the circular ones.

FIGURE 17.10 **Details of an Antibody Molecule.**

IgE plays a major role in allergic reactions by sensitizing cells to certain antigens, as we discuss below. Both the function and significance of IgD are unclear, but research evidence indicates that the antibody is a cell surface receptor on the B cells. For this reason, it is called a membrane antibody.

Antigen-Antibody Interactions. In order for specific resistance to develop, antibodies must interact with antigens in such a way that the antigen is altered. The alteration may result in the death of the microbe having the antigen or an increased susceptibility of the antigen to other body defenses. This interaction is the crux of antibody-mediated immunity. For example, antibodies called neutralizing antibodies react with viral capsids and prevent viruses from entering their host cells, while encouraging phagocytosis. Moreover, other neutralizing antibodies provide a vital mechanism of defense against toxins. These antibodies, known as antitoxins, alter the toxin molecules near their active sites and mask their toxicity.

Other antibodies called agglutinins react with antigens on the surface of bacteria. This action causes clumping (or agglutination) of the microbes and enhances phagocytosis. Moreover, movement of microbes is inhibited if antibodies react with antigens on their flagella. And the reaction of antibodies with antigens of the pili (Chapter 5) prohibits attachment of a bacterium to the tissues. Still other antibodies called precipitins react with dissolved antigens and convert them to solid precipitates. In this form, antigens are usually inactive and more easily phagocytized, as FIGURE 17.11 shows.

A final example of antigen-antibody interaction involves the complement system. Originally described in 1895 by Jules Bordet at the Pasteur Institute, the complement system is a series of over 20 proteins that function in a cascading set of reactions. The pathway of reactions is set into motion by the interaction of antigen and antibody molecules. This interaction usually takes place on the surface of a cell, such as a bacterium. It results in several substances, one of which increases the permeability of the cell membrane and induces the cell to undergo lysis through the leakage of fluid from its cytoplasm. Another substance attracts phagocytes, while a third binds the cell to the phagocyte.

Types of Immunity. The word "immune" is derived from the Latin *immuno*, meaning "safe" or "free from." In its most general sense, the term implies a condition under which an individual is protected from disease. This does not mean, however, that the person is immune to all diseases, but rather to a specific disease or group of diseases.

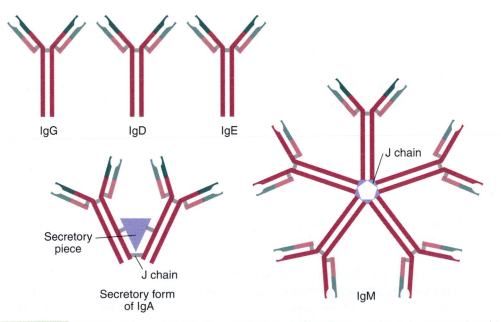

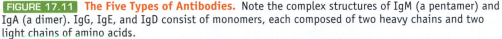

FIGURE 17.11 **The Five Types of Antibodies.** Note the complex structures of IgM (a pentamer) and IgA (a dimer). IgG, IgE, and IgD consist of monomers, each composed of two heavy chains and two light chains of amino acids.

Two general types of immunity are recognized: innate immunity and acquired immunity. Innate immunity is an inborn capacity for resisting disease. It begins at birth and is synonymous with the nonspecific resistance we discussed earlier. Acquired immunity, by contrast, begins after birth. It depends on the activity of T cells, antibodies, and other factors originating in the immune system. Acquired immunity is termed active immunity if it develops when the immune system responds to antigens and forms antibodies or passive immunity if it develops when antibodies enter the body from an outside source. Active immunity takes several hours or days to develop fully, but it remains for a long period of time, often throughout a person's life. Passive immunity, by comparison, comes about immediately when antibodies enter the body, but it lasts for only several days or weeks.

Active and passive immunity are further subdivided into four types. The first is naturally acquired active immunity. This usually follows a bout of illness. Memory T or B cells residing in the lymphoid tissues remain active for many years and produce IgG immediately if the pathogen later enters the host. Artificially acquired active immunity, by comparison, develops after an exposure to antigens in a vaccine. The antigens may be toxoids, inactivated viruses, synthetic viral parts, bacterial parts, or other components we note in Chapters 15 and 19. Vaccines promote a long-term immune response in the form of memory T or B cells and IgG antibodies. Many are effective for a lifetime, but some need to be readministered periodically as booster injections to restimulate the immune response.

The third type of immunity is naturally acquired passive immunity, also called congenital immunity. This immunity develops when IgG antibodies pass from the mother's bloodstream into the fetal circulation via the placenta and umbilical cord. The antibodies remain with the child for approximately 3 to 6 months after birth and fade as the child's immune system becomes fully functional. Maternal antibodies, predominantly IgA type, also pass to the newborn through the colostrum. Artificially acquired passive immunity, the fourth type, arises from the injection of antibody-rich serum into the patient's circulation, as FIGURE 17.12 illustrates. This form of therapy is used for serious viral diseases and toxin-related diseases such as botulism and tetanus. The serum injected is often referred to as gamma globulin. Such injections must be used with caution because in many individuals, the immune system recognizes the foreign serum proteins as antigens and synthesizes antibodies against them.

Destructive Immune Processes

Modern scientists realize that not all immune processes protect the body, for there are many immune system activities that are destructive. Among these are a set of disorders commonly known as hypersensitivity reactions, because they arise from an overreaction of the immune system to an antigen's presence.

One example of a hypersensitivity reaction is anaphylactic shock, a process that begins with the entry of an antigen into the body. This antigen may be any of diverse chemicals, such as bee venom, serum proteins, or drugs (for example, penicillin). The immune system responds, and B cells revert to plasma cells, which produce an antibody of type IgE. This antibody fixes itself to the surface of mast cells (a type of connective tissue cells) and white blood cells known as basophils. Over a period of a week or more, millions of molecules of IgE attach to thousands of mast cells and basophils, and the person becomes highly sensitive to the antigen.

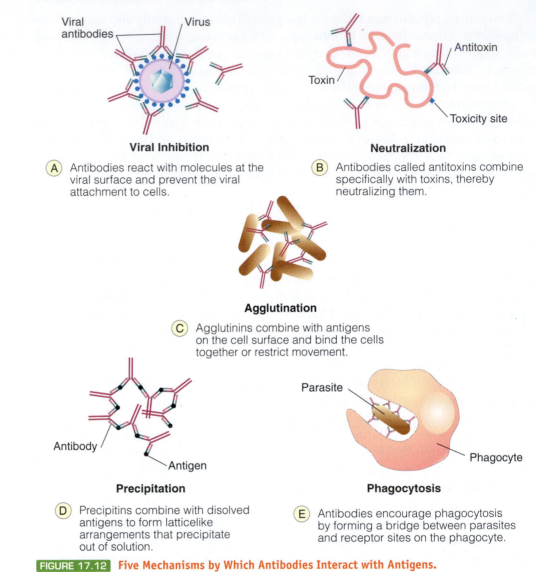

Viral Inhibition

Ⓐ Antibodies react with molecules at the viral surface and prevent the viral attachment to cells.

Neutralization

Ⓑ Antibodies called antitoxins combine specifically with toxins, thereby neutralizing them.

Agglutination

Ⓒ Agglutinins combine with antigens on the cell surface and bind the cells together or restrict movement.

Precipitation

Ⓓ Precipitins combine with disolved antigens to form latticelike arrangements that precipitate out of solution.

Phagocytosis

Ⓔ Antibodies encourage phagocytosis by forming a bridge between parasites and receptor sites on the phagocyte.

FIGURE 17.12 **Five Mechanisms by Which Antibodies Interact with Antigens.**

The symptoms of anaphylaxis occur rapidly during a later exposure to the antigen. After entering the body, the antigen molecules unite with IgE molecules on the surfaces of the mast cells and basophils, as **FIGURE 17.13** pictures. As this takes place, the cells swell, and cytoplasmic granules quickly move out of the cell. As the granules flow into the outside fluid, they emit a number of chemical substances having substantial effect on the body. For example, one substance called histamine contracts the smooth muscles in the body with numerous effects: The skin swells around the eyes, wrists, and ankles, and a hivelike rash develops, along with burning and itching. Contractions along the gastrointestinal and respiratory tracts lead to sharp cramps and shortness of breath, respectively. And the small veins and capillary pores expand, forcing fluid out and leading to further swelling and shock. Death from suffocation may occur in 10 to 15 minutes if prompt action is not taken. Epinephrine, antihistamines, and smooth muscle relaxers are among the first priorities of treatment.

(A) Naturally acquired active immunity arises from an exposure to antigens and often follows a disease.

(B) Artificially acquired active immunity results from an inoculation of toxoid or vaccine.

(C) Naturally acquired passive immunity stems from the passage of IgG across the placenta from the maternal to the fetal circulation.

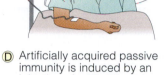

(D) Artificially acquired passive immunity is induced by an injection of antibodies taken from the circulation of an animal or another person. (short-term)

FIGURE 17.13 **The Four Types of Acquired Immunity.**

The hypersensitivity reaction need not result in whole-body involvement. Indeed, the vast majority of these reactions are accompanied by limited production of IgE antibodies and sensitization in localized areas of the body. This phenomenon is called allergy. An example is hay fever, a condition that develops from springtime inhalations of tree and grass pollens, and summer and fall exposures to grass and weed pollens. The hypersensitivity reaction brings on sneezing, tearing, swollen membranes, and other well-known symptoms. Similar reactions may be caused by house dust, mold spores, dust mites, and detergent enzymes. Food allergies, another possibility, are accompanied by symptoms in the GI tract, including swollen lips, abdominal cramps, nausea, and diarrhea. Foods causing these symptoms include chocolate, strawberries, oranges, and crabmeat. According to public health estimates, almost 20% of Americans have some type of allergy (see A Closer Look on page 396).

The immune system disorders include another form of hypersensitivity reaction called cytotoxic hypersensitivity. This is a cell-damaging reaction that occurs when IgG antibodies react with antigens on the surfaces of body cells. The complement system is often activated, and IgM antibodies may be involved. IgE antibodies do not participate, nor is there any degranulation of mast cells or basophils. A well-known example of cytotoxic hypersensitivity is the transfusion reaction arising from the mixing of incompatible blood types. Another example of cytotoxic hypersensitivity is hemolytic disease of the newborn, or Rh disease, a problem that develops when an Rh (+) man and an Rh (−) woman have a child who is Rh (+).

Other examples of cytotoxic hypersensitivity are less familiar. One example, called thrombocytopenia, results from IgG antibodies the body produces against such drugs as aspirin, certain antibiotics, or antihistamines. The antibodies combine with antigens and drug molecules (as usual), but they also bind to antigens found on the surface of thrombocytes (blood platelets). The thrombocytes break down and disintegrate. Impaired blood clotting soon develops, and hemorrhages may occur on the skin and in the mouth.

A third form of hypersensitivity reaction is immune complex hypersensitivity. This problem develops when IgG antibodies combine excessively with antigens dissolved in the body fluids. So much antigen-antibody reaction takes place that visible masses of material, the so-called immune complexes, soon accumulate along blood vessels or on tissue surfaces. Systemic lupus erythematosus (SLE) is an example of this form of hypersensitivity. When immune complexes accumulate in the skin and body organs, the patient experiences a butterfly rash (a skin condition appearing across the nose and cheeks), as well as damage to the blood-rich organs such as the spleen and kidneys. In patients with rheumatoid arthritis, another example of this form of hypersensitivity, immune complexes form in the joints and damage occurs here.

Cellular hypersensitivity, the final immune disorder we consider, is an exaggeration of the process of cell-mediated immunity. This is a delayed reaction whose max-

A CLOSER LOOK

The Peanut Dilemma

Americans love peanuts—salted, unsalted, oil-roasted, dry-roasted, Spanish, honey-crusted, in shells, out of shells, and on and on. But as declared peanut fanatics know, there is also the risk of a rather nasty allergic reaction: An individual can break out in hives, develop a serious headache, experience a racing heartbeat, or double over with intestinal cramps.

And, if that's not bad enough, research reported in 1996 indicates that feeding peanut butter to very young children increases their sensitivity to the legumes when they reach adulthood. Moreover, a 1997 report in the *New England Journal of Medicine* indicates that peanut-specific antibodies can be transferred from an organ donor to an organ recipient. (In the recipient, the skin reaction is not threatening, but it does complicate matters.)

The answer to all those miseries can be best summed up as "V plus V." The first V is for vigilance. Vigilance means avoiding peanuts or peanut butter; it also means being cautious about egg roll wrappers, chili fillers, and protein extenders in cake mixes, all of which may contain peanuts in one form or another. And vigilance means that manufacturers must be required to label their products clearly and to perform peanut-detection tests routinely.

The second V is for vaccine. In 1999, investigators from Johns Hopkins University tested a peanut vaccine and showed that it protects sensitized mice against peanut proteins. The vaccine consists of DNA segments that encode the peanut proteins. Encased in protective molecules and delivered orally, the vaccine decreased the mice's ability to produce peanut-related IgE. The developers postulated that the vaccine elicits so-called blocking antibodies that bind to the peanut antigens before they reach the animal's immune system.

So does that mean we can expect health officials to distribute vaccine where we buy peanut butter, peanut brittle, or beer nuts? Not likely, say the researchers, at least not in the immediate future. But check back soon. You never know.

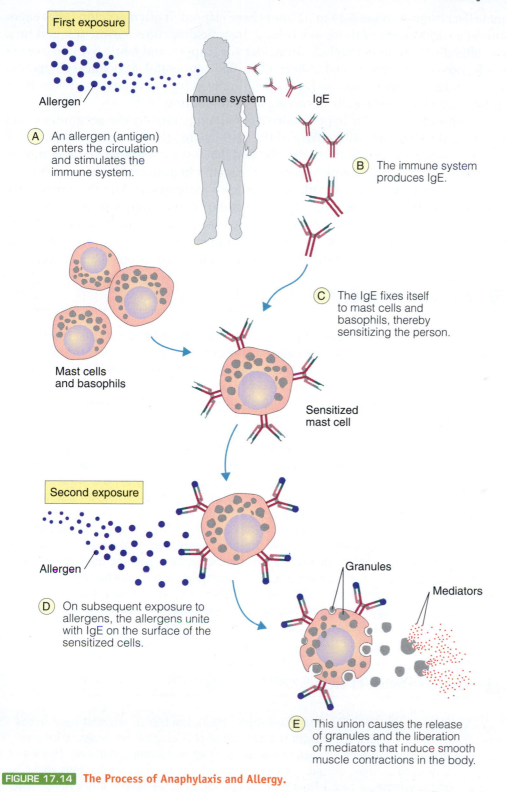

First exposure

Allergen

Immune system   IgE

(A) An allergen (antigen) enters the circulation and stimulates the immune system.

(B) The immune system produces IgE.

(C) The IgE fixes itself to mast cells and basophils, thereby sensitizing the person.

Mast cells and basophils

Sensitized mast cell

Second exposure

Allergen

Granules

Mediators

(D) On subsequent exposure to allergens, the allergens unite with IgE on the surface of the sensitized cells.

(E) This union causes the release of granules and the liberation of mediators that induce smooth muscle contractions in the body.

FIGURE 17.14 **The Process of Anaphylaxis and Allergy.**

imal effect is not seen until 24 to 72 hours have elapsed. It often develops after exposure to antigens such as those in clothing, insecticides, coins, cosmetics, and furs. The offending substances include formaldehyde, copper, and fabric dyes. In one example, poison ivy, the antigen has been identified as urushiol, a chemical compound found on the surface of poison ivy leaves. A poison ivy rash consists of very itchy pinhead-sized blisters usually occurring in a straight row.

An episode of cellular hypersensitivity is accompanied by the accumulation of T cells in the skin and overactivity of the cells against the offending antigen. The T cells soon begin to attack the body cells, and that results in drying of the skin, with reddening and scaling. Examples occur on the scalp when allergenic shampoo is used, on the hands when contact is made with detergent enzymes, and on the wrists when the person is allergic to chemicals in watchbands. Immunologists are at a loss to explain why the immune system, the fundamental defensive system of the body, should be related to these harmful effects. They point out, however, that immunology is a relatively new science that at the present time has more questions than answers.

◼ A FINAL THOUGHT

It may have occurred to you that an episode of disease is much like a war. First, the invading microbes must penetrate the natural barriers of the body; then, they must escape the phagocytes and other chemicals that constantly patrol the body's circulation and tissues. Finally, they must elude the antibodies or T cells that the body sends out to combat them. How well the body does in this battle will determine whether the individual survives the disease.

Obviously, our bodies do very well, because most of us are in good health today. As it turns out, most infectious organisms are stopped at their point of entry to the body. Cold viruses, for example, get no farther than the upper respiratory tract. Many parasites come into the body by food or water, but they rarely penetrate beyond the intestinal tract. The staphylococci in a wound may cause inflammation at the infection site, but this is usually the extent of the problem.

To be sure, drugs and medicines help in those cases where diseases pose life-threatening situations. In addition, sanitation practices, vector control, care in the preparation of food, and other public health measures prevent microbes from reaching the body in the first place. However, in the final analysis, body defense represents the bottom line in protection against disease, and, as history has shown, it works very well. Indeed, as Lewis Thomas has suggested in *The Lives of a Cell*, a microbe that catches a human is in considerably more danger than a human who catches a microbe.

◼ QUESTIONS TO CONSIDER

1. In his classic book *The Mirage of Health* (1959), the French scientist René Dubos develops the idea that health is a balance of physiological processes, a balance that takes into account such things as nutrition and living conditions. (Such a view opposes the more short-sighted approach of locating an infectious agent and developing a cure.) From your experience, describe several other things that Dubos might add to his list of "balancing agents."

2. The transparent windows placed over salad bars are commonly called "sneeze bars" because they help prevent nasal droplets from reaching the salad items. What other suggestions might you make to prevent disease transmission via a salad bar?

3. Between 1982 and 1992, the number of Americans dying of infectious diseases rose 58%. That rise continues to this day. Population shifts, modern travel patterns, and microbial evolution are three of the many reasons given for the emergence and reemergence of infectious diseases. How many other reasons can you name?

4. An environmental microbiologist has created a stir by maintaining that a plume of water is aerosolized when a toilet is flushed, and that the plume carries bacteria to other items in the bathroom, such as toothbrushes. Assuming this is true, what might be two good practices to follow in the bathroom? Further, suppose you are a microbe hunter assigned to make a list of the ten worst "hot zones" in your home. The title of your top-ten list will be "Germs, Germs Everywhere." What places will make your list, and why?

5. It has been said that no other system in the human body relies so heavily on signals as the immune system. What evidence can you offer to support or reject this claim?

6. The ancestors of modern humans lived in a sparsely settled world where communicable diseases were probably very rare. Suppose one of those individuals was magically thrust into the contemporary world. How do you suppose he or she would fare in relation to infectious disease? What is the immunological basis for your answer?

7. Some years ago, a novel entitled *Through the Alimentary Canal with Gun and Camera* appeared in bookstores. The book described a fictitious account of travels through the human body. What perils, microbiologically speaking, would you encounter if you were to take such an adventure?

KEY TERMS

Informative facts are necessary for the expression of every concept, and the information for a concept is founded in a set of key terms. The following terms form the basis for the concepts of this chapter. On completing the chapter, you should be able to explain and/or define each one:

acquired immunity
active immunity
agglutinins
allergy
anaphylactic shock
antibodies
antibody-mediated immunity (AMI)
antigenic determinant
antigens
antitoxins

B cell
carrier
cell-mediated immunity (CMI)
cellular hypersensitivity
cytotoxic hypersensitivity
cytotoxic T cell
disease
droplets
endemic
endotoxins

epidemic	normal microflora
exotoxins	opportunistic microbe
fomites	pandemic
gamma globulin	parasite
helper T cells	passive immunity
immune complex hypersensitivity	pathogenicity
infection	pathogenicity islands
inflammation	perforin
innate immunity	phagocytosis
invasiveness	plasma cells
lymphokines	portal of entry
lymphopoietic cells	reservoir
lysozyme	precipitins
macrophages	somatic recombination
mechanical vector	T cell
memory B cells	thymus
memory T cells	toxoid
MHC proteins	virulence

http://microbiology.jbpub.com/book/microbes

The site features **eLearning**, an online review area that provides quizzes and other tools to help you study for your class. You can also follow useful links for in-depth information, read more stories of microbiology, or just find out the latest microbiology news.

Viral Diseases of Humans: AIDS to Zoster

18

Improperly sterilized equipment, such as tattoo needles, could spread the hepatitis B virus (inset), which can cause hepatic cancer.

▌ Looking Ahead

The viral diseases of humans cover a broad spectrum of illnesses and include such mild diseases as measles, mumps, and chickenpox, as well as such serious maladies as smallpox, rabies, and AIDS. These diseases will be surveyed in this chapter.

On completing this chapter, you should be able to . . .

- appreciate the broad scope of viral diseases that affect virtually all parts of the human body.
- summarize the wide variety of signs and symptoms that can accompany viral diseases and understand how they can be used to help identify a disease.
- specify the varied methods for transmitting viral diseases among humans and explain how to avoid being exposed to viruses.
- define some of the mechanisms by which viral diseases result in damage to the cells, tissues, and organs of the body.
- summarize the salient characteristics of some traditional viral diseases and be aware of the viral diseases of contemporary times.

- recognize some misconceptions regarding viral diseases and gain new insights about these public health problems.
- develop a vocabulary of terms relating to viral diseases and be able to discuss them confidently.

It appeared in England in 1986. First one cow showed the symptoms, then another and another. All of them were apprehensive and twitchy, overreacting to a sound or touch. Soon the whole herd developed a peculiar, high-stepping, swaying gait with an unsteady lurch. Some cows became overly aggressive, and soon the local residents were talking about the day the cows went mad.

By 1990, cows were becoming ill at the rate of hundreds per week, and cartoon writers were having a field day drawing mad cows descending on farmhouses and milk factories. To health officials, however, the situation was more foreboding than mirthful. Were the milk and meat supplies contaminated, and would humans be next? Indeed, there were reports of mad cats, and speculation arose that cat food made from bovine parts was to blame.

Through all the months, virologists had been drawing a parallel between the unknown disease and scrapie, a disease in which sheep develop neurological symptoms. Under the microscope, scrapie-infected tissue looks spongy, with tiny fluid-filled holes, exactly like the cows' brain tissue. But how was the agent responsible for scrapie transferred to cows? Veterinary researchers eventually discovered that young calves are often fed a protein-rich feed made from the carcasses of sheep.

Today, the term "mad cow disease" is used primarily by journalists. The proper name for the disease is bovine spongiform encephalopathy, or BSE ("bovine" refers to cows, "spongiform" to the brain's spongy appearance, and "encephalopathy" to a brain disease). Most investigators consider BSE in the same terms as scrapie, and a human equivalent of the disease is currently being researched.

BSE is one of the contemporary viral diseases we shall study in this chapter. We shall discuss the diseases according to the part of the body affected, a method used in many books. However, the discussion will not be traditional. Rather, in this chapter and in Chapter 19, we discuss the diseases in the form of a conversation between you and a clinical microbiologist. Microbial disease is surrounded by many questions, and this format explores many of them. (If you have a question that is not included, please let us know and perhaps we can include it in the next edition.)

■ Viral Diseases of the Skin

The viral diseases of the skin are a diverse collection of human maladies. Certain ones, such as herpes simplex, remain epidemic in our time; others, such as measles, mumps, and chickenpox, are being brought under control through effective vaccination programs. The diseases are generally transmitted by contact with an infected individual, and at least some of the symptoms are seen on the skin tissues. With these diseases, we begin our "microbial conversation."

Herpes Simplex

Someone told me that herpes simplex is one of the oldest diseases known. Is that so?

The sores and blisters of herpes simplex have been known for centuries. In ancient Rome, an epidemic was so bad that the Emperor Tiberius banned kissing; in

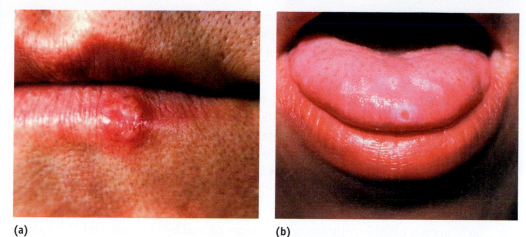

(a)   (b)

FIGURE 18.1 **Two Manifestations of Herpes Simplex.** (**a**) The cold sores (fever blisters) of herpes simplex erupting as tender, itchy papules and progressing to vesicles that burst, drain, and form scabs. Contact with the sores accounts for spread of the virus. (**b**) Gingivostomatitis involving the oral mucosa, tongue, and cheeks with sores and blisters.

Romeo and Juliet, in the 1500s, Shakespeare wrote of "blisters o'er ladies lips"; and in the 1700s, genital herpes was common enough for French prostitutes to consider it a vocational disease.

Is herpes simplex the same as genital herpes?

Yes, it is. Herpes simplex is not one disease, but an array of viral diseases caused by a large icosahedral virus containing DNA. Among them are herpes simplex, the cold sores that occur around the mouth (); genital herpes, the sexually transmitted disease; herpes encephalitis, a viral disease of the brain; and herpes keratitis, a herpes infection of the eyes. Antiviral drugs such as acyclovir (Zovirax) and idoxuridine can be used to treat these infections.

What's genital herpes like?

Genital herpes is estimated to affect between 10 and 20 million Americans yearly, of whom about 500,000 are new cases. Signs generally appear within a few days of sexual contact, often as itching or throbbing in the genital area. This is followed by reddening and swelling of a small area where painful blisters erupt. The blisters crust over and the sores disappear, usually within about 3 weeks. In the majority of cases, however, the symptoms reappear, often in response to stressful triggers, such as sunburn, fever, menstruation, or emotional disturbance.

Can pregnant women give herpes to their newborns?

Yes. In a pregnant woman, herpes simplex viruses sometimes pass to the fetus via the placenta and cause neurological problems and/or mental impairment in the newborn. The virus can also be picked up during childbirth. For that reason, obstetricians recommend that women with active genital herpes consider giving birth by cesarean section.

What are the two herpes types I've heard of?

In the 1960s, scientists learned that the herpes simplex virus has two different forms: type I and type II. For reasons that remain unclear, type I virus generally infects areas above the waist and is the cause of most cold sores, while type II virus appears prevalent below the waist. Type II herpes simplex virus is especially worrisome because it is associated with cervical cancer, a disease that strikes over 15,000 American women annually.

Chickenpox (Varicella)

Why do they call it chickenpox?

Chickenpox got its name because it makes the skin look like the skin of a freshly plucked chicken. In the centuries when pox diseases regularly swept across Europe, people had to contend with the Great Pox (syphilis), the smallpox, the cowpox, and the chickenpox.

Is chickenpox easy to catch?

Chickenpox is a highly communicable disease caused by an icosahedral DNA virus that is usually transmitted by skin contact. Viruses localize in the skin and in nerves close to the skin, where they trigger the formation of hundreds of small, teardrop-shaped, fluid-filled vessels (varicella, the other name for chickenpox, is the Latin word for "little vessel"). The vessels develop over 3 or 4 days in a succession of crops. They itch intensely and eventually break open to yield highly infectious virus-laden fluid. The drug acyclovir lessens the symptoms of chickenpox and hastens recovery.

Can chickenpox be prevented?

Yes, it can. In 1995, the Food and Drug Administration licensed a vaccine for children that contains attenuated (weakened) chickenpox viruses to help prevent the disease. This vaccine has been largely responsible for the dramatic decrease in the incidence of chickenpox in the United States.

Is chickenpox the same as shingles?

Herpes zoster, or shingles, is an adult disease caused by the same virus that causes chickenpox. For this reason, the virus is often referred to as the varicella-zoster (VZ) virus. In adults, the viruses multiply in nerve tissue along the spinal cord and travel down the nerves to the skin. Here they cause blisters with blotchy patches of red that often appear to encircle the trunk (*herpes* is Greek for "creeping," and *zoster* is Greek for "girdle"). Many sufferers also experience headaches, as well as facial paralysis and sharp "ice-pick" pains. The condition can occur repeatedly and is linked to emotional and physical stress, as well as to a suppressed immune system or aging.

Measles (Rubeola)

I heard that you get measles from someone coughing in your face. Why then is it a skin disease?

Measles (or rubeola) is a highly contagious disease usually transmitted by respiratory droplets (bits of mucus and saliva) coughed out during the early stages of disease. But its recognizable symptoms appear on the skin, and so we consider it a skin disease. A helical RNA virus is responsible.

What's it like to have measles?

Measles symptoms commonly include a hacking cough, sneezing, nasal discharge, eye redness, sensitivity to light, and a high fever. The characteristic red rash of measles soon appears, beginning as pink-red pimplelike spots and developing into a rash that

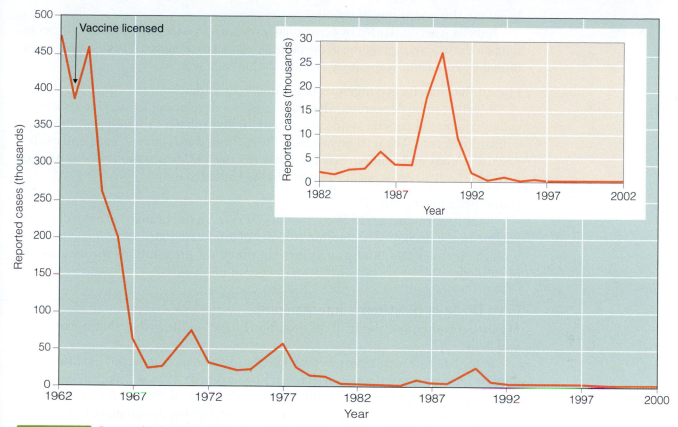

FIGURE 18.2 **Reported U.S. Cases of Measles (Rubeola) From 1962 to 2000.** Note the sharp dropoff in cases after licensing of the MMR vaccine in the mid-1960s. Unfortunately, the immunity from this vaccine was not long-lasting, and a new epidemic of measles broke out in the late 1980s. The inset shows the rise from a low of 1400 cases in 1983 to over 27,000 cases in 1990. The recent reduction in cases is partly due to renewed efforts to revaccinate susceptible individuals.

breaks out at the hairline, covers the face, and spreads to the trunk and extremities. Within a week, the rash turns brown and fades.

How common is measles in the United States?

In past decades, measles often occurred in epidemic form, but in 2002, there were only 60 cases in the United States (FIGURE 18.2). The cornerstone of the campaign to control measles is immunization of all school-age children with attenuated measles viruses in the **measles-mumps-rubella (MMR)** vaccine. Immunization of children is now mandatory in all 50 of the United States.

Rubella (German Measles)
What's the difference between measles and German measles?

For generations, rubella was thought to be a mild form of measles. But, in 1829, Rudolph Wagner, a German physician, noted the difference in symptoms and suggested that the two diseases were separate (the name German measles was soon adopted). Scientists now know that different viruses cause the two diseases. The name rubella ("small red") was suggested in the 1860s because the disease is accompanied by a slightly red rash. An icosahedral RNA virus causes rubella.

What are the symptoms of rubella?

Rubella is accompanied by occasional fever with a variable, pale-pink rash beginning on the face and spreading to the body trunk and extremities. The disease is transmitted by contact or respiratory droplets and is usually mild.

I heard that rubella is dangerous in a fetus. Why is that?

Rubella can be dangerous when it occurs in a developing fetus. This condition, called congenital rubella syndrome, occurs when rubella viruses pass across the woman's placenta. The fetal organs most often affected are the eyes, ears, and cardiovascular organs, and children may be born with cataracts, glaucoma, deafness, or heart defects. About 50 cases occur each year in the United States.

Can rubella be prevented?

Since its introduction in 1969, the rubella vaccine has brought about a dramatic decline in the disease. The vaccine consists of attenuated viruses and is combined with the measles and mumps vaccines (in the MMR preparation) for inoculation of children.

Fifth Disease (Erythema Infectiosum)

My little brother had the fifth disease, but I have no idea what that means. Can you help?

In the late 1800s, numbers were assigned to diseases accompanied by skin rashes. Disease I was measles, II was scarlet fever, III was rubella, IV was Duke's disease (also known as roseola), and V was the fifth disease. The agent of the fifth disease remained a mystery until the modern era, when a strain of parvovirus designated B19 was identified as the cause. The virus contains DNA and has icosahedral symmetry. Transmission appears to be by respiratory droplets.

How do doctors identify the fifth disease?

Most cases of fifth disease occur in children. The outstanding feature of the illness is a fiery red rash on the cheeks and ears (hence, another name is slapped-cheek disease). The rash may spread to the trunk and extremities, but it fades within several days, leaving a lacy pattern of red on the skin. No immunization is available.

Mumps

How did mumps get its unusual name?

Mumps takes its name from the English "to mump," meaning to be sullen or to sulk. The characteristic sign of the disease is enlarged jaw tissues arising from swollen salivary glands, especially the parotid glands. Epidemic parotitis is an alternate name for the disease. A helical RNA virus is the cause.

Why do the glands swell in mumps?

When a person has mumps, obstruction of the ducts leading from the parotid glands retards the flow of saliva and causes the characteristic swelling. The skin overlying the glands becomes taut and shiny, and patients experience pain when the glands are touched.

When I was a kid, I had the mumps and my father and uncles were told to stay away from me until I got over the disease. Why was that?

In adult males, the mumps virus may infect the testicles and cause a lowering of the sperm count. This condition is called orchitis. To avoid that possibility, adult males are told to avoid children with mumps.

Can you get immunized to mumps?

Most certainly. The mumps vaccine, developed in 1967, consists of attenuated mumps viruses and is usually combined with the measles and rubella vaccines (MMR). Although the campaign against mumps never attained the fame of the campaigns against measles or rubella, the reduction of mumps cases has been equally notable.

Smallpox

I thought I read about smallpox in my history book. Is that possible?

You probably came across smallpox more than once in your reading. Smallpox has ravaged people since prebiblical times. It moved swiftly across Europe and Asia, often doubling back on its path, and it was apparently brought to the New World in the 1500s by troops serving under the Spanish conquistador Hernando Cortez. There it killed 3.5 million Native Americans and contributed to the collapse of the Inca and Aztec civilizations. Few people escaped the pitted scars that accompanied the disease. Indeed, children were not considered part of the family until they had survived smallpox.

Is smallpox a serious disease?

Yes, smallpox is a horrible disease. The earliest signs are pink-red spots, called macules. Soon they become large fluid-filled vessels (the disease is also called variola from *varus*, meaning "vessel"). The vesicles become deep pustules, which break open and emit pus, as shown in FIGURE 18.3A. If the person survives the onslaught, the pustules leave pitted scars, or pocks. These are generally smaller than the lesions of syphilis (the Great Pox) or varicella (the chickenpox).

Who was Jenner, and why is his name linked to smallpox?

Edward Jenner was an English physician. In 1798, he noted that milkmaids contracted a mild form of smallpox named cowpox, or vaccinia (*vacca* is Latin for "cow"). Anyone who experienced cowpox apparently did not develop smallpox. Jenner therefore got the idea of inoculating people with material from a cowpox lesion to give them vaccinia. He reasoned that they would not then get smallpox. After much research, he established the method of vaccination (FIGURE 18.3B). His method was so successful that Napoleon ordered his entire army vaccinated in 1806. The effort to vaccinate the American population was led by President Thomas Jefferson. A century passed before scientists understood that antibodies produced against the mild cowpox virus were neutralizing the more deadly smallpox virus as well. A brick-shaped DNA virus is now known to cause smallpox.

How common is smallpox now?

At present, there is no smallpox in the world. In 1966, the World Health Organization (WHO) began the global eradication of smallpox. Surveillance and containment methods were used to isolate every known pox victim, and all contacts were

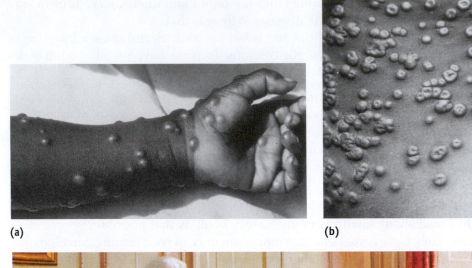

(a)   (b)

(c)

FIGURE 18.3 **Smallpox.** (**a**) Smallpox lesions are raised, fluid-filled vesicles (left) similar to those of chickenpox. For this reason, cases of chickenpox have been misdiagnosed as smallpox. (**b**) Later, the lesions become pustules and then form pitted scars, the pocks. (**c**) A painting by Robert Thom showing Edward Jenner vaccinating a young boy. The woman at the right is holding her wrist at the spot from which cowpox material was taken.

vaccinated. The eradication was aided by the fact that smallpox viruses do not exist anywhere in nature except in humans. On October 26, 1977, health care workers reported isolation of the last smallpox case.

But does that mean that the smallpox viruses are gone as well?

No. Smallpox viruses still remain in two laboratories, one at the CDC in Atlanta, the other in Russia. The World Health Organization has recommended that stocks of smallpox viruses be destroyed, especially because scientists have deciphered the base sequence of the virus' genome. Not all scientists agree, however.

Could smallpox viruses by used by bioterrorists?

Yes, unfortunately they could. Smallpox is a very dangerous disease, it spreads with relative ease, and as a bioweapon, it could cause a calamity of unprecedented proportions. Smallpox viruses, however, are very difficult to cultivate (as compared to bacteria), and thus are not accessible to most bioterrorist groups.

What can individuals do to protect themselves against a bioterrorist attack with smallpox viruses?

Probably the best defense is to be immunized against the disease, a possibility that may exist by the time you read this book.

Warts

What are warts, and what do they have to do with viruses?

Warts are small, usually benign skin growths commonly due to viruses. Plantar warts occur on the soles of the feet. Genital warts develop on the genital organs and are often transmitted in sexual contact; they are usually moist and pink. An estimated 4 million Americans are believed to be infected by genital warts.

What causes warts?

One of the primary causes of warts is human papilloma viruses, a collection of over two dozen types of icosahedral DNA viruses. In most cases, the skin warts they cause are a minor problem. Because the warts caused by the papilloma virus include genital warts, this virus is considered one of the many sexually transmitted viruses. Furthermore, some strains of the papilloma virus cause cervical and vaginal cancer as well as malignancies of the penis and anus, which also are the result of sexual transmission. Fortunately, in 2006, a vaccine known as Gardasil became available that provides protection against 70% of the cancer-causing papilloma virus strains.

Viral Diseases of the Respiratory Tract

A number of viral diseases occur within the human respiratory system. Certain ones, such as influenza, affect the upper regions of the system, while others, such as respiratory syncytial disease, involve the lungs. Also included in this group are the common colds and head colds that affect millions of Americans annually.

Influenza and Parainfluenza

How did influenza get its name? Does it have anything to do with "influence"?

Yes, you are close. Influenza is believed to take its name from the Italian word for "influence," perhaps a reference to the influence of heavenly bodies; alternately, the disease may have been named *influenza de freddo*, meaning "influence of the cold." Influenza is an acute, contagious disease of the upper respiratory tract transmitted by droplets.

Is there anything unique about the influenza virus?

The influenza virus is quite unusual. It is composed of eight single-stranded segments of RNA wound helically and associated with protein to form a nucleocapsid. The viral envelope contains a series of projections called spikes. The spikes contain the enzymes hemagglutinin (H) and neuraminidase (N), which assist the entry of the virus into its host cell. Chemical changes occur periodically in these two enzymes, thereby yielding new strains of virus. These changes have practical consequences because the antibodies produced during last year's attack of influenza fail to recognize this year's strain, and individuals suffer another bout of disease.

I've heard of type A and type B flu viruses. What does that mean?

Three "types" of influenza virus are recognized: type A, which causes most pandemics; type B, which is less widespread than type A; and type C, which is rare. The nomenclature for influenza viruses is based on the type and the H and N strain. For example, influenza virus A(H3N2) predominated in the United States in 1999. Emergence of an avian strain in humans caused much concern in the 1990s, as A Closer Look on page 411 recounts.

What's it like when you develop the flu?

The onset of influenza is abrupt, with sudden chills, fatigue, headache, and pain that is most pronounced in the chest, back, and legs. Over a 24-hour period, the body temperature often rises to 40°C (104°F), and a severe cough develops. Despite these severe symptoms, influenza is normally short-lived and has a favorable prognosis. However, secondary complications may occur if bacteria such as staphylococci invade the damaged respiratory tissue.

Are there any drugs to treat influenza?

Cases of influenza may be treated with amantadine, a drug believed to interfere with viral uncoating in the replication cycle. A new drug that neutralizes neuraminidase is also available.

How about the flu vaccine? Does it work?

High-risk individuals, such as the elderly and very young, may be immunized with the influenza vaccine and receive protection against the disease The vaccine is prepared with inactivated influenza viruses of the type and strain predicted for the impending flu season. For example, the vaccine for 1999 consisted of A(H3N2) virus and two others predicted to be the causes of flu that year.

Is parainfluenza the same as influenza?

Parainfluenza resembles influenza, but it's not the same. For one thing, the viruses are different (parainfluenza is caused by an RNA helical virus). And, although as widespread as influenza, parainfluenza is a much milder disease often referred to as a cold. Bronchitis (inflammation of the respiratory tubes) and croup (hoarse coughing) may accompany the disease, which is most often seen in children under the age of 6.

Adenovirus Infections

Do adenoviruses have anything to do with adenoids?

Adenoviruses are a group of DNA-containing icosahedral viruses that take their name from the adenoid tissue from which they were first isolated in 1953. They are among the most frequent causes of upper respiratory diseases collectively called the common cold. Adenoviral colds are distinctive because the fever is substantial, the throat is very sore, and the cough is usually severe.

A CLOSER LOOK

Carnage

In May 1997, a worried mother in Hong Kong brought her 3-year-old son to a doctor for treatment for influenza. The doctor gave the boy aspirin and an antibiotic and sent him home. But soon the boy's symptoms worsened, and he was brought to Queen Elizabeth Hospital with a massive lung infection. Then his lungs collapsed, his liver shut down, and his kidneys failed. The boy died soon thereafter.

Baffled by the illness, doctors sent samples of the boy's tissues to laboratories in The Netherlands and the United States. Three months later, the diagnosis was complete: The boy's death was due to influenza virus A(H5N1), a virus fiercely pathogenic in chickens, but unknown to occur in humans—until then, that is.

Tension quickly developed among public health officials around the world. They recalled the 20 million who died of influenza in 1918 and 1919; the 100,000 dead of Asian flu in 1957; and the 36,000 who died of Hong Kong flu in 1968. They watched for other cases of the avian virus in humans (even as a nervous Hong Kong community filled hospital emergency rooms with symptoms of routine respiratory diseases). And they monitored the epidemic as it spread among the population of chickens: One moment a chicken stood contentedly in its cage; the next moment it leaned over and fell dead, blood oozing from its orifices. By November, there were four more human cases; by Christmas, another thirteen.

On December 28, 1997, Hong Kong's Chief Executive took decisive action: "No chicken will be allowed to walk free in the territory," he declared. Knives flashed, blood splattered, and the carnage began, as a million-and-a-half chickens were slaughtered. Thousands of workers were mobilized to stuff dead chickens into garbage bags and haul them to landfills. Carcasses rotted, and rats picked apart bags left by the road-side. A thousand tons of chickens disappeared from the Hong Kong landscape.

In January 1998, the Chinese New Year came and went without the traditional poultry dishes. There were no live chickens or chicken dishes to offer to the gods or to honor Chinese ancestors. To be sure, it was a strange holiday. But there were no more deaths from influenza. It had been a close call.

Most recently, between 2005 and the summer of 2006, the H5N1 bird flu reemerged and infected the human population in Asia, Southeast Asia, and the Middle East. In 2003, only four cases of H5N1 were observed world-wide; these cases were limited to China and Vietnam, and all lead to death. In 2005, the frequency rose to 95 cases of H5N1 with 41 deaths. This increase in the H5N1 infection rate rose further, with 90 human cases of the H5N1 bird flu reported across 9 countries within the first eight months of 2006. Again, the greatest concern is that human-to-human spread of H5N1, of which there has only been a handful of cases since 2003, could lead to a world-wide influenza pandemic like those seen in 1918, 1957, and 1968.

Respiratory Syncytial Disease

Is respiratory syncytial disease common? I've never heard of it.

Since 1985, respiratory syncytial (RS) disease has been the most common lower respiratory tract disease affecting infants and children under 2 years of age. Infection takes place in the bronchioles and air sacs of the lungs, and the disease is often described as viral pneumonia. RS disease occasionally breaks out in adults, usually as an upper respiratory disease with influenzalike symptoms. Ribavirin has been used for treatment.

Where does respiratory syncytial disease get its unusual name?

When the RS virus (a helical RNA virus) infects tissue cells, the cells tend to fuse and form giant cells called syncytia (sing., syncytium). The syncytia cannot function normally.

Rhinovirus Infections

Does the word "rhinovirus" have anything to do with "nose"?

Yes, good call. Rhinoviruses are a group of over 100 different icosahedral RNA viruses that take their name from the Greek *rhinos*, meaning "nose" and referring to the infection site. The viruses are among the major causes of the upper respiratory infections called head colds. Because so many different viruses are involved, the prospects for developing a vaccine for head colds is not promising.

What's a head cold?

A head cold involves a regular sequence of symptoms beginning with headache, chills, and a dry, scratchy throat. A "runny nose" and obstructed air passageways are the dominant symptoms, but the cough is variable and fever is often absent or slight. Antihistamines can be used to relieve the symptoms because these are often due to histamines released from damaged host cells. **FIGURE 18.4** shows the seasonal variations of this and other viral respiratory diseases.

Viral Diseases of the Nervous System

Some serious viral diseases affect the human nervous system. This fragile system suffers substantial damage when viruses replicate in its tissues. Rabies, a highly fatal and well-known disease, is characteristic of these diseases, as we shall see as our conversation continues.

Rabies

I've heard that rabies is a really dangerous disease. Is that so?

Absolutely. Rabies is notable for having the highest mortality rate of any human disease, once the symptoms have fully materialized. Few people in history have recovered from rabies.

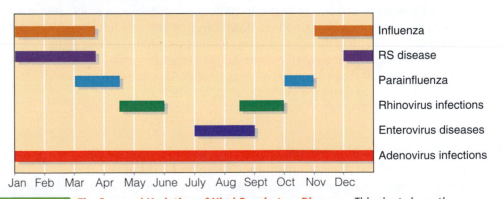

FIGURE 18.4 **The Seasonal Variation of Viral Respiratory Diseases.** This chart shows the seasons associated with various viral diseases of the respiratory tract. Note that the seasons for influenza and RS disease overlap; these are the primary viral respiratory diseases of the winter season. Enteroviruses cause diseases of the gastrointestinal tract as well as respiratory disorders and are usually acquired from the environment.

In what animals does rabies occur?

Rabies can occur in most warm-blooded animals, including dogs and cats, horses and rats, skunks and bats. The helical RNA virus enters the tissue through a skin wound contaminated with the saliva, urine, blood, or other fluid from an infected animal. The incubation period (the time from viral entry to symptom emergence) may vary from days to years, depending on such things as the amount of virus entering the tissue and the wound's proximity to the central nervous system.

What's it like to have rabies?

Rabies is a gruesome disease. Early signs are abnormal sensations such as tingling, burning, or coldness at the site of the bite. Then, increased muscle tension develops, and the individual becomes alert and aggressive. Soon there is paralysis, especially in the swallowing muscles, and saliva drips from the mouth. Brain degeneration, together with an inability to swallow, increases the violent reaction to the sight, sound, or thought of water. The disease has therefore been called hydrophobia—literally, "fear of water." Death follows from respiratory paralysis.

Can anything be done to prevent rabies from developing?

A person who is bitten or has had contact with a rabid animal should be given rabies vaccine via four or five injections in the arm muscle. Because of the long incubation period of the disease, there is time for the immune system to produce protective antibodies. The number of human cases of rabies in the United States is usually less than 5 per year, due in large measure to immunizations, as illustrated by the case in FIGURE 18.5 .

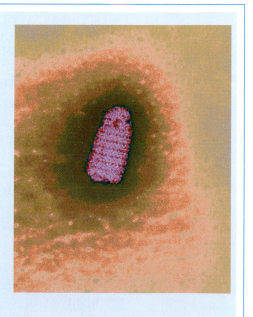

1 Twenty-seven guests attended a wedding reception in rural Columbia County, New York, on November 13, 1994. The bride and groom sparkled on their wedding day.

2 During the reception, the guests were invited to stroll around the farm. A young calf was the center of attention, and most of the guests had an opportunity to pet the animal.

3 During the next 2 weeks, the calf behaved erratically, with twitching movements and paralysis developing first in its limbs, then in its breathing muscles. The veterinarian could do little to help. On November 27, the calf died.

4 Because there were numerous reports of rabies in raccoons in New York, the veterinarian suggested that the calf's head be sent to the lab for analysis. The lab report was positive for rabies.

5 The family members and all 27 guests were contacted, and all received post-exposure immunizations to prevent their getting the disease.

FIGURE 18.5 **An Unusual Case of Rabies.** This case was unusual because rabies is not usually associated with calves. Transmission had probably occurred during contact with a wild animal.

Polio

How did polio gets its name?

Polio is a shortened form of poliomyelitis, a word derived from the Greek *polios* for "gray " and *myelon* for "matter." The gray matter is the nerve tissue of the spinal cord and brain, which are affected in the disease. Viruses that cause polio are among the smallest known.

How do you get polio?

Polio viruses contain RNA and have icosahedral symmetry. They usually enter the body with contaminated water or food. They then pass into the bloodstream and cause inflammation of the meninges. Paralysis of the arms, legs, and body trunk may result. In the most severe form of polio, the viruses infect the medulla of the brain. Swallowing is difficult, and paralysis develops in the tongue, facial muscles, and neck. Paralysis of the diaphragm muscle causes labored breathing and may lead to death.

I've heard that there are two polio vaccines. What are they made of?

In 1955, a team led by Jonas Salk cultivated large quantities of polio viruses and inactivated them with chemicals to produce the first polio vaccine. The viruses in this vaccine cannot multiply in the body and must be injected to be effective as a vaccine. Albert Sabin's group subsequently developed a vaccine containing attenuated (weakened) polio viruses. This vaccine can be taken orally. The Salk vaccine is currently the vaccine of choice in the United States because there is less possibility of developing polio from its inactivated viruses.

Is it true that polio may one day be eradicated from the Earth?

Polio still breaks out occasionally in various parts of the world. However, in North America, the attempt at its eradication by the Pan American Health Organization has apparently been successful, and no case of polio due to a " wild " (environmental) virus has occurred since August 23, 1991. Worldwide eradication in the twenty-first century is a goal of public health officials.

Slow Virus (Prion) Diseases

What's the relationship between slow viruses and prions?

Certain viruses appear to cause slow-developing degenerative diseases in which many years elapse between the infection and any detectable symptoms. These viruses, known as slow viruses, have not yet been visualized with the electron microscope. Moreover, no nucleic acid has been detected in material that transmits the diseases. Many investigators support the theory that slow viruses are really **prions**, the proteinaceous infectious particles we discuss in Chapter 6. To accommodate both groups, we shall refer to the agents as slow viruses (prions).

Which diseases are caused by the slow viruses (prions)?

Virologists have related at least three diseases to slow viruses (prions). The first is kuru, a fatal nerve disorder found in villages of the Fore people living in the remote highlands of New Guinea. The second is scrapie, a disease of sheep and goats that interferes with nervous system coordination, making the animals unable to walk or stand. The third is bovine spongiform encephalopathy (BSE), also called mad cow disease. BSE affects the brain tissue in cattle, causing their brains to develop a "spongy" appearance with large numbers of deposits called plaques. The animals become apprehensive and unable to stand on their feet; they behave erratically and soon die (FIGURE 18.6). The opening to this chapter describes some of the concern the disease has evoked.

Is BSE related to any human disease?

In 1996, British researchers announced a possible link between BSE and a human malady called Creutzfeldt Jakob disease (CJD). This disease has been recognized for decades in humans and is much like kuru in its effects on the body. Patients suffer from nervousness, bizarre behavior, memory loss, wobbly walk, lethargy, and hunched posture, and, eventually, death.

Arboviral Encephalitis

What's the difference between encephalitis and arboviral encephalitis?

The term encephalitis refers to an acute inflammation of the brain. Used in the general sense, encephalitis may refer to any brain disorder. Here, we use the term to mean a number of viral disorders that are *ar*thropod*b*orne (hence, arboviral). Arboviral encephalitis is caused by a series of icosahedral RNA viruses. Patients suffer sudden, very high fever, a severe headache, stupor, and possibly a series of convulsions before lapsing into a coma.

Are there different kinds of arboviral encephalitis?

Yes, there are many forms and many vectors of the disease. St. Louis encephalitis, California encephalitis, La Crosse encephalitis, Japanese B encephalitis, and West Nile encephalitis are forms transmitted by mosquitoes. Russian encephalitis and Louping ill encephalitis are transmitted by ticks. Arboviral encephalitis is also a serious problem in horses, and birds are natural reservoirs of the viruses and spread them during annual migrations.

FIGURE 18.6   Bovine Spongiform Encephalopathy. During the 1990s, an outbreak of BSE—"mad cow disease"—brought a vigorous response from public health officials in England. The epidemic subsided after extreme measures were taken to destroy all animals that might possibly be infected.

Viral Diseases of the Visceral Organs

Viral diseases of the viscera affect the blood and such organs as the liver, spleen, and small and large intestines. To reach these organs, the viruses are generally introduced into the body tissues by arthropods or by contaminated food and drink, as we shall see in the discussions that follow.

Yellow Fever

I think I read about yellow fever in my history class. Is that possible?

Yes, it is possible. In 1803, for example, emperor Napoleon of France sent troops to quell an uprising in Haiti, but yellow fever killed thousands of his men. Napoleon soon came to think of the Americas as a fever-ridden land, and when President Thomas Jefferson sent emissaries to negotiate the purchase of the New Orleans region, Napoleon offered the entire Louisiana territory at a bargain price. You may also have read about the need to control yellow fever during the building of the Panama Canal (see A Closer Look on page 416).

A CLOSER LOOK

"For the Cause of Humanity . . ."

During the Spanish American War, the U.S. government became acutely aware of infectious disease because more soldiers were dying from disease than from bullet wounds. Yellow fever was particularly bad in Cuba, where it exacted a heavy toll. When the war was over, Cuba remained under U.S. control, and the Surgeon General sent a commission of four men to study the disease. Led by Major Walter Reed, the group included three assistant surgeons: James Carroll, Jesse W. Lazear, and Aristides Agaramonte. On June 25, 1900, the four men assembled in Cuba and began their work.

At first the commission devoted its energy to isolating a bacterium, but none could be found. As the weeks wore on, the investigators were impressed with the peculiar way the disease jumped from house to house, even when there was no contact with infected persons or contaminated objects. They also visited Carlos J. Finlay, a physician from Havana who insisted that mosquitoes were involved in transmission. If his theory was true, then the disease could be interrupted by simply killing the mosquitoes.

By now 2 months had passed, and Reed had been called back to Washington. Carroll, Lazear, and Agaramonte pushed forward and bred mosquitoes from eggs given them by Finlay. They allowed the mosquitoes to feed on patients with established cases of yellow fever, and then they applied the insects to the skin of volunteers, including themselves. The results were inconclusive: Some volunteers got yellow fever, but others did not. At that point, two accidents then saved the research. The first was fortunate, the second tragic:

Carroll decided to feed an "old" mosquito some of his own blood, lest it die. Three days later, he was ill with the fever, but he survived. Lazear's notebook recorded that the insect had fed "*twelve days* before on a yellow fever patient, who was then in his second day of the disease." This, they discovered, was the proper combination of two factors necessary for a successful transmission. The disease could be reproduced over and over again, if this procedure was followed.

Then came tragedy. Lazear was working at the bedside of a yellow fever patient when a stray mosquito settled on his wrist. For reasons not clear, Lazear let the insect drink its fill. Five days later, he developed yellow fever; on the seventh day of his illness, he died. Lazear had been bitten previously, but apparently by uninfected mosquitoes. This time, the mosquito was infected.

Reed returned to Cuba in October, and a new set of experiments was planned to prove once and for all that clothing and other objects could not transmit yellow fever. The experiments were gruesome: Some volunteers slept in blood-soaked and vomit-stained garments of disease victims; others allowed themselves to be mercilessly bitten by mosquitoes; still others came forward to be injected with blood from yellow fever patients. By late 1900, there was no doubt that mosquitoes were the carriers of yellow fever.

It has been said that the experiments performed in Cuba are among the noblest in the history of medicine. Two volunteers, Private John R. Kissinger and clerk John J. Moran, were asked why they were agreeing to such life-threatening experiments. "We volunteer," they replied, "for the cause of humanity and in the interest of science."

But why is yellow fever so named?

In cases of yellow fever, mosquitoes inject the viruses into the bloodstream, and infection of the liver causes an overflow of bile pigments into the blood, a condition called jaundice; the complexion becomes yellow (the disease is often called "yellow jack"). The gums bleed, the stools turn bloody, and the delirious patient often vomits blood. Patients die of internal bleeding, and mortality rates are very high. An icosahedral RNA virus is responsible.

Can anything be done for patients?

Except for supportive therapy, no treatment exists for yellow fever. However, the disease can be prevented by immunization with either of two vaccines.

Dengue Fever

What's dengue fever?

Dengue fever is a viral disease due to an icosahedral DNA virus that multiplies in white blood cells. High fever and severe prostration are early signs of dengue fever. These are followed by sharp pains in the muscles and joints, and patients often report sensations that their bones are breaking. Like the yellow fever virus, the dengue fever virus is transmitted by mosquitoes.

Is the disease dangerous like yellow fever?

Death is uncommon with dengue fever, but if another strain of the virus later enters the body, a condition called dengue hemorrhagic fever may occur. This condition is dangerous enough that dengue viruses are considered a possible weapon of bioterrorists. Yellow fever viruses are also a potential bioweapon.

Infectious Mononucleosis

Isn't infectious mononucleosis also called "mono"?

Yes, it is. Infectious mononucleosis (or "mono" in the vernacular) is a familiar term to young adults because the disease is common in this age group. It is sometimes called the "kissing disease" because it can be spread by contact with saliva.

What kind of disease is mono?

Infectious mononucleosis is a blood disease, especially affecting the antibody-producing B cells of the immune system. Enlargement of the lymph nodes ("swollen glands") is accompanied by a sore throat, fever, enlarged spleen, and a high count of damaged mononuclear white blood cells. Those who recover usually become carriers for several months and shed the viruses into their saliva.

Does the Epstein-Barr virus have anything to do with mono?

A substantial body of evidence indicates that the mononucleosis virus is identical to the Epstein-Barr (EB) virus (an icosahedral DNA virus named for British virologists M. Anthony Epstein and Yvonne M. Barr). This virus has also been detected in patients who have Burkitt's lymphoma, a tumor of the connective tissues of the jaw that is prevalent in areas of Africa.

Hepatitis A

What exactly is hepatitis?

Hepatitis is an acute inflammatory disease of the liver caused by any of several viruses. Symptoms include anorexia, nausea, vomiting, and low-grade fever. Discomfort

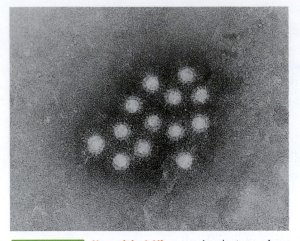

FIGURE 18.7 **Hepatitis A Viruses.** An electron micrograph of hepatitis A viruses. The particles were coated with antibodies to assist staining. The coating accounts for the halo around the viruses.

in the upper-right quadrant of the abdomen follows, as the liver enlarges. Considerable jaundice usually follows the onset of symptoms (the urine darkens, as well).

How is hepatitis A transmitted?

Hepatitis A (or infectious hepatitis) is the form most commonly transmitted by contaminated food or water or by infected food handlers. The disease is often transmitted by raw shellfish such as clams and oysters, since these animals filter and concentrate the viruses from contaminated seawater. A small, icosahedral RNA virus that is resistant to the environment outside the body causes the disease (FIGURE 18.7).

I heard that there is a new vaccine available for hepatitis A. Is that right?

Yes, it is. The Food and Drug Administration (FDA) has licensed a vaccine composed of inactivated viruses. Use of the vaccine should reduce the incidence of hepatitis A, estimated to be over 20,000 annually in the United States.

Is there any treatment if you get hepatitis A?

There is no treatment for hepatitis A except for prolonged rest and care of symptoms. In people exposed to the virus, it is possible to prevent development of the disease by administering hepatitis A immune globulin within 2 weeks of infection. This preparation consists of antiviral antibodies obtained from blood donors who have had the disease.

Hepatitis B

What's the difference between hepatitis A and hepatitis B?

Both hepatitis A and hepatitis B occur in the liver, and the symptoms are similar. But there are important differences. For one thing, the hepatitis B virus is a very fragile, complex DNA virus. Second, transmission of hepatitis B usually involves contact with an infected body fluid such as blood or semen. For example, transmission may occur via blood-contaminated needles used in hypodermic syringes or for tattooing, acupuncture, or ear-piercing. Sexual contact can also lead to transmission (FIGURE 18.8).

Are there any other differences?

Yes, there are. Hepatitis B (or serum hepatitis) has a longer incubation period than hepatitis A, and the vaccine components are different. For hepatitis B, the vaccine contains protein fragments produced by genetically engineered yeast cells. Also, long-term liver damage may occur as a result of hepatitis B, including a liver cancer called hepatocarcinoma.

Can anything be done if you get hepatitis B?

Injections of the antiviral drugs interferon and lamuvidine (3TC) can influence the course of hepatitis B. Moreover, if you are exposed to the virus, injections of hepatitis B immune globulin can help prevent development of the disease.

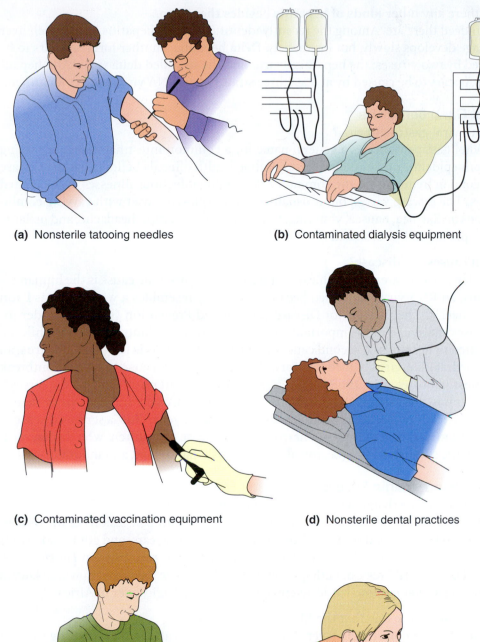

(a) Nonsterile tatooing needles

(b) Contaminated dialysis equipment

(c) Contaminated vaccination equipment

(d) Nonsterile dental practices

(e) Contaminated drug needles

(f) Nonsterile body piercing equipment

FIGURE 18.8   **Some Methods for the Transmission of Hepatitis B.**

Are there any other kinds of hepatitis besides these two?

Indeed there are. Among the recently identified ones is hepatitis C, in which liver damage develops slowly, but insidiously. Delta hepatitis, another form, appears to be caused by two viruses: the hepatitis B virus and the so-called delta virus. And hepatitis E appears to be caused by at least three strains of an RNA virus not yet identified.

Viral Gastroenteritis

What's viral gastroenteritis?

Viral gastroenteritis is a general name for a common intestinal illness occurring in both epidemic and endemic forms. Public health officials believe that the disease is second in frequency to the common cold, among infectious illnesses in the United States. The disease varies, but it usually has an explosive onset with varying combinations of diarrhea, nausea, vomiting, low-grade fever, cramps, headache, and malaise. Some people call it "stomach flu."

What causes the disease?

Many viruses are responsible for viral gastroenteritis. One cause is the human rotavirus, an RNA virus so-named because it vaguely resembles a wheel (*rota* is Latin for "wheel"). The Centers for Disease Control and Prevention (CDC) consider rotaviruses the single most important cause of diarrhea in infants and young children admitted to hospitals. A second cause of viral gastroenteritis is the Norwalk virus, an icosahedral RNA virus named for Norwalk, Ohio, where it caused a notable outbreak of intestinal disease in 1968 (**FIGURE 18.9** chronicles an outbreak in New Jersey). Another possible cause is the Coxsackie virus, another icosahedral RNA virus, named for a town in New York. Coxsackie virus also causes myocarditis, a serious disease of the heart muscle and valves, and herpangina, a disease of children, who display vesicles on the soft palate, tongue, tonsils, and hands ("hand, foot, and mouth disease").

Viral Hemorrhagic Fevers

Why were hemorrhagic fevers given that name?

Certain viral fevers are accompanied by severe hemorrhagic lesions of the tissues. Physicians report blood spouting from patients' eyes, nose, ears, and gums and organs turning to liquid. They tell a horror story similar to the one recounted in *The Hot Zone*, a novel by Richard Preston. Perhaps you've seen the movie *Outbreak*, in which Dustin Hoffman and Rene Russo battle an epidemic of hemorrhagic fever in Africa.

Is this disease the one that the Ebola virus causes?

Yes. Ebola hemorrhagic fever has occurred sporadically in Africa during the past decade. Its symptoms are typical of hemorrhagic fever. Other examples are Lassa fever and Marburg disease, which have similar symptoms. They are caused by a filovirus, a long, threadlike RNA virus (*filum* is Latin for "thread") that often takes the shape of a fishhook or U.

What about that disease that broke out in Arizona some years ago? Was that a hemorrhagic fever?

Yes, it was. In the summer of 1993, a brief epidemic occurred among Native Americans living in the southwestern United States. It was named the "four-corners disease" for the place where four states come together. The disease is a rapidly developing flulike illness whose symptoms include hemorrhaging blood and respiratory failure. The dis-

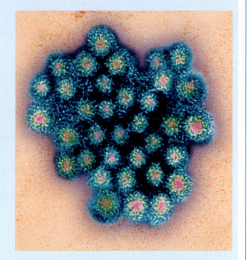

1 A tourist resort in the Gulf of Taranto, Italy has 456 guest rooms in 19 buildings. The buildings are situated around a central area where a restaurant, a swimming pool, and the resort management office are located.

2 This resort accommodates 1,000 guests, who generally arrive to start their vacations on Saturday and depart 1–2 weeks later, resulting in a 50% turnover of guests each weekend.

3 In July 2000, an outbreak of gastroenteritis occurred. Gastroenteritis is defined as the occurrence of diarrhea (three or more loose stools in any 24-hour period) or vomiting (at least one episode) or both, in the same period. A total of 344 people, 69 of whom were staff members, met the case definition.

4 The norovirus was found in 22 of 28 stool specimens tested. The source of the virus was thought to be contaminated drinking water. Illness rates were increased in staff and guests who reported being involved in water sports, using the beach showers, and consuming drinks with ice. Although Italy has no surveillance system for nonbacterial gastroenteritis, no outbreak caused by a norovirus had been described previously in the country.

FIGURE 18.9   A Case of Norovirus Gastroenteritis at an Italian Tourist Resort. The Norwalk virus and the Norovirus are members of related virus strains that together are the most frequent cause of gastroenteritis outbreaks worldwide. The norovirus accounts for 96% of viral gastroenteritis outbreaks in the United States.

ease was formerly called Korean hemorrhagic fever, but it has been renamed hantavirus pulmonary syndrome because the viruses responsible for it are a group of RNA viruses called hantaviruses. CDC investigators identified airborne viral particles from the dried urine and feces of rodents (especially deer mice) as the responsible agents.

Cytomegalovirus Disease

Why was the cytomegalovirus given that name?

The cytomegalovirus (CMV) takes its name from the enlarged cells ("cyto-megalo") found in infected tissues. When a person is infected, fever, malaise, and, in some cases, an enlarged spleen develop, but few other signs of disease are observed. Most patients recover uneventfully. An icosahedral DNA virus is the cause.

Is cytomegalovirus ever dangerous?

Yes. If a woman is pregnant, the virus can pass into the fetal bloodstream and damage the fetal tissues. The cytomegalovirus has also demonstrated its invasive tendency in patients who have AIDS. Up to one-third of AIDS patients experience CMV-induced retinitis, a serious infection of the retina of the eye that can lead to blindness. The drug ganciclovir is useful for therapy.

HIV Infection and AIDS

When did the AIDS epidemic begin?

In 1981, physicians described a syndrome involving a deficiency of the immune system. This clinical entity included the development of certain opportunistic infections,

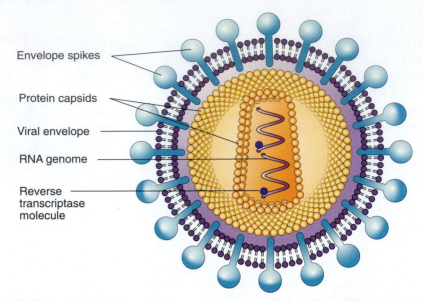

Envelope spikes

Protein capsids

Viral envelope

RNA genome

Reverse
transcriptase
molecule

FIGURE 18.10 **The Human Immunodeficiency Virus (HIV).** The virus consists of two molecules of RNA and molecules of reverse transcriptase. A protein capsid surrounds the genome, and an envelope with spikes of protein lies outside the capsid.

as well as an unusual type of skin cancer called Kaposi's sarcoma. The most plausible factor was a virus, but a definitive agent was not identified until 1984, when a French group led by Luc Montagnier isolated it. In 1986, the virus was given its current name of human immunodeficiency virus (HIV). By that time, the disease was well-known as acquired immune deficiency syndrome (AIDS).

What makes HIV unique?

The HIV is an RNA-containing icosahedral virus with an envelope that contains spikes, as **FIGURE 18.10** shows. Within its genome, HIV contains molecules of an enzyme called reverse transcriptase. When the RNA is released in the cytoplasm of a host cell, the enzyme synthesizes a molecule of DNA utilizing the genetic message in the RNA as a template. (This reversal of the usual mode of genetic information transfer is why the virus is called a retrovirus.) The synthesized DNA molecule, termed a provirus, integrates into the host's DNA, and from that location, it transcribes its genetic message into new particles of HIV. The viral particles then "bud" from the host cell and infect other cells. This individual now has HIV infection.

What are the host cells for HIV?

The normal host cell for HIV is a cell of the immune system called a helper T cell. These cells participate in cell-mediated immunity (Chapter 17) where they assist in the development of cytotoxic T cells and the production of antibodies. Infection by HIV leads to the failure of this immune response and places the host at risk for opportunistic pathogens and cancers, similar to those in patients with other immunodeficiencies.

Why is the helper T cell a target for HIV?

The helper T cell bears the CD4 receptor site (the helper T cell is also called a CD4+ cell because of this site) as well as other coreceptors to which HIV binds during its replication cycle. For a period of time, the body keeps up with the virus and replaces infected cells as they are destroyed. Eventually, however, the individual suffers a decline from the normal 1000 T cells per microliter of blood to less than 100 per microliter. Infections due to opportunistic pathogens follow.

How prevalent is AIDS in the United States and the world?

As of January 2004, a total of over 900,000 cases of AIDS had been reported to the CDC since the epidemic's beginning. Over 500,000 deaths have occurred from complications due to opportunistic pathogens. It is important to note that AIDS is the end result of HIV infection. As many as 1.0 million Americans are believed to have HIV infection. Worldwide, scientists estimate that over 35 million people are infected with HIV.

What is it like to have AIDS?

A person with HIV infection manifests an ever-weakening condition. Fever, diarrhea, rash, swelling of the lymph nodes, night sweats, malaise, and fatigue may be present. These symptoms occur sporadically and last several weeks at a time. There is also severe depression from the never-ending series of illnesses. This stage is sometimes called AIDS-related complex. The progression to AIDS has occurred when the individual experiences neurological disease, including dementia, memory loss, mood swings, and other nerve-related pathologies. AIDS also is present when the individual develops a wasting syndrome with excessive diarrhea and loss of muscle mass. However, the most widespread evidence of AIDS is the presence of opportunistic illnesses such as Kaposi's sarcoma, *Pneumocystis carinii* pneumonia, cytomegalovirus infection, cryptosporidiosis, *Candida albicans* infection, and cryptococcosis. Tuberculosis and other mycobacterial diseases may also be significant problems.

■ *Pneumocystis carinii*
nü-mō-sis′tis kär-i′nē-ī
(or kär-i′nē-ē)

In what ways is HIV transmitted?

Transmission of HIV occurs primarily by infected blood or semen. Intimate sexual contact, including anal intercourse, is a common method of transmission, particularly when rectal tissues bleed and give the virus access to the blood stream. Unprotected vaginal intercourse is also a high-risk sexual activity, especially if lesions, cuts, or abrasions of the vaginal tract exist. (The use of condoms has been shown to decrease the transmission of HIV significantly.) The sharing of blood-contaminated needles by injection drug users also transmits HIV. Moreover, transplacental transfer from mother to child is an acknowledged method of transmission. In very rare cases, transmission can occur through contacts with blood products used for medical purposes, so health care workers should always practice the established infection-control procedures known as universal precautions.

How can a doctor detect if you've been exposed to HIV?

A number of diagnostic tests can be used to determine whether a person has been exposed to HIV. For example, HIV antibodies can be detected by the sophisticated ELISA and Western blot tests. A newer test called the viral load test detects the RNA of HIV and is also used to assess the extent of infection. With this test, almost 100% of infected individuals can be detected. Gene probes and PCR amplification methods (Chapter 14) can also be used.

What about the new treatments for people who have AIDS?

Treatments for HIV infection and AIDS have been researched for many years. The drug azidothymidine, commonly known as AZT, interferes with reverse transcriptase activity and inhibits DNA synthesis. Other anti-HIV agents approved for use include

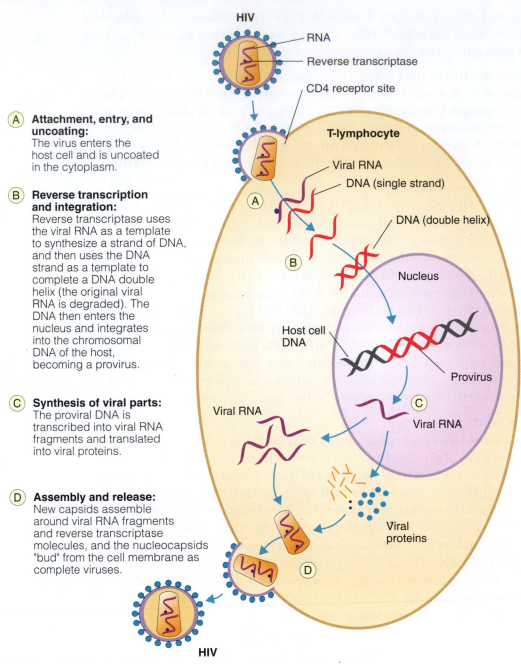

A **Attachment, entry, and uncoating:**
The virus enters the host cell and is uncoated in the cytoplasm.

B **Reverse transcription and integration:**
Reverse transcriptase uses the viral RNA as a template to synthesize a strand of DNA, and then uses the DNA strand as a template to complete a DNA double helix (the original viral RNA is degraded). The DNA then enters the nucleus and integrates into the chromosomal DNA of the host, becoming a provirus.

C **Synthesis of viral parts:**
The proviral DNA is transcribed into viral RNA fragments and translated into viral proteins.

D **Assembly and release:**
New capsids assemble around viral RNA fragments and reverse transcriptase molecules, and the nucleocapsids "bud" from the cell membrane as complete viruses.

FIGURE 18.11 **The Replication Cycle of the Human Immunodeficiency Virus (HIV).**

dideoxyinosine (ddI) and dideoxycytidine (ddC). Both are similar to AZT, but have fewer side effects. Another agent is lamivudine (3TC), an inhibitor of DNA synthesis. Still another group of anti-HIV agents are the protease inhibitors. These drugs interfere with protease, the enzyme used to form the viral capsid during the replication process (**FIGURE 18.11**). Several protease inhibitors are now in use, including saquinavir (Invirase), indinavir (Crixivan), and ritonavir (Norvir). They are used with AZT and

3TC in the highly reactive antiretroviral therapy (HAART) that has been largely responsible for a drop in the death rate associated with AIDS.

Will there ever be an AIDS vaccine?

An anti-HIV vaccine is being actively researched. Research evidence indicates that inactivated whole viruses may protect susceptible individuals, but there is reluctance to use such vaccines because of the possibility that active HIV may be present. Other vaccines use synthetic viral fragments, such as the gp41 and gp120 proteins found in the envelope spikes of HIV (Chapter 14). Problems arise because animal models are not available for testing. Moreover, HIV tends to mutate in the body and avoid the antibody response, and the pool of volunteers for testing candidate vaccines is limited. Nevertheless, vaccine research continues, and while a vaccine is being developed, emphasis continues to be placed on preventing the disease. Indeed, public health officials emphasize that at the present time, the best vaccine is education.

A FINAL THOUGHT

In the remote tropics of South America, Africa, and Asia, there lurks a coterie of viruses that infect animals but seldom bother humans. Occasionally, however, humans stumble into their paths, with results horrifying enough to mark the annals of medicine. The hemorrhagic fever viruses, for example, make the internal organs bleed and rot, and patients ooze contagious, virus-laden blood from their eyes, ears, nose, and other orifices. Marburg virus appeared in 1967, then Lassa fever virus in the 1970s, followed by Ebola virus and hantavirus in the years thereafter. In 1994, another hemorrhagic fever virus, the Sabia virus, escaped from a high-security laboratory at Yale University and roamed the streets of New Haven and Boston before felling its human host with symptoms (the patient recovered, and none of his 80 contacts developed the disease).

Where have these viruses come from, and what do they portend? These "new" pathogens are believed to be ancient organisms that lacked the opportunity to attack until humans blundered into their habitat. Lassa virus, Marburg virus, Ebola virus, hantavirus, HIV—all were probably confined for millennia to isolated human groups or animals. Then civilization encroached on the forest and they broadened their range; all that was needed was for nature to trigger an expansion of the host organism. Prior to the 1993 hantavirus epidemic, for instance, heavy rains had disturbed the desert and mountain ecology of the southwestern United States, leading to an abundance of piñon nuts and grasshoppers. Suddenly, the deer mouse population exploded, and hantaviruses had a ready mode of transport. In other cases, humans luck out before the viruses can find an alternate host: Ebola virus kills its victims so quickly that it does not have time to spread. And it does not have an animal host—not yet, at least.

As the world shrinks to a global village, a great biological soup is emerging. It is a soup into which dangerous microorganisms from long-isolated ecologies are being mixed. Air travel provides an excellent opportunity for an ill person to carry viruses across the oceans in a matter of hours. Military bases in far corners of the globe and increased trade with new nations promote the exposure of foreigners to exotic viruses. This has occurred in the past, of course. Europeans encountered dangerous new microbes in the New World and brought them back home to Europe. Now their descendants are repeating that experience in their own New Worlds.

■ QUESTIONS TO CONSIDER

1. In February 1992, the CDC reported an outbreak of measles at an international gymnastics competition in Indianapolis, Indiana. A total of 700 athletes and numerous coaches and managers from 51 countries were involved. Although the potential for a disastrous international epidemic was high, it never materialized. What steps do you think the local health agencies took to quell the spread of the disease?

2. Thomas Sydenham, the "English Hippocrates," was a London physician in the 1600s. In 1661, he differentiated measles from scarlet fever, smallpox, and other fevers, and set down the foundations for studying these diseases. How would a modern Thomas Sydenham go about distinguishing the variety of look-alike skin diseases discussed in this chapter?

3. A child experiences "red bumps" on her face, scalp, and back. Within 24 hours, they have turned to tiny blisters and become cloudy, some developing into sores. Finally, all become brown scabs. New "bumps" keep appearing for several days, and her fever reaches 39°C (102°F) by the fourth day. Then the blisters stop coming and the fever drops. What disease does she have?

4. In 1996, a medical school student at the University of Maryland was assigned a case for a seminar on undetermined diagnoses in difficult cases. His subject was identified only as E. P., a gentleman who lived in the mid-1800s and was a well-known poet and animal lover. The man was taken to a hospital with delirium and tremors. He was confused and combative, and he refused to drink any water or other liquid. He died on October 7, 1849. The traditional diagnosis had been alcoholism, but the student made a different diagnosis. What do you think the diagnosis was? When told the man's name, the student thought "Nevermore!" Who was the man?

5. Walt Disney World uses a series of sentinel chickens strategically placed on the grounds to detect any signs of viral encephalitis. Why do you suppose they use chickens? Why is Disney World particularly susceptible to outbreaks of viral encephalitis? And what recommendations might be offered to tourists if the disease broke out?

6. Sicilian barbers are renowned for their skill and dexterity with razors (and sometimes their singing voices). In 1995, French researchers studied a group of 37 Sicilian barbers and found that 14 had antibodies against hepatitis C, despite never having been sick with the disease. By comparison, when a random group of 50 blood donors was studied, none had the antibodies. What might account for the high incidence of exposure to hepatitis C among the barbers?

7. With many diseases, the immune system overcomes the infectious agent, and the person recovers. With certain diseases, the infectious agent overcomes the products of the immune system, and death follows. Compare this broad overview of disease and resistance to what is taking place with AIDs, and explain why AIDS is probably unlike any other human disease.

KEY TERMS

All the names of the viral diseases represent the key terms of this chapter. To use the names as a study guide, discuss each one for a few moments, indicating what sort of disease it is, where it occurs, whether a treatment or vaccine is available, and other information that is relevant to a concise view of the disease.

adenovirus infections	measles (rubeola)
AIDS	mumps
arboviral encephalitis	parainfluenza
chickenpox (varicella)	polio
cytomegalovirus disease (CMV)	rabies
dengue fever	respiratory syncytial disease
fifth disease (erythema infectiosum)	rhinovirus infections
hepatitis A	rubella (German measles)
hepatitis B	slow virus (prion) diseases
herpes simplex	smallpox
HIV infection	viral gastroenteritis
infectious mononucleosis	viral hemorrhagic fevers
influenza	warts
jaundice	yellow fever

http://microbiology.jbpub.com/book/microbes

The site features **eLearning**, an online review area that provides quizzes and other tools to help you study for your class. You can also follow useful links for in-depth information, read more stories of microbiology, or just find out the latest microbiology news.

19

Bacterial Diseases of Humans: Slate-Wipers and Current Concerns

In 2006, spinach contaminated with *E. coli 0157:H7* caused a nationwide outbreak of *E. coli* poisoning. The spinach was contaminated in the fields, and the *E. Coli* survived the processing steps all the way to the grocery store shelves. Investigators couldn't determine the definitive source of contamination, but they were able to narrow it down to either the manure used as fertilizer or a handler's hands. The FDA concluded that more thorough washing from field, to handlers to food preparers is essential to prevent further such outbreaks.

■ Looking Ahead

Bacterial diseases are among the most well-known of human afflictions. In some epidemics, such diseases have carved out great swathes through humanity as they coursed over the continents of Earth. In other instances, the diseases continue to pose serious challenges to the public health community. We shall see both cases as we pass through the pages of this chapter.

On completing this chapter, you should be able to . . .

- appreciate the broad scope of bacterial diseases both historically and today.
- recognize the different ways in which bacterial diseases can be spread among human populations.
- summarize the variety of bacteria that can cause human disease.
- conceptualize the numerous methods by which bacteria can cause disease in the body.
- identify the high mortality rates associated with certain serious bacterial diseases and understand why these are called "slate-wipers."
- explain the methods of treatment available for enhancing recovery from particular bacterial diseases.
- communicate knowledgeably with others about bacterial diseases.

Throughout history, bacterial diseases have posed a formidable challenge to humans and have often swept through populations virtually unchecked. In the 1700s, for example, the first European visitors to the South Pacific found the islanders robust, happy, and well-adapted to their environment. But the explorers introduced syphilis, tuberculosis, and pertussis (whooping cough), and soon these diseases spread like wildfire. Hawaii was struck with unusually terrible force. The population of the islands was about 300,000 when Captain Cook landed there in 1778; by 1860, it had been reduced to fewer than 37,000 people.

With equally devastating results, the Black Death came to Europe from the Orient, and cholera spread eastward from India. Together with tuberculosis, diphtheria, and dysentery, these bacterial diseases ravaged European populations for centuries and wove themselves insidiously into daily life. Infant mortality was particularly shocking: England's Queen Anne, who reigned in the early 1700s, lost 16 of her 17 babies to disease; and until the mid-1800s, only half the children born in the United States reached their fifth year.

Today, humans are better able to cope with bacterial diseases. Though credit is often given to wonder drugs, the major health gains have resulted from an understanding of disease processes and the body's resistance mechanisms, coupled with modern sanitary methods that prevent microbes from reaching their targets. Furthermore, immunization has also played a key role in preventing disease. Indeed, very few people in our society die of the bacterial diseases that once accounted for the majority of human deaths.

In this final chapter of the book, we study the bacterial diseases of humans according to the major mode of transmission of the disease. Some of the diseases are of historical interest and are currently under control. But just as a garden always faces new onslaughts of weeds and pests, so, too, the human body is continually confronted with newly emerging diseases. In this regard, the modern era is no different from the past. On the fundamental level, disease has not changed. Only the pattern of diseases has changed.

Airborne Bacterial Diseases

The bacterial diseases of the respiratory tract can be severe, as several diseases we consider in this section illustrate. Moreover, the respiratory tract is a portal of entry to the blood, and from there, a disease can spread and affect the more sensitive internal organs. As in Chapter 18, we shall use the question-answer format in the discussion. Imagine you are conversing with a microbiologist.

Streptococcal Diseases

I've heard of streptococci in a number of contexts such as strep throat, tooth decay, and even yogurt. Are all the streptococci the same?

Not really. Streptococci are a large and diverse group of encapsulated Gram-positive bacteria occurring in chains (FIGURE 19.1). One species, *Streptococcus pyogenes*, causes streptococcal sore throat, popularly known as strep throat. Patients experience a high fever, coughing, swollen lymph nodes and tonsils, and a fiery red "beefy" appearance to pharyngeal tissues owing to tissue erosion. Scarlet fever is strep throat

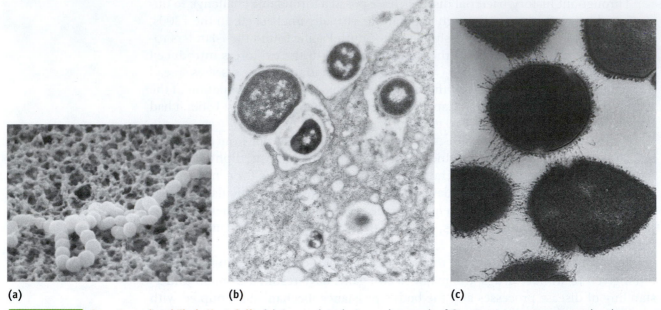

(a)   **(b)**   **(c)**

FIGURE 19.1   Streptococci and Their Host Cell. (**a**) A scanning electron micrograph of *Streptococcus mutans*, a species that causes many cases of tooth decay. (**b**) A transmission electron micrograph of *Streptococcus pyogenes* attaching to cells cultured from the human pharynx. Surface proteins, especially the streptococcal M protein, function in this attachment. A streptococcal cell has been internalized and is seen within the cytoplasm of the pharyngeal cell (×24,300). (**c**) A closeup view of *S. pyogenes* cells, showing strands of M protein. The M protein enhances virulence by encouraging attachment to the host cell.

accompanied by a skin rash. Streptococci are also involved in tooth decay and childbed fever of the newborn, and yes, they are among the important "active cultures" in a cup of yogurt. So you see, they can be quite dangerous or quite harmless, depending on the species.

Suppose streptococcal diseases are not treated? Can they become complicated?

Streptococci usually respond to antibiotics such as penicillin. However, untreated streptococcal disease can be dangerous. An important complication is rheumatic fever, a condition characterized by fever and inflammations of the small blood vessels. The most significant long-range effect is permanent scarring and distortion of the heart valves, a condition called rheumatic heart disease.

Do streptococci have anything to do with the "flesh-eating bacteria"?

Yes, they are one and the same. What happens is this: Sometimes, invasive streptococci infect the fatty tissue (called fascia) lying over the muscles and beneath the skin. The streptococcal toxins destroy the fascia and cause a condition called necrotizing fasciitis, with an unsightly degeneration of the skin tissues. The disease is not new by any means, but tabloid newspapers and TV programs had a field day with it in the 1990s as they heralded an outbreak of the "flesh-eating bacteria."

Diphtheria

"Diphtheria" is an unusual word. How did the disease get its name?

Diphtheria was first recognized as a clinical entity in 1826 by French pathologist Pierre F. Bretonneau. Bretonneau named the disease *la diphtérie*, from the Greek *diph-*

thera for "membrane," a reference to the suffocating membranes that appear in the throats of patients.

What causes diphtheria, and why is it uncommon in the United States?

Diphtheria is caused by *Corynebacterium diphtheriae*, a Gram-positive rod that is acquired by inhaling respiratory droplets. The bacteria produce a potent exotoxin that kills cells lining the respiratory tract. The toxin is chemically altered to produce the diphtheria toxoid contained in the diphtheria-tetanus-acellular pertussis (DTaP) vaccine, which is given to all children in the United States. The toxoid induces the immune system to produce antibodies (antitoxins) that circulate in the bloodstream throughout the person's life and protect vaccine recipients.

■ *Corynebacterium*
kôr′ē-ne-bak-ti-rē-um

Pertussis

My father had pertussis as a child, but then it was called "whooping cough." Why is that?

Pertussis is one of the more dangerous diseases of childhood years. Disintegrating cells and mucus accumulate in the airways and cause labored breathing. Children experience violent spells of rapid-fire coughing in one exhalation, followed by a forced inhalation over a partially closed windpipe. The rapid inhalation results in the characteristic "whoop" (hence, the name "whooping cough").

Which bacterium causes pertussis, and can the disease be prevented?

Pertussis is caused by *Bordetella pertussis*, a small Gram-negative rod. The bacilli are spread by droplets, and treatment of pertussis is generally successful when antibiotics are administered.

■ *Bordetella*
bor-de-tel′lä

Can the disease be prevented?

The declining incidence of pertussis stems partly from use of a pertussis vaccine. The older vaccine (diphtheria-pertussis-tetanus, or DPT) contained killed *B. pertussis* cells and was considered risky because a few recipients suffered high fevers and seizures. Public health officials now recommend the cell-free ("acellular") vaccine, the "aP" in DTaP.

Bacterial Meningitis

What is meningitis, and which bacteria cause it?

The term meningitis refers to several diseases of the meninges, the three membranous coverings of the brain and spinal cord. Patients suffer pounding headaches, neck paralysis, and numbness in the extremities. A particularly dangerous form of meningitis is meningococcal meningitis. This disease is caused by *Neisseria meningitidis*, a small Gram-negative diplococcus commonly called the meningococcus.

■ *Neisseria meningitides*
nī-se′rē-ä
me-ninji′ti-dis

What's meningococcal meningitis like?

Meningococcal meningitis begins as an influenzalike upper respiratory infection. The infection then spreads to the bloodstream, where bacterial toxins may overwhelm the body in as little as 2 hours and cause death. This condition is called meningococcemia. In survivors, a rash also appears on the skin, beginning as bright red patches, which progress to blue-black spots. Paralytic symptoms soon appear. FIGURE 19.2 shows the progress of the disease.

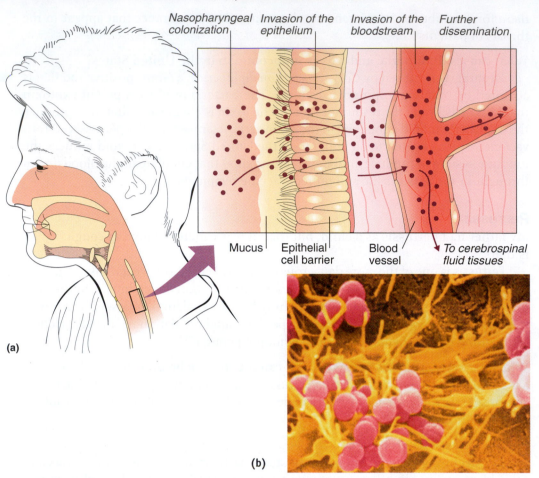

FIGURE 19.2 **Meningococcal Meningitis.** Meningococcal meningitis caused by the Gram-negative diplococcus *Neisseria meningitidis*. (**a**) The bacteria colonize the cell layers of the throat, causing respiratory distress. They pass into the blood stream, where they induce meningococcemia. Finally, they disseminate to tissues near the spinal cord, causing inflammation and meningitis. (**b**) A scanning electron micrograph of *N. meningitides* at the surface of epithelial cells of the respiratory tract. Numerous hair-like pili can be seen. The pili bind the meningococci to the epithelial cells and encourage infection.

How does a doctor find out if you have the disease, and what can be done?

Early diagnosis and treatment of meningococcal meningitis can prevent irreversible nerve damage or death. A principal criterion for diagnosis is the observation of Gram-negative diplococci in samples of spinal fluid obtained by a spinal tap. Aggressive treatment with antibiotics is recommended.

Is this the same type of meningitis that occurs primarily in children?

No, it's not. Another form of meningitis, caused by *Haemophilus influenzae* type b, occurs primarily in children between the ages of 6 months and 2 years (the disease is sometimes called Hib disease). Symptoms include stiff neck, severe headache, and other evidence of neurological involvement, such as listlessness, drowsiness, and irritability. Rifampin and other drugs are used for treatment.

There's a vaccine for this childhood meningitis, right?

Yes, there is, and it's been used quite successfully. In 1986, when the vaccine was first used, about 18,000 cases of *Haemophilus* meningitis were occurring in the United States annually, but by 2002, there were less than 200 cases each year. The Hib vaccine consists of polysaccharides from the organism's capsule.

■ *Haemophilus*
hē-mä'fil-us

Tuberculosis

Which bacterium is responsible for tuberculosis?

Tuberculosis is caused by *Mycobacterium tuberculosis*, the tubercle bacillus first isolated by Robert Koch in 1882. It is a small rod that enters the respiratory tract in droplets. In its cell wall, *M. tuberculosis* contains a layer of waxy material that greatly enhances resistance to environmental pressures and staining. Once stained, however, the microbes resist decolorization, even when subjected to a 5 percent acid-alcohol solution. Thus, the bacilli are said to be acid-resistant, or acid-fast.

Why is tuberculosis called by that name?

Tuberculosis patients experience chronic cough, chest pain, and high fever. The body responds to the disease by forming a wall of white blood cells, calcium salts, and fibrous materials around the organisms. As these materials accumulate in the lung, a hard nodule called a tubercle arises (hence the name "tuberculosis"), as FIGURE 19.3 indicates. This tubercle may be visible in a chest X ray. Tubercle bacilli produce no discernible toxins, but growth is so unrelenting that the tissues are literally consumed, a factor that gave tuberculosis its other common name, consumption.

How do the diagnostic tests for tuberculosis work?

The acid-fast test is an important screening tool used on the patient's sputum. Early detection of tuberculosis is also aided by the tuberculin test, a procedure that begins with the application of a purified protein derivative (PPD) of *M. tuberculosis* to the skin. If the patient has been exposed to tubercle bacilli, the skin becomes thick, and a raised, red welt develops within 48 to 72 hours.

Which drugs are used to treat tuberculosis?

Physicians treat tuberculosis with such drugs as isoniazid (INH), pyrazinamide, and rifampin. The recent appearance of multidrug-resistant *M. tuberculosis* has necessitated the use of two or more drugs to help delay the emergence of resistant strains. In addition, drug therapy is intensive and must be extended over a period of 6 to 9 months or more, partly because the organism multiplies at a very slow rate. In the United States, about 20,000 new cases are identified annually, with almost 3000 fatalities.

Is there a vaccine for tuberculosis?

Immunization to tuberculosis may be rendered by injections of an attenuated (weakened) strain of *Mycobacterium bovis*. This species causes tuberculosis in cows and rarely in humans. The attenuated strain is called bacille Calmette Guérin, or BCG, after Albert Calmette and Camille Guérin, the two French investigators who developed it in the 1920s. Many scientists oppose its use in the United States, pointing to the success of early detection and treatment and to the vaccine's occasional side effects.

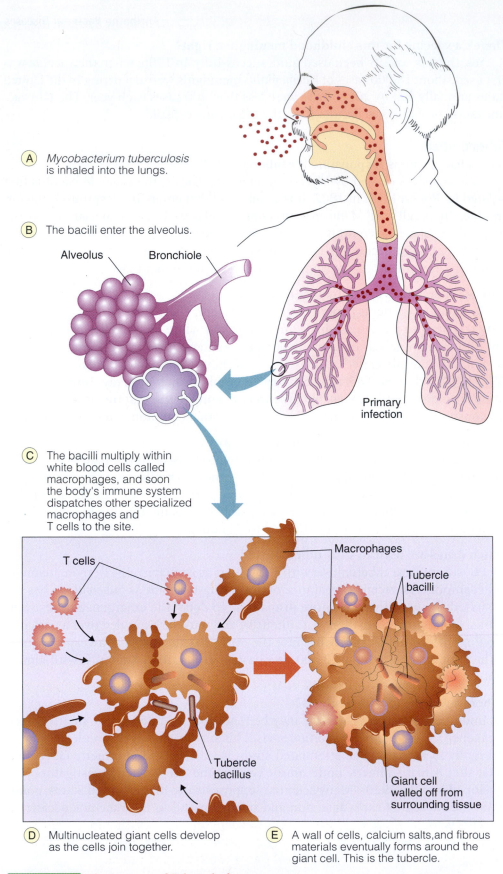

(A) *Mycobacterium tuberculosis* is inhaled into the lungs.

(B) The bacilli enter the alveolus.

Alveolus

Bronchiole

Primary infection

(C) The bacilli multiply within white blood cells called macrophages, and soon the body's immune system dispatches other specialized macrophages and T cells to the site.

T cells

Macrophages

Tubercle bacilli

Tubercle bacillus

Giant cell walled off from surrounding tissue

(D) Multinucleated giant cells develop as the cells join together.

(E) A wall of cells, calcium salts, and fibrous materials eventually forms around the giant cell. This is the tubercle.

FIGURE 19.3 The Progress of Tuberculosis.

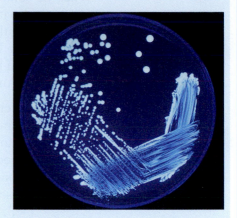

1 On June 25, 1994, the cruise ship *Horizon* set sail from New York City with hundreds of passengers bound for Bermuda. Among the attractions that awaited them on deck were a set of Jacuzzi whirlpool baths. Passengers found them to be an exhilarating way to relax, the warm spray covering their faces.

2 The ship docked at Hamilton, Bermuda, where passengers enjoyed a three-day visit. On the return trip, the whirlpools were busy once again. The cruise ended on July 2, 1994.

3 Beginning on July 15, the New Jersey State Department of Health began receiving reports of passengers who had coughing, fever, and pneumonia. Some cases were quite severe, and some patients required hospitalization.

4 Cultures of respiratory secretions taken from the patients yielded *Legionella pneumophila*, the agent of Legionnaires' disease. Antibody tests confirmed the diagnosis.

5 Health investigators were drawn to the whirlpool baths, since most ill patients had used them. In the sand filters, they located the same strain of *L. pneumophila* as in the patients. Breathing the spray had apparently transmitted the bacteria.

FIGURE 19.4 **Legionnaires' Disease on a Cruise Ship.** This outbreak demonstrates how warm water can provide an incubator for infectious microorganisms.

Legionnaires' Disease (Legionellosis)

How did Legionnaires' disease get its name?

From July 21 to July 24, 1976, the Bellevue-Stratford Hotel in Philadelphia was the site of the 58th annual convention of Pennsylvania's chapter of the American Legion. Toward the end of the convention, 140 conventioneers and 72 other people in or near the hotel became ill with fever, coughing, and pneumonia. Eventually, 34 individuals died of the disease, which came to be called Legionnaires' disease. In January 1977, CDC investigators announced the isolation of a bacterium from the lung tissue of one of the patients. They called it *Legionella pneumophila*.

■ *Legionella pneumophila*
lē-jä-nel'lä nü-mo'fi-lä

Does Legionnaires' disease spread like other airborne diseases?

Not quite. *L. pneumophila* exists where water collects, and it apparently becomes airborne in wind gusts and breezes. Cooling towers, industrial air-conditioning units, humidifiers, stagnant pools, and puddles of water have been identified as sources of the bacteria. Humans breathe the contaminated droplets into the respiratory tract, and disease develops a few days later. The symptoms of Legionnaires' disease include fever, a dry cough with little sputum, and lung infection. **FIGURE 19.4** describes an outbreak on a cruise ship.

Bacterial Pneumonia

What do doctors mean by "pneumonia," and which bacteria cause it?

The term "pneumonia" refers to microbial disease of the bronchial tubes and lungs. A wide spectrum of organisms, including many different viruses and bacteria, may cause pneumonia. Over 80 percent of bacterial cases are due to *Streptococcus pneumoniae*, a Gram-positive chain of diplococci traditionally known as the pneumococcus. The disease is commonly called pneumococcal pneumonia.

What are some symptoms of pneumococcal pneumonia?

Patients with pneumococcal pneumonia experience high fever, sharp chest pains, difficulty breathing, and rust-colored sputum resulting from blood seeping into the air sacs of the lungs. The involvement of an entire lobe of the lung is called lobar pneumonia. If both left and right lungs are involved, the condition is called double pneumonia. Antibiotics are used in therapy.

Is there a vaccine for pneumococcal pneumonia?

In 1983, the FDA licensed a vaccine for immunization to 23 of the over 80 strains of the bacterium. Since these strains are responsible for 87% of cases, there is hope that pneumococcal pneumonia may be controlled in high-risk patients. Pneumococcal pneumonia exists in all age groups, but the mortality rate is highest among the elderly and those with underlying medical conditions.

Is there a milder form of bacterial pneumonia?

Yes. A second type of bacterial pneumonia is termed primary atypical pneumonia "primary" because it occurs in previously healthy individuals (pneumococcal pneumonia usually develops in people who are already ill); "atypical" because the organism differs from the typical pneumococcus and because symptoms are unlike those in pneumococcal disease. The patient experiences fever, fatigue, and a characteristic dry, hacking cough.

Which bacteria cause this type of pneumonia?

The agent of primary atypical pneumonia is *Mycoplasma pneumoniae*, one of the smallest bacteria causing human disease. Mycoplasmas have no cell wall, and therefore they have no Gram reaction or sensitivity to penicillin. Often the disease is called *Mycoplasma* pneumonia.

Other Respiratory Diseases

Which other respiratory diseases are caused by bacteria?

There are several other respiratory diseases caused by bacteria. Among them is Q fever, which is caused by *Coxiella burnetii*, a small bacterium known as a rickettsia. Patients experience severe headache, high fever, a dry cough, and occasionally, lesions on the lung surface. Another bacterial respiratory disease is psittacosis, a flulike disease caused by a chlamydia called *Chlamydia psittaci*. This disease affects humans as well as birds such as parrots, canaries, pigeons, and chickens (FIGURE 19.5). Humans acquire the disease by inhaling airborne dust or dried droppings of infected birds. A final disease in this group is chlamydial pneumonia, caused by *Chlamydia pneumoniae*. The disease is clinically similar to primary atypical pneumonia.

■ *Chlamydia psittaci*
kla-mi'dē-ä sit'tä-sē

■ Foodborne and Waterborne Bacterial Diseases

In this section, we move to the bacterial diseases transmitted by food and water. We shall encounter two types of such diseases: intoxications, where a toxin produced outside the body is the prime consideration in the disease; and infections, where microbes grow in the body and cause the illness. When an intoxication occurs, there is a brief time between the entry of the toxin into the body and the appearance of symp-

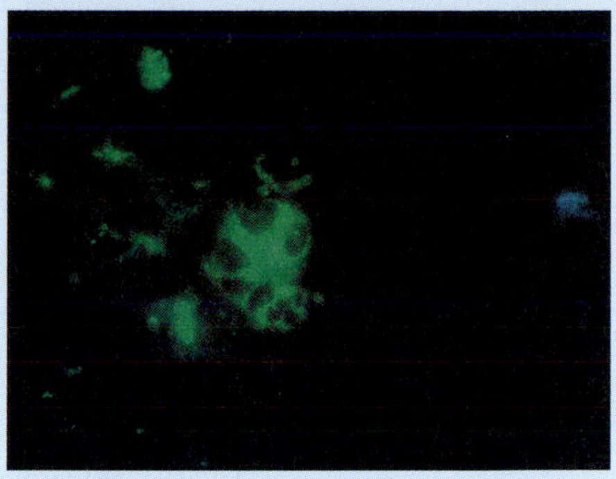

1 On February 13, a distributor in Mississippi shipped a supply of parakeets and cockatiels to retail pet stores in Massachusetts and Tennessee. The birds had been supplied to him by domestic breeders.

2 Four days later, on February 17, a man purchased one of the parakeets from the Massachusetts store. The man noted that the bird became very "tired-looking" some days after he brought it home.

3 On February 19, a family from Tennessee purchased a cockatiel sent by the distributor. The bird was taken home and kept in a birdcage where all family members could enjoy it. Some days later, family members observed that the bird was "irritable."

4 On March 1, the man from Massachusetts was hospitalized with fever, sore throat, and pneumonia. Two other members of his family were also sick.

5 Also on March 1, six members of the Tennessee family were experiencing fever, cough, and sore throat.

6 In both families a diagnosis of psittacosis was made. All recovered.

FIGURE 19.5   Two Outbreaks of *Psittacosis*. These 1992 outbreaks of *psittacosis* were traced to birds distributed by a single supplier. Coincidence linked the disease to the distributor's birds, even though there was no evidence of widespread disease among his stock.

toms (i.e., the incubation period), and the situation also resolves in a relatively brief period. For infections, the incubation period is longer, and the disease takes longer to resolve. As before, we continue our conversational mode of discussion.

Botulism

Is botulism a problem to be taken seriously?

Absolutely. Of all the foodborne intoxications in humans, none is more dangerous than botulism. The causative agent, *Clostridium botulinum*, produces a toxin so powerful that 1 pint of the pure material dispersed evenly over the world could eliminate the entire population; 1 ounce would kill all the people in the United States. The toxin is thus a potent weapon for bioterrorism.

What's the cause of botulism?

Botulism is caused by *Clostridium botulinum*, a Gram-positive anaerobic bacillus that forms spores. The spores exist in the soil, and when they enter the anaerobic environment of cans or jars, they germinate to multiplying bacilli that produce and release the toxin (technically called an exotoxin).

How does one recognize botulism in patients?

The symptoms of botulism develop within hours of consuming toxin-contaminated food. Patients suffer blurred vision, slurred speech, difficulty swallowing and chewing, labored breathing and developing paralysis. These symptoms result from a complex process in which the toxin penetrates the ends of nerve cells and inhibits the release of the neurotransmitter acetylcholine into the junctions between nerves and muscles. Without acetylcholine, nerve impulses cannot pass into the muscles, and the muscles do not contract. Death follows within a day or two.

Can anything be done for patients?

Because botulism is a type of foodborne intoxication, antibiotics are of no value as a treatment. Instead, large doses of special antibodies called antitoxins must be administered to neutralize the toxins. Life-support systems such as respirators are also used. The disease can be avoided by heating foods before eating them, because the toxin is destroyed by exposure to temperatures of 90°C (194°F) for 10 minutes. At present, there is no general-use vaccine against botulism, but the possibility that *C. botulinum* toxin may be used as a bioweapon has spurred research into the development of such a vaccine.

I think I read in a newspaper that botulism toxin is used to smooth out wrinkles on the face. Is that true?

Yes, some cosmetologists are using the toxin to smooth out wrinkles on the face by paralyzing the muscles. Botox, as the material is known, is also used to treat strabismus (misalignment of the eyes, commonly known as "cross eye"), and it may be valuable in relieving stuttering, uncontrolled blinking, musician's cramp (the bane of the violinist), and spastic closure of the anal and urinary sphincter.

Staphylococcal Food Poisoning

Is staph food poisoning the same as ptomaine poisoning?

Years ago, it was common for people to complain of ptomaine poisoning shortly after eating contaminated food. Modern microbiologists, however, have placed the blame for most food poisonings on the Gram-positive bacterium *Staphylococcus aureus*, often obtained from skin wounds or the nose. Today, staphylococcal (staph) food poisoning ranks as the second most often reported of all types of foodborne disease (*Salmonella*-related illnesses are first).

Is a toxin involved in staph food poisoning?

Like botulism, staphylococcal food poisoning is caused by an exotoxin secreted into foods and consumed by unsuspecting individuals. Patients experience abdominal cramps, nausea, vomiting, prostration, and diarrhea as the toxin encourages the release of water. The symptoms last for several hours, and recovery is usually rapid and complete.

How can you tell that it's staphylococcal food poisoning and not due to something else, such as *Salmonella*?

One telltale sign of staph food poisoning is the short incubation period, a brief 1 to 6 hours. *Salmonella* infections have a longer incubation period, and the symptoms are different (see below). Foods containing *S. aureus* toxin lack an unusual taste, odor, or appearance, and the only clues to possible contamination are factors such as moisture content, low acidity, and improper heating.

Salmonellosis

Is *Salmonella* an intoxication or a poisoning?

"Salmonella" as you call it, is an abbreviation for salmonellosis, a foodborne infection. It is caused by hundreds of species of *Salmonella*, all Gram-negative rods (**FIGURE 19.6**). The bacteria grow in the body for 1 to 3 days, then the patient experiences fever, nausea, vomiting, diarrhea, and severe abdominal cramps. The symptoms may last a week or more, and if dehydration occurs, fluid replacement may be necessary.

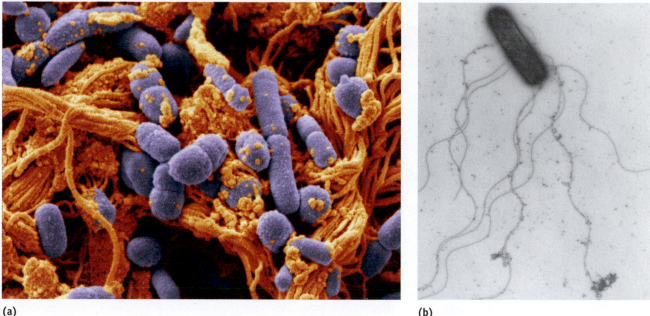

(a)   (b)

FIGURE 19.6 *Salmonella* **Species.** Two views of *Salmonella* involved in salmonellosis. (a) *S. typhimurium* observed on the collagen fibers of muscle tissue from an infected chicken. (b) A transmission electron micrograph of a Salmonella species covered with flagella (×40,000). Note the long length of the flagella relative to the cell size.

Is salmonellosis the disease I've heard about in chickens?

Yes, it is. Poultry products are particularly notorious because *Salmonella* species commonly infect chickens and turkeys. The microbes may be consumed directly from the poultry or in poultry products such as chicken salad or cold cuts made from chicken. With over 4 billion chickens and turkeys consumed annually in the United States (an average of 75 pounds per American), the possibilities for acquiring *Salmonella* are plentiful.

How about eggs? Are they dangerous too?

Eggs can be another source of salmonellosis when used in foods such as custard pies, cream cakes, egg nog, ice cream, and mayonnaise. Researchers believe that *Salmonella* species infect the ovary of the hen and pass into the egg before the shell forms. Consumers should store eggs in the main compartment of the refrigerator, refrigerate leftover egg dishes quickly in small containers (to accelerate cooling), and avoid "runny" or undercooked eggs.

Are *Salmonella* species useful as a bioweapon?

Although *Salmonella* species cause an incapacitating illness, they do not possess the lethal potential of other bioweapons such as anthrax bacilli and smallpox viruses. Nevertheless, *Salmonella* were used in a bioterrorist incident in Oregon in 1984. Members of a religious cult sprayed the bacteria on the salad bars in several restaurants. Over 750 people became ill, but none died.

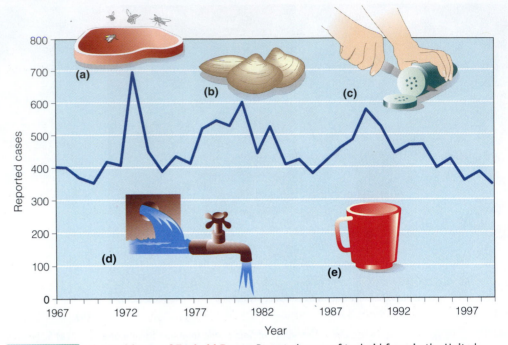

FIGURE 19.7   **The Incidence of Typhoid Fever.** Reported cases of typhoid fever in the United States by year, 1967 to 1999. For 1999, 346 cases were reported. In about half the cases, the disease was acquired during foreign travel. Portrayed on the graph are the five Fs important in the transmission of typhoid fever: (**a**) flies, (**b**) food, (**c**) fingers, (**d**) feces, and (**e**) fomites.

Typhoid Fever

Is typhoid fever as serious as salmonellosis?

Typhoid fever is among the classical diseases (the "slate-wipers") that have ravaged human populations for generations. *Salmonella typhi*, the Gram-negative rod that causes it, displays high resistance to environmental conditions outside the body. This factor enhances its ability to remain alive for long periods of time in water, sewage, and certain foods. About 300 to 400 cases occur each year in the United States as **FIGURE 19.7** shows.

What's typhoid fever like?

During cases of typhoid fever, *S. typhi* invades the tissues of the small intestine, causing deep ulcers and bloody stools. Blood invasion follows, and after a few days, the patient experiences mounting fever, lethargy, and delirium. The abdomen becomes covered with **rose spots**, an indication that blood is hemorrhaging in the skin.

How is typhoid fever treated?

Treatment of typhoid fever is generally successful with antibiotics, but about 5% of recoverers become carriers and continue to harbor and shed the organisms for a year or more. The experiences of one of history's most famous carriers, Typhoid Mary, are recounted in A Closer Look on page 376.

Shigellosis

I know someone who had bacterial dysentery. What's that?

Bacterial dysentery is technically known as shigellosis. It is manifested by waves of intense abdominal cramps and frequent passage of small-volume, bloody, mucoid stools. About 25,000 cases occur annually in the United States.

What causes the disease?

Shigellosis may be caused by any of four species of *Shigella*. Humans ingest the organisms in contaminated water or foods. *Shigella* bacilli usually penetrate the cells lining the intestines, and, after 2 to 3 days, they produce sufficient toxins to encourage water release. Antibiotics are sometimes effective, but many strains of *Shigella* are resistant because of R factors, as the story in Chapter 10 recounts. The dysentery can be quite indisposing. Indeed, historians report that at the Battle of Crécy in 1346, the English army was racked with dysentery. When the French attacked, they literally caught the English with their pants down.

Cholera

I've heard that the diarrhea of cholera is quite bad. Is that so?

Yes, it is. No diarrhea can compare with the extensive diarrhea associated with cholera. In the most severe cases, a patient may lose up to 1 liter of colorless, watery fluid every hour for several hours. The patient's eyes become gray and sink into their orbits; the skin is wrinkled, dry, and cold, and muscular cramps occur in the arms and legs; the blood thickens, urine production ceases, and the sluggish blood flow to the brain leads to shock and coma.

So, is cholera another of those "slate-wipers"?

Most definitely. In untreated cases, the mortality rate for cholera may reach 70%. Cholera has been observed for centuries in human populations, and seven pandemics have been documented. Travelers to regions of the world where cholera is active are immunized with preparations of dead *V. cholerae*, thereby obtaining protection for about 6 months.

What's the cause of cholera?

Cholera is caused by *Vibrio cholerae*, a curved Gram-negative rod first isolated by Robert Koch in 1883 (FIGURE 19.8). The bacilli enter the intestinal tract in contaminated water or food. As they move along the intestinal epithelium, they secrete a toxin that stimulates the unrelenting loss of fluid. Antibiotics kill the bacteria, but the key treatment is restoration of the body's water balance with oral or intravenous (IV) rehydration solutions.

E. Coli Diarrheas

I thought that *Escherichia coli* was a harmless member of the intestinal microflora. Does it also cause disease?

Unfortunately, there are pathogenic strains of *Escherichia coli*. One such strain induces diarrhea in infants when it invades the intestinal lining and produces powerful toxins that cause water loss. Other strains cause traveler's diarrhea, a term usually applied to a disease in which a traveler experiences diarrhea within 2 weeks of visiting a tropical location; the diarrhea lasts up to 10 days.

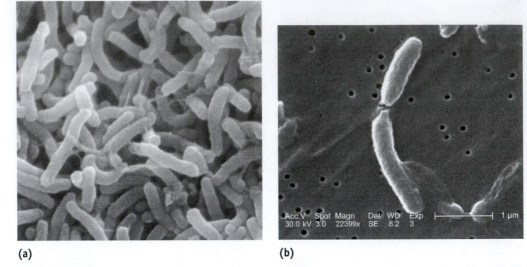

(a)   (b)

FIGURE 19.8   **The Bacilli of Cholera.** Two views of *Vibrio cholerae*, the cholera bacillus. (**a**) A colony of *Vibrio cholerae* on a solid culture medium. Many cells are seen in various stages of division, and their irregular arrangement is visible. (**b**) *Vibrio cholerae* (×22,399).

What about that *E. coli* in foods that I've read about?

It is possible to contract a rather serious *E. coli* diarrhea due to *E. coli* O157:H7. This strain has been acquired from hamburger meat and other foods. When confined to the large intestine, the strain causes grossly bloody diarrhea, a complication known as hemorrhagic colitis. When the disease involves the kidneys, it can lead to kidney failure and is called hemolytic uremic syndrome (HUS). Seizures, coma, colon perforation, and liver disorder have been associated with HUS.

How widespread is this *E. coli* disease?

Epidemiologists estimate that approximately 20,000 cases of disease due to *E. coli* O157:H7 occur annually in the United States, with about 250 deaths. The prevailing wisdom is that the bacteria exist in the intestines of cattle but cause no disease in these animals. Contamination during slaughtering brings *E. coli* to beef products, and excretion to the soil accounts for transfer to plants and fruits.

Peptic Ulcer Disease

Did I read somewhere that peptic ulcers are now considered an infectious disease?

■ *Helicobacter pylori*
hĕ′lik-ō-bak-tĕr pī′lō-rē

Yes, you did. One of the more remarkable discoveries of the modern era is that many cases of peptic (stomach) ulcers are caused by the bacterium *Helicobacter pylori*. This Gram-negative curved rod is apparently transmitted by contaminated food and water. Doctors have revolutionized the treatment of ulcers by prescribing antibiotics such as tetracycline. They have achieved cure rates of up to 90%, and relapses are uncommon.

But isn't the stomach very acidic? How does the microbe manage to survive?

How *H. pylori* manages to survive in the intense acidity of the stomach is interesting. Apparently, the bacterium twists its way through the mucus coating of the stomach lining and attaches to the stomach wall. There it secretes the enzyme ure-

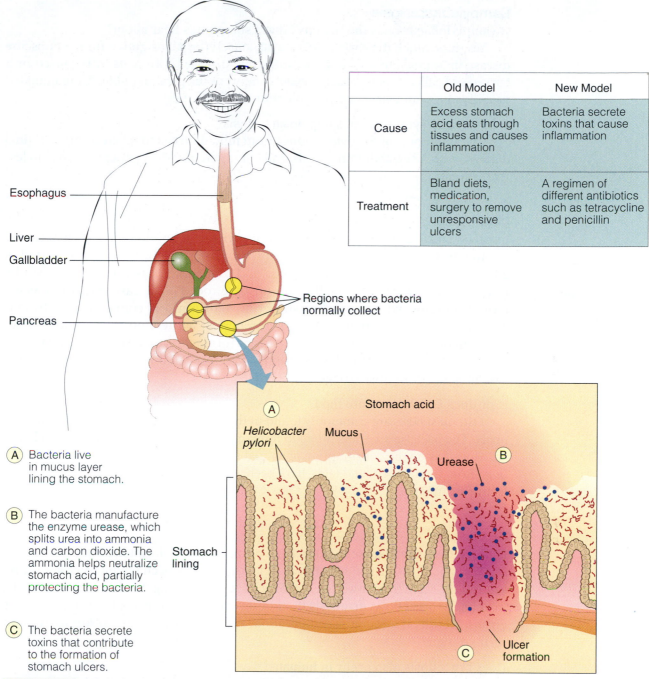

Esophagus

Liver

Gallbladder

Pancreas

	Old Model	New Model
Cause	Excess stomach acid eats through tissues and causes inflammation	Bacteria secrete toxins that cause inflammation
Treatment	Bland diets, medication, surgery to remove unresponsive ulcers	A regimen of different antibiotics such as tetracycline and penicillin

Regions where bacteria normally collect

(A) Bacteria live in mucus layer lining the stomach.

(B) The bacteria manufacture the enzyme urease, which splits urea into ammonia and carbon dioxide. The ammonia helps neutralize stomach acid, partially protecting the bacteria.

(C) The bacteria secrete toxins that contribute to the formation of stomach ulcers.

Stomach acid

(A) *Helicobacter pylori* Mucus Urease (B)

Stomach lining

Ulcer formation (C)

FIGURE 19.9 **The Progression of Peptic Ulcers.** Scientists now believe that the majority of peptic ulcers are caused by the bacterium *Helicobacter pylori*. This figure illustrates how the bacteria cause such an ulcer and highlights the old and new views of ulcers' cause and treatment.

ase, which digests urea in the area and produces ammonia as an end-product. The ammonia neutralizes stomach acid, and the organism begins its destruction of the tissue, supplemented by digestive enzymes normally found in the stomach tissue. FIGURE 19.9 illustrates the process.

Campylobacteriosis

Is camplylobacteroisis the "campy" that I sometimes hear about?

Yes, it is. Since the early 1970s, campylobacteriosis has grown from an obscure disease in animals to a widespread intestinal disease in humans. It is caused by a curved rod called *Campylobacter jejuni* "campy" to the general public. Contaminated dairy products and water are possible sources of infection.

What are the symptoms of campylobacteriosis?

The symptoms of the disease range from mild diarrhea to severe gastrointestinal distress, with fever, abdominal pains, and bloody stools. Most patients recover in less than a week.

Listeriosis

I watched a *20-20* segment on TV about listeriosis in hot dogs. What's listeriosis?

In late 1998 and early 1999, close to 100 cases of listeriosis were reported in 22 states, all cases linked to hot dogs and deli meats. Listeriosis is caused by *Listeria monocytogenes*, a small Gram-positive rod. The disease may occur as listeric meningitis, with headaches, stiff neck, delirium, and coma. Another form is a blood disease accompanied by high numbers of white blood cells called monocytes (hence the organism's name). A third form is characterized by infection of the uterus, with vague flulike symptoms. If contracted during pregnancy, the disease may result in miscarriage.

■ *Listeria monocytogenes* lis-te′rē-ä mo-nō-sī-tô′je-nēz

How does *Listeria* get into food?

Listeria species are commonly found in the soil and in the intestines of many animals. The bacteria are transmitted to humans by food contaminated with fecal matter, as well as by the consumption of animal foods. Cold cuts, as well as soft cheeses (e.g., Brie, Camembert, and feta), have been associated with cases. Antibiotics are effective treatments.

Brucellosis

Is brucellosis the disease that has been occurring in Yellowstone Park?

Yes, since 1997, a major outbreak of brucellosis has been taking place among bison and elk herds in Yellowstone National Park in Wyoming, Montana, and Colorado. In animals, brucellosis manifests itself in several organs, especially the reproductive organs. Sterility is a common complication, and pregnant animals are known to abort their young.

Does brucellosis also occur in humans?

Yes, but the major focus of brucellosis in humans is the blood-rich organs, such as the spleen and lymph nodes. Patients experience flulike weakness, as well as backache, joint pain, and a high fever (with drenching sweats) in the daytime and low fever (with chills) in the evening. This fever pattern gives the disease its alternate name, undulant fever. Brucellosis is caused by Gram-negative rods belonging to the genus *Brucella*. Transmission can occur by contact with animals and by the consumption of milk and other dairy products.

Other Intestinal Diseases

Are there any other bacterial pathogens of the intestine that I should know about?

Yes, there are a couple of important ones. *Vibrio parahaemolyticus*, for example, is a Gram-negative rod that is a cause of foodborne infections where seafood is the main staple of the diet. *Vibrio vulnificus* occurs naturally in brackish water and seawater, where oysters and clams live and is transmitted to humans by them. *Bacillus cereus* is a Gram-positive sporeforming bacillus that causes food poisoning, frequently experienced after consuming cooked rice. And *Yersinia enterocolitica* is a Gram-negative rod that causes fever, diarrhea, and abdominal pain.

■ vulnificus
vul-ni′fi-kus

■ enterocolitica
en′-tėr-ō-kōl-it-ik-ä

Soilborne Bacterial Diseases

Soilborne diseases are those whose bacterial agents are transferred from the soil to the unsuspecting individual. To remain alive in the soil, the bacteria must resist environmental extremes, and often they form spores, as these three diseases illustrate.

Anthrax

Is anthrax the disease they mean when they talk about biological warfare?

Yes, anthrax is considered a threat in biological warfare and in bioterrorism because it is a serious disease caused by a sporeformer, and the spores can be aerosolized in microscopic droplets and distributed to large areas to infect huge numbers of people (see A Closer Look on page 447). For example, an aerosol of spores could be released unobtrusively and could drift through a large city without being detected. The seriousness of the bioterrorism risk has been underscored by a public health official who called anthrax "the poor man's atomic bomb."

But what exactly is anthrax?

Anthrax is a blood disease that occurs in humans as well as large animals such as cattle, sheep, and goats. The disease is caused by *Bacillus anthracis*, a Gram-positive sporeforming rod (FIGURE 19.10). Patients inhale the spores or come into contact with them in the air or ingest them in meat, and soon their organs fill with bloody, black, infected fluid. On the skin, there are boil-like lesions covered with a black crust. Violent dysentery with bloody stools accompanies the intestinal form.

Is anthrax common in humans?

No, but humans can acquire anthrax by shearing infected sheep, processing wool, consuming infected meat, or coming in contact with such animal products as violin bows, shaving bristles, goatskin drums, and leather jackets. Penicillin and other antibiotics such as ciprofloxacin are used for therapy. In untreated cases, the mortality rate is more than 80%. In the 1990s, the total number of anthrax cases in the United States was less than a dozen, primarily because of the testing of imported animal products.

What conditions would have to be met for anthrax to be useful as a bioweapon?

To use anthrax as a bioweapon, bioterrorists would have to obtain anthrax cultures, grow the bacilli in huge amounts in fermentation tanks, encourage them to form spores, find a way to produce a biodust that would keep the spores from clumping,

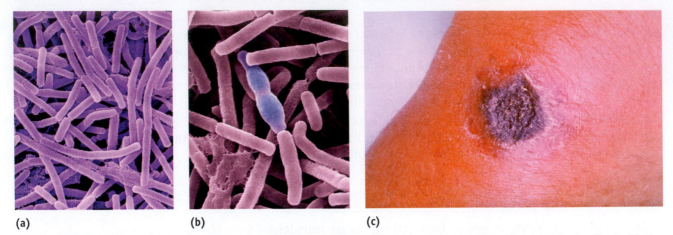

(a)   (b)   (c)

FIGURE 19.10 *Bacillus anthracis.* *Bacillus anthracis* is the cause of anthrax. (**a**) Vegetative cells of *B. anthracis* visualized with the scanning electron microscope. Note the oval shape of the spores and the typical rodshape of the vegetative cells. (**b**) Anthrax spores in the process of germinating. The spore coat of the spore in the center of the photograph has divided and is beginning to separate. (**c**) This cutaneous lesion is a result of infection with anthrax bacilli. Lesions like this one develop when anthrax spores contact the skin, germinate to vegetative cells, and multiply.

and develop a method for distributing the bacteria-laced dust. The anthrax attacks of the autumn of 2001 used a biodust in letters sent through the postal service.

Is there a vaccine available to protect against anthrax?

At this writing, a vaccine is available, but it is used only for high-risk individuals such as medical workers, researchers, and the military. The vaccine consists of antigens derived from *B. anthracis* and is administered in five or six injections over a period of 18 months. Research is currently underway to develop a more efficient vaccine.

Why would anyone mount a bioterrorist attack?

Bioterrorists operate outside the laws and morals of society. They spread pathogenic microbes that are silent, stealthy, invisible, and slow-acting, and they generate levels of anxiety approaching hysteria. By unleashing the microbes of disease, they hope to gain a political advantage. As the events of 2001 showed, however, they also elicit a rapid and reassuring response based on the collective technological advances of human culture.

Tetanus

Is tetanus the disease you get when you step on a nail?

Yes, it is possible to get tetanus through a puncture wound from an soil-contaminated nail or a piece of glass, a needle, a thorn, or an animal bite. *Clostridium tetani*, the anaerobic Gram-positive rod that causes the disease, occurs everywhere in the environment. Its spores enter a wound and revert to multiplying bacilli that produce the second most powerful toxin known to science (after the botulism toxin). The toxin provokes sustained and uncontrolled contractions of the muscles, and spasms occur throughout the body.

Can anything be done for patients with tetanus?

Tetanus patients are treated with sedatives and muscle relaxants and are placed in quiet, dark rooms. Physicians prescribe penicillin to destroy the bacteria and tetanus antitoxin to neutralize the toxin. The United States has had a steady decline in the incidence of tetanus, with approximately 100 cases confirmed annually.

A CLOSER LOOK
The Legacy of Gruinard Island

In 1941, the specter of airborne biological warfare hung over Europe. Fearing that the Germans might launch an attack against civilian populations, British authorities performed a series of experiments to test their own biological weapons. The spot chosen was Gruinard Island, a mile-long patch of land off the coast of Scotland. Investigators placed 60 sheep on the island and exploded a bomb containing anthrax spores overhead. Within days, all the sheep were dead.

Warfare with biological weapons never came to reality in World War II, but the contamination of Gruinard Island remained. A series of tests in 1971 showed that anthrax spores were still alive at and below the upper crust of the soil, and that they could be spread by earthworms. Officials posted signs warning people not to set foot on the island, but did little else.

Then a strange protest occurred in 1981. Activists demanded that the British government decontaminate the island. They backed their demands with packages of soil taken from the island. Notes led government officials to two 10-pound packages of spore-laden soil, and the writers threatened that 280 pounds were hidden elsewhere.

Partly because of the protests, the British government instituted a decontamination of the island in 1986. Technicians used a powerful brushwood killer, combined with burning and treatment with formalin in seawater. Finally, they managed to rid the soil of anthrax spores. By April 1987, sheep were once again grazing on the island. However, people were somewhat reluctant to return. Gruinard Island remains a monument of sorts to the effects of biological warfare.

What does a tetanus shot do for you?

Immunization to tetanus is accomplished by injections of tetanus toxoid in the diphtheria-tetanus-acellular pertussis (DTaP) vaccine. The toxoid induces the immune system to produce protective antitoxins (antibodies). Booster injections of tetanus toxoid in the Td vaccine (a "tetanus shot") are recommended every 10 years to keep the level of immunity high. (A Closer Look on page 448 explains what can happen if the immunity is not present.)

Gas Gangrene
Is gas gangrene the same as gangrene?

No, gangrene is self-destruction of the muscle tissues brought about by enzymes from partially repaired tissue. Gas gangrene occurs when soilborne *Clostridium perfringens* and other *Clostridium* species invade the dead, anaerobic tissue of a wound. These Gram-positive rods ferment the muscle carbohydrates and putrefy the muscle proteins, producing large amounts of gas that tear the tissue apart.

What are the symptoms of gas gangrene?

The symptoms of gas gangrene include intense pain and swelling at the wound site, as well as a foul odor. The site initially turns dull red, then green, and finally blue-black. Treatment consists of antibiotic therapy as well as debridement (excision), amputation, or exposure in a hyperbaric oxygen chamber. The disease spreads rapidly, and death frequently results.

A CLOSER LOOK

Shoelaces

The untied shoelaces told the story—that and the difficult swallowing and distorted facial features. She had left the hospital the day before, but the symptoms were not gone—not yet, at least.

It began on Sunday, July 5, 1992. The 4th of July weekend was hot in Rutland, Vermont, and most of her friends were recovering from the previous day's celebrations. But she decided to catch up on her gardening. The ground was warm, and she took off her shoes to walk about the garden—nothing like good fertile soil, she must have been thinking. Then it happened. "Ouch!" A splinter entered the base of her right big toe. No matter. Take out the splinter, and get on with the gardening. Gone and forgotten.

But it was not forgotten. Three days later, the pain on the left side of her face necessitated a visit to her family doctor. Probably a facial infection, she was told. Take the amoxicillin, and it should resolve.

It did not resolve—it worsened. Now it was July 12, and her jaw was so tight that she had not eaten for three days. The muscle spasms in her face were intense, and her friends nervously suggested that it looked like lockjaw. When she arrived at the hospital's emergency room, an alert doctor recognized the classic risus sardonicus (the grinning expression caused by spasms of the facial muscles) and the trismus (the lockjaw caused by spasms of the chewing muscles). No question—she had tetanus.

Treatment was swift and aggressive—3250 units of tetanus antitoxin, intravenous penicillin, and a tetanus booster, plus removing the traces of wood still in the wound. She was placed in a quiet room and given muscle relaxants. Fifteen days would pass before she was discharged. She was well on her way to recovery . . . except she still could not tie those darn shoelaces.

■ Arthropodborne Bacterial Diseases

Arthropods transmit diseases to or among humans usually by taking a blood meal from an infected animal or person and themselves becoming infected. Then they pass the microbes to another individual during the next blood meal. Arthropod-related diseases occur primarily in the blood stream, and they are often accompanied by a high fever and a body rash.

Plague

I've heard that plague had a powerful influence on the course of Western civilization. Is that true?

Yes, it is. Few diseases have had the terrifying history of bubonic plague, nor can any match the array of social, economic, and religious changes it has brought about. The first documented pandemic of plague probably began in Africa in 542, during the reign of the Roman emperor Justinian. It lasted 60 years, killed millions, and contributed to the down-fall of Rome. The second pandemic was known as the Black Death because of the purplish-black splotches on victims and the terror it evoked in the 1300s. The Black Death decimated the world and, by some accounts, killed an estimated 40 million people in Europe, almost one-third of the population.

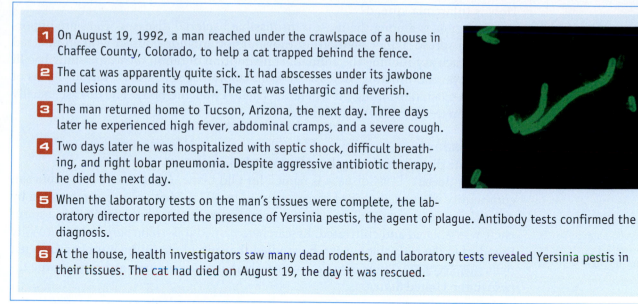

1. On August 19, 1992, a man reached under the crawlspace of a house in Chaffee County, Colorado, to help a cat trapped behind the fence.

2. The cat was apparently quite sick. It had abscesses under its jawbone and lesions around its mouth. The cat was lethargic and feverish.

3. The man returned home to Tucson, Arizona, the next day. Three days later he experienced high fever, abdominal cramps, and a severe cough.

4. Two days later he was hospitalized with septic shock, difficult breathing, and right lobar pneumonia. Despite aggressive antibiotic therapy, he died the next day.

5. When the laboratory tests on the man's tissues were complete, the laboratory director reported the presence of Yersinia pestis, the agent of plague. Antibody tests confirmed the diagnosis.

6. At the house, health investigators saw many dead rodents, and laboratory tests revealed Yersinia pestis in their tissues. The cat had died on August 19, the day it was rescued.

FIGURE 19.11   An Episode of Plague. This incident demonstrates that plague occurs in modern times in the United States.

Has plague ever occurred in the United States?

Yes, again. In the late 1800s, Asian warfare facilitated the spread of plague, and migrations brought infected individuals to China and Hong Kong. During this epidemic in 1894, the causative organism was isolated by Alexandre Yersin and Shibasaburo Kitasato. Plague first appeared in the United States in San Francisco in 1900, carried by rats on ships from the Orient. The disease spread to ground squirrels, prairie dogs, and other wild rodents, and it is now endemic in the southwestern states, where it is commonly called sylvatic plague (FIGURE 19.11 describes a case of human infection).

What's the cause of plague?

Plague is caused by the Gram-negative rod *Yersinia pestis* (named for Yersin). The bacteria are transmitted by the rat flea, and they localize in the lymph nodes, especially those of the armpits, neck, and groin, where hemorrhaging causes substantial swellings called buboes (hence the name bubonic plague). From there, the microbes spread to the bloodstream, where they cause septicemic plague; then to the lungs, where they cause called pneumonic plague (spread by respiratory droplets). There is extensive coughing and hemorrhaging, and many patients suffer cardiovascular collapse. Mortality rates for pneumonic plague approach 100% unless antibiotic therapy is instituted.

Tularemia

I've never heard of tularemia. What is it?

Tularemia is a plaguelike disease, although it is not as serious as plague. It is caused by *Francisella tularensis*, a small Gram-negative rod. The disease occurs in a broad variety of wild animals, especially rodents, and it is particularly prevalent in rabbits (where it is known as rabbit fever).

■ *Francisella tularensis*
fran'sis-el-lä
tü'lä-ren-sis

How is tularemia spread?

Tularemia is among the most transmissible bacterial diseases. Ticks are important to its spread. Other methods of transmission include contact with an infected animal (such as touching the animal), consumption of rabbit meat, splashing bacilli into the eye, and inhaling bacilli, even from laboratory specimens. Various forms of tularemia exist, depending on where the bacilli enter the body. The disease usually resolves on treatment with antibiotics.

Lyme Disease

Does Lyme disease have anything to do with limes?

■ *Borrelia burgdorferi*
bôr-rel'ē-ä
burg-dôr'fĕr-ē

No, it doesn't. Lyme disease is named for Old Lyme, Connecticut, the suburban community where a cluster of cases occurred in 1975. That year, researchers led by Allan C. Steere traced the disease to deer ticks. Several years would pass before a spirochete was isolated and cultivated. Researchers named it *Borrelia burgdorferi* for Willy Burgdorfer, the microbiologist who studied the spirochete in the gut of an infected tick (FIGURE 19.12). Lyme disease is currently the most common arthropodborne illness in the United States.

Isn't there a rash associated with Lyme disease?

Yes, there is rash at the site of the tick bite. It expands slowly and is called erythema (red) chronicum (persistent) migrans (expanding), or ECM. It has an intense red border and a red center, and it resembles a bull's eye. About one-third of patients do not develop ECM. Fever, aches and pains, and flulike symptoms usually accompany the rash.

Can the disease be treated with antibiotics?

In the rash stage, effective treatment can be rendered with penicillin, tetracycline, or other antibiotics. Left untreated, some cases enter a second stage where the patient experiences pain, swelling, and arthritis in the large joints, especially the knee, shoul-

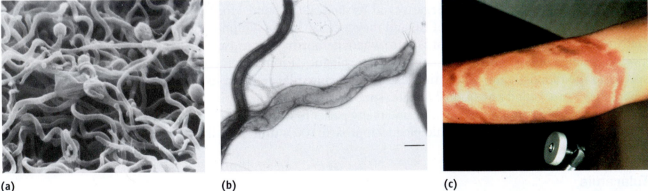

(a)   (b)   (c)

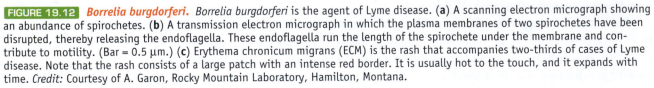

FIGURE 19.12   *Borrelia burgdorferi.* *Borrelia burgdorferi* is the agent of Lyme disease. (a) A scanning electron micrograph showing an abundance of spirochetes. (b) A transmission electron micrograph in which the plasma membranes of two spirochetes have been disrupted, thereby releasing the endoflagella. These endoflagella run the length of the spirochete under the membrane and contribute to motility. (Bar = 0.5 μm.) (c) Erythema chronicum migrans (ECM) is the rash that accompanies two-thirds of cases of Lyme disease. Note that the rash consists of a large patch with an intense red border. It is usually hot to the touch, and it expands with time. *Credit:* Courtesy of A. Garon, Rocky Mountain Laboratory, Hamilton, Montana.

der, ankle, and elbow joints. In some individuals a third stage occurs, and the arthritis is complicated by damage to the cardiovascular and nervous systems.

How does the doctor detect Lyme disease and is there a vaccine to prevent it?

A diagnosis of Lyme disease is initially based on symptoms. A blood sample may also be taken to test for spirochetal antibodies or spirochetal DNA. A vaccine has been approved by the FDA, but the manufacturer has recently suspended production because of a low usage rate.

Relapsing Fever
Did relapsing fever get its name the way I think it did?

Yes, the name says it all. Cases of relapsing fever are characterized by substantial fever, shaking chills, headache, prostration, and drenching sweats. The symptoms last for a couple of days, then disappear, then reappear up to ten times during the following weeks.

What causes the disease?

Relapsing fever is caused by a spirochete referred to as *Borrelia recurrentis*, which is transmitted by lice and ticks. When lice are present, relapsing fever can occur in epidemics. One such epidemic descended on Napoleon's soldiers during the Russian campaign. Serious losses from it and other bacterial diseases decimated the French army, and the balance tipped in favor of the Russians. Peter Ilyich Tchaikovsky wrote the *1812 Overture* to celebrate the great Russian victory over Napoleon's forces.

Rocky Mountain Spotted Fever
Does Rocky Mountain spotted fever happen only in the Rocky Mountains?

Interestingly, very few cases of the disease happen in the Rocky Mountains. Most cases occur in southeastern and Atlantic Coast states where ticks transmit *Rickettsia rickettsii*, the responsible agent.

What are the signs and symptoms of the disease?

The hallmarks of Rocky Mountain spotted fever are a high fever lasting for many days and a skin rash reflecting damage to the small blood vessels. The rash begins as pink spots and progresses to pink-red pimplelike spots that fuse to form a maculopapular rash. The rash generally begins on the palms of the hands and soles of the feet and progressively spreads to the trunk. Tetracycline and other antibiotics are therapeutically useful.

Typhus Fever
Is typhus fever the same as typhoid fever?

No, the two diseases are quite different, although both are accompanied by extremely high fever. Typhus fever is one of the most notorious of all bacterial "slate-wipers." It is a prolific killer of humans and has altered the course of history on several occasions, for example, when it helped decimate the Aztec population in the 1500s. It also helped the Russians defeat Napoleon's army.

What causes typhus fever?

Typhus fever is caused by *Rickettsia prowazekii*. The bacteria are transmitted to humans by head and body lice, which flourish where sanitation measures are lacking and hygiene is poor. The characteristic fever and rash of rickettsial disease are evident in epidemic typhus. There is a maculopapular rash, but unlike the rash of Rocky

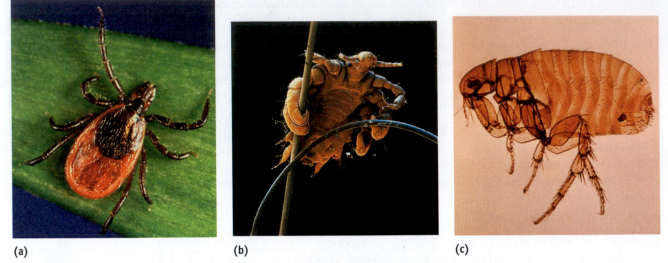

(a)   (b)   (c)

FIGURE 19.13 **Three Arthropods That Transmit Bacterial Disease.** (**a**) The tick (Ixodes). (**b**) The body louse *(Pediculus)*. (**c**) The flea (Xenopsylla).

Mountain spotted fever, it appears first on the trunk of the body and progresses to the extremities, accompanied by intense fever.

Other Rickettsial Diseases

Are there any other rickettsial diseases I should be aware of?

Several other rickettsial diseases occur sporadically. Endemic typhus, for example, is a mild form of typhus fever transmitted by fleas (**FIGURE 19.13**). The disease is usually characterized by a mild fever, persistent headache, and a maculopapular rash. Rickettsialpox resembles chickenpox and is transmitted by mites. Trench fever, a rickettsial illness, was the most widespread disease among troops during World War I. An estimated 1 million soldiers are thought to have been infected. Transmission is by head and body lice.

I've recently heard of cases of ehrlichiosis. Is that a new disease?

Ehrlichiosis was first described in humans in 1986. Formerly believed to be confined to dogs, ehrlichiosis has been recognized in two forms in humans: human monocytic ehrlichiosis (HME), which affects the body's monocytes (a type of white blood cell); and human granulocytic ehrlichiosis (HGE), which affects the body's neutrophils (another type of white blood cell). Both forms are transmitted by ticks, and both are accompanied by headache, malaise, and fever, with some liver disease and, infrequently, a maculopapular rash. Antibiotics are effective as therapy.

Sexually Transmitted Diseases

The sexually transmitted diseases (STDs) belong to a broad category of diseases transmitted by contact. The contact in this case is with the reproductive organs. Person-to-person transmission is necessary for bacterial survival because the microbes usually cannot remain alive outside the body tissues. The list of STDs is diverse, as the fol-

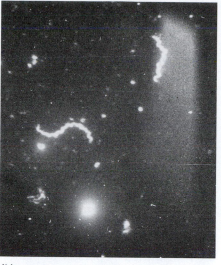

(a)   (b)

FIGURE 19.14 *Treponema pallidum.* Two views of *Treponema pallidum*, the agent of syphilis.
(**a**) A scanning electron micrograph view. The thin threads are axial filaments extending along the periphery of the spirochete. (**b**) A dark-field microscope view of the spirochetes seen in a sample taken from the chancre of a patient. Courtesy of the CDC.

lowing examples will demonstrate. Incidentally, STDs are what used to be called VD, short for "venereal disease" (after Venus, the goddess of love). Health agencies made the switch to STD to call attention to the mode of transmission for the diseases and to avoid the connotation of love.

Syphilis

In some books, syphilis is called the Great Pox. Why is that?

Over the centuries, Europeans have had to contend with four pox diseases: chickenpox, cowpox, smallpox, and the Great Pox, a disease now known as syphilis. The first European epidemic was recorded in the late 1400s, shortly after the conquest of Naples by the French army. Syphilis resembles smallpox, but the pocks are larger and more disfiguring.

What causes syphilis, and what's the disease like?

Syphilis is caused by *Treponema pallidum*, literally, the "pale spirochete" (depicted in **FIGURE 19.14**). The spiral bacteria penetrate the skin surface and cause a disease that progresses in three stages: Primary syphilis, the first stage, is characterized by the chancre, a painless, hard, circular, purplish ulcer often on the genital organs. It persists for 2 to 6 weeks, then disappears. Several weeks later, the patient experiences secondary syphilis, with lesions over the entire body surface, fever, rash, and a patchy loss of hair on the head. Recovering patients bear pitted scars from the lesions and remain "pockmarked." Tertiary syphilis is next. Its hallmark is the gumma, a soft, gummy granular lesion that weakens the blood vessels, causing them to bulge and burst. In the nervous system, gummas lead to paralysis and insanity.

Can syphilis be treated?

Yes, penicillin is the drug of choice for the primary and secondary stages of the disease, but antibiotics are ineffective in tertiary syphilis because gummas appear to be an immunological response to the spirochetes. The cornerstone of syphilis control is the identification and treatment of the sexual contacts of patients. In the United States, about 45,000 people are afflicted with the disease annually. Syphilis is also a serious problem in pregnant women because the spirochetes penetrate the placental barrier.

Gonorrhea

How common is gonorrhea in the United States?

As of 2001 gonorrhea was the second most frequently reported microbial disease in the United States (after chlamydia), with several hundred thousand cases reported annually. The disease is sometimes called "the clap," from the French *clappoir* for "brothel."

What's the cause of gonorrhea?

Gonorrhea is caused by *Neisseria gonorrhoeae*, a small Gram-negative diplococcus commonly known as the gonococcus. The great majority of cases of gonorrhea are transmitted by person-to-person contact during sexual intercourse.

Is gonorrhea a deadly disease?

Not really. In women, the gonococci invade the cervix and the urethra. Patients often report a discharge, abdominal pain, and a burning sensation on urination. In some women, gonorrhea also spreads to the Fallopian tubes, and these thin passageways become riddled with adhesions, causing a difficult passage for egg cells. In men, gonorrhea occurs primarily in the urethra. Onset is usually accompanied by a tingling sensation in the penis, followed in a few days by pain when urinating. There is also a thin, watery discharge at first, and later a whitened, thick fluid that resembles semen.

Can gonorrhea occur in other organs?

Gonorrhea does not restrict itself to the urogenital organs. Gonococcal pharyngitis, for example, may develop in the pharynx; in infants born to infected women, gonococci may cause a disease of the eyes called gonococcal ophthalmia. To preclude the blindness that may ensue, most states have laws requiring that the eyes of newborns be treated with silver nitrate or antibiotics. In adults, gonorrhea therapy consists of antibiotic treatment.

Chlamydia

I've heard that chlamydia is the most reported infectious disease in the United States. Is that true?

As of 2001, over 500,000 cases of chlamydia were occurring in Americans annually (FIGURE 19.15). Chlamydia is a gonorrhealike disease transmitted by sexual contact. Its symptoms are remarkably similar to those of gonorrhea, although somewhat milder. The causative agent is *Chlamydia trachomatis*, a species of chlamydiae.

■ *Chlamydia trachomatis*
kla-mi′dē-ä
trä-kō′mä-tis

What are the typical symptoms of chlamydia?

Women suffering from chlamydia often note a slight vaginal discharge, as well as inflammation of the cervix. Burning pain is also experienced on urination. In complicated cases, the disease may spread to the Fallopian tubes, causing adhesions that block the passageways. In men, chlamydia is characterized by painful urination and a discharge that is more watery and less copious than that of gonorrhea. Tingling sensations in the penis are generally evident, and inflammation of the epididymis (the tube leading from the testis) may result in sterility.

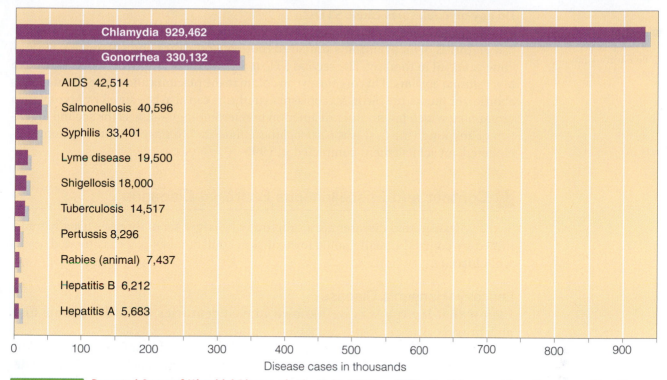

FIGURE 19.15 **Reported Cases of Microbial Diseases in the United States, 2004.** For this year, chlamydia was considerably more prevalent than the next most common reportable disease, gonorrhea.

Is chlamydia a problem in newborns?

Yes, a newborn may contract *C. trachomatis* from an infected mother and develop a disease of the eyes known as chlamydial ophthalmia. Chlamydial pneumonia may also develop in a newborn. Health officials estimate that each year in the United States, over 75,000 newborns suffer chlamydial ophthalmia and 30,000 newborns experience chlamydial pneumonia. The disease may be successfully treated with tetracycline or erythromycin.

Other Sexually Transmitted Diseases

Which other sexually transmitted diseases bear remembering?

A number of other STDs occur sporadically and should be mentioned. Chancroid, for example, is common in tropical climates. The causative agent is *Haemophilus ducreyi*, a small Gram-negative rod. It causes a painful, shallow, saucer-shaped ulcer that bleeds easily. Lymphogranuloma venereum (LGV) is caused by a variant of *Chlamydia trachomatis*. The disease is accompanied by fever, malaise, and swelling and tenderness in the inguinal lymph nodes. Granuloma inguinale remains an endemic problem in tropical and subtropical areas of the world. It is caused by *Calymmatobacterium granulomatis*, a small Gram-negative bacillus that produces a granular ulcer that bleeds easily. Vaginitis is a general term for various mild infections of the vagina and sometimes the vulva. One cause of vaginitis is the Gram-negative rod *Gardnerella vaginalis*. A foul-smelling discharge is the most prominent symptom of this illness.

■ *Calymmatobacterium*
ka-lim-mä-to-bak-ti′rē-um

■ *Ureaplasma urealyticum*
ū-rē-ă-plaz′mä
ū-rē-ă-lit′i-kum

Do any mycoplasmas cause STDs?

Yes, two STDs are related to mycoplasma. *Ureaplasma urealyticum* causes ureaplasmal urethritis, with symptoms similar to those of gonorrhea and chlamydia; tetracycline is useful in therapy. The second STD is mycoplasmal urethritis, a disease due to *Mycoplasma hominis*. This organism can colonize the placenta and cause spontaneous abortion or premature birth. Mycoplasmal urethritis can also be caused by *Mycoplasma genitalium*, which has another distinction in microbiology—it was the second organism of any kind (the first was *Haemophilus influenzae*) to have its entire genome deciphered, an achievement completed in 1995.

■ Contact and Miscellaneous Bacterial Diseases

Sexually transmitted diseases are a subgroup of a larger set of diseases usually transmitted by contact. Usually, some form of skin contact takes place, as these diseases will illustrate.

Leprosy (Hansen's Disease)

I know that leprosy was well-known in past centuries. How did people look on leprosy?

For many centuries, leprosy was considered a curse of the damned. It did not kill, but neither did it seem to end. Instead, it lingered for years, causing the tissues to degenerate and deforming the body. In biblical times, the afflicted were required to call out "Unclean! Unclean!" and usually were ostracized from the community. Among the more heroic stories of medicine is that of the work of Father Damien de Veuster, the Belgian priest who in 1870 established a hospital for leprosy patients on the island of Molokai in Hawaii.

Which bacterium causes leprosy?

The agent of leprosy is *Mycobacterium leprae*, an acid-fast rod. It is referred to as Hansen's bacillus, and leprosy is commonly called Hansen's disease. Ironically, *M. leprae* has not yet been cultivated in artificial laboratory media, and live animals must be used to propagate it.

How is leprosy spread, and what do patients experience?

Leprosy is spread by multiple skin contacts, as well as by droplets from the upper respiratory tract. The disease has an unusually long incubation period of 3 to 6 years. Patients with leprosy experience disfiguring of the skin and bones, twisting of the limbs, and curling of the fingers. The largest number of deformities develop from the loss of pain sensation due to nerve damage. A sulfur compound known commercially as Dapsone is often used to treat the disease.

Staphylococcal Skin Disease

What causes staphylococcal skin disease, and what are the various forms?

Staphylococcus aureus, the grapelike cluster of Gram-positive cocci, is the species usually involved in staphylococcal skin disease (FIGURE 19.16). The hallmark of disease is the abscess, a circumscribed pus-filled lesion. (A boil is a skin abscess.) Deeper skin abscesses, called carbuncles, develop when the staphylococci work their way

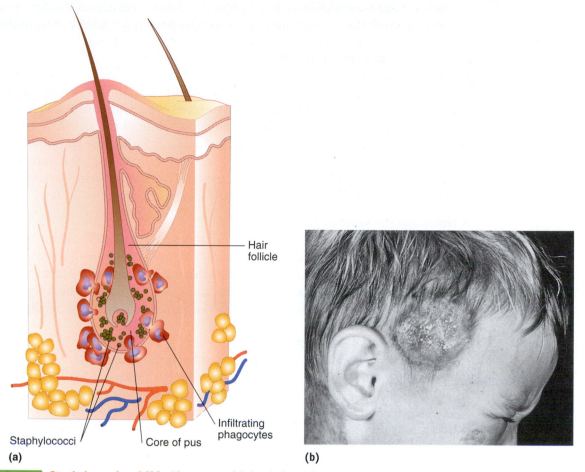

FIGURE 19.16 **Staphylococci and Skin Abscesses.** (**a**) Staphylococci at the base of a hair follicle during the development of a skin lesion. Phagocytes (white blood cells that engulf bacteria) have begun to collect at the site as pus accumulates, and the skin has started to swell. (**b**) A severe abscess on the head of a young boy. Abscesses often begin as trivial skin pimples and boils, but they can become serious as staphylococci penetrate to the deeper tissues.

into the tissues below the skin. In impetigo contagiosum, the infection is more superficial and involves patches of epidermis just below the outer skin layer.

How is staphylococcal skin disease spread, and can it be treated?

Skin contact with an infected individual is the usual mode of transmission. Staphylococcal diseases are commonly treated with penicillin, but resistant strains of *S. aureus* are well known. This problem was brought to public awareness during the 1990s, when reports of antibiotic resistance in staphylococci became widespread. Multidrug-resistant *Staphylococcus aureus* (MRSA) soon appeared in many hospitals, as we discuss in Chapter 10.

Is toxic shock syndrome a staphylococcal disease?

Toxic shock syndrome (TSS) is caused by a toxin-producing strain of *S. aureus*. The earliest symptoms of disease include a rapidly rising fever, accompanied by vomiting and watery diarrhea. Patients then experience a sore throat, severe muscle aches, and a sun-burnlike rash with peeling of the skin, especially on the palms of the hands and soles of the feet. A sudden drop in blood pressure also occurs, leading to shock and heart failure. High-absorbency tampons have been implicated in past outbreaks.

Conjunctivitis and Trachoma

Is conjunctivitis an eye disease?

Yes, it is. Conjunctivitis is a disease of the conjunctiva, the thin membranous covering of the cornea and inner eyelid. When infected, the membrane becomes inflamed, a factor that imparts a brilliant pink color to the white of the eye (hence the name "pinkeye"). Several microbes cause conjunctivitis, including the Gram-negative rod *Haemophilus aegyptius*, which is transmitted by face-to-face contact and airborne droplets, as well as by contaminated instruments. The disease normally runs its course in about 2 weeks. Therapy with antibiotics hastens recovery.

■ *aegyptius*
ē-jip′tē-us

Is trachoma the same as glaucoma?

No, the names only sound similar. Glaucoma is a physiological disease caused by a buildup of fluid in the eye chambers, while trachoma is an infectious disease caused by a variant of *Chlamydia trachomatis*. The chlamydiae multiply in the conjunctiva. A series of tiny, pale nodules forms on this membrane, giving it a rough appearance. Fingers, towels, optical instruments, and face-to-face contact are possible modes of transmission.

Animal Contact Diseases

How important are animals in the spread of bacterial disease?

Actually, quite important. Public health officials estimate that each year in the United States, about 3.5 million people are bitten by animals. Most of these wounds heal without complications, but in certain cases, bacterial disease may develop.

Which animal diseases can you get from a bite?

An important cause of bite infections is *Pasteurella multocida*, a Gram-negative rod. This organism is a common inhabitant of the pharynx of cats and dogs, where it causes a disease called pasteurellosis. In humans, the symptoms of pasteurellosis develop rapidly, with local redness, warmth, swelling, and tenderness at the site of the bite wound. In addition, two different species of bacteria can cause rat-bite fever. One is *Streptobacillus moniliformis*, a Gram-negative rod that occurs in long chains. The other is *Spirillum minor*, a rigid spiral bacterium with polar flagella. Patients experience a lesion at the site of the bite, then a typical triad of fever, arthritislike pain in the large joints, and skin rash.

■ *moniliformis*
mon-i-li-FORM-is

How about cat-scratch disease?

Although cats transmit few diseases to humans, a notable problem is cat-scratch disease (also known as cat-scratch fever). The disease affects an estimated 20,000 Americans each year. Symptoms include a papular or pustular lesion at the site of entry, followed by headache, malaise, and low-grade fever. Swollen lymph glands, generally on the side of the body near the bite, accompany the disease, and antibiotics hasten recovery. The causative agent of cat-scratch disease has not yet been identified with certainty, but a leading candidate is *Bartonella henselae*, a rickettsia.

■ *Bartonella*
bar-to-NEL-ah

Can you also get animal diseases from water?

Yes. One example is leptospirosis, which affects household pets such as dogs and cats, as well as rats, mice, and barnyard animals. Humans acquire it by contact with these animals or from water contaminated with their urine. The agent of leptospirosis is *Leptospira interrogans*, a small spirochete. Patients experience flulike symptoms, such as fever, aches, and muscle weakness. The mortality rate is low, and antibiotics are generally used with success.

Dental Diseases

I know it's not a contact disease, but what about dental caries?

Scientists estimate that there are between 50 billion and 100 billion bacteria in the adult mouth at any time. (To put the number in perspective, consider that about 78 billion people have lived on Earth since the beginning of time.) Dental plaque is essentially a deposit of dense gelatinous material consisting of protein, polysaccharides, and an enormous mass of bacteria. In order for dental caries to develop, three elements must be present: a caries-susceptible tooth with a buildup of plaque; dietary carbohydrates, usually in the form of sucrose (sugar); and acidogenic (acid-producing) plaque bacteria. One of the primary bacterial causes of caries is acidogenic *Streptococcus mutans*. This Gram-positive coccus has a high affinity for the smooth surfaces, pits, and fissures of a tooth. The microbes then ferment dietary carbohydrates to lactic acid, along with smaller amounts of acetic acid, formic acid, and butyric acid. The acids dissolve the calcium compounds of the tooth enamel, and protein-digesting enzymes break down any remaining organic materials. Soon dental caries develops.

How can I protect against dental caries?

Preventing dental caries has three principal thrusts: protecting the tooth, modifying the diet, and combating the plaque bacteria. Tooth protection may be accomplished by the ingestion and topical application of fluorides. These compounds displace hydroxyl ions in a component of tooth enamel called hydroxyapatite, thus reducing the solubility of the enamel. Teeth can also be protected by applying polymers to cover pits and fissures in their surfaces, thereby preventing bacterial adhesion. Diet modification requires minimizing sucrose in foods. Efforts to eliminate the bacteria focus on eliminating the streptococci, as well as stimulating antibody production against the bacteria.

■ hydroxyapatite
HI-drax-e-AP-ah-tite

What about gum disease?

Caries is not the only form of dental disease. The teeth are surrounded by tissues that provide essential support. These tissues, called the periodontal tissues, may be the site of a periodontal disease called acute necrotizing ulcerative gingivitis (ANUG). ANUG is characterized by punched-out ulcers that first appear along the gingival margin and inter-dental papillae and then spread to the soft palate and tonsil areas. A foul odor and bad taste come from gases produced by anaerobic bacteria. As the periodontal tissues decay, the teeth may become loosened and eventually dislodged completely. ANUG has long been associated with conditions such as malnutrition, viral infection, excessive smoking, poor oral hygiene, and mental stress. The disease was common among soldiers in World War I and was known at that time as trench mouth.

Urinary Tract Disease

I heard that urinary tract disease is among the most common diseases in the United States. Is that so?

Yes, it is. According to a study reported in the late-1990s, about 7 million episodes of urinary tract infections occur annually in the United States and this number has risen to 8 million episodes in 2006. Urinary tract infections are the second most frequent cause of visits to the doctor's office (after respiratory tract infections). Sufferers report abdominal discomfort, burning pain on urination, and frequent urges to urinate. Most infections develop in the bladder, and women are apparently infected more often than men.

What causes urinary tract disease, and how can it be avoided?

Among the major causes of these infections are *Proteus mirabilis* and *Escherichia coli*, both Gram-negative rods. Such practices as avoiding tight-fitting clothes and urinating soon after sexual intercourse can reduce the possibility of infection. Studies have also shown that cranberry juice and vitamin C may inhibit the bacteria by increasing the acidity of the urinary tract. Unfortunately, if urinary tract disease is left untreated, the infection can involve the entire bladder or spread to the kidneys.

■ A FINAL THOUGHT

As we conclude this chapter of *Microbes and Society*, you should take a moment to reflect on the vast body of knowledge you have acquired over the many pages of this book. You have undoubtedly become better aware of the microbes and their profound influence on your life. The last two chapters have focused on the negative aspects, but you will certainly also remember the ways in which they affect your life for the better. You have learned an alphabet of microbes and a language of microbiology that will allow you to keep learning the rest of your life. No longer will you bypass a news headline referring to an emerging disease, a biotechnology breakthrough, or a new drug. Now you will be able to read and digest the contents of the story. Then, in using and sharing your new knowledge, you will become a better citizen. And that's what education is all about. Congratulations, you are now a microbiologist.

■ QUESTIONS TO CONSIDER

1. It is a paradox that technology gives us appliances that improve our lives but often provides breeding grounds for microorganisms. For example, *Legionella* probably lived undisturbed for millennia in ponds and lakes; but with the rise of air conditioning, powerful vents swept air over reservoirs of water and picked up droplets containing *Legionella* to disperse to unsuspecting people who breathed the air. How many other examples of "technology" can you relate to the problem of Legionnaires' disease?

2. In 1997, the incidence of pneumonia was recorded in various states of the United States. Massachusetts had a very high rate, but so did Georgia. North Dakota had the third lowest rate, Florida the lowest. The highest rate was in California. Suppose you were an analyst of infectious diseases. What explanation(s) might you offer for these observations?

3. The story is told of a doctor in New York City in the early 1800s who was an expert at diagnosing typhoid fever even before the symptoms of disease appeared. He would go up and down the rows of hospital beds, feeling the

tongues of patients and announcing that a patient was in the early stages of typhoid. Sure enough, a few days later the symptoms would surface. What do you think was the secret to his success?

4. You have volunteered to do the supermarket shopping for the upcoming class barbecue. What are some precautions you can take to ensure that the event is remembered for all the right reasons?

5. Although the tetanus toxin is second in potency to the botulism toxin, many physicians consider tetanus to be a more serious threat than botulism. Would you agree? Why?

6. In Chapter 9 of the Bible, in the Book of Exodus, the sixth plague of Egypt is described in this way: "Then the Lord said to Moses and Aaron, 'Take a double handful of soot from a furnace, and in the presence of Pharaoh, let Moses scatter it toward the sky. It will then turn into a fine dust over the whole land of Egypt and cause festering boils on man and cattle throughout the land.'" Which bacterial disease is possibly being described?

7. A column by Ann Landers carried the following letter: "I am a 34-year-old married woman who is trying to get pregnant, but it doesn't look promising. . . . When I was in college, I became sexually active. I slept with more men than I care to admit. . . . Somewhere in my wild days, I picked up an infection that left me infertile. . . . The doctor told me I have quite a lot of scar tissue inside my Fallopian tubes." The woman, who signed herself "Suffering in St. Louis," went on to implore readers to be careful in their sexual activities. What advice do you think the woman gave to readers?

◼ KEY TERMS

All the names of the bacterial diseases represent the key terms of this chapter. To use the names as a study guide, talk about each one for a few moments, indicating what sort of disease it is, what causes it, how it is transmitted, where it occurs, whether a treatment or vaccine is available, and other information that is relevant to a concise view of the disease.

anthrax
bacterial meningitis
bacterial pneumonia
botulism
brucellosis
campylobacteriosis
cat-scratch disease
chancre
chancroid
chlamydia
cholera
conjunctivitis
dental caries

diphtheria
E. coli diarrheas
ehrlichiosis
gas gangrene
gonorrhea
granuloma inguinale
Legionnaires' disease (legionellosis)
leprosy
leptospirosis
listeriosis
Lyme disease
mycoplasmal urethritis
pasteurellosis

peptic ulcer disease
pertussis
plague
rat-bite fever
relapsing fever
Rocky Mountain spotted fever
salmonellosis
shigellosis
staphylococcal food poisoning
staphylococcal skin diseases
streptococcal disease

syphilis
tetanus
trachoma
tuberculosis
tularemia
typhoid fever
typhus fever
ureaplasmal urethritis
urinary tract infections
vaginitis

http://microbiology.jbpub.com/book/microbes

The site features **eLearning**, an online review area that provides quizzes and other tools to help you study for your class. You can also follow useful links for in-depth information, read more stories of microbiology, or just find out the latest microbiology news.

GLOSSARY

A

abscess   A circumscribed pus-filled lesion characteristic of staphylococcal skin disease; also called a boil.

acid   A substance that releases hydrogen ions (H$^+$) in solution. *See also* base.

acid-fast technique   A staining process in which certain bacteria resist decolorization with acid alcohol.

acidic dye   A negatively charged colored substance in solution that is used to stain an area around cells.

acidophile   A microorganism that grows at acidic pHs below 4.

acquired immunity   A response to a specific immune stimulus that involves immune defensive cells and frequently leads to the establishment of host immunity.

acquired immunodeficiency syndrome (AIDS)   A serious viral disease caused by the human immunodeficiency virus (HIV) in which the T (CD4) lymphocytes are destroyed and opportunistic illnesses occur in the patient.

actinomycete   A soil bacterium that exhibits fungus-like properties when cultivated in the laboratory.

active immunity   The immune system responds to antigen by producing antibodies and specific lymphocytes.

active site   The region of an enzyme where the substrate binds.

acyclovir   A drug used as a topical ointment to treat herpes simplex and injected for herpes encephalitis.

adenoid   A mass of secondary lymphoid tissue at the back of the throat that helps in immune system activities.

adenosine diphosphate (ADP)   A molecule in cells that is the product of ATP hydrolysis.

adenosine triphosphate (ATP)   A molecule in cells that provides most of the energy for metabolism.

adhesin   A protein in bacterial fimbriae that assists in attachment to the surface molecules of cells.

aerobe   An organism that uses oxygen gas (O$_2$) for metabolism.

aerobic respiration   The process for transforming energy to ATP in which the final electron acceptor in the electron transport chain is oxygen gas (O$_2$).

agar   A polysaccharide derived from marine seaweed that is used as a solidifying agent in many microbiological culture media.

alcoholic fermentation   A catabolic process that forms ethyl alcohol during the reoxidation of NADH to NAD$^+$ for reuse in glycolysis to generate ATP.

alga (pl. algae)   An organism in the kingdom Protista that performs photosynthesis.

algaecide   A chemical that kills algae.

algal bloom   An excessive growth of algae on or near the surface of water, often the result of an oversupply of nutrients from organic pollution.

alpha helix   The spiral structure of a polypeptide consisting of amino acids stabilized by hydrogen bonds.

amantadine   A drug for the treatment of influenza, the virus which causes the flu.

amino acid   An organic acid containing one or more amino groups; the monomers that build proteins in all living cells.

aminoglycoside   An antibiotic that contains amino groups bonded to carbohydrate groups that inhibit protein synthesis; examples are gentamicin, streptomycin, and neomycin.

amoeba   A protozoan that undergoes a crawling movement by forming cytoplasmic projections into the environment.

amylase   A group of enzymes that break down starch.

anabolism   An energy-requiring process involving the synthesis of larger organic compounds from smaller ones. *See also* catabolism.

anaerobe   An organism that does not require or cannot use oxygen gas (O$_2$) for metabolism.

anaerobic respiration   The production of ATP where the final electron acceptor is an inorganic molecule other than oxygen gas (O$_2$); examples include nitrate and sulfate.

anaphylactic shock   A life-threatening allergic reaction resulting from to the release of mediators that cause contractions of smooth muscle throughout the body and a drop in blood pressure.

anthrax   An infection with the bacterium *Bacillus anthracis*.

antibiotic   A substance naturally produced by bacteria or fungi that inhibits or kills bacteria.

antibody   A highly specific protein produced by the body in response to a foreign substance, such as a bacterium or virus, and capable of binding to the substance.

anticodon   A three-base sequence on the tRNA molecule that binds to the codon on the mRNA molecule during translation.

antigen   A chemical substance that stimulates the production of antibodies by the body's immune system; also called immunogen.

antigen binding site   The region on an antibody that binds to an antigen.

antigenic determinant   A section of an antigen molecule that stimulates antibody formation and to which the antibody binds; also called epitope.

antihistamine   A drug that blocks cell receptors for histamine, preventing allergic effects such as sneezing and itching.

antimicrobial agent (drug)   A chemical that inhibits or kills the growth of microorganisms.

antisepsis   The use of chemical methods for eliminating or reducing microorganisms on the skin.

antiseptic   A chemical used to reduce or kill pathogenic microorganisms on a living object, such as the surface of the human body.

antiserum (pl. antisera)   A blood-derived fluid containing antibodies and used to provide immunity.

antitoxin   An antibody produced by the body that circulates in the bloodstream to provide protection against toxins by neutralizing them.

antiviral protein   A protein made in response to interferon and that blocks viral replication.

apicomplexan   A protozoan containing a number of organelles at one end of the cell that are used for host penetration; no motion is observed in adult forms.

Archaea   The domain of living organisms that excludes the *Bacteria* and *Eukarya*.

archaebacterium   The former term for a unicellular organism in the domain *Archaea*.

arthrospore   An asexual fungal spore formed by fragmentation of a septate hypha.

ascocarp (pl. ascomata)   The fruiting body of an ascomycete fungus.

ascomycete   A division of fungi which produce spores in a distinctive type of sporangium or sac called an ascus.

ascospore   A sexually produced fungal spore formed by members of the ascomycetes.

ascus (pl. asci)   A saclike structure containing ascospores; formed by the ascomycetes.

asexual reproduction   The form of reproduction that maintains genetic constancy while increasing cell numbers.

asthma   A period of wheezing and stressed breathing resulting from a type I hypersensitivity reaction taking place in the respiratory tract.

asymptomatic   Without obvious indications of infection or disease.

atom   The smallest portion into which an element can be divided and still enter into a chemical reaction.

atomic nucleus   The positively charged core of an atom, consisting of protons and neutrons that make up most of the mass.

attenuate   To reduce the ability of a bacterium or virus to do damage to the exposed individual.

autoclave   An instrument used to sterilize microbiological materials by means of high temperature using steam under pressure.

autotroph   An organism that uses carbon dioxide (CO_2) as a carbon source; *see also* chemoautotroph and photoautotroph.

avirulent   Referring to an organism that is not likely to cause disease.

axial filament   *See* endoflagellum.

B

bacillé Calmette Guérin (BCG)   A strain of attenuated *Mycobacterium bovis* used for immunization against tuberculosis and, on occasion, leprosy.

bacillus (pl. bacilli)   (1) Any rod-shaped prokaryotic cell. (2) When referring to the genus *Bacillus*, it refers to an aerobic or facultatively anaerobic, rod-shaped, endospore-producing, Gram-positive bacterial cell.

bacitracin   An antibiotic derived from a *Bacillus* species, effective against Gram-positive bacteria when used topically.

Bacteria   The domain of living things that includes all organisms not classified as *Archaea* or *Eukarya*.

bacteriocin   One of a group of bacterial proteins toxic to other bacterial cells.

bacteriology   The scientific study of prokaryotes; originally used to describe the study of bacteria.

bacteriophage (phage)   A virus that infects and replicates within bacterial cells.

bacteriostatic   Any substance that prevents the growth of bacteria.

bacterium (pl. bacteria)   A single-celled microorganism lacking a cell nucleus and membrane-enclosed compartments, and often having peptidoglycan in the cell wall.

bacteroides   Gram-negative anaerobic rod-shaped bacteria which are unusual because their plasma membranes contain sphingolipids.

barophile   A microorganism that lives under conditions of high atmospheric pressure.

basal body   A structure at the base of a bacterial flagellum consisting of a central rod and set of enclosing rings.

base   A chemical compound that accepts hydrogen ions (H^+) in solution. *See also* acid.

basic dye   A positively charged colored substance in solution that is used to stain cells.

basidiocarp   The fruiting body of a basidiomycete fungus.

basidiomycete   A division of fungi that produce spores in a club-shaped structure called a basidium.

basidiospore   A sexually-produced fungal spore formed by members of the basidiomycetes.

basidium (pl. **basidia**)   A club-like structure containing basidiospores; formed by the basidiomycetes.

basophil   A type of white blood cell with granules that functions in allergic reactions.

B cell   *See* B lymphocyte.

benign   Referring to a tumor that usually is not life threatening or likely to spread to another part of the body.

beta-lactam nucleus   The chemical group central to all penicillin antibiotics.

binary fission   An asexual process in prokaryotic cells by which a cell divides to form two new cells while maintaining genetic constancy.

binomial system   The method of nomenclature that uses two names (genus and specific epithet) to refer to organisms.

biochemical oxygen demand (BOD)   A number referring to the amount of oxygen used by the microorganisms in a sample of water during a 5-day period of incubation.

biofilm   A complex community of microorganisms that form a protective and adhesive matrix that attaches to a surface, such as a catheter or industrial pipeline.

bioinformatics   The use of computers and statistical techniques to manage and analyze biological information, especially nucleotide sequences in genes.

biological vector   An infected arthropod, such as a mosquito or tick, that transmits disease-causing organisms between hosts. *See also* mechanical vector.

bioluminescence   The emission of light by a living organism in which the light is produced by an enzyme-catalyzed chemical reaction during which chemical energy is converted to light energy.

bioreactor   A large fermentation tank for growing microorganisms used in industrial production; also called a fermentor.

bioremediation   The use of microorganisms to degrade toxic wastes and other synthetic products of industrial pollution.

biosphere   The areas of the earth that are inhabited by living organisms.

biotechnology   The commercial application of genetic engineering using living organisms.

bioterrorism   The intentional or threatened use of biological agents to cause fear in or actually inflict death or disease upon a large population.

blanching   A process of putting food in boiling water for a few seconds to destroy enzymes.

bloom   An excessive growth of algae or cyanobacteria on or near the surface of water, often the result of an oversupply of nutrients from organic pollution.

B lymphocyte (B cell)   A white blood cell that matures into memory cells and plasma cells that secrete antibody.

boil   *See* abscess.

bone marrow   A soft reddish substance inside some bones that is involved in the production of blood cells.

booster shot   A repeat dose of a vaccine given some years after the initial course to maintain a high level of immunity.

Bordeaux mixture   A combination of hydrated lyme and copper sulfate, first developed in the vineyards of the Bordeaux region of France, and used mainly to control agricultural infestations of fungus.

bright-field microscope   An instrument that magnifies an object by passing visible light directly through the lenses and object. *See also* light microscope.

broad spectrum   Referring to an antimicrobial drug useful for treating many groups of microorganisms, including Gram-positive and Gram-negative bacteria; *see also* narrow-spectrum.

broth   A liquid containing nutrients for the growth of microorganisms.

bubo   A swelling of the lymph nodes due to inflammation.

budding   An asexual process of reproduction in fungi, in which a new cell forms as a swelling at the border of the parent cell and then breaks free to live independently.

buffer   (1) A compound that minimizes pH changes in a solution by neutralizing added acids and bases. (2) Refers to a solution containing such a substance.

C

cancer   A disease characterized by the radiating spread of malignant cells that reproduce at an uncontrolled rate.

capsid   The protein coat that encloses the genome of a virus.

capsomere   Any of the protein subunits of a capsid.

capsule   A layer of polysaccharides and small proteins covalently bound some prokaryotic cells.

carbohydrate   An organic compound consisting of carbon, hydrogen, and oxygen that is an important source of carbon and energy for all organisms; examples include sugars, starch, and cellulose.

carbolic acid   *See* phenol.

carbon cycle   A series of interlinked processes involving carbon compound exchange between living organisms and the nonliving environment.

carbon-fixing reactions   The stage of photosynthesis where electrons and ATP are used to reduce carbon dioxide gas (CO_2) to sugars.

carbuncle   An enlarged abscess formed from the union of several smaller abscesses or boils.

carcinogen   Any physical or chemical substance that causes cancer.

carrier   An individual who has recovered from a disease but retains the infectious agent in the body and continues to shed them.

casein   The major protein in milk.

catabolism   An energy-liberating process in which larger organic compounds are broken down into smaller ones. *See also* anabolism.

cation   A positively charged ion. *See also* anion.

cell culture   The process of growing cells (either prokaryotic or eukaryotic) in the control conditions of a laboratory.

cell envelope   The cell wall and cell membrane of a prokaryotic cell.

cell line   A group of identical cells in culture and derived from a single cell.

cell-mediated immune response   The body's ability to resist infection through the activity of T-lymphocyte recognition of antigen peptides presented on macrophages and dendritic cells and on infected cells.

cell membrane   A thin bilayer of phospholipids and proteins that surrounds the prokaryotic cell cytoplasm. *See also* plasma membrane.

cell theory   The tenet that all organisms are made of cells and arise from preexisting cells.

cellulase   An enzyme produced by fungi, bacteria, and protozoans that is used to degrade and metabolize cellulose.

cellulose   A polysaccharide carbohydrate composed of beta-glucose subunits. It forms the primary structural component of cell walls.

cell wall   A carbohydrate-containing structure surrounding fungal, algal, and most prokaryotic cells.

Centers for Disease Control and Prevention (CDC)   An agency of the U.S. Department of Health and Human Services that, among other things, focuses on infectious disease prevention and control to improve the health of the citizens of the United States.

central dogma   The doctrine that DNA codes for RNA through transcription and RNA is converted to protein through translation.

cephalosporin   An antibiotic derived from the mold *Cephalosporium* that inhibits cell wall synthesis in Gram-positive bacteria and certain Gram-negative bacteria.

chancre   A painless, circular, purplish hard ulcer with a raised margin that occurs during primary syphilis.

chemical bond   A force between two or more atoms that tends to bind those atoms together.

chemical element   Any substance that cannot be broken down into a simpler one by a chemical reaction.

chemical reaction   A process that changes the molecular composition of a substance by redistributing atoms or groups of atoms without altering the number of atoms.

chemiosmosis   The use of a proton gradient across a membrane to generate cellular energy in the form of ATP.

chemoautotroph   An organism that derives energy from inorganic chemicals and uses the energy to synthesize nutrients from carbon dioxide gas (CO_2).

chemoheterotroph   An organism that derives energy from organic chemicals and uses the energy to synthesize nutrients from carbon compounds other than carbon dioxide gas (CO_2).

chemolithotroph   A microorganism which uses an inorganic molecules (usually of mineral origin) to gain the energy to synthesize organic molecules (e.g. carbon dioxide fixation).

chitin   A polymer of acetylglucosamine units that provides rigidity in the cell walls of fungi.

chlamydia (pl. **chlamydiae**)   A very small, round pathogenic bacterium visible only with the electron microscope and cultivated within living cells.

chloramphenicol   A broad-spectrum antibiotic derived from a *Streptomyces* species that interferes with protein synthesis.

chlorination   The process of treating water with chlorine to kill harmful organisms.

chlorophyll   A green or purple pigment in algae and some bacterial cells that functions in capturing light for photosynthesis.

chloroplast   A double membrane-enclosed compartment in algae that contains chlorophyll and other pigments for photosynthesis.

chocolate agar   A culture medium consisting of a nutritious base and whole blood; the medium is heated to disrupt the blood cells and release the hemoglobin.

chromosome   A structure in the nucleoid or cell nucleus that carries hereditary information in the form of genes.

chronic disease   A disease that develops slowly, tends to linger for a long time, and requires a long convalescence.

ciliate   A protozoan that moves with the aid of cilia.

cilium (pl. **cilia**)   A hair-like projection on some eukaryotic cells that along with many others assist in the motion of some protozoa and beat rhythmically to aid the movement of a fluid past cell of the respiratory epithelial cells in humans.

citric acid cycle   *See* Krebs cycle.

clone   A population of cells genetically identical to the parent cell.

coagulase   An enzyme produced by some staphylococci that catalyzes the formation of a fibrin clot.

coccus (pl. **cocci**)   A spherical-shaped prokaryotic cell.

codon   A three-base sequence on the mRNA molecule that specifies a particular amino acid insertion in a polypeptide.

coenocytic   Referring to a fungus containing no septa (cross-walls) and multinucleate hyphae.

coenzyme   A small, organic molecule that forms the nonprotein part of an enzyme molecule; together they form the active enzyme.

coenzyme A (CoA)   A small, organic molecule of cellular respiration that functions in release of carbon dioxide gas (CO_2) and the transfer of electrons and protons to another coenzyme.

cold sore   A herpes-induced blister that may occur on the lips, gums, nose, and adjacent areas.

coliform bacterium   A Gram-negative, non-spore-forming, rod-shaped cell that ferments lactose to acid and gas and usually is found in the human and animal intestine; high numbers in water is an indicator of contamination.

colony   A visible mass of microorganisms of one type.

colostrum   The yellowish fluid rich in antibodies secreted from the mammary glands of animals or humans prior to the production of true milk.

commercial sterilization   A canning process to eliminate the most resistant bacterial spores.

communicable disease   A disease that is readily transmissible between hosts.

competence   Referring to the ability of a cell to take up naked DNA from the environment.

complement   A group of blood proteins that functions in a cascading series of reactions with antibodies to recognize and help eliminate certain antigens or infectious agents.

compound   A substance made by the combination of two or more different chemical elements.

compound microscope   *See* light microscope.

conidiophore   The supportive structure on which conidia form.

conidium (pl. **conidia**)   An asexually produced fungal spore formed on a supportive structure without an enclosing sac.

conjugation   (1) In prokaryotes, a unidirectional transfer of genetic material from a live donor cell into a live recipient cell during a period of cell contact. (2) In the protozoan ciliates, a sexual process involving the reciprocal transfer of micronuclei between cells in contact.

conjugation pilus (pl. **pili**)   A hollow projection for DNA transfer between the cytoplasms of donor and recipient bacterial cells.

conjunctivitis   A general term for disease of the conjunctiva, the thin mucous membrane that covers the cornea and forms the inner eyelid; also called pinkeye.

contagious   Referring to a disease whose agent passes with particular ease among hosts.

contractile vacuole   A membrane-enclosed structure within a cell's cytoplasm that regulates the water content by absorbing water and then contracting to expel it.

contrast   In microscopy, to be able to see an object against the background.

corticosteroid   A synthetic drug used to control allergic disorders by blocking the release of chemical mediators.

covalent bond   A chemical bond formed by the sharing of electrons between atoms or molecules.

critical control point   In the food processing industry, a place where contamination of the food product could occur.

culture medium   A mixture of nutrients in which microorganisms can grow.

cyanobacterium (pl. **cyanobacteria**)   An oxygen-producing, pigmented bacterium occurring in unicellular and filamentous forms that carries out photosynthesis.

cyst   A dormant and very resistant form of a protozoan and multicellular parasite.

cystitis   An inflammation of the urinary bladder.

cytochrome   A compound containing protein and iron that plays a role as an electron carrier in cellular respiration and photosynthesis; *see also* electron transport chain.

cytokine   Small proteins released by immune defensive cells that affects other cells and the immune response to an infectious agent.

cytoplasm   The complex of chemicals and structures within a cell; in plant and animal cells excluding the nucleus.

cytoskeleton   (1) The structural proteins in a prokaryotic cell that help control cell shape and cell division. (2) In a eukaryotic cell, the internal network of protein filaments and microtubules that control the cell's shape and movement.

cytosol   The fluid, ions, and compounds of a cell's cytoplasm excluding organelles and other structures.

cytotoxic T cell   The type of T lymphocyte that searches out and destroys infected cells.

D

dander   Particles of animal skin, hair, or feathers that may contain materials that cause allergic reactions.

dapsone   A chemotherapeutic agent used to treat leprosy patients.

dark-field microscopy   An optical system on the light microscope that scatters light such that the specimen appears white on a black background.

deamination   A biochemical process in which amino groups are enzymatically removed from amino acids or other organic compound.

decline phase   The final portion of a bacterial growth curve in which environmental factors overwhelm the population and induce death; also called death phase.

decomposer   An organism, such as a bacterium or fungus, that recycles dead or decaying matter.

dedifferentiation   A cellular process in which a cell reverts to an earlier developmental stage.

dehydration synthesis   A process of bonding two molecules together by removing the products of water and joining the open bonds.

denaturation   A process caused by heat or pH in which proteins lose their function due to changes in their molecular structure.

denitrification   The process of reducing nitrate and nitrite into gaseous nitrogen, which is far less accessible to life forms but makes up the bulk of our atmosphere.

deoxyribonucleic acid (DNA)   The genetic material of all cells and many viruses.

dermatophyte   A pathogenic fungus that affects the skin, hair, or nails.

desensitization   A process in which minute doses of antigens are used to remove antibodies from the body tissues to prevent a later allergic reaction.

desiccation   The process through which things are made to be extremely dry by removing water.

detergent   A synthetic cleansing substance that dissolves dirt and oil.

diabetes   A disorder that causes the body to produce an excessive amount of urine; type I is an autoimmune disorder caused by a lack of insulin production.

diarrhea   Excessive loss of fluid from the gastrointestinal tract.

diatomaceous earth   Filtering material composed of the remains of diatoms.

diatom   One of a group of microscopic marine algae that performs photosynthesis.

differential medium   A growth medium in which different species of microorganisms can be distinguished visually.

differential stain technique   A procedure using two dyes to differentiate cells or cellular objects based on their staining; *see also* simple stain technique.

dikaryon   A fungal cell in which two genetically different haploid nuclei closely pair.

dikaryotic   The condition of possessing a nuclear feature which is unique to the fungi, in which two nuclei cohabitate the cytoplasm of two fused cells in which the paired nuclei divide synchronously during cell growth.

dimorphic   Referring to pathogenic fungi that take a yeast form in the human body and a filamentous form when cultivated in the laboratory.

dinoflagellate   A microscopic photosynthetic marine alga that forms one of the foundations of the food chain in the ocean.

diplococcus (pl. diplococci)   A pair of spherical-shaped prokaryotic cells.

diplomonad   A protozoan that contains four pair of flagella, two haploid nuclei, and live in low oxygen or anaerobic environments; most members are symbiotic in animals.

direct contact   The form of disease transmission involving close association between hosts; *see also* indirect contact.

disaccharide   A sugar formed from two single sugar molecules; examples include sucrose and lactose.

disease   Any change from the general state of good health.

disinfectant   A chemical used to kill or inhibit pathogenic microorganisms on a lifeless object such as a tabletop.

disinfection   The process of killing or inhibiting the growth of pathogens.

DNA   *See* Deoxyribonucleic acid

DNA fingerprinting   A technique used to distinguish between and identify individuals of the same species using samples of their DNA. DNA fingerprinting is used in forensic science, to match samples of blood, hair, saliva or semen found at crime scenes to specific suspects.

DNA ligase   An enzyme that binds together DNA fragments.

DNA polymerase   An enzyme that catalyzes DNA replication by combining complementary nucleotides to an existing strand.

DNA probe   A short segment of single stranded DNA used to locate a complementary strand among many other DNA strands.

DNA replication   The process of copying the genetic material.

DNA vaccine   A preparation that consists of a DNA plasmid containing the gene for a pathogen protein.

domain   (1) The most inclusive taxonomic level of classification; consists of the *Archaea, Bacteria*, and *Eukarya*. (2) A loop of DNA consisting of about 10,000 bases.

double helix   The structure of DNA, in which the two complementary strands are connected by hydrogen bonds between complementary nitrogenous bases and wound in opposing spirals.

droplet   An airborne particle of mucus and sputum from the respiratory tract that contains disease-causing microorganisms.

dwarfism   A pathological condition in which the physical size of an organism is well below normal.

dysentery   A condition marked by frequent, watery stools, often with blood and mucus.

E

ecosystem   The collection of living and nonliving components and processes that comprise, and govern the behavior of some defined subset of the biosphere (the outermost part of the earth's shell which includes air, land, surface rocks and water.

electron   A negatively charged particle with a small mass that moves around the nucleus of an atom.

electron microscope   An instrument that uses electrons and a system of electromagnetic lenses to produce a greatly magnified image of an object. *See also* transmission electron microscope *and* scanning electron microscope.

electron transport system   A series of proteins that transfer electrons in cellular respiration to generate ATP.

electrophoresis   A laboratory technique involving the movement of charged organic molecules through an electrical field; used to separate DNA fragments in diagnostic procedures.

ELISA   *See* enzyme-linked immunosorbent assay.

encapsulated   Referring to a prokaryotic cell surrounded by a capsule.

encephalitis   Inflammation of the tissue of the brain or infection of the brain.

endemic   Referring to a disease that is constantly present in a specific area or region.

endocarditis   An inflammation of the membranous lining of the heart's cavities.

endoflagellum   A microscopic fiber located along cell walls in certain species of spirochetes; contractions of the filaments yield undulating motion in the cell; also called axial filament.

endonucleases   A class of enzymes that cleave the phosphodiester bond within a polynucleotide chain of DNA and as a result cut the DNA strand. Restriction endonucleases (Restriction Enzymes) cut DNA at specific sites in the nucleotide sequence.

endoplasmic reticulum   A network of membranous plates and tubes in the eukaryotic cell cytoplasm responsible for the synthesis and transport of materials from the cell.

endospore   An extremely resistant dormant cell produced by some Gram-positive bacteria.

endosymbiosis theory   This theory explains the origins of organelles in eukaryotic cells as the result of bacteria engulfing other prokaryotic cells through endophagocytosis. These cells with the bacteria trapped inside them entered a symbiotic relationship (symbiosis).

endotoxin   A metabolic poison, produced chiefly by Gram-negative bacteria, that are part of the bacterial cell wall and consequently are released on cell disintegration; composed of lipid-polysaccharide-peptide complexes.

enriched medium   A growth medium in which special nutrients must be added to get an species to grow.

enterotoxin   A toxin that is active in the gastrointestinal tract of the host.

envelope   The flexible membrane of protein and lipid that surrounds many types of viruses.

enzyme   A reusable protein molecule that brings about a chemical change while itself remaining unchanged.

enzyme-linked immunosorbent assay (ELISA)   A serological test in which an enzyme system is used to detect an individual's exposure to a pathogen.

enzyme-substrate complex   The association of an enzyme with its substrate at the active site.

epidemic   Referring to a disease that spreads more quickly and more extensively within a population than normally expected.

epidemiology   The scientific study of the source, cause, and transmission of disease within a population **epitope**. *See* antigenic determinant.

ergotism   A disease caused by the transfer of a toxin produced by the fungus *Claviceps purpurea* from rye grain to humans.

erythema   A zone of redness in the skin due to a widening of blood vessels near the skin surface.

erythema chronicum migrans (ECM)   An expanding circular red rash that occurs on the skin of patients with Lyme disease.

Eukarya   The taxonomic domain encompassing all eukaryotic organisms.

eukaryote   An organism whose cells contain a cell nucleus with multiple chromosomes, a nuclear envelope, and membrane-bound compartments; *see also* prokaryote.

eukaryotic   Referring to a cell or organism containing a cell nucleus with multiple chromosomes, a nuclear envelope, and membrane bound compartments.

exon   Any region of a gene that is transcribed and retained in the final messenger RNA product, in contrast to an intron, which is spliced out from the transcribed RNA molecule.

exotoxin   A bacterial metabolic poison composed of protein that is released to the environment; in the human body, it can affect various organs and systems.

extreme halophile   An archaeal organism that grows at very high salt concentrations.

extremophile   A microorganism that lives in extreme environments, such as high temperature, high acidity, or high salt.

F

facultative   Referring to an organism that grows in the presence or absence of oxygen gas (O_2).

FAD   *See* flavin adenine dinucleotide.

fermentation   A metabolic pathway in which carbohydrates serve as electron donors, the final electron acceptor is not oxygen gas (O_2), and NADH is reoxidized to NAD^+ for reuse in glycolysis for generation of ATP; *see also* industrial fermentation.

fever   An abnormally high body temperature that is usually caused by a bacterial or viral infection.

flagellates   Cells with one or more whip-like organelles called flagella.

flagellin   One of the proteins forming the filament of the bacterial flagellum.

flagellum (pl. **flagella**)   A long, hair-like appendage composed of protein and responsible for motion in microorganisms; found in some bacteria, protozoa, algae, and fungi.

flash method   A treatment in which milk is heated at 71.6°C for 15 seconds and then cooled rapidly to elimi-

nate harmful bacteria; also called HTST ("high temperature, short time") method.

flavin adenine dinucleotide (FAD)   A coenzyme that functions in electron transfer during oxidative phosphorylation.

fluid mosaic model   The representation for the cell (plasma) membrane where proteins "float" within or on a bilayer of phospholipid.

fluorescence   The emission of one color of light after being exposed to light of another wavelength.

fluorescence microscopy   An optical system on the light microscope that uses ultraviolet light to excite dye-containing objects to fluoresce.

folic acid   The organic compound in bacteria whose synthesis is blocked by sulfonamide drugs.

folliculitis   An inflammation of one or more hair follicles, producing small boils.

fomite   An inanimate object, such as clothing or a utensil, that carries disease organisms.

food vacuole   A membrane-enclosed compartment in some eukaryotic that results from the intake of large molecules, particles, or cells, for digestion.

foraminiferan   A shell-containing amoeboid protozoan having a chalky skeleton with window-like openings between sections of the shell.

formalin   A solution of formaldehyde used as embalming fluid, in the inactivation of viruses, and as a disinfectant.

freeze drying   *See* lyophilization.

fruiting body   The general name for a reproductive structure of a fungus from which spores are produced.

Fungi   One of the five kingdoms in the Whittaker classification of living organisms; composed of the molds and yeasts.

fungicide   Any agent that kills fungi.

G

gamma globulin   A general term for antibody rich serum.

gangrene   A physiological process in which the enzymes from wounded tissue digest the surrounding layer of cells, inducing a spreading death to the tissue cells.

gastroenteritis   Infection of the stomach and intestinal tract often due to a virus.

gene   A segment of a DNA molecule that provides the biochemical information for a function product.

gene linkage map   A genetic map which tells you the positions of genes in relation to each other based on the frequency of crossing overs.

gene probe   A short piece of single stranded DNA which is radioactively labeled that will hybridize with or bind to a complementary target DNA or RNA strand.

generalized transduction   A process by which a bacteriophage carries a bacterial chromosome fragment from one cell to another. *See also* specialized transduction.

generation time   The time interval for a cell population to double in number.

genetic code   The specific order of nucleotide sequences in DNA or RNA that encode specific amino acids for protein synthesis.

genetic engineering   The use of bacterial and microbial genetics to isolate, manipulate, recombine, and express genes.

genetic recombination   The process of bring together different segments of DNA.

genome   The complete set of genes in a virus or an organism.

genomics   The study of an organism's gene structure and gene function in viruses and organisms.

genus (pl. **genera**)   A rank in the classification system of organisms composed of one or more species; a collection of genera constitute a family.

germ theory   The principle formulated by Pasteur and proved by Koch that microorganisms are responsible for infectious diseases.

gibberellins   First identified in fungi, these are plant growth hormones that promote stem elongation and mobilization of food reserves in seeds and other processes.

glycocalyx   A viscous polysaccharide material covering many prokaryotic cells to assist in attachment to a surface and impart resistance to desiccation. *See also* capsule *and* slime layer.

glycolysis   A metabolic pathway in which glucose is broken down into two molecules of pyruvate with a net gain of two ATP molecules.

golgi   A stack of flattened, membrane-enclosed compartments in eukaryotic cells involved in the modification and sorting of lipids and proteins.

gonococcus   A colloquial name for *Neisseria gonorrhoeae*.

Gram negative   Referring to a bacterial cell that stains red after Gram staining.

Gram-negative cell wall   Bacterial cell walls composed of little peptidoglycan and an outer membrane.

Gram positive   Referring to a bacterial cell that stains purple after Gram staining.

Gram-positive cell wall   Bacterial cell walls composed of a thick layer of peptidoglycan and teichoic acid.

Gram stain   A staining procedure used to identify bacterial cells as Gram-positive or Gram-negative.

granulocyte   A white blood cell with visible granules in the cytoplasm; includes neutrophils, eosinophils, and basophils. *See also* agranulocyte.

granuloma   A small lesion caused by an infection

griseofulvin   An antifungal drug used against infections of the skin, hair, and nails.

growth curve   The plotted or graphed measurement of the size of a population of bacteria as a function of time.

Guillain-Barré syndrome (GBS)   A complication of influenza and chickenpox, characterized by nerve damage and polio-like paralysis.

gumma   A soft, granular lesion that forms in the cardiovascular and/or nervous systems during tertiary syphilis.

H

HAART   *See* highly active antiretroviral therapy.

habitat   The place where a particular species lives and grows.

HACCP   *See* hazard analysis critical control point.

halogen   A chemical element whose atoms have seven electrons in their outer shell; examples include iodine and chlorine.

halophile   An organism that lives in environments with high concentrations of salt.

hapten   A small molecule that combines with tissue proteins or polysaccharides to form an antigen.

hay fever   A type I hypersensitivity reaction resulting from the inhalation of tree and grass pollens.

Hazard Analysis Critical Control Point (HACCP)   A set of federally enforced regulations to ensure the dietary safety of seafood, meat, and poultry.

heat fixation   The use of warm temperatures to prepare microorganisms for staining and viewing with the light microscope.

heavy metal   A chemical element often toxic to microorganisms; examples include mercury, copper, and silver.

heliozoa   Also known as sun animalcules, these amoeba are roughly spherical have many stiff projections, called axopods, radiating outward from the cell surface. These projections give them the sun-like appearance for which they are named.

helix (pl. **helices**)   A twisted shape such as that seen in a spring, screw or a spiral staircase.

helminth   A term referring to a multicellular parasite; includes roundworms and flatworms.

helper T lymphocyte   A T lymphocyte that enhances the activity of B lymphocytes and stimulate destruction of macrophages infected with bacteria.

hemagglutination   The formation of clumps of red blood cells.

hemagglutinin   (1) An enzyme composing one type of surface spike on influenza viruses that enables the viruses to bind to the host cell. (2) An agent such as a virus or an antibody that causes red blood cells to clump together.

hemoglobin   The red oxygen-carrying pigment in erythrocytes.

hemolytic disease of the newborn   A type II hypersensitivity reaction disease in which Rh antibodies from a pregnant woman combine with Rh antigens on the surface of fetal erythrocytes and destroy them; also called Rh disease and erythroblastosis fetalis.

hemophilia A   A blood clotting disorder resulting from a mutation in the gene for factor VIII, an important protein in the process of blood clotting.

hemorrhagic colitis   Bloody diarrhea associated with infection by *E. coli* O157:H7.

hemorrhagic fever   Any of a series of viral diseases characterized by high fever and hemorrhagic lesions of the throat and internal organs.

heterotroph   An organism that requires preformed organic matter for its energy and carbon needs. *See also* photoheterotroph *and* chemoheterotroph.

Hfr cell   A bacterial cell containing an F factor incorporated into the bacterial chromosome.

high efficiency particulate air (HEPA) filter   A type of air filter that removed particles larger than 0.3 micrometers.

highly active antiretroviral therapy (HAART)   The combination of several (typically three or four) antiretroviral drugs for the treatment of infections caused by retroviruses, especially the human immunodeficiency virus.

histamine   A mediator in type I hypersensitivity reactions that is released from the granules in mast cells and basophils and causes the contraction of smooth muscles.

histone   One of several proteins that organize and pack the DNA in a chromosome.

homeostasis   The tendency of an organism to maintain a steady state or equilibrium.

host   An organism on or in which a microorganism lives and grows, or a virus replicates.

human immunodeficiency virus (HIV)   The retrovirus that causes acquired immunodeficiency syndrome (AIDS).

humoral immune response   The immune reaction of producing antibodies directed against antigens in the body fluids.

hydrocarbon   An organic molecule containing only hydrogen and carbon atoms that are connected by a sharing of electrons.

hydrogen bond   A weak chemical bond between a positively charged hydrogen atom (covalently bonded to oxygen or nitrogen) and a covalently bonded, negatively charged oxygen or nitrogen atom in the same or separate molecules.

hydrogen peroxide   An unstable liquid that readily decomposes in water and oxygen gas (O_2).

hydrophilic   Referring to a substance that dissolves in or mixes easily with water; *see also* hydrophobic.

hydrophobia   An emotional condition ("fear of water") arising from the inability to swallow as a consequence of rabies.

hydrophobic   Referring to a substance that does not dissolve in or mixing easily with water; *see also* hydrophilic.

hypha (pl. hyphae)   A microscopic filament of cells representing the vegetative portion of a fungus.

I

icosahedron   A symmetrical figure composed of 20 triangular faces and 12 points; one of the major shapes of some viral capsids.

IgA (*immunoglobulin A*)   The class of antibodies found in respiratory and gastrointestinal secretions that help neutralize pathogens.

IgD (*immunoglobulin D*)   The class of antibodies found on the surface of B cells that act as receptors for binding antigen.

IgE (*immunoglobulin E*)   The class of antibodies responsible for type I hypersensitivities.

IgG (*immunoglobulin G*)   The class of antibodies abundant in serum that are major diseases fighters.

IgM (*immunoglobulin M*)   The first class of antibodies to appear in serum in helping fight pathogens.

imidazole   An antifungal drug that interferes with sterol synthesis in fungal cell membranes; examples are miconazole and ketoconazole.

immune complex   A combination of antibody and antigen capable of complement activation and characteristic of type III hypersensitivity reactions.

immune complex hypersensitivity   Type III hypersensitivity, in which antigens combine with antibodies to form aggregates that are deposited in blood vessels or on tissue surfaces.

immune deficiency   The lack of an adequate immune system response.

immunity   The body's ability to resist infectious disease through innate and acquired mechanisms.

immunization   The process of making an individual resistant to a particular disease by administering a vaccine. *See also* vaccination.

immunocompetent   The ability of the body to develop an immune response in the presence of a disease-causing agent.

immunoglobulin (Ig)   The class of immunological proteins that react with an antigen; an alternate term for antibody.

immunology   The scientific study of how the immune system works and responds to nonself agents.

incubation period   The time that elapses between the entry of a pathogen into the host and the appearance of signs and symptoms.

indicator organism   A microorganism whose presence signals fecal contamination of water.

indirect contact   The mode of disease transmission involving nonliving objects. *See also* direct contact.

induced mutation   A change in the sequence of nucleotide bases in a DNA molecule arising from a mutagenic agent used under controlled laboratory conditions.

industrial fermentation   Any large scale industrial process, with or without oxygen gas (O_2), for growing microorganisms. *See also* fermentation.

inert gas   A element with filled electron shells; examples include helium and neon.

infection   The relationship between two organisms and the competition for supremacy that takes place between them.

infectious disease   A disorder arising from a pathogen invading a susceptible host and inducing medically significant symptoms.

inflammation   A nonspecific defensive response to injury; usually characterized by redness, warmth, swelling, and pain.

innate immunity   An inborn set of the pre-existing defenses against infectious agents; includes the skin, mucous membranes, and secretions.

interferon   An antiviral protein produced by body cells on exposure to viruses and which trigger the synthesis of antiviral proteins.

intoxication   The presence of microbial toxins in the body.

intron   A section of DNA that is removed after transcription by splicing the mRNA transcript.

invasiveness   The ability of a pathogen to spread from one point to adjacent areas in the host and cause structural damage to those tissues.

iodophor   A complex of iodine and detergents that is used as an antiseptic and disinfectant.

ion   An electrically charged atom.

ionizing radiation   A type of radiation such as gamma rays and X rays that causes the separation of atoms or a molecule into ions.

isomer   A molecule with the same molecular formula but different structural formula.

isoniazid (INH)   An antimicrobial drug effective against the tubercle bacillus.

isotope   An atom of the same element in which the number of neutrons differs.

J

jaundice   A condition in which bile seeps into the circulatory system, causing the complexion to have a dull yellow color.

K

Kaposi's sarcoma   A type of cancer in immunocompromised individuals, such as AIDS patients, where cancer cells and an abnormal growth of blood vessels form solid lesions in connective tissue.

Koch's postulates   A set of procedures by which a specific organism can be related to a specific disease.

Krebs cycle   A cyclic metabolic pathway in which carbon from acetyl-CoA is released as carbon dioxide; the reactions yield ATP as well as protons and high-energy electrons that are transferred to coenzymes; also called citric acid cycle.

L

lactone   A chemical compound used as a flavoring ingredient in foods and beverages.

lactose   A milk sugar composed of one molecule of glucose and one molecule of galactose.

lag phase   A portion of a bacterial growth curve encompassing the first few hours of the population's history when no growth occurs.

larva (pl. **larvae**)   A sexually immature stage in the life cycle of a multicellular parasite.

latency   A condition in which a virus integrates into a host chromosome without immediately causing a disease.

leukemia   Cancer of the white blood cells.

leukocyte   Any of a number of types of white blood cells.

lichen   An association between a fungal mycelium and a cyanobacterium or alga.

light microscope   An instrument that uses visible light and a system of glass lenses to produce a magnified image of an object; *also* called a compound microscope.

lipid   A nonpolar organic compound composed of carbon, hydrogen, and oxygen; examples include triglycerides and phospholipids.

lobar pneumonia   An inflammation involving an entire side or lobe of the lung.

lockjaw   *See* trismus.

logarithmic (log) phase   The portion of a bacterial growth curve during which active growth leads to a rapid rise in cell numbers.

lymph   The tissue fluid that contains white blood cells and drains tissue spaces through the lymphatic system.

lymph node   A bean-shaped organ located along lymph vessels that is involved in the immune response and contains phagocytes and lymphocytes.

lymphocyte   A type of white blood cell that functions in the immune system.

lymphoid progenitor   A bone marrow cell that gives rise to lymphocytes. *See also* myeloid progenitor.

lymphokines   Soluble growth factors and signaling molecules that are produced by cells of the immune system.

lymphopoietic cells   *See* lymphoid progenitor

lyophilization   A process in which food or other material is deep frozen, after which its liquid is drawn off by a vacuum; also called freeze-drying.

lysis   The rupture of a cell and the loss of cell contents.

lysogenic cycle   The events of a bacterial virus that integrates its DNA into the bacterial chromosome.

lysosome   A membrane-enclosed compartment in many eukaryotic cells that contains enzymes to degrade or digest substances.

lysozyme   An enzyme found in tears and saliva that digests the peptidoglycan of Gram-positive bacterial cell walls.

lytic cycle   A process by which a bacterial virus replicates within a host cell and ultimately destroys the host cell.

M

macronucleus   The larger of two nuclei in most ciliates that is involved in controlling metabolism. *See also* micronucleus.

macrophage   A large cell derived from monocytes that is found within various tissues and actively engulfs foreign material, including infecting bacteria and viruses.

macule   A pink-red skin spot associated with infectious disease.

maculopapular rash   A rash consisting of pink-red spots that later become dark red before fading; occurs in rickettsial diseases.

major histocompatibility complex (MHC)   A set of genes that controls the expression of MHC proteins; involved in transplant rejection.

major histocompatibility complex (MHC) protein Any of a set of proteins at the surface of all body cells that identify the uniqueness of the individual.

malignant   Referring to a tumor that invades the tissue around it and may spread to other parts of the body.

malting   Referring to the process when barley begins to germinate by being soaked in water to produce simpler carbohydrates.

mashing   A fermentable mixture of hot water and barley grain from which alcohol is distilled.

mass   The amount of matter an object contains.

mast cell   A type of cell in connective tissue which release histamine during allergic attacks.

maternal antibodies   The IgG antibodies that cross the placenta from the maternal to the fetal circulation and protect the newborn for the first few months of life.

matrix protein   A protein shell found in some viruses between the genome and capsid.

matter   Any substance that has mass and occupies space.

mechanical vector   A living organism, or an object, that transmits disease agents on its surface. *See also* biological vector.

membrane filter technique   A method to test water quality by identifying any coliform bacteria trapped on a filter.

memory B cells   *See* memory cell.

memory cell   A cell derived from B lymphocytes or T lymphocytes that reacts rapidly upon re-exposure to antigen.

memory T cells   *See* memory cell.

meninges   The covering layers of the brain and spinal cord.

meningitis   A general term for inflammation of the covering layers of the brain and spinal cord due to any of several bacteria, fungi, viruses, or protozoa.

meningococcus   A common name for *Neisseria meningitidis*.

mesophile   An organism that grows in temperature ranges of 20°C to 40°C.

messenger RNA (mRNA)   An RNA transcript containing the information for synthesizing a specific polypeptide.

metabolic pathway   A sequence of linked enzyme-catalyzed reactions in a cell.

metabolism   The sum of all biochemical processes taking place in a living cell. *See also* anabolism and catabolism.

metastasize   Referring to a tumor that spreads from the site of origin to other tissues in the body.

methanogen   An archaeal organism that lives on simple compounds in anaerobic environments and produces methane during its metabolism.

MHC proteins   *See* Major histocompatability complex.

microaerophile   An organism that grows best in an oxygen-reduced environment.

microbe   *See* microorganism.

microbicide   Any agent that kills microbes.

microbiology   The scientific study of microscopic organisms and viruses, and their roles in human disease as well as beneficial processes.

micrometer (μm)   A unit of measurement equivalent to one millionth of a meter; commonly used in measuring the size of microorganisms.

micronucleus   The smaller of the two nuclei in most ciliates that contains genetic material and is involved in sexual reproduction. *See also* macronucleus.

microorganism (microbe)   A microscopic form of life including bacterial, archaeal, fungal, and protozoal cells.

microtubule   A hollow protein tube making up part of the cytoskeleton, flagella, and cilia in eukaryotic organisms.

mitochondrion (pl. mitochondria)   A double membrane-enclosed compartment in eukaryotic cells that carries out aerobic respiration.

mold   A type of fungus that consists of chains of cells and appears as a fuzzy mass in culture.

molecule   Two or more atoms held together by a sharing of electrons.

monera   An obsolete biological kingdom that comprised most living things with a prokaryotic cell organization.

monobactam   A synthetic antibacterial drug used to inhibit cell wall synthesis in Gram-negative bacteria.

monocyte   A circulating white blood cell with a large bean-shaped nucleus that is the precursor to a macrophage.

monomer   A simple organic molecule that can join in long chains with other molecules to form a more complex molecule. *See also* polymer.

monosaccharide   A simple sugar that cannot be broken down into simpler sugars; examples include glucose and fructose.

monospot test   A method used to detect the presence of heterophile antibodies, which is indicative of infectious mononucleosis.

mucosa   *See* mucous membrane.

mucous membrane   A moist lining in the body passages of all mammals that contains mucus-secreting cells and is open directly or indirectly to the external environment.

mutagen   A chemical or physical agent that causes a mutation.

mutant   An organism carrying a mutation.

mutation   A change in the characteristic of an organism arising from a permanent alteration of a DNA sequence.

mutualism   A close and permanent association between two populations of organisms in which both benefit from the association.

mycelium (pl. mycelia)   A mass of fungal filaments from which most fungi are built.

mycoplasma   One of a group of tiny submicroscopic bacteria that lacks cell walls and is visible only with an electron microscope.

mycorrhiza (pl. mycorrhizae)   A close association between a fungus and the roots of many plants.

myocarditis   Infection of the heart muscle, often due to Coxsackie virus.

myxobacteria   A group of soil-dwelling bacteria that exhibit multicellular behaviors.

N

nanometer   (nm) A unit of measurement equivalent to one billionth of a meter; the unit is often used in measuring viruses and the wavelength of energy.

nematode   A common name for the roundworm.

neuraminidase   An enzyme composing one type of surface spike of influenza viruses that facilitates viral release from the host cell.

neurotoxin   A toxin that is active in the nervous system of the host.

neutron   An uncharged particle in the atomic nucleus.

neutrophil   The most common type of white blood cell; functions chiefly to engulf and destroy foreign material, including bacteria and viruses that have entered the body.

nicotinamide adenine dinucleotide (NAD$^+$)   A coenzyme that transfers and transports electrons during oxidative phosphorylation and fermentation reactions.

nitrification   The biological oxidation of ammonia into nitrates and is performed in two steps by two different bacteria both collectively known as nitrifying bacteria.

nitrogen cycle   The processes that convert nitrogen gas (N_2) to nitrogen-containing substances in soil and living organisms, then reconverted to the gas.

nitrogen fixation   The chemical process by which microorganisms convert nitrogen gas (N_2) into ammonia.

nitrogenous base   Any of five nitrogen-containing compounds found in nucleic acids, including adenine, guanine, cytosine, thymine, and uracil.

nomenclature   A method of assigning specific identifying names.

nucleic acid   A high-molecular-weight molecule consisting of nucleotide chains that convey genetic information and are found in all living cells and viruses. *See* DNA *and* RNA.

nucleocapsid   The combination of genome and capsid of a virus.

nucleoid   The chromosomal region of a prokaryotic cell.

nucleotide   A component of a nucleic acid consisting of a carbohydrate molecule, a phosphate group, and a nitrogenous base.

nucleus (pl. **nuclei**)   (1) The portion of an atom consisting of protons and neutrons. (2) A membrane-enclosed compartment in eukaryotic cells that contains the chromosomes.

O

objective lens   The lens or mirror in a microscope that receives the first light rays from the object being observed.

Okazaki fragment   A segment of DNA resulting from discontinuous DNA replication.

oncogene   A segment of DNA that can induce uncontrolled growth of a cell if permitted to function.

oncogenic   Referring to any agent such as viruses that can cause tumors.

oncology   The scientific study of tumors and cancers.

operator   A sequences of bases in the DNA to which a repressor protein can bind.

operon   The unit of bacterial DNA consisting of a promoter, operator, and a set of structural genes.

ophthalmia   Severe inflammation of the eye.

opportunist   A microorganism that invades the tissues when body defenses are suppressed.

opportunistic infection   A disorder caused by a microorganism that does not cause disease but that can become pathogenic or life-threatening if the host has a low level of immunity.

orchitis   A condition caused by the mumps virus in which the virus damages the testes.

organelle   A specialized compartment in eukaryotic cells that has a particular function.

organic compound   A substance characterized by chains or rings of carbon atoms that are linked to atoms of hydrogen and sometimes oxygen, nitrogen, and other elements.

origin of replication   The fixed point on a DNA molecule where copying of the molecule starts.

osmosis   The net movement of water molecules from where they are in a high concentration through a semipermeable membrane to a region where they are in a lower concentration.

osmotic pressure   The force that must be applied to a solution to inhibit the inward movement of water across a membrane.

outbreak   A small, localized epidemic.

outer membrane   A bilayer membrane forming part of the cell wall of Gram-negative bacteria.

oxidation   A chemical change in which electrons are lost by an atom. *See also* reduction.

oxidation lagoon   A large pond in which sewage is allowed to remain undisturbed so that digestion of organic matter can occur.

P

pandemic   A worldwide epidemic.

papule   A pink pimple on the skin.

para-aminobenzoic acid (PABA)   A precursor for folic acid synthesis.

parasite   A type of heterotrophic organism that feeds on live organic matter such as another organism.

parasitemia   The spread of protozoa and multicellular worms through the circulatory system.

parasitism   A close association between two organisms in which one (the parasite) feeds on the other (the host) and may cause injury to the host.

passive immunity   The temporary immunity that comes from receiving antibodies from another source.

pasteurization   A heating process that destroys pathogenic bacteria in a fluid such as milk and lowers the overall number of bacteria in the fluid.

pathogen   A microorganism or virus that causes disease in a host organism.

pathogenicity   The ability of a disease-causing agent to gain entry to a host and bring about a physiological or anatomical change interpreted as disease.

pathogenicity island   A set of adjacent genes that encode virulence factors.

pébrine   A protozoal disease of silkworms studied by Louis Pasteur.

pellicle   A flexible covering layer typical of the protozoan ciliates.

penicillin   Any of a group of antibiotics derived from *Penicillium* species or produced synthetically; effective against Gram-positive bacteria and several Gram-negative bacteria by interfering with cell wall synthesis.

penicillinase   An enzyme produced by certain microorganisms that converts penicillin to penicilloic acid and thereby confers resistance against penicillin.

peptide bond   A linkage between the amino group on one amino acid and the carboxyl group on another amino acid.

peptidoglycan   A complex molecule of the bacterial cell wall composed of alternating units of N-acetylglucosamine and N-acetylmuramic acid cross linked by short peptides.

perforin   A protein secreted by cytotoxic T lymphocytes and natural killer cells that forms holes in the plasma membrane of an infected cell.

period of convalescence   The phase of a disease during which the body's systems return to normal.

period of decline   The phase of a disease during which symptoms subside.

periplasmic space   A metabolic region between the cell membrane and outer membrane of Gram-negative cells.

pH   An abbreviation for the hydrogen ion concentration [H^+] of a solution.

phage   *See* bacteriophage.

phagocyte   A white blood cell capable of engulfing and destroying foreign materials or cells, including bacteria and viruses.

phagocytosis   A process by which foreign material or cells are taken into a white blood cell and destroyed.

pharyngitis   An inflammation of the pharynx; commonly called a sore throat.

phase-contrast microscopy   An optical system on the light microscope that uses a special condenser and objective lenses to examine cell structure.

phenol   A chemical compound that has one or more hydroxyl groups attached to a benzene ring and derivatives are used as an antiseptic or disinfectant; also called carbolic acid.

phospholipid   A water-insoluble compound containing glycerol, two fatty acids, and a phosphate head group; forms part of the membrane in all cells.

photosynthesis   A biochemical process in which light energy is converted to chemical energy, which is then used for carbohydrate synthesis.

physical map   A reference, in genetics, to how much DNA separates the location of two genes and is measured in nucleotide base pairs, as opposed to a genetic map that tells you the positions of genes in relation to each other based on the frequency of crossing-overs.

phytoplankton   Microscopic free-floating communities of cyanobacteria and unicellular algae.

pilus   A hairlike extension of the plasma membrane found on the surface of many bacteria and is used for cell attachment and anchorage. Some bacteria possess a specialized larger pilus (the F-pilus or conjugation pilus) that is used for conjugation; the passage of genetic information between bacteria.

plaque   (1) A clear area on a lawn of bacteria where viruses have destroyed the bacterial cells. (2) The gummy layer of gelatinous material consisting of bacteria and organic matter on the teeth.

plasma   The fluid portion of blood remaining after the cells have been removed. *See also* serum.

plasma cell   An antibody-producing cell derived from B lymphocytes.

plasma membrane   The phospholipid bilayer with proteins that surrounds the eukaryotic cell cytoplasm. *See also* cell membrane.

plasmid   A small, closed-loop molecule of DNA apart from the chromosome that replicates independently and carries non-essential genetic information.

pneumococcus   A common name for *Streptococcus pneumoniae*.

pneumonia   An inflammation of the bronchial tubes and one or both lungs.

polyadenylation   The covalent linkage of a polyadenosine (poly-A) tail to messenger RNA after transcription. The poly-A tail protects the mRNA molecule from degradative enzymes.

polymer   A substance formed by combining smaller molecules into larger ones. *See also* monomer.

polymerase chain reaction (PCR)   A technique used to replicate a fragment of DNA many times.

polymyxin   An antibiotic derived from a *Bacillus* species that disrupts the cell membrane of Gram-negative rods.

polypeptide   A chain of linked amino acids.

polysaccharide   A complex carbohydrate made up of sugar molecules linked into a branched or chain structure; examples include starch and cellulose.

polysome   A cluster of ribosomes linked by a strand of mRNA and all translating the mRNA.

portal of entry   The site at which a pathogen enters the host.

portal of exit   The site at which a pathogen leaves the host.

pox   Pitted scars remaining on the skin of individuals who have recovered from smallpox.

precipitation   A type of antigen–antibody reaction in which thousands of molecules of antigen and antibody cross-link to form visible aggregates.

primary structure   The sequence of amino acids in a polypeptide.

prion   An infectious, self-replicating protein involved in human and animal diseases of the brain.

prodromal   The phase of a disease during which general symptoms occur in the body.

product   A substance or substances produced in a chemical reaction.

prokaryote   A microorganism in the domain *Bacteria* or *Archaea* composed of single cells having a single chromosome but no cell nucleus or other membrane-bound compartments. *See also* eukaryote.

prokaryotic   Referring to cells or organisms having a single chromosome but no cell nucleus or other membrane-bound compartments.

promoter   The region of a template DNA strand or operon to which RNA polymerase binds.

prontosil   A red dye found by Domagk to have significant antimicrobial activity when tested in live animals, and from which sulfanilamide was later isolated.

protease inhibitor   A compound that breaks down the enzyme protease, inhibiting the replication of some viruses, such as HIV.

proteases   Are enzymes that use water to hydrolyze and break peptide bonds between amino acids of proteins.

protein   A chain or chains of linked amino acids used as a structural material or enzyme in living cells.

protein synthesis   The process of forming a polypeptide or protein through a series of chemical reactions involving amino acids.

proteomics   The study of proteins, particularly their structures and functions and localization.

Protista   One of the five kingdoms in the Whittaker classification of living things, composed of the protozoa and unicellular algae.

proton   A positively charge particle in the atomic nucleus.

proto-oncogene   A region of DNA in the chromosome of human cells; they are altered by carcinogens into oncogenes that transform cells.

protozoan (pl. **protozoa**)   A single-celled eukaryotic organism that lacks a cell wall and usually exhibits chemoheterotrophic metabolism.

provirus   The viral DNA that has integrated into a eukaryotic host chromosome and is then passed on from one generation to the next through cell division.

pseudopod   A projection of the plasma membrane that allows movement in members of the amoebozoans.

psychrophile   An organism that lives at cold temperature ranges of 0°C to 20°C.

psychrotroph   An organism that lives at cold temperature ranges of 0°C to 30°C.

pure culture   An accumulation or colony of microorganisms of one species.

pus   A mixture of dead tissue cells, leukocytes, and bacteria that accumulates at the site of infection.

pustule   A raised bump on the skin that contains pus.

pyruvic acid   The end product of the glycolysis metabolic pathway.

R

radiolarian   A single-celled marine organism with a round silica-containing shell that has radiating arms to catch prey.

reactant   A substance that interacts with another in a chemical reaction.

recombinant DNA molecule   A DNA molecule containing DNA from two different sources.

red tide   A brownish-red discoloration in seawater caused by increased numbers of dinoflagellates. *See also* algal bloom.

reduction   The gain of electrons by a molecule.

regulatory gene   A DNA segment that codes for a repressor protein.

rennin   An enzyme that accelerates the curdling of protein in milk.

replication fork   The point where complementary strands of DNA separate and new complementary strands are synthesized.

repressor protein   A protein that when bound to the operator blocks transcription.

restriction enzyme   A type of enzyme that splits open a DNA molecule at a specific restricted point; important in genetic engineering techniques.

reverse transcriptase   An enzyme that synthesizes a DNA molecule from the code supplied by an RNA molecule.

reverse transcriptase inhibitor   A compound that inhibits the action of reverse transcriptase, preventing the viral genome from being replicated.

revertant   Referring to a mutant organism or cell that has reacquired its original phenotype or metabolic ability.

rheumatoid arthritis   An autoimmune disorder characterized by immune complex formation in the joints.

ribonucleic acid (RNA)   The nucleic acid involved in protein synthesis and gene control; also the genetic information in some viruses.

ribosomal RNA (rRNA)   An RNA transcript that forms part of the ribosome's structure.

ribosome   A cellular component of RNA and protein that participates in protein synthesis.

rickettsia (pl. rickettsiae)   A very small bacterial cell generally transmitted by arthropods; most rickettsiae are cultivated only within living tissues.

rifampin   An antibiotic prescribed for tuberculosis and leprosy patients and for carriers of *Neisseria* and *Haemophilus* species.

RNA   *See* Ribonucleic acid

RNA polymerase   The enzyme that synthesizes an RNA polynucleotide from a DNA template.

rose spots   Bright red skin spots associated with diseases such as typhoid fever and relapsing fever.

ruminant   Any hooved animal that digests its food in two steps. The first step is eating the raw material and then regurgitating the semi-digested form of the food known as cud. The second step is then eating the cud, a process called ruminating. Ruminants include cattle, goats, sheep, camels, llamas, giraffes, bison, buffalo, and deer.

S

Sabin vaccine   A type of polio vaccine prepared with attenuated viruses and taken orally.

Salk vaccine   A type of polio vaccine prepared with inactivated viruses and injected into the body.

sanitization   To remove microbes or reduce their populations to a safe level as determined by public health standards.

saprobe   A type of heterotrophic organism that feeds on dead organic matter, such as rotting wood or compost.

sarcina (pl. sarcinae)   (1) A packet of eight spherical-shaped prokaryotic cells. (2) A genus of Gram-positive, anaerobic spheres.

sarcoma   A tumor of the connective tissues.

saturated fat   A water-insoluble compound that cannot incorporate any additional hydrogen atoms. *See also* unsaturated fat.

scanning electron microscope (SEM)   The type of electron microscope that allows electrons to scan across an object, generating a three-dimensional image of the object.

secondary structure   The region of a polypeptide folded into an alpha helix or pleated sheet.

selective medium   A growth medium that contains ingredients to inhibit certain microorganisms while encouraging the growth of others.

semiconservative replication   The DNA copying process where each parent (old) strand serves as a template for a new complementary strand.

sensitizing dose   The first exposure to an allergy-causing antigen.

sepsis   The growth and spreading of bacteria or their toxins in the blood and tissues.

septic shock   A collapse of the circulatory and respiratory systems caused by an overwhelming immune response.

septicemia   A growth and spreading of bacterial in the bloodstream.

septum (pl. septa)   A cross-wall in the hypha of a fungus.

serum (pl. sera)   The fluid portion of the blood consisting of water, minerals, salts, proteins, and other organic substances, including antibodies; contains no clotting agents. *See also* plasma.

sexually transmitted disease (STD)   A disease such as gonorrhea or chlamydia that is normally passed from one person to another through sexual activity.

simple stain technique   The use of a single cationic dye to contrast cells. *See also* differential stain technique.

slime layer   A thin, loosely bound layer of polysaccharide covering some prokaryotic cells. *See also* capsule *and* glycocalyx.

slime molds   Peculiar amoeboid protists that can under certain conditions transform into a multicellular slug and migrate to better growth conditions.

sludge   The solids in sewage that separate out during sewage treatment.

soap   A compound made by potassium or sodium hydroxide reacting with fatty acids.

sour curd   The acidification of milk, causing a change in the structure of milk proteins.

spawn   A mushroom mycelium used to start a new culture of the fungus.

specialized transduction   The transfer some bacterial genes by a bacterial virus that carries the genes to another bacterial cell. *See also* generalized transduction.

species   The fundamental rank in the classification system of organisms.

specific epithet   The second of the two scientific names for a species. *See also* genus.

spike   A protein projecting from the viral envelope or capsid that aids in attachment and penetration of a host cell.

spirillum (pl. **spirilla**)   (1) A bacterial cell shape characterized by twisted or curved rods. (2) A genus of aerobic, helical cells usually with many flagella.

spirochete   A twisted bacterial rod with a flexible cell wall containing axial filaments for motility.

spleen   An organ in the left upper abdomen of humans that helps to destroy pathogens.

spontaneous generation   The doctrine that nonliving matter could spontaneously give rise to living things.

sporangium (pl. **sporangia**)   A protective sac containing asexually produced fungal spores.

spore   (1) A reproductive structure formed by a fungus. (2) A highly resistant dormant structure formed from vegetative cells in several genera of bacteria, including *Bacillus* and *Clostridium. See also* endospore.

sporulation   The process of spore formation.

sputum   Thick, expectorated matter from the lower respiratory tract.

staphylococcus (pl. **staphylococci**)   (1) An arrangement of bacterial cells characterized by spheres in a grapelike cluster. (2) A genus of facultatively anaerobic, nonmotile, non-spore-forming, Gram-positive spheres in clusters.

stationary phase   The portion of a bacterial growth curve in which the reproductive and death rates of cells are equal.

stem cell   An undifferentiated cell from which specialized cells arise.

sterile   Free from living microorganisms, spores, and viruses.

sterilization   The removal of all life forms, including bacterial spores.

sterol   A organic solid containing several carbon rings with side chains; examples include cholesterol.

streptobacillus (pl. **streptobacilli**)   (1) A chain of bacterial rods. (2) A genus of facultatively anaerobic, nonmotile, Gram-negative rods.

streptococcus (pl. **streptococci**)   (1) A chain of bacterial cocci. (2) A genus of facultatively anaerobic, nonmotile, non-spore-forming, Gram-positive spheres in chains.

streptogramin   An antibiotic that block protein synthesis.

streptokinase   An enzyme that dissolves blood clots; produced by virulent streptococci.

structural formula   A chemical diagram representing the arrangement of atoms and bonds within a molecule.

structural gene   A segment of a DNA molecule that provides the biochemical information for a polypeptide.

substrate   The substance or substances upon which an enzyme acts.

sulfonamide   A synthetic, sulfur-containing antibacterial agent; also called sulfa drug.

sulfur cycle   The processes by which sulfur moves through and is recycled in the environment.

symbiosis   An interrelationship between two populations of organisms where there is a close and permanent association.

symptom   An indication of some disease or other disorder that is experienced by the patient. *See also* sign.

syncytium (pl. **syncytia**)   A giant tissue cell formed by the fusion of cells infected with respiratory syncytial viruses.

syndrome   A collection of signs or symptoms that together are characteristic of a disease.

T

tapeworm   *See* cestode.

taxon   Subdivisions used to classify organisms.

taxonomy   The science dealing with the systematized arrangements of related living things in categories.

T cell   *See* T lymphocyte.

teichoic acid   A negatively charged polysaccharide in the cell wall of Gram-positive bacteria.

tertiary structure   The folding of a polypeptide back on itself.

tetanospasmin   An exotoxin produced by *Clostridium tetani* that acts at synapses, thereby stimulating muscle contractions.

tetracycline   An antibiotic characterized by four benzene rings with attached side groups that blocks protein synthesis in many Gram-negative bacteria, rickettsiae, and chlamydiae.

tetrad   An arrangement of four bacterial cells in a cube shape.

thermoacidophile   An archaeal organism living under high temperature and high acid conditions.

thermoduric   Referring to an organism that tolerates the heat of the pasteurization process.

thermophil   An organism that lives at high temperature ranges of 40°C to 90°C.

three-domain system   The classification scheme placing all living organisms into one of three groups based, in part, on ribosomal RNA sequences.

thymus   A flat, bilobed organ where T lymphocytes mature.

tincture   A substance dissolved in ethyl alcohol.

tinea   Any of a group of fungal infections of the skin, feet, or scalp.

tissue plasminogen activator   A secreted protease (an enzyme which cuts proteins) that converts cuts plasminogen and thereby converts it to plasmin.

T lymphocyte (T cell)   A type of white blood cell that matures in the thymus gland and is associated with cell-mediated immunity.

tonsil   One of two small oval masses secondary lymphoid tissue that is important for the body's immune system.

total magnification   The enlargement of an object using the ocular lens and a specific objective lens of the microscope.

toxin   A poisonous chemical substance produced by an organism.

toxoid   A preparation of a microbial toxin that has been rendered harmless by chemical treatment but that is capable of stimulating antibodies; used as vaccines.

transcription   The biochemical process in which RNA is synthesized according to a code supplied by the bases of a gene in the DNA molecule.

transduction   The transfer of a few bacterial genes from a donor cell to a recipient cell via a bacterial virus.

transfer RNA (tRNA)   A molecule of RNA that unites with amino acids and transports them to the ribosome in protein synthesis.

transformation   (1) The transfer and integration of DNA fragments from a dead and lysed donor cells to a recipient cell's chromosome. (2) The conversion of a normal cell into a malignant cell due to the action of a carcinogen or virus.

transgenic   Referring to organisms containing DNA from another source.

translation   The biochemical process in which the code on the mRNA molecule is translated into a sequence of amino acids in a polypeptide.

transmission electron microscope (TEM)   The type of electron microscope that allows electrons to pass through the object, resulting in a detailed view of the object's structure.

triclosan   A phenol derivative incorporated as an antimicrobial agent into a wide variety of household products.

trichocyst   A structure of some ciliate and flagellate protozoans that consists of a cavity that contains long, thin threads that are ejected in response to the correct stimuli.

trismus   A sustained spasm of the jaw muscles, characteristic of the early stages of tetanus; also called lockjaw.

trophozoite   The feeding form of a microorganism, such as a protozoan.

tubercle   A hard nodule that develops in tissue infected with *Mycobacterium tuberculosis*.

tuberculin test   A procedure performed by applying purified protein derivative from *Mycobacterium tuberculosis* to the skin and noting if a thickening of the skin with a raised vesicle appears within a few days; used to establish if someone has been exposed to the bacterium.

tumor   An abnormal uncontrolled growth of cells that has no physiological function.

U

ultrapasteurization   A treatment in which milk is heated at 140°C for 3 seconds to destroy harmful bacteria.

ultraviolet (UV) light   A type of electromagnetic radiation of short wavelengths that damages DNA.

uncoating   Referring to the loss of the viral capsid inside an infected eukaryotic cell.

universal precautions   Using those measures to avoid contact with a patient's bodily fluids; examples include wearing gloves, goggles, and proper disposal of used hypodermic needles.

urethritis   An inflammation of the urethra.

V

vaccination   Inoculation with weakened or dead microbes, or viruses, in order to generate immunity. *See also* immunization.

vaccine   A preparation containing weakened or dead microorganisms or viruses, treated toxins, or parts of microorganisms or viruses to stimulate immune resistance.

vaginitis   A general term for infection of the vagina.

vancomycin   An antibacterial drug that inhibits cell wall synthesis and is used in treating diseases caused by Gram-positive bacteria, especially staphylococci.

vector   (1) An arthropod that transmits the agents of disease from an infected host to a susceptible host. (2) A plasmid used in genetic engineering to carry a DNA segment into a bacterium or other cell.

vesicle   (1) A fluid-filled skin lesion, such as that occurring in chickenpox. (2) A small, membrane-enclosed compartment found in many eukaryotic cells.

vibrio   (1) A prokaryotic cell shape occurring as a curved rod. (2) A genus of facultatively anaerobic, Gram-negative curved rods with flagella.

viral load test   A method used to detect the RNA genome of HIV.

virion   A completely assembled virus outside its host cell.

viroid   An infectious RNA segment associated with certain plant diseases.

virology   The scientific study of viruses.

virulence   The degree to which a pathogen is capable of causing a disease.

virulence factor   A molecule possessed by a pathogen that increases its ability to invade or cause disease to a host.

virulent   Referring to a virus or microorganism that can be extremely damaging when in the host.

virus   An infectious agent consisting of DNA or RNA and surrounded by a protein sheath; in some cases, a membranous envelope surrounds the coat.

W

wart   A small, usually benign skin growth commonly due to a virus.

wheal   An enlarged, hive-like zone of puffiness on the skin, often due to an allergic reaction; *see also* flare.

whey   The clear liquid remaining after protein has curdled out of milk.

white blood cell   *See* leukocyte.

wort   A sugary liquid produced from crushed malted grain and water to which is added yeast and hops for the brewing of beer.

Y

yeast   (1) A type of unicellular, nonfilamentous fungus that resembles bacterial colonies when grown in culture. (2) A term sometimes used to denote the unicellular form of pathogenic fungi.

Z

zooplankton   Small crustaceans and other animals that inhabit the water column of oceans, seas, and bodies of fresh water.

zoospore   An asexual spore that uses a flagellum for locomotion.

Zygomycota   A phylum of fungi whose members have coenocytic hyphae and form zygospores, among other notable characteristics.

zygospore   A sexually produced spore formed by members of the Zygomycota.

Index

Photo Acknowledgments

47(4): 647–625.); 8.12a: Reprinted with permission from the American Society for Microbiology (*ASM News.* December 1998, Vol. 64(12): 693–698). Photo courtesy of Doctor Stuart Levitz, University of Massachusetts Medical Center; 8.12b: Reprinted with permission from the American Society for Microbiology (Dykstra, M.A., Friedman, L. and Murphy, J.W.; *Infect. Immun.* 1977 May; 16(1): 129–135.); 8.13: © Medical-on-Line/Alamy Images.

Chapter 9 Opener: Courtesy of the Soil Biology Primer/USDA; 9.1a: Courtesy of David McKay/NASA; 9.1b: Courtesy of Dr. John Bradley, Institute for Geophysics and Planetary Physics, Livermore, CA.; 9.3a: © Jones and Bartlett Publishers. Photographed by Kimberly Potvin; 9.3b: © Photodisc; 9.3c: Courtesy of J Schmidt/Yellowstone National Park/NPS; 9.4: Reprinted with permission from the American Society for Microbiology (Tao, L., Tanzer, J.M and MacAlister, T.J.; *J. Bacteriol,* 1987 June; 169(6): 2543–2547.); 9.5: © Brian Chase/ShutterStock, Inc.

Chapter 10 Opener: © Birgit Sommer/ShutterStock, Inc.; Opener Insert: © Medical-on-Line/Alamy Images; 10.1a: © H. Potter-D. Dressler/Visuals Unlimited; 10.3: Courtesy of the CDC; 10.9b: © National Library of Medicine.

Chapter 11 Opener: © Photos.com; Opener Insert: © Dr Kari Lounatmaa/Photo Researchers, Inc.; 11.1a-b: Reprinted with permission from the American Society for Microbiology (Chou, F.I. and Tan, S.T.; *J. Bacteriol,* 1991 May; 173(10): 3184–3190.); 11.5b: Courtesy of Glenn Howard, Pall Biomedical Products Corporation, Glen Cove, NY; 11.6: Collection of the University of Michigan Health System, Gift of Pfizer, Inc.; 11.9: Collection of the University of Michigan Health System, Gift of Pfizer, Inc.; 11.11a-b: © National Library of Medicine; 11.13a-b: Reprinted with permission from the American Society for Microbiology (Thomas, C.J. and McMeekin, T.A.; *J. Bacteriol,* 1986 December; 168(3): 1476–1478.)

Chapter 12 Opener: © matka_Wariatka/ShutterStock, Inc.; Opener Insert: © SciMAT/Photo Researchers, Inc.; 12.1: © Jones and Bartlett Publishers. Photographed by Kimberly Potvin; 12.2a: Courtesy of Pabst Brewing Company; 12.2b: © David M. Phillips/Visuals Unlimited; 12.2c: © Photodisc; 12.3: © Lorraine Kourafas/ShutterStock, Inc.; 12.4a-f: Photos courtesy of Switzerland Cheese Marketing, Rosemarie Schilt; 12.5: © Photodisc; 12.6: © Pacific Press Service/Alamy Images; 12.7a-b: Courtesy of M.A. Dawschel and H.P. Flemming/USDA ARS; 12.8: © The Print Collector/Alamy Images.

Chapter 13 Opener: © Mark Breck/ShutterStock, Inc.; Opener Insert: © Phototake/Alamy Images; 13.1: Courtesy of Michael Otto Ph.D., NIAID/Rocky Mountain Laboratories; 13.3: Reprinted with permission from the American Society for Microbiology (Whelan, K.F., Sherburne, R.K. and Taylor, D.E.; *J. Bacteriol,* 1997 January; 179(1): 63–71.); 13.4: Reprinted with permission from the American Society for Microbiology (Thomas, C.J. and McMeekin, T.A.; *Appli. Environ. Microbiol,* 1981 February; 41(2): 492–503.); 13.5a-b: Reprinted with permission from the American Society for Microbiology (NG, L.K., Sherburne, R., Taylor, D.E. and Stiles, M.E.; *J. Bacteriol,* 1985 October; 164(1): 338–343.); 13.7: © Phototake/Alamy Images; 13.8: © Robert Longuehaye, NIBSC/SPL/Photo Researchers, Inc.; 13.9: © Jones and Bartlett Publishers. Photographed by Kimberly Potvin; 13.10: Photo courtesy of Institute of Food Research, Norwich Research Park, Colney.

Chapter 14 Opener: © Teo/ShutterStock, Inc.; Opener Insert: © George Musil/Visuals Unlimited; 14.1: © Photodisc; 14.6: © kavram/ShutterStock, Inc.; 14.8: Courtesy of Keith Weller/USDA ARS; 14.9a-c: © Photodisc; 14.9d: © Andi Berger/ShutterStock, Inc.; 14.10a: Courtesy of Pfizer, Inc.; 14.10b: Reprinted with permission from the American Society for Microbiology (Kudo, N., Kimura, M, Beppu, T. and Horinuchi, S.; *J. Bacteriol,* 1995 November; 177(22): 6401–6410.)

Chapter 15 Opener: © Alex Melnick/ShutterStock, Inc.; Opener Insert: © Dr. Dennis Kunkel/Visuals Unlimited; 15.1a: Courtesy of Harold Evans; 15.1b: © Medical-on-line/Alamy Images; 15.4: © Robert J. Breyers II/ShutterStock, Inc.; 15.5a: © melissa bouyounan/ShutterStock, Inc.; 15.5b: © Photos.com; 15.6: © Photodisc; 15.7b: Courtesy of Lian Bruno; 15.8a-c: Reprinted with permission from the American Society for Microbiology (López-Meza, J.E. and Ibarra, J.E.; *Appl. Environ. Microbiol,* 1996 April; 62(4): 1306–1310). Photo courtesy of Doctor J.E. López-Meza.

Chapter 16 Opener: © jeff gynane/ShutterStock, Inc.; Opener Insert: © Simko/Visuals Unlimited; 16.3: Courtesy of D. Hardesty/USGS; 16.4: © Clayton Thacker/ShutterStock, Inc.; 16.6: Reprinted with permission from the American Society for Microbiology (Nickel, J.C., Ruseska, J.B., Wright, J.B. and Closterton, J.W.; *Antimicrob. Agents Chemother,* 1985 April; 27(4): 619–624.) Photo courtesy of Doctor J. Curtis Nickel; 16.8: © Reuters/Mike Blake/Landov.

Chapter 17 Opener: Courtesy of James Gathany/CDC; Opener Insert: © Omikron/Photo Researchers, Inc.; 17.4: Reprinted with permission from the American Society for Microbiology (*ASM News.* 1992, Vol. 58: 486). Photo courtesy of Doctor Brett Finlay.

Chapter 18 Opener: © Scott Kapich/ShutterStock, Inc.; Opener Insert: © LSHTM/Photo Researchers, Inc.; 18.1a: Courtesy of Dr. Hermann/CDC; 18.1b: © Dr. Milton Reisch/Corbis; 18.3a: Courtesy of World Health Organization; Diagnosis of Smallpox Slide Series/CDC; 18.3b: Courtesy of James Hicks/CDC; 18.3c: Collection of the University of Michigan Health System, Gift of Pfizer, Inc.; 18.5: © Phototake/Alamy Images; 18.6: © Sinclair Stammers/SPL/Photo Researchers, Inc., 18.7: Courtesy of Betty Partin/CDC, 18.9: © Science VU/CDC/Visuals Unlimited.

Chapter 19 Opener: © Corbis/age footstock; Opener Insert: © CAMR/A. Barry Dowsett/Photo Researchers, Inc.; 19.1a: Reprinted with permission from the American Society for Microbiology (Tao, L., Tanzer, J.M and MacAlister, T.J.; *J. Bacteriol,* 1987 June; 169(6): 2543–2547.); 19.1b: Reprinted with permission from the American Society for Microbiology (Fluckiju, U. and Fischetti, V.A.; *Infect. and Immun,* 1998 March; 66(3): 974–979.) Photo courtesy of Doctor Vincent A. Fischetti; 19.1c: Courtesy of Vincent A. Fischetti, Ph.D., Head of the Laboratory of Bacterial Pathogenesis at Rockefeller University; 19.2b: Reprinted with permission from the American Society for Microbiology (*ASM News.* July 1998, Vol. 64, Cover.); 19.4: Courtesy of James Gathany/CDC; 19.5: Courtesy of Dr. Vester Lewis; 19.6a: © Scimat/Photo Researchers, Inc.; 19.6b: Courtesy of Dr. W.L. Dentler; 19.8a: Courtesy of Louisa Howard, Dartmouth College, Electron Microscope Facility; 19.8b: Courtesy of Janice Carr/CDC; 19.10a-b: © Phototake/Alamy Images; 19.10c: Courtesy of CDC/James H. Steele; 19.11: Courtesy of Larry Stauffer, Oregon State Public Health Laboratory/CDC; 19.12a: Courtesy of A. Garon, Rocky Mountain Laboratory/NIAID; 19.12b: Reprinted with permission from the American Society for Microbiology (Brusca, J.S., McDowall, A.W., Norgard, M.V. and Radolf, J.D.; *J. Bacteriol,* 1991 December; 173(24): 8004–808). Photo courtesy of Doctor Justin D. Radolf; 19.12c: © Science VU/Visuals Unlimited; 19.13a: Courtesy of Jim Gathany/CDC; 19.13b: © Oliver Meckes/Photo Researchers, Inc.; 19.13c: Courtesy of World Health Organization/CDC; 19.14a: Courtesy of the CDC; 19.14b: Courtesy of Susan Lindsley/CDC; 19.16b: Courtesy of CDC.